URBAN GEOGRAPHY
A Text-Book

URBAN GEOGRAPHY

A Text-Book

R.B. MANDAL

University Prof. of Geography

R.D. & D.J. College, Munger-811201,

CONCEPT PUBLISHING COMPANY, NEW DELHI-110059

ISBN 81-7022-794-1 (HB)

First Published 2000

Reprint : 2023

Ram Bahadur **Mandal** (*b.* 1948)

Published and Printed by
Ashok Kumar Mittal
Concept Publishing Company
A/15-16, Commercial Block, Mohan Garden
NEW DELHI-110059 (India)

Phones : 5648039, 5649024
Fax : 091-11-5648053

PREFACE

URBANIZATION and urban development are the catchwords not only in the western world but even in the orient as well. The process of population growth in the urban areas along with the commercial, industrial and transport preferment have favoured the recent spurt of urban development throughout the world. This has been accelerated with the overwhelming growth of population in urban areas through migrational accretion and natural growth. These have given rise to systems of central places, problem of slums and squatter settlements besides enhancing the linkages of industrial centres thereby increasing the entropy of urban places. All these have forced the governments to think about change in the urban policies and planning prospects.

This book will highlight some of these points as a text-book for university students. It has been written after consulting the syllabus of different universities of India. I hope, this book will help the university students upto farthest limit of their requirements.

I am extremely grateful to my urban geography teacher, Prof. L.N. Ram, Head of the Department of Geography, Patna University and my favourite teacher of this subject Prof. B. Thakur, University of Delhi, for suggestions and help.

I am also thankful to Dr. G.N. Singh who provoked me to complete this work within shortest possible time. I request the readers to point out the mistakes found in this book besides giving suggestions for further modification of this work in the next edition.

Munger, 1998 RAM BAHADUR MANDAL

CONTENTS

PART-III : URBAN PLANNING

LIST OF FIGURES

LIST OF TABLES

PART-I

URBANIZATION

1

MEANING AND SCOPE OF URBAN GEOGRAPHY

Introduction

URBAN GEOGRAPHY, as a distinct study, is a recently developed branch of human geography, dealing with the study of compact non-agricultural settlements, mainly towns and cities. It concerns with the spatial dimensions of urban centres, i.e., their origin, location, siting, growth, functions and relationship with each other within and outside its surrounding areas or zone of influence.

The term 'Urban' refers to towns or cities having marked secondary and tertiary functions along with a municipality or notified area committee. It concerns with the geography of towns, their situations, patterns, layout, housing and hierarchical organization of population concentration and service areas. Urbanization on the other hand, is a process of population increase in urban areas. Where ever population does not increase, we may say that deurbanization has taken place. Urbanization on the one hand, indicates the growth of secondary and tertiary activities like manufacturing, trading etc. and on the other threatens the growth of agricultural areas. Urban Geography discusses the town as a dwelling place whose residents are mainly engaged in industry, retail trade, wholesale trade and transport activities. The Western or the European civilization is regarded as modern civilization, the Asian is regarded as an ancient civilization and the current is metropolitan civilization based on industrialization and urbanization. The civilization of the world has synchronized to urban life.

As a science of human settlement, urban geography deals with the complex urban areas which possess sharp internal differentiation. It concerns about delineation of urban activities which are expressed in characteristic association of intensive landuse and human occupance features. Thus man is the pivotal point in urban geography and the study itself comes under the cultural environment made by man.

The term 'urban centre' has been defined by different scholars and institutions in different ways. On the basis of minimum population size, UNO defines an urban place as "a permanent settlement with not less than 20,000 inhabitants" (Brian, 1979, p. I). But in Albania, an urban place is a settlement which has about 400 people, and 250 inhabitants in Denmark, 2000 in France and about 20,000 in the United States of America are considered for an urban place. However, in some countries like Israel, an urban place is regarded as a settlement occupied by people who are not engaged in agricultural activities.

The most conspicuous feature of today's accelerated world population growth is its even greater rapidity of urbanization. In history, cities have grown but the tempo and dimensions of its growth in recent years have never been equalled. The many ramifications of the growth have special significance. It is associated with growth of non-agricultural activities, spread of popular education, material and cultural aspirations, etc. These changes have caused migration into cities in search of cash earnings.

According to Thompson[1] urbanization is characterized by movement of people from Small Communities concerned solely with agriculture to other Larger Communities whose activities are primarily centred in government, trade, manufacture and allied interests.

Hauser and Duncan[2] characterize urbanization as a change in the pattern of population distribution. It involves an increase in the size of urban population, and growth in number and size of urban places with increasing concentration of population of such places.

Urbanization has been systematically treated by Hope Tisdale Eldridge.[3] He argued that "the process of population concentration" involves two elements :

1. Multiplication of the points of concentration; and
2. Increase in the size of individual concentration.

It means an increasing shift from agrarian to industrial services and distributive occupations is a process by which the number of points of population concentration increases involving a shift from agarian to non-agrarian occupations.

There are great differences among the various countries regarding the relative number of urban units in a specified range of population. In some countries, the number of larger agglomerations is dominant over the small urban places, while in others the smaller towns are more prevalent than the larger ones. For example, more than 50 per cent of urban units in India fall under the category of urban places having a population of more than 10,000, while in the Unied States more than 70 per cent of total urban units fall in the population range of less than 10,000.[4]

In major parts of the world urbanization has been accompanied more by an increase in the size of existing urban units than the multiplication of points of concentration. The rate of increase in the size of population varies from one range of towns to another range in the same country and from one country to another in the same range of towns.

The characteristics of urbanization differ in advanced countries from those in the developing countries. Factors responsible for this difference are the following :

1. Difference in forces making for urbanization.
2. Difference in the ratio of population to resources and levels of living.
3. Difference in basic outlook and value systems.
4. Difference in the technological advancment.

The net migration is the major component of urbanization and is the chief mechanism by which all the great urbanization trends in the world's have been accomplished.[5]

Factors contributing to migration and therefore, to urbanization may be divided into two categories.

1. Factors at home, in the rural village which push the individual towards urban places.
2. Factors in the urban places themselves which exert pull mechanism in the urban places.

Data in different surveys on the causes of migration due to urbanization tend to highlight "pull" as the main motivating factor in the migration. Both push as well as pull factors are accountable to the process of urbanization. Push factors include—low level productivity, and agricultural lands, tenant factors, improve in transportation, improved communication and higher education facilities in urban areas and so on.

The plight of agriculture sector, high density of population on arable land, excessive rural indebtness, the progressive sub-division and fragmentation of holdings, recurrence of drought and crop failure, inefficient methods of cultivation, uncertainty of foreign and local markets, all conspire to make the peasant's life hard. It is the misery and poverty and lack of technological change in agriculture and the existence of risk and uncertainty in agriculture which push the people to urban centres. There is another argument that it is prosperity and revolutionary change in agriculture which encourage urbanization.[6]

Urbanization is so closely bound up with technology that we can say without qualification technology is the *sin-qua-non* of urbanization.[7]

Among the pull factors the principal causes of urbanization are industrial developments including the progress of inventions and discoveries. The growth of urban centres is closely associated with commercial and industrial trends. Revolutionary improvements in the system of communication and transportation are another conditions for urbanization.

Other non-economic factors causing urbanization include educational facilities, standard of living and revolutionary developments in medical science, etc.

What is a Town?

The study of towns is a pivotal point of human geography. It has been defined by various geographers in numerous ways, which vary from each other in one way or the other. In fact, town has been defined on the basis of various characteristics such as, their size, population, lay-out, functions etc. Emrys Jones (in *Towns and Cities*) has rightly stated that "a town seems to be all things to all men : a physical agglomeration of streets and houses, a centre of commerce and administration, a kind of society, even a cultural frame of mind of urbanite or a way of life."[8]

The term 'town' refers to a compact settlement, larger than a village, having marked secondary and tertiary functions with some form of municipal administration or a notified area committee, with a community pursuing an urban way of living. Towns have also been defined by taking into account a fixed minimum number of inhabitants. For instance, in India a place having a population of more than 5,000 and a density of 386 persons per square kilometre is called town in case it has a municipality. But the minimum population size for a town varies from country to country. It may range from only 250 inhabitants in Denmark to 3,000 in Japan.

A town is dominated by distinctive secondary and tertiary functions which include manufacturing, trading and other professional services.

At least two-thirds of urban inhabitants follow non-agricultural occupations. They have a variety of shops selling varying goods, a bank, a cinema hall, a school, a hotel, a hospital and such other urban facilities.

Meaning or Definitions by various Geographers

"Urban geography includes the site, evolution, pattern and classification of towns"—*G. Taylor.*[9]

"Urban geography is in fact the intensive study of towns and their development in all their geographical aspects"—*L.D. Stamp.*[10]

"Urban geography is concerned with interpreting the patterns and relationships, that exist within urban areas on the one hand and betwe urban areas on the other"—*H.M. Mayer.*[11]

"Urban geography is the geographic study of urban places which evolve, grow and exist as service centres for their surrounding areas"

—*G.S. Gosal.*[12]

"Urban Geographic investigations are concerned primarily with the areal variations within and among cities, including the relationships between urban and non-urban areas, and the forces of development and change that are shaping the urban landscape"—*Pater Sholar.*[13]

"Urban geography deals with spatial aspects of urban development and non-urban areas as they relate to cities. The concern is with determining the areal patterns associated with urban centres and in explaining their arrangements"—*R.E. Murphy.*[14]

According to R.M. Northam,[15] the study of urban geography concerns with four types of relations with man and the land (Fig. 1.1).

(a) First of all it is related with man and the land, in terms of spatial variations of population distribution.

(b) Secondly, the relationship is provided by two places—which explains the morphological patterns, spatial distribution and spacing of urban settlements,

(c) Third relationship is linked with more than two places as concentration of population. It explains the economic and functional characteristics of population found in various urban centres.

(d) Fourth point explains the relationship found internally at one place. In this respect, it explains the urban landuse and commercial structure of population found inside the commercial ribbon.

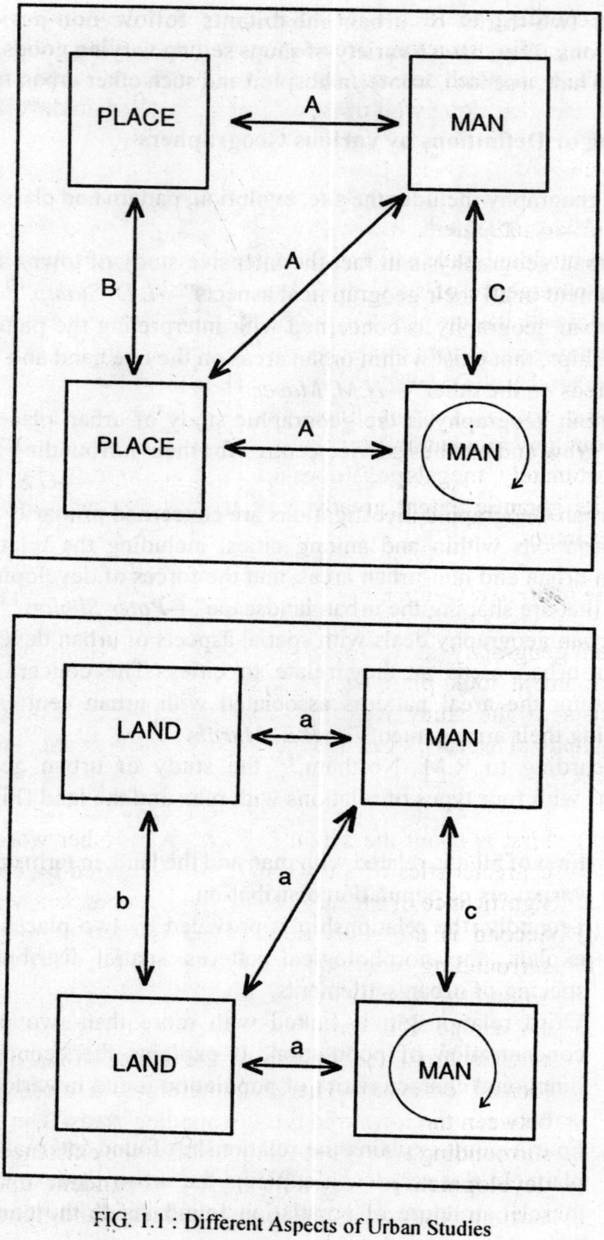

FIG. 1.1 : Different Aspects of Urban Studies

According to R.E. Dickinson[16] in urban geography studies are made regarding structure, process and stages of urban settlements as the home of man which determine its physical structure, social and economic characteristics along with traditional process of urban development.

Relations of Man with Land in Urban Areas

The urban settlements today serve unique functions in the organization of international trade and in industrial innovation and management. The rapid emergence of such a massive units of settlement has, however, brought with it acute planning problems. The essential need is to anticipate the long-term effects at the outset, so that efficient solutions can be formulated to the problems caused by metropolitan growth before they become critical. But the finest statistical analysis will be of no use without government machinery to implement the indicated policies. Unfortunately, megalopolis does not respect administrative boundaries and its gigantic extent greatly exceeds any existing regional unit of government.

Scope of Urban Geography

Urban geography as a distinct subject is of recent origin and its development took place within last five decades. Urban geography consists of the study of towns and their development in all their geographical aspects. The field of urban geography covers two types of studies :

(a) First is about the actual town itself. In other words the chief characteristics of a town have to be studied by analyzing the significance of towns and linkages of urban landscape.

(b) Second is the study about the influence of towns over its surrounding countryside. In fact, there is an extensive area around the town which is always influenced by the town with respect to economy, culture and political make-up. The surrounding areas have also the same influence over town located centrally. There is, therefore, a close relationship between the town and the surrounding areas. The study of the surrounding area is necessary because the existence and further development of town is highly dependent upon it. This surrounding area is often known as umland, hinterland, city region, zone of influence and command area.

According to Mayer and Kohn "The study of urban geography is largely a product of the 20th century. Early writers devoted much of their attention to the physical sites of urban places and to their situation (Fig. 1.2). It is closely related with other branches of geography in terms of location and structure of specific cities and the land they occupy."

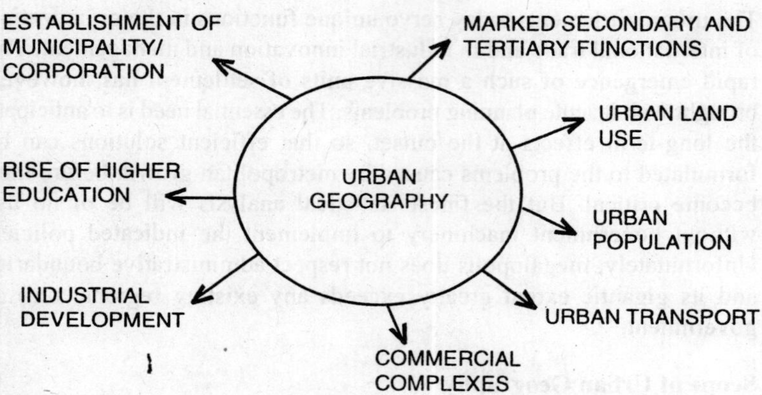

FIG. 1.2 : Subject Matter of Urban Geography

Today the urban geography commonly approaches this study from three points of view.

Firstly it considers city as a discrete phenomena in the general fabric of settlement of the entire earth; and a major part of it. As such it can be analyzed geographically in more or less the same way as any other phenomena similarly distributed. Concepts and theories may be framed regarding their distribution, size, function and rates of growth.

Secondly, area served by urban places should be delineated and the spatial interactions between urban places, inside or outside the delimited tributary areas, may be studied. Within the past few years several papers in urban geography have contributed the theoretical notions regarding the functional base of central places, the size and spacing of cities within selected areas, nodal regions besides other geographical considerations.

Thirdly, urban geography gives us the clue of cities in terms of their morphology besides origin, growth and functions. The studies which have focused on these aspects of urban geography have given rise to the concepts and generalizations related to the character and intensity of landuse within the city and to the spatial interactions of one part of the city with another. These interactions are reflected in the amount, direction, time and character of the movements of persons and goods between various functional areas that together make up the urban agglomeration. The study of urban transport, as a result, has become a major concern of geography besides the increasing number of urban places in recent years, especially the extension of the urban agglomeration into the adjacent rural areas. As a result of their interest in morphology, some geographers have recently turned their attention to problems involved in guiding urban growth and development.

Harold M. Mayer, in his article, "Geography and Urbanism" outlines[17] the major problems with which modern geography deals. Among these, he stresses the geographer's interest in the areal association of activities within urban places. The economic base of cities, the areas which cities serve in one way or another and patterns of distribution of cities are important subject matter of the study of urban geography. He further describes the growing interest in describing and understanding the spatial frictions existing in cities because of their morphology and the need for planned development of urban centres. Robert E. Dickinson,[18] in his paper, "The Scope and Status of Urban Geography : An Assessment," elaborates several problems, especially on the geographic approach to the study of a specific urban settlement. Dickinson states that the task of a geographer in all urban study is to determine the characteristics of site and situation of urban settlement, its historical development and the limits which it has currently attained. The article is also concerned with comparative functional and morphological studies and presents a valuable review of the significant papers that have contributed to a better understanding of the location, spacing and size of cities along with their layout and build.

In social surveys and town planning schemes it is invariably assumed that the geographer is primarily concerned with the physical ground plan on which the city rests. It has further conceded that the geographer follows such appraisal by tracing the effects of physical ground plan on which the growth and character of the city depend. But as a rule the investigation of these particular aspects is handled by the economist, sociologist, historian and architect. There are, however, numerous thorough studies of cities by geographers which indicate clearly that urban geography has a well-established scope and developed techniques for its study.

Economist, historian, sociologist and geographer study the urban settlement from different angles and it would be better at the outset to define the geographer's conception of urban settlement. The urban settlement is regarded by a geographer as a man-made habitat on the earth where secondary and tertiary functions dominate the scene. The problem of exact definition lies in the smallest urban settlements in any area. With such cases it will be found that size and administrative status are not essential criteria of true urban character. Function and form are the essential criteria of true urban character. The word 'urban' as opposed to 'rural' implies an activity that divorced from the cultivation of the soil which carried out in close association with commercial activities at fixed places. These activities in the broadest sense, are cultural, commercial, industrial, administrative and residential. Farming, however is not excluded from the occupations of the human communities. A farming element is particularly characteristic feature of the towns on the Mediterranean lands and of east central Europe.. It was important in origin of the early medieval towns of western Europe.

Specialization of urban functions in one place or another is a marked feature in the lands of western civilization. The development of transport facilities help the location of an isolated factory, a group of worker's houses in the open country, a mining camp and a inland resort. All these may be described as urban settlements. Whenever, these activities occur in combination in a permanent compact settlement with some measure of community organization, the place assumes the character of a town. A city is king among the towns enjoying leadership over its neighbours. A fundamental trait of both town and city in all ages has been they are institutional centres (commercial, cultural and administrative) for their surrounding territory. It is only in recent years that industry has become a primary cause of urban development.

By virtue of its distinct activities, the layout of buildings of urban settlements are also distinct from the rural settlement. The urban activities are accommodated in shop, workshop, office, warehouses and public building and their mutual and outside contacts by street, road, river and more recently by rail. The dwellings of the workers normally commands the entire build-up area.

The larger the town, the more the buildings. The various wings of the town spread from the nucleus, along the main streets. In large towns the urban functions are so numerous and varied that they have to compete for space as near as possible to the nucleus so as to form a central business district that throng with activities in the day and acts as dead heart during night hours.

The Limits of the Urban Settlement

The urban settlement cannot be interpreted adequately as a mass of materials or in terms of dead patterns. It must be interpreted as an organic part of a social group in terms of its four functions.[19]

(i) Dwelling,
(ii) Work Place,
(iii) Recreation, and
(iv) Transport.

Every urban settlement forms part of the economic, social, cultural and political whole, upon which its development depends and all these relations have geographic expressions. In urban geography studies have been made about town itself and its surrounding areas, and we must have the clear cut ideas about these two.

Finally, these characteristics of urban settlements—their size, functions, spacing, layout and build, vary regionally with the physical environment, the density of population and with the character of the human economies and cultures which they serve and represent.

Review of Past Literature

Urbanization deals[20] with the land as well as machine production. It is closely linked with the concentration of people at one place through migration from the fringe and surrounding areas with an ambitious hope for future materialistic self-development. But due to limited resource, the polarization of transportation networks are found just near urban centres which help in urban development on multi-level whether the same may be an urban expansion in a zig-zag crab like fashion or enlargement of the zone of influence by administrative set up and commercial development.

Hence urbanization is a component of regional economic development because urban centres provide *inter-alia*, a variety of centralized services for the surrounding zone of influence, e.g. marketing of agricultural surplus, products of cottage industries, including the supply of fertilizer, engineering goods, pumping sets, medicines and specialized skills in a wide variety of situations which are necessary for regional development.

There are three words which are generally used in urban geography such as urban centre, urbanism, and urbanization. Among these, urban centre is a place where the concentration of population takes place with marked secondary and tertiary occupations having a municipality and notified area committee. Urbanism, on the other hand, *is a social behaviour in* urban areas, e.g., there is a marked differentiation in social structure as the people of diverse regions, religion and income live close together which could not form a part and parcel of good neighbourhood. In urban areas one person uses to meet with thousand persons in a day and hence, it may not be possible for an urbanite to remember all of them for a longer period. Thus urbanization, on the other hand, is a process of social transformation from rural to urban areas. In this sense one place may be more urbanized in comparison with the other whether the same may be a village or town. A certain percentage of people in urban areas are found to be engaged in agricultural activities and similarly the urban component of secondary and tertiary populations are found in villages. In this way, urbanization in one place is a comparative element which may continue, decline, stop, found in lesser degree and there may be a high level of urbanization.

It is quite evident that due to urbanization the merger of villages into towns has been observed. Due to this process, the small towns developed into major urban centres. This also helps in transforming rural way of life into urban economic and social organization.

The transformation of rural service centres in both time and space, due to concentration of commercial, transportation, and professional services, give rise to the development of urban centres. The increasing intensity of such processes over a landscape leads to urbanization. The growth of urban centres also leads to changes in infrastructure facilities which in turn affect regional development, industrialization, transportation linkages, population distribution, and the entire rural-urban continuum. At the same time the dynamics of regional economics, population growth, increasing employment in the tertiary sector and the continuation of these processes for longer period of time and over increasing dimensions of space are signs of regional development. Thus *urbanization is a polarization technique* which have been adopted for regional development.

Urbanization usually brings with it regional prosperity as the provision of infrastructure facilities. It stimulates the development of locally available resources, increasing regional income and the level of employment. Hence, urbanization and regional development are closely related and cannot be separated. However, exceptions to the general pattern of regional prosperity abound.

Urban slums, e.g., are areas of regional, economic, cultural, political and social deprivation, which have not reaped the rewards of urbanization and regional development.

Urbanization provides, new economic opportunities through providing employment in new industries, including the provision of several other infrastructural facilities, e.g., electricity, irrigation, provision of higher education, transport and communication lines and the better opportunity of employment in urban areas. Besides these, urban areas also provide the facility of new political features of urban, state and national level, marketing of produce, diversification of occupations, commercialization of agriculture and change in consumption pattern. The problem of urbanization and regional economic development are basic issues of investigation all over the world. Almost all countries developed and developing continue to witness the process of urban development with its specialization of services, industrialization, multifarious modes of transportation linkages and allied functions. Urban development, of course is in different stages of development in different countries.

The process of urbanization includes the change of economy which may be from forest to agriculture, agriculture to manufacturing oriented production centres and centres of exchange of goods. During ancient period urbanization was the basic way of life in the valley of Nile, Greece and Indus valley's especially of Mohanjodaro and Harrappa. After industrial revolution world wide preferment of urbanization was quite noticeable when most of the lower order production centres have developed into urban centres. This has also engulfed the productive bari-Land of the country side, which favoured the concentration of industries and opened a new vista of employment opportunities at urban centres. This gave a great impetus to the growth of urban population. Later on with the establishment of government offices, the opening of higher educational institutions and the modes of transportation, nowadays urbanization becomes the focus of civilization. This is because most of the administrative, political, educational, religious and policy decisions are centred in the urban areas in each and every part of the globe. Hence, urban agglomeration have become focal points in the modern world.

City dwellers now control most of the large scale institutions that dominate the economic life and the well aspirations as generated from the offices located in cities.

Relationship between Urbanization and Economic Development

The study about the relationship between urban economic development is a nice subject to investigate in social sciences. The process of development acts like organic growth of a human body which is not a reversible process because one cannot revert the development which has already taken place in one time and space, but a decline may be the possibility in later stage. One can see the prospect of urbanization through rural-urban migration, urban social structure and rural-urban economic order inclusive of urban politics.

Due to urbanization the development of a region takes place in the following way :

Urbanization has a capacity to raise the intensity of rural-urban migration in each and every part of the globe. This is due to the changing social order by the people of various caste and class, such as Brahmins migrated to the urban centres partially for maintaining the traditional occupation, inclusive of secured jobs in different departments of the government. In this way, only a certain group of people have dominance in urban areas. Proportionately, the other dominant residents of urban areas are that of businessmen and industrial labourers.

Urbanization is the effect of survival of people from the beginning of human civilization when wandering group were formed and they settled on land with permanent agricultural occupation. In this way, agriculture which now the chief occupation of non-urban people, was one of the earliest steps in the long process of urbanization.

A city is necessarily made of labourers working in non-agricultural occupations. Growth of urbanization is closely linked with the agricultural productivity by releasing some of the workers from the necessity of working on land. In 1937 one farm family was feeding seven urban families. This factor is a powerful force which make possible a large number of population to live in cities. There must be a motivation and means for cities development. These were provided by a combination of several factors among which are the survival complex, advantages of group living and others. Other conditions attributed to the growth of cities are technological development, commercial revolution, increased efficiency of transportation and demographic revolution in combination with other economic and social development of a society.

The Concept of Urbanization

This concept is generally defined in terms of involving a situation in which larger proportion of population live in urban places than their degrace of economic development justifies.[21] This phenomenon has been noted with particular interest and concern because of the common observations. In all developing countries with few exceptions there appears to be larger population in urban areas to be supported by the employment, services, and facilities available from urban areas. This reflects the observation of western countries as compared to newly developing countries that a similar degree of participation of the labour force in occupations that were urban in character, i.e., non-agricultural is taking place at an usually rapid rate. However, it may be considered over urbanization from one point of view but may not be the same from another point of view.

Significance of Urbanization in Regional Development

There is a close liasion between urbanization and regional economic development. It exercises perceptible influence on the process of capital formation, saving, investment and industrialization.

The early stages of urbanization adversely affects savings. Because Low Saving Capacity exists in urban areas due to high cost of living and high standard of living. Although increase in income also takes place due to transfer and use of disguised unemployed labourers. Social structure and habits amount to more expenditure by the urban people but it is more compensated by the increase in per capita income resulting from the application of the improved methods of production which accompany urbanization.

Urbanization enhances the magnitude of investment in social overhead capital and directly productive activities. Social overhead capital comprises those basic services without which primary, secondary and tertiary productive activities cannot function. It includes all public services from law and order, education, and public health to transportation, communication, power and water supply as well as such agricultural overhead capital as irrigation and drainage systems.

Directly productive investment is primary in the production of goods in industries. With rapid urbanization the demand for the investment in social overhead capital increases in relation to the investment in directly productive activities. This distorts the pattern of investment in the sectors that might be conceived as the most efficient from the general economic development and employment points of view.[22]

Urbanization and industrialization have a very close relationship. Modern large scale industries cannot develop unless there are adequate economic and social overhead. Economics of scale are available only in towns and cities. Urbanization is accompanied by a shift in the occupation of people. It implies transfer of workers from agriculture to manufacturing services and distributive occupations. It implies a change in the occupational diversification and change in the occupational structure of the working force towards non-agricultural activities.

In this way, urbanization accelerates the pace of regional economic development by increasing the rate of capital formation, augmenting the magnitude of investment, enhancing the potentiabilities and possibilities of saving by expanding the level of income and by bringing occupational and structural transformation.

Concepts of Urban-Regional Development Growth Foci

According to R.P. Misra there are four tiers of the concept of "growth foci," which help in urbanization and regional develop in India.

Service Centres of the Local Level

Such centres are called *urban centre* or larger villages of lowest rank in urban development having a population of 5,000 to 10,000 concentrated at one nucleated settlement or detached into a number of tiny hamlets. The service centres of this category have the functional units of grocery shops, general merchandise shops, minor repair facilities, tailor shops, barbar shops, restaurants, primary and middle schools, post-office, co-operatives, community centres and other necessary infrastructural facilities. Such centres are generally the headquarters of *Mukhia* or *Sarpanch* with a few agricultural extension workers of the government.

Growth Points

These form the second higher category of service centres in any region. For the sake of regional development growth points play a crucial role. Such growth points will serve at least 10 to 20 smaller service centres. The population of such centres will range from 50,000 to 100,000. It is just like a major urban centre of higher market place in India.

Growth Centres

In the hierarchy of growth foci the third level centres are called growth centres. Growth centres are necessarily the production centres and manufacturing is the chief function. In this way, it is a centre of marked secondary and terriary functions. They will serve a population of 1,00,000 to 2,00,000 having a wider zone of influence. Hence, it is just the functional character of such a development node which are liable to be called growth centres. Thus it is the industrial hub of the area which it serves. It has also the facility of grain collection, storage, grading including fertilizer, pesticide and agri-implements, besides college and technological institutions.

Growth Poles

At highest level of the growth foci hierarchy lies the growth poles. Such centres may have a population of 5,00,000 to 25,00,000. In growth poles tertiary population have higher share in comparison with secondary and primary services. They are the focal points of regional economic development having predominance of financial, educational, technological and industrial sectors. In this way, growth pole hypothesis is indispensable for national, regional and local planning processes.

The Theory of Development Poles

It was first propounded by Francis Perroux but it has got an important place in French regional economy just like the balanced and steady growth theories. Perroux approach holds that analysis of sustained growth of total production should concentrate on the process by which various functions appear, grow and decline in temporal dimensions. It also emphasizes that growth rates which vary considerably from one sector to another and entrepreneurial innovation is a dominant factor in explaining the growth process, which takes the form of a succession of dynamic sectors or poles in time. Here regional interest has been aroused in the pole concept by emphasizing that growth is concentrated in various spatial foci as well as in leading industrial functional zones useful in analyzing the various consequences of alternative choice of location. According to Perroux "growth does not appear every where at the same time," it shows itself in points or poles with variable intensities. It spreads by different channels and with variable final effects for the economy as a whole.[23] This did not consider that development pole is equivalent to a key industry, an economic base, an industrial zone, or even some geographically concentrated phenomenon. Rather Perroux interpreted it essentially in economic and functional sense.

The Concept of Dominance

It first of all appeared in French economic literature and it explains a dynamic interpretation of economic activity. For Perroux the effect of domination "Consists of an irreversible or partially reversible influence exercised by 94 unit or another. An economic unit exercises this effect by reasons of its dimension, its negotiating strength, the nature of its activity or because it belongs to a zone of dominant activity.[24] The effect of domination has both economic and geographic dimensions, abstracted from any consideration of geographic space. Besides, as soon as any inequality among firms appears the breach is opened by which cumulative effect of domination insinuates itself.[25] In terms of geographic space dominant and propulsive industries make the cities in which they are located and poles of their regions.

Conflicting Tendencies in Spatial Development

Hirschman emphasized the spatial concentration of economic activity within the framework of his theory of unbalanced growth. He argued that development strategies should concentrate on a relatively few sectors rather than on widely dispersed projects. The key sectors would be determined by measuring backward linkage and forward linkage effects in terms of input-output Maxima.[26] In such a situation the growth is communicated from the leading sectors of the economy to the followers, from one firm to the other. The advantage of this approach over balanced growth where every activity expends perfectly in step with every other is that it leaves considerable scope to induce investment decisions and economizes over principal scar resource namely "genuine decision-making."

Myrdal's concepts of "back wash and spread" effects involve the workings of population migration, trade and capital movements, he emphasized about the selective nature of migration from the hinterland to the growth centres, where specially young are more prone to migrate. He considers the higher fertility rates of poor races and their impact on the working age group to total population ratio, whose effects may be unfavourable in the hinterland. Similarly, capital and income flow to growth centres will continue due to induced investment. Myrdal recognized the critical significance of non-economic factors to the cumulative process of maintaining poverty in the hinterland. The inability to support adequate health and education facilities and other conservative outlook, all are detrimental to the experimental and rational orientation of an economically well off society. The spread effect are meant for increased outlets of the hinterland agricultural products and tendency for technical advance to diffuse from the growth centres. The spread effects will be stronger at the higher level of economic development. The attractiveness of a growth centre is also weakened by using external technological diseconomies and higher labour costs. Ultimately the national and central governments are also likely to initiate policies directed towards greater regional equality.

Friedman's Core-Periphery Model

This model was initially put forward in a study of Venezuelan regional development policy by Friedman.[27] He felt that development is the unfolding of the creative in a society through a successive series of structural transformations by the process of innovation. Development taken place from small number of "centre of change" located at the points of highest potential interaction within a communication field. Innovations diffuse from these centres to areas of lower potential interaction. Such 'core regions' are major centres of innovative change, while all other territory consists of "peripheral regions" which are dependent on core regions for development. It is controlled by the following processes of feed back effects of core regions growth.

(a) The weakining of the periphery by resource transfer to the core;
(b) increased interaction and innovation in the core;
(c) a higher rate of innovation due to higher expectations and lower risks;
(d) the modernization effect;
(e) social and institutional change favouring innovation;
(f) linkage effects or the tendency of innovations to induce at another innovations; and
(g) production effect which increase scale and agglomeration of economics.

In India State capitals may be known as core regions within a nested hierarchy of spatial systems. The peripheral regions are dependent on core regions by virtue of supply and consumer relations, including administrative organization.

Lasuen Growth Poles and Organizational Space

The growth centre concept has immensely help in studying the problems of regional development through industrial organization. Lasuen believes that the "diffusion channels of specific information will become less and less selective and the spread of adoptions more extensive over sectoral and geographic space, and economic development faster and less polarized."[28] He proposes two-faced strategy to avoid problems of regional—economic disparities. At this stage, it is required to create a stronger interactions between all points across all topological spaces. This can be achieved by consolidating all plants into large multi-plant, multi-product, multi-city of firms by fostering inter-firms linkages. These should also be a very flexible policy to increase the relative attractiveness of lagging regions. Capital, credit, training, service, infrastructure and other firms of assistance to public and private firms should take account of the different priorities of firms at various stages of organization in case they are to be effective.[29]

Urban Development and Regional Planning in India

According to the policy of the government in the coming years the major thrust is to slow down the growth of large centres in order to increase the growth rate of the small and medium sized towns. It is possible with the provision of infrastructure facilities to small town in order to equip them to act as growth centres for the hinterland around each and every town. For this purpose increased investment proposed for housing, water supply, transportation and communication facilities, besides education, medical care and recreation. In large urban centres emphasis will also have to be given in improving the conditions of urban poor, slum dwellers and raising the civic services upto acceptable level.

In the past, the policy was to help in improving the conditions of congested metropolitan centres so that they would continue to grow at their current pace indefinitely.

Regional development has also assumed a new dimension in Indian planning. Towns, villages, blocks and districts are conceived as spatial units for micro-level planning on various levels. The hierarchy of spatial units relates mainly to the development of rural and urban areas. The Planning Commission has appointed a committee on block level planning with a view to prepare proper guide lines for planning from below in conformity with national goals.

(a) The basic objectives are optimum utilization of potential resources to increase income and employment;
(b) ensuring the benefits accrued from the development programme to the weaker sections of the population in particular; and

(c) Reorientation of the institutional set-up and building of appropriate organizational frame to protect the rights and privileges of the poor with progressive promotion of an egalitarian structure of ownership of national assets and utilization of the basic resources at area level.

Hypothesis Testing

In hypothesis testing the author tries to search the real world situation with the help of data available for different elements of urbanization after careful examination of their traits and modality in a particular region. In North Bihar the following hypothesis regarding the growth of urbanization and regional development may be tested to highlight the following points:

1. In an urban centre the increase in functional units is closely related with the growth of urban population.
2. The number of persons engaged in trade and commerce is also related with the areal extent of the city.
3. From the nucleus of an urban centre towards its periphery the decline in the density of population is almost a rule.
4. In case the ground plan of an urban centre is circular in shape, in such a situation the circular ring of settlements are found and in each such strata population having different income groups reside. In case the site is linear then central business area shows a ribbon development which is followed by light industrial, residential, slum, urban blight in the zone of refuge which lies almost on the rural urban fringe.
5. In the hinterland of urban centres the increase in infrastructure facilities like the development of roads, railways, ferry service, electricity development, construction of embankment and communication lines are available as these are the emblem of regional development. Urbanization helps in increasing the basic and non-basic income and functions of a city.
6. The urban centre having larger zone of influence attracts consumers from bigger area where as smaller centres attract only smaller number of consumers from the surrounding zone of influence.
7. With the growth of urban development and formation of social overhead capital and growth of directly productive activities taken place.

8. Urbanization leads to industrialization and consequently various linkage effects.
9. Urbanization enhances employment opportunities, expands level of income and encourages optimum exploitation of natural resources.
10. Urbanization accelerates rural-urban migration.
11. It helps in building the focus of civilization and nodal centre for the convergence of transportation routes.
12. Urbanization helps in making the urban centre the seat of local administration and place for higher education.

In this way, the author conclude that several parameters of urbanization are accelerating the pace of regional development as the above illustrations highlight the point.

In Table 1.1 the relationship between the urban population of North Bihar and the length of roads and railways have been demonstrated for testing the hypothesis.

TABLE 1.1 : Relationship between Population and Length of Roads and Railways in North Bihar

District	*Urban Population in '000 (X)*	*Length of Road Railways (Y)*	*XY*	X^2	Y^2
Saran	169,428	400	67,699	28,561	160,000
Siwan	78,252	325	25,350	6,084	105,625
Gopalganj	68,788	312	21,216	4,624	97,344
Paschim Champaran	145,559	485	40,325	21,025	235,225
Purbi Champaran	112,853	523	59,099	12,769	273,529
Sitamarhi	87,282	176	15,312	7,569	30,976
Muzaffarpur	190,765	412	78,280	36,100	169,744
Vaishali	108,640	300	32,400	11,664	90,000
Begusarai	154,689	315	48,510	23,716	99,225
Samastipur	88,026	360	31,680	7,744	309,600
Darbhanga	176,879	555	97,680	30,976	308,025
Madhubani	72,307	400	28,800	5,184	160,000
Saharsa	169,929	399	67,431	28,561	159,201
Purnea	286,018	514	147,004	81,796	264,196
Katihar	134,286	370	49,580	17,956	136,900
Khagaria	44,115	157	6,908	1,936	24,649
North Bhagalpur	23,231	136	3,128	529	18,496
Total	2,103	6,139	850,303	326,794	2,642,735

The hypothesis shows the relationship in between two variables. Thus a statistical hypothesis is a formidable answer to a question about relationship. Statistical hypothesis is usually classified into Null Hypothesis and Alternative Hypothesis; and test of such a hypothesis consists of testing the Null Hypothesis against alternative or research hypothesis.

The hypothesis $\overline{X} = \overline{X}_1$ is tested against the alternative $\overline{X} \neq \overline{X}_1$, where $\overline{X}$ and $\overline{X}_1$ are the means of two observed variables, where as $\neq$ is a sign of not equal to. Hypothesis originate from the observations of events. They provide a bridge between the theory and observed real world situation. This is because researchers generally ask questions about the relationship between variables in light of the existing theory. After the formulation of hypothesis, it is advanced for testing and drawing inferences.

The testing of hypothesis may provide us drawing deductive or inductive logic of inferences.

In deductive research events are explained by citing conclusions which are based on theoretical principles, where as in deductive research relationship between two variables is illustrated on the basis of data gathered through experience. This is because we have new principles and laws in geography and, therefore, a geographical research is mostly inductive. Both theoretically and by common knowledge it could be suggested that roads and railways length in kilometres have close relationship with the urban population.

The problem can be restated by the hypothesis that the greater number of urban population (independent variable) from 1981 to 1991 is dependent on the increase in Roads and Railways length (dependent variable). This simple relationship is illustrated in scatter diagram (Fig. 1.3) with the dependent variables on vertical axis and the independent on the horizontal axis. To know the distribution and magnitude of one variable with the other comparisons may be made to show that their correspondence could be estimated.

In order to test the hypothesis an example has been given below :

$$\overline{X} = 123.70, \overline{Y} = 361.11 = 17$$

To get the correlation value the formula for Pearsonian correlation co-efficient is as follows :

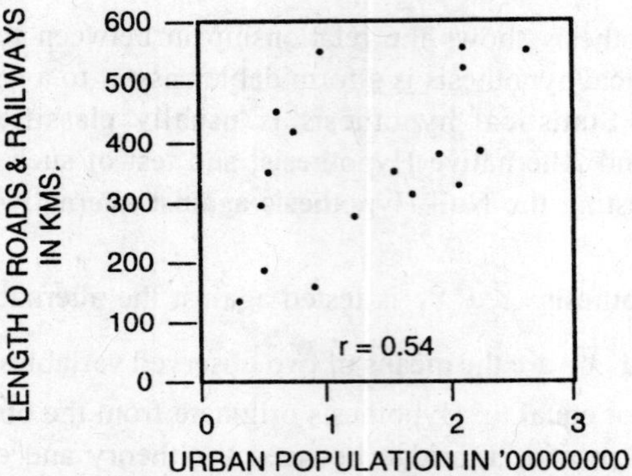

FIG. 1.3 : Relationship between Population and Length of Roads and Railways in North Bihar

$$r = \frac{\dfrac{\Sigma XY - \{(X)(Y)\}}{N}}{\sqrt{\left\{\Sigma X^2 - \left(\dfrac{(\Sigma X)^2}{N}\right)\right\}\left\{\Sigma Y^2 - \left(\dfrac{(\Sigma Y)^2}{N}\right)\right\}}}$$

where

r = Pearsonian correlation co-efficient,
ΣX = Sum of all values of independent variables,
ΣX^2 = Sum of squares of all independent variables,
ΣY = Sum of all values of dependent variables,
ΣY^2 = Sum of squares of all dependent variables;
ΣXY = Sum of products of all X and Y values;
N = Number of units.

With the help of above formula the product moment correlation co-efficient calculation is as follows :

$$= \frac{\dfrac{850303 - (2103 \times 6139)}{17}}{\sqrt{\left\{326794 - \dfrac{(2103)^2}{17}\right\}\left\{2642935 - \dfrac{(6139)^2}{17}\right\}}}$$

$$= \frac{90873}{168457} = 0.54 \ (r)$$

The effect of X and Y could be established with the help of linear regression, i.e.,

$$Y = a + bX \pm Sey$$

where $$b = \frac{XY - \frac{X.Y}{N}}{X^2 - \frac{(X)^2}{N}}$$

$$= \frac{90873}{66641} = 1.36$$

where $$a = \overline{Y} - \overline{X} - b$$

$$a = \frac{6139}{17} - 1.36.\frac{2103}{17}$$

$$= 361.11 - 1.36 \times 123.70 = 193.8780$$

Therefore, $$Y = 193.87 + 1.36\,X \pm Sey$$

Calculating from the above, we get the value of r = 0.54. This shows that the relationship is very close. The result has been tested by student '*t*' test by the formula:

$$t = \sqrt{\frac{r^2(n-2)}{1-r^2}} = \sqrt{\frac{.2916\,(17-2)}{1-.2916}}$$

$$= \sqrt{\frac{4.3740}{.7084}}$$

$$= \sqrt{6.17} \quad = 2.48$$

Ho According to Null Hypothesis (*Ho*) there is no correlation between the distribution of urban population and the length of roads and railways in 1991 in North Bihar.

Hi According to Alternative Hypothesis (*Hi*) there is positive correlation between the distribution of urban population and the length of roads and railways in 1991 in North Bihar.

Table value of '*t*' at 5 per cent level of significance and $df\,(n-2)$ 15 is 1.75 where as calculated value of *tc* is = 2.48.

Since the calculated value of '*t*' is greater than the table value at 5 per cent level of significance and degree of freedom 15. Therefore, Null Hypothesis is rejected in favour of Alternative Hypothesis that there is positive correlation between the distribution of urban population and the length of roads and railways in 1991 in North Bihar.

FOOTNOTES

1. W.S. Thompson, *Urbanization in Encyclopaedia of Social Sciences*, Vol. XV, Macmillan, 1935, p. 189.
2. Philip M. Hauser and Otis Dudley Duncan (eds.), *The Study of Population: An Inventory and Appraisal*, The University of Chicago Press, 1954, p. 34.
3. Hope Tisdale Eldridge, "The Process of Urbanization," in J.J. Spengler and O.D. Duncan (eds.) *Demographic Analysis*, Glence, III: Free Press, 1956, p. 338.
4. U.S. Census of Population, 1950.
5. R.B. Vance and Demcrath, N.J., *Op. cit*, pp. 22-23.
6. Spengler and Duncan (eds.), *Op. cit.*, p. 343.
7. O. Hirschman Albert, *The Strategy of Economic Development*, New Haven and London ,Yale University Press, p. 83.
8. G. Brian, *The Economics of Urban Area*, Pergamon Press, London, 1979.
9. G. Taylor, "Towns and Townships in Southern Ontario," *Economic Geography*, XXI, 1945, pp. 86-96.
10. L.D. Stamp, *Applied Geography*, Middlesex, 1960.
11. H.M. Mayer, Urban Geography, in *American Geography: Inventory and Prospect*, Syracuse University Press, 1954.
12. G.S. Gosal, "The Occupational Structure of India's Rural Population : A Regional Analysis." *The National Geographical Journal of India*, 1959, Vol. 5, pp. 137-48.
13. Pater Sholar, "The Scope and Status of Urban Geography," *Land Economics*, 1948, pp. 221-38.
14. K.E. Murphy, *The American City : An Urban Geography*, McGraw Hill Book Company, New York, 1966.
15. R.M. Northam, *Urban Geography*, (Second Edition), John Wiley & Sons, New York, 1979.
16. R.E. Dickinson, *The West European City*, London, 1951.
17. H.M. Mayer, *Readings in Urban Geography*, Central Book Depot, Allahabad, 1967.
18. R.E. Dickinson, "The Scope and Status of Urban Geography," *Land Economics*, 1948, pp. 221-38.
19. W. Winid, "The Scope of Urban Geography," Comptes Rendu du Congress International De Geographie, *Varsovle*, Vol. 3, 1934, pp. 171-83.
20. R.P. Misra, *Growth Poles and Growth Centres in Urban and Regional Planning in India*, University of Mysore, 1971, pp. 19-24.
21. H.V. Sovani, *Economic Development and Cultural Change*, 12 January, 1964, pp. 113-22.
22. G.R. Friedman, *A Central Theory of Polarized Development*, 1967, pp. 29-30.
23. F. Perroux, *Leconomic du XX Siecle*, Universitaries de France, 1964, pp. 142-53.
24. *Ibid.*, p. 85.
25. *Ibid.*, p. 40.
26. A.O. Hirschman, *The Strategy of Economic Development*, New Haven, Yale University Press, 1958, pp. 100-17.
27. *Ibid.*, pp. 62-63.
28. J.R. Lasuen, "Urban—Hierarchy Stability and Spatial Polarization: A Rejoinder," *Urban Studies*, 1970, Vol. 7, No. 1, pp. 86-87.
29. N.M. Hausen, *Intermediate Size Cities As Growth Centres*, Praeger Publishers, New York, 1971, p. 30.

2

ORIGIN AND EVOLUTION OF TOWNS

THE ORIGIN of urban centres or towns has been traced to about 5000 B.C. when the growth and development of several river valley civilizations in the world led to the origin of towns at favourable locations as permanent human settlements. For instance in Egypt Memphes developed about 4000 B.C. besides Indus Valley civilization of Mohanjodaro and Harrappa in Pakistan, the Chechar civilization in North Bihar and the Maya and Inka civilizations of North America. Taylor has traced the cyclic character of urban development. According to him, a town reaches the end stage of one cycle with the decay of previous civilization of which it was a product and it begins a new life with the imposition of a new civilization. Thus the present character and cultural background of an urban centre cannot be fully comprehended without knowing its past history of origin and evolution. The origin of towns may be caused by various factors which are in action for its site, situation and development (Fig. 2.1).

Origin and Growth of Towns

Ancient Period

In the Greek-Roman world of Europe, the city development started in 600 B.C. to 400 A.D. The early urban growth in western Europe was associated with Greek and Roman civilizations.[1] The city of Babylone evolved in 600 B.C. in Yunan which shows its typical circular structure and is based on religious sentiments (Fig. 2.2). The Roman conquest brought common culture to scattered hamlets in the villages of western Europe besides the road building, development of agriculture, mining and the creation of towns on favourable sites, for example gap, strategic and nodal towns, e.g., Stransbourg and St. Albans. The full potentialities of the ancient world to support a large city were realized only with the

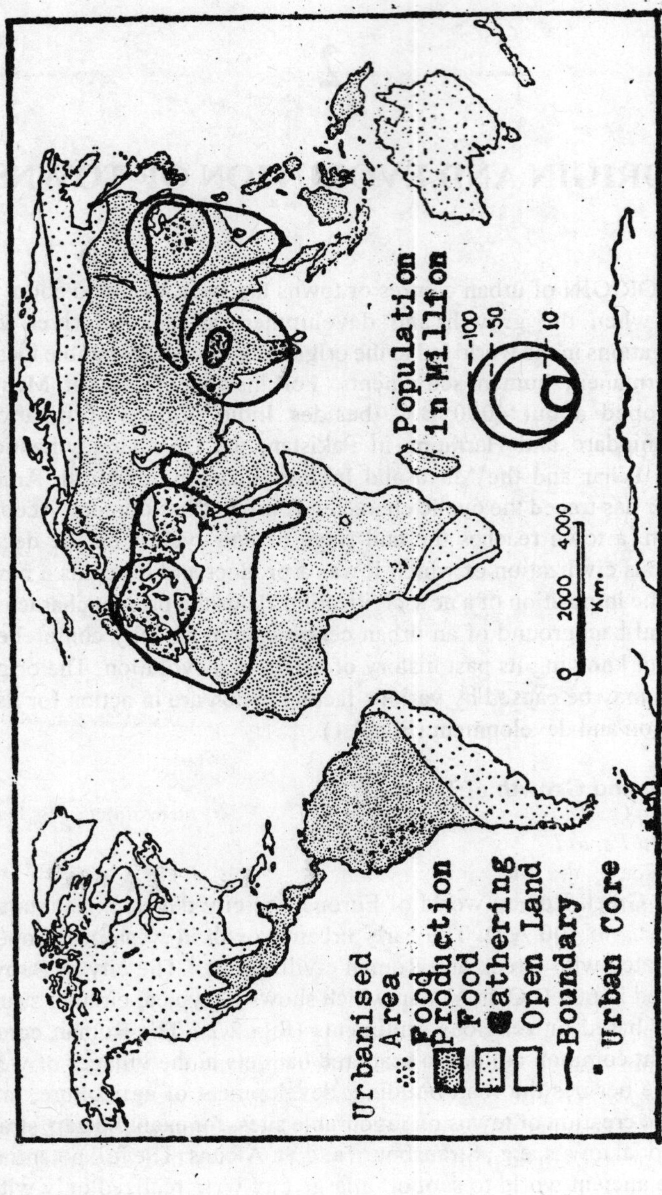

FIG 2.1 : 2000 Years Ago Distribution of Urbanized Area in the World along with Food Production and Collection Centres

Romans. In Western Europe, many of the modern trading cities originated during the Roman peaceful rule, London is a case in point (Fig. 2.3).

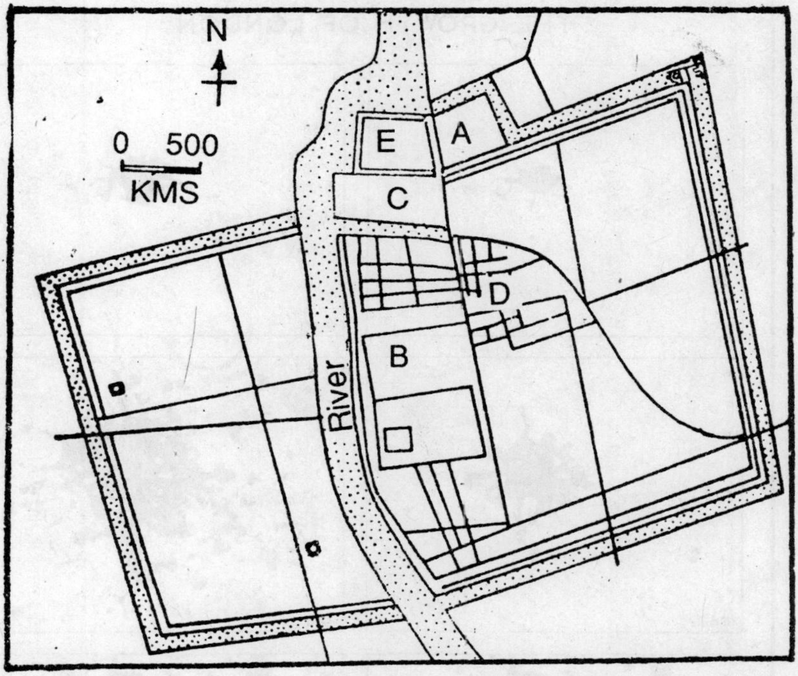

FIG. 2.2 : The City of Babylone in 600 B.C.

A. Isthar Gate, B. Temple, C. Hanging Garden, D. Old Pattern of Roads, E. Fort

In Spain, the Moors founded many towns in the course of their occupation. They include Gordoba and Granada towns which still posses Moorish architecture. In northern Spain, this was less affected by the Arab sway, indigenous fortress towns were established during the Spanish conquest of the peninsula, especially along the Fbro, Douro and Tagus.

In India the tradition of urban living and town planning goes back to 5000 years. The cities of the Indus valley civilization, Mohanjodaro and Harrappa, which flourished about 5000 years ago, were large and well-planned.[2] The other towns of this period are Patliputra of Chandragupta Maurya period, Ujjain of the Guptas period followed by Kannanj, Banaras and Mathura.

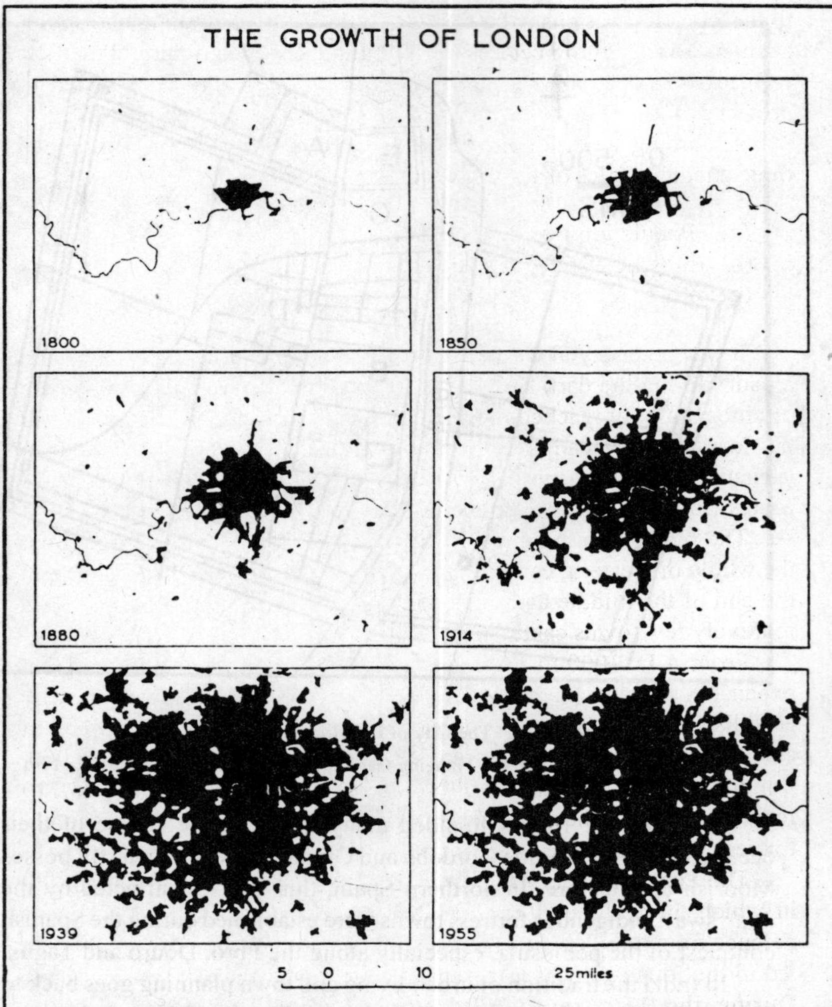

FIG. 2.3 : The Growth of London

The Growth and Spread of the Medieval Towns

During the medieval period, some of the villages in western Europe due to their nodal positions were found to be suitable for trading and socio-economic and cultural centres for their hinterland. Agricultural markets were established, churches were built and walls were constructed to protect the valuable workshops and communities of urban places.

According to H. Pirenue, a medieval town of western Europe had three characteristics of formation:

(a) Population was engaged in industry and commerce;
(b) it had a distinct legal constitution and institution; and
(c) it was a centre of administration and a fortress.

It was precisely in western Europe, where urbanization had reached a nadir during the dark age. Starting at the zero point, the development of cities not only reached the stage that the ancient world had achieved but kept going ahead. It kept going on the basis of improvement in agriculture and transport, the opening of new lands and new trade routes, and above all, the rise in production activity.

The overwhelming majority of the urban places today, throughout the whole of western, central and southern Europe, were in existence at the end of the middle ages. Thereafter, from about A.D. 1500 to 1800 relatively few towns came into being.

From A.D. 1000 to 1400 in western Europe there occurred a great expansion of urban places. By A.D. 1150 the concept of town was fully established. During the medieval period many urban places were of small population, as for example, Venice in the year 1422 had 190,000, London in 1337 A.D. had 30,000 and Frankfurt in 1440 A.D. had 8,719 residents. Most of these urban places developed along the site of church, fortress, administrative nucleus and business establishments (Fig. 2.4).

The Scottish rulers developed an independent pattern of urban life in which the walls and castles were important elements as means of defence from turbulent neighbours. The conquest of wales by the English led to the foundation of many small new towns adjacent to the castles. During the 13th and 14th centuries, urban life experienced its florescence—great commercial, trade and industrial emporium grew. It was under such circumstances that the large cities of western Europe developed, their size being proportional to the importance of the trade routes that centered on them. From A.D. 1500 to 1800 with the dominance of water and wood complex, relatively few new towns came into being

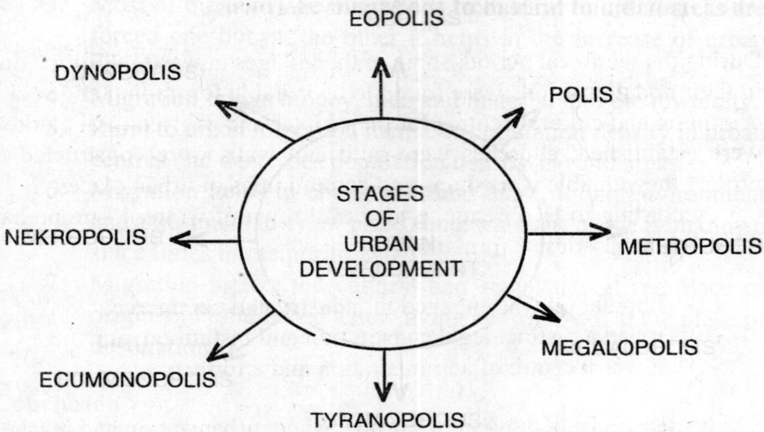

FIG. 2.4 : Lewis Mumfords Stages of Urban Development

especially in Scandinavea and the middle and lower Danube plains. This period was one of the relative stability, the most marked growth being in the capital, port and court cities. At the end of the middle ages, the Roman empire with twelve million people had some 10 to 15 per cent living in towns. The entire A.D. 1821 to 1851 was a period of concentration of cities in which case there was the rise of sixteen great cities and the expansion of London.

In south India, the great cities were built in the medieval period by the Chalukyas, the Rashtrakutas, the Cholas, the Hoysalas and others. Vijayanagar flourished during 15th and 16th centuries. In India our metropolitan port cities like Calcutta, Mumbai and Chennai are the examples of British contribution. The cities of Delhi and Bangalore present a synthesis of different traditions of the Hindus, Muslim and British.

Modern Urban Growth: The Industrial Revolution

The industrial revolution had its beginning in England and in the adjacent countries in the later half of the 18th century. In the middle of the 19th century had began the modern phase of coal and iron, the railway and the steam driven machines. In this period urban revolution began as the substantial proportion of population started living in towns and cities. By 1901 not less than 35 per cent of the population of England and Wales was living in cities of 100,000 or more and 58 per cent was living in cities of 20,000 or more.

Urbanization had tended to reach its highest point wherever economic productivity had been greatest, that is where the economy is industrialized and revolutionalized. This explains why urbanization is so closely associated with north-west Europeans and their culture, since they were mainly responsible for the industrial revolution.

In India, after independence, the economic and industrial development gave rise to a large number of mining, industrial, commercial and administrative towns. An urban centre having a population of 50,000 is known as small town, more than 50,000 is known as major town, population of 100,000 is known as city and having 10,000,000 is known as metropolis.

Types of Origin

Political Origin

In the entire globe, the capital cities and the stronghold of emperors during historical periods are the urban centres of political origin. Because, such human settlements have favourable ground for defence, development of commercial activities, transportation node, the facilities of fortification and the construction of moat have immensely helped in political origin of a town. For example, Samastipur is a district headquarter of North Bihar and it has been named after Shamsuddin Illias, the then ruler during Muslim period who established it. Similarly, Aurangabad originated in the name of Aurangzeb, a Muslim ruler in medieval India.[3]

Village Origin

A considerable number of towns transformed from village. They are known as overgrown villages, small towns or urban settlements or quasi-urban centres. For example, Jhanjharpur in Madhubani district was a village in 1961 census but it had been counted as an urban centre in the 1971 census. Similarly, Bariarpur in Munger district was a village in 1971 census but in 1981 census it had developed into a town. Some of the villages developed into an urban centre due to expansion of urban boundary. For example in 1971 census Hajipur urban centre had amalgamated 67 villages of the surrounding areas into its urban boundary.

Religious Origin

The religious character of some towns along with their association to some saints, deities and the presence of a number of temples, mosques and churches favoured the idea of their religious origin. In India the origin of most of the religious towns like Varanasi, Allahabad, Ayodhya, Gaya, Deoghar, Mathura and Haridwar are supposed to have religious history behind their growth and development. During Muslim period the urban centres of Fatehpursikri and Dehradun originated. During later period with the massing of devotees, the traders and artisans flocked to these centres to serve the periodic visit of pilgrims to these religious centres.

Industrial Origin

Many towns have developed due to industrial development. For example, before 1968 Jamshedpur was known as Sakchi village but with the establishment of iron and steel industry now it has developed into a city level centre. In European countries this momentum had gained long ago due to industrial revolution in 1917 but in developing countries of the world the industrial origin of town have gained momentum only in the last five decades. Other examples are Barauni, Sindri, Durgapur, Chitaranjan, Digboi and Bokaro.

Commercial Origin

Most towns and cities in the world have developed due to establishment of business nodes. For example, Mumbai, Calcutta, Lakhisarai and Rosra are some of the greatest business marts in India besides Chennai, Trivandrum and others.

Transport and Communication Origin

The connection point of roads, railways besides land and sea routes are the places of urban development. For example, Banka is a town in Bhagalpur district which originated as a road side town, while on the sea-shore the development of ports of Mangalore, Paradeep, Tutikorin and Cochi are towns of port origin.

Resort Towns

In the later half of the nineteenth century quite a large number of towns had came into existence as health resorts on hill stations and as military sanitoria especially in the Himalayan region. Such towns are Murrey, Mussoorie, Chakrata, Nainital, Darjeeling and Jammu. Even in interior parts of India, Ranchi developed as the queen of hills, Panchmadhi, Mount Abu and Ottakamund have originated as hill resorts.

Origin Due to Whim of Rulers

It has been said that 'God made the countryside and man made the town'. During historical periods the rulers had developed some towns for political and strategic reasons. Sometimes, they made towns to suit their pleasure and to satisfy their ego or whim. For example, in U.P. Jaunpur was made by Firoz Shah on the bank of Gomati river. Similarly, Asafaddulah left Faizabad in favour of Farukhabad. But the favourable geographical factors of defence, good site and productive upland were always there with the establishment of these urban centres.

Origin Due to Resettlement

After partition of India and Pakistan the towns developed were Faridabad and Nilokheri. Similarly, the partition of Punjab and Haryana gave rise to a planned urban settlement of Chandigarh.

The origin and history of urban development is essential for tracing its evolution. For example, Kanpur originated as commercial and industrial town whereas Lucknow originated as a capital town, and hence, they are quite different in morphology and evolutionary pattern.

The characteristic response in many parts of the world is poverty, degraded environmental conditions and lack of opportunities for personal advancement in rural areas. This has increased migration to the towns. This happened during the industrial revolution in the newly developed countries (where it still continues to a lesser extent) and it is happening today in many developing countries on a massive and frightening scale, bringing with it even greater environmental problems and dangers than the migrant leaves behind them. The migrant is not motivated solely, or even predominantly, by environmental considerations the dominant factor has been and is chiefly economic.

The principal influences on population movements and on the location and growth of large settlements have been the industrial revolution, with its accompanying advances in transport and communications, and other aspects of modern technology and organization. Underlying and reinforcing these technical trends are human and cultural factors—notably the principle of economic determinism, according to which those options which are seen to yield the greatest economic return are given priority. This principle, a legacy of man's long struggle against scarcity, which has contributed to some of the most severe and unnecessary environmental ills of the twentieth century.

In the less industrialized parts of the world economic considerations are not the only factor. Other factors such as desire for better educational opportunities and access to public utilities and health and welfare services also play a significant role in stimulating migration. In these regions, in fact, urbanization is now proceeding with greater momentum than industrialization, with the result that the very amenities which attract the migrant become incapable of meeting the increased demands made upon them. The mass movement to the cities, together with the natural growth of urban population, has placed a severe strain on the fabric and facilities of many older cities and thus created some of the most squalid and degrading human environments.[4]

The environmental side-effects of individuals' and society's desire for economic and social benefits are sometimes only a temporary phenomenon but in many rapidly growing cities, the problem is acute and permanent.

FOOTNOTES

1. Emrys Jones, *Social Geography of Belfast*, London, 1960.
2. G.L. Possehl, *Ancient Cities of the Indus*, Vikas Publishing House Private Ltd., New Delhi, 1979.
3. E. Ahmad, *Geographical Essays on India*, Kusum Prakashan, Patna, 1954.
4. *Human Settlements: The Environmental Challenge*, United Nations, Stockholm Conference, 1972, The Macmillan Press Ltd., 1975.

3

LOCATION AND SITING OF TOWNS

THE LOCATION of town in an area shows its latitudinal and longitudinal position and its significance and importance. Location shows the importance of towns and highlights about the surrounding environment. For example, it may be said that Chennai is located at sea shore of Bay of Bengal.

On the basis of locational aspects the towns may be classified into the following.[1]

LOCATION OF TOWNS

Riverside Towns

These are the towns located on river banks. Location of riverside towns is favoured by following reasons:

(a) Rivers have always been the source of drinking water supply for the human being especially for large population centres during the historical periods;
(b) Riverside location has religious importance for the sages and pilgrims. For this fact sanctity is always attached with the riverside location of towns;
(c) The migrant people who successively came to India from drier part of central Asia since the Aryan period upto the present were always in search of water frontage for which riverside is the best location for the towns;
(d) During the historical periods the problem of defence was an important consideration in the location of towns. The levees along the river embankment provided commanding sites for forts and towns. The protection was given by rivers which was considered as an important advantage;
(e) Before the construction of metalled road and railway the rivers were the main lines of transport and communication; and

(f) Rivers also create barriers for land routes and tracks for which crossings began to be used at ferry point and this favoured the location of towns such as at Mokameh in South Bihar (Table 3.1) and Calcutta along the Hooghly river (Fig. 3.1).

TABLE 3.1 : Location of Riverside Towns/Cities of India (1998)

Town/City	*Riverside Location*
Agra	Yamuna
Allahabad	Ganga, Yamuna and Saraswati
Ayodhya	Sarju
Badrinath	Mandakini Ganga
Calcutta	Hooghly
Cuttack	Mahanadi
Delhi	Yamuna
Dibrugarh	Brahmaputra
Ferozpur	Sutlej
Gauhati	Brahmaputra
Haridwar	Ganga
Jabalpur	Narmada
Kanpur	Ganga
Patna	Ganga
Munger	Ganga
Bhagalpur	Ganga
Varanasi	Ganga
Surat	Tapti
Vijayawada	Krishna
Sri Nagar	Jhelum
Nasik	Godavari
Lucknow	Gomati
Kurnool	Tungbhadra
Kota	Chambal
Sekobaghat	Brahmaputra
Shrirangpatham	Kaveri

Source : Collected by Author, 1998.

Interfluvial Location

The interfluve in between two rivers favoures the location of towns because they have the advantage of the convergence of land routes, commanding topography and a central location with respect to a productive region. Sometimes, interfluvial towns are also associated with administrative functions besides the development of trade and commerce.[2]

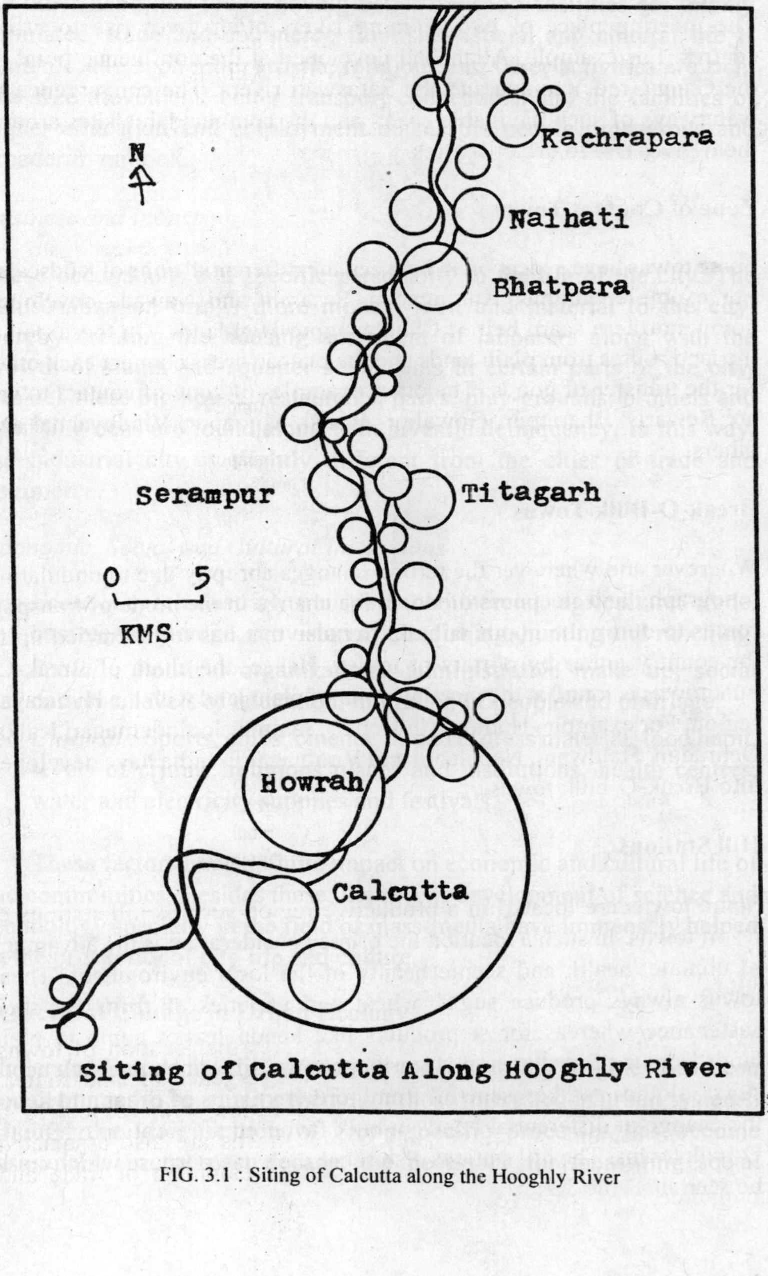

FIG. 3.1 : Siting of Calcutta along the Hooghly River

Confluence Towns

The meeting place of two or more rivers often gives rise to urban centres. For example, Allahabad developed at the confluence point of the Ganga, the Yamuna and the Saraswati rivers. The convergence of waterways of such navigable rivers and the commandable sites around them gives rise to urban centres.

Zone of Contact Towns

Some towns have a location in between the differential units of landscape. For example, Jammu, Aurangabad, Sasaram and Nawada developed below northern scarp belt of Chhotanagpur Highlands. On these points the land routes from plain lands and the plateau areas connect each other for the transfer of goods. The other examples of zone of contact towns are Rewari,[3] Bharatpur, Gowalior, Jhansi, Mirzapur, Vindhyachal and others.

Break-O-Bulk Towns

Wherever and whenever the terrain changes abruptly due to undulating topography and steepness of slope, the change in the mode of transport forces to dump the goods enbulk, because that has to be carried up to the country either by porters or mules. Hence, the chain of Break-O-bulk towns is found at the meeting point of plain land with the Himalayan region. For example, Hathout (in Nepal), Jammu, Jogindernagar, Kalka, Dehradun, Haldwani, Nepalganj and Viratnagar in India have developed into Break-O-bulk towns.

Hill Stations

Some towns are located in a productive region such as hill stations or resort towns. In such a location the prime consideration is the advantage of climate, health and scenic beauty of the local environment. These towns always produce sugar, wheat and varieties of fruits for their sustenance whereas forest products like kendu leaves come to plain lands from the forest area of Hazaribagh town. In turn Hazaribagh needs sugar, clothes and kerosene oil from northern plains of Bihar and hence the change of differential goods in bulk favoured the location of Break-O-bulk towns. The hill stations also serve the same purpose which could be seen in Table 3.2.

TABLE 3.2 : Mountainous Towns/Cities of India

Towns/Cities	*State*	*Height MSL (metres)*	*Characteristics*
Almora	U.P. (Kumaon)	1,538	Kath Godam station
Nainital	U.P. (Kumaon)	1,950	Naina lake
Lense Down	U.P. (Pauri Garhwal)	1,524	Kotdwar station
Mukteshwar	U.P. (Kumaon)	1,828	Veterinary Research Institute
Mussoorie	U.P. (Dehradun)	2,288	Queen of hills
Simla	Himachal Pradesh	2,136	Kalka station
Mahabaleshwar	Maharashtra	1,371	Western ghat
Kasauli	Himachal Pradesh	2,193	Kalka station
Utakamand	Tamilnadu	2,288	Ooti lake
Dalhousi	Himachal Pradesh	2,400	Pathankot
Srinagar	J&K	1,900	Dal lake
Gulmarg	J&K	2,515	48 kms. from Srinagar
Cherapunji	Meghalaya	1,358	Annual highest rainfall in the world (1,142 cms)
Darjeeling	West Bengal	2,177	Everest and Kanchanjanga can be seen
Kalimpong	West Bengal	1,219	Access through Siliguri
Ranchi	Bihar	640	Queen of hills
Shillong	Meghalaya	1,524	Jaintia hills
Kunjur	Tamilnadu	2,053	Nilgiri hills
Panchmadhi	M.P.	1,066	Hill station

Source : Collected by the Authors, 1998.

In IGC abstract volume held in New York Wong Pob Poh (1992) emphasized about appropriate Coastal Sites for resort development (p. 684).

The sea beach resorts of Southeast Asia are at various stages of development. Existing models of beach resort development do not consider adequately the coastal site; the coast is often assumed to be a single entity and the beach is treated as a linear sector. A classification of the coastal sites for Southeast Asian beach resorts is proposed :

(a) Hard coast beaches
(b) Gravel beaches
(c) Low sandy beaches
  (i) linear
  (ii) seaward concave: crescentic, zetaform, horseshoe
  (iii) seaward convex

(d) High sandy beaches
(e) Coral beaches
  (i) no moat/lagoon
  (ii) with moat/lagoon
  (iii) cay
(f) Estuarine beaches
(g) Artificial beaches.

Significant aspects of the coastal sites will be discussed with reference to examples given above. Although some overlapping of the categories is inevitable, the classification is simple to use. A better understanding of the coastal sites would help to establish an improved model of beach resort development.

Frontier Guard Towns

Some towns of strategic value are located in frontier areas. They are popularly known as frontier guard towns, e.g., Peshawar, Siliguri and Jalandhar. Agra had developed as frontier town for the Mughal homeland against rivalry of Malwa chieftains. Saharanpur developed as a frontier town for Delhi Sultunate in between A.D. 1206 to 1556.

Nodal Towns

The convergence points of rivers, roads and railways give rise to nodal towns. These towns are not supposed to serve the surrounding territory, and their source of support are the passengers passing through the route. New York is an example of nodal town.[4]

Fall Line Towns

The abrupt change of topography gives rise to waterfall, hydro-electricity generation stations and the resultant fall line towns. Such towns are Mohara and Muzaffarabad on the Jhelum river, Jammu on the Chenab, Jogindernagar on the Beas, Bahadurabad on the upper Ganges canal followed by Panchet and Tilaiya on the Damodar and the several fall line towns can be observed on the eastern side of Appalachian Mountains in the U.S.A.

Raw Materials or Mining Towns

In case of industrial towns the nearness of the availability of raw material is a prime factor in its location. For example, Jamshedpur is located in between the coalfield of Jharia, iron ore of Gua, limestone of Hathibari, water from Subarnarekha and labourers from North Bihar. The cement factory of Dehri-on-Sone is located close to the limestone deposits of Kaimur highlands; Ahmedabad and Sholapur are located amidst cotton belt of the Punjab and Maharashtra Kodarma on the mica belt of Hazaribagh, Kolar on the gold mines of Karnataka and Kalgurdi in the gold mines of west Australian desert. A list of industrial towns in India is shown in Table 3.3.

TABLE 3.3 : Industrial Towns of India (1998)

Towns	*State*	*Industry*
Jamshedpur	Bihar	Iron and Steel
Barauni	Bihar	Oil Refinery
Bokaro	Bihar	Iron & Steel
Durgapur	West Bengal	Iron & Steel
Bhilai	M.P.	Iron & Steel
Bhadrawati	Karnataka	Iron & Steel
Aligarh	U.P.	Locks
Anand	Gujarat	Cream, milk
Bangalore	Karnataka	Telephone machine tools
Chitaranjan	West Bengal	Railway workshop
Chandigarh	Punjab/Haryana	Locomotive
Calcutta	West Bengal	Fan, bulb, lamp
Delhi	Delhi	D.D.T.
Kanpur	U.P.	Cotton textile
Munger	Bihar	Gun factory
Dindigul	Tamilnadu	Cigar and tobacco
Baroda	Gujarat	Glass factory
Ujjain	M.P.	Nylon thread
Suratgarh	Rajasthan	Agricultural machine
Hatia	Bihar	Heavy machine
Pinjour	Haryana	Machine tools
Peramvoor	Tamilnadu	Coach factory
Koyali	Gujarat	Petro-chemicals
Khetri	Rajasthan	Copper processing
Gwalior	M.P.	Utensils of China clay
Gomia	Bihar	Explosive goods
Rourkela	Orissa	Iron & Steel

Source : Collected by Author, 1998.

Seaboard and Bridgehead Location

Quite a large number of towns have sea-board or bridgehead location. The sea-board towns are quite often the convergence points of land and sea routes in the form of port towns. It may be a naval port and sea resort. The bridgehead towns have the facility of river crossing through a bridge. Such towns may be exemplified by Calcutta, London, Yorkshire and Hajipur. Some of the major port towns of India are listed in Table 3.4.

TABLE 3.4 : Port Towns of India (1998)

Port	*Location*	*Characteristics*
Allepy	West coast	Kerala
Vatkal	West coast	Karnataka
Okha	West coast	Gujarat
Goa	West coast	Goa (Island)
Mangalore	West coast	Run with tides
Malpo	West coast	Fishing port
Chennai	East coast	Cotton textile
Kandla	West coast	Gujarat
Dhanushkodi	East coast	Railway terminal
Cochi	West coast	Kerala
Kojikode	West coast	Kerala
Calcutta	East coast	Run with tide
Mumbai	West coast	Maharashtra
Bhawnagar	West coast	Gujarat
Paradweep	East coast	Orissa
Surat	West coast	Maharashtra
Tutikorin	East coast	Tamil Nadu
Vishakhapatnam	East coast	Andhra Pradesh

Source : Collected by Author, 1998.

SITING OF TOWNS

Site is the actual space occupied by the town including its immediate environment. In plateau and mountainous regions the rocky structure of the site, the underlying rock formations, their resistance to erosion and suitability for foundation should be carefully studied. In a plain land of alluvial formation the mellow sediments lying horizontally consist of clay, loam and sand do not provide a stable siting. In case the ground is full of grit or pebbles the site is especially considered suitable for the growth and development of towns. Varanasi, Jaipur and Delhi are examples of such a siting of towns. Delhi is sited on Delhi ridge, especially in the fringe area of Peninsular India or on the northern flank of Aravalli range. The drainage lines of a town is wholly guided by the topography of the land. It is, therefore, said that site guides the morphology of a town whereas its situation guides its importance. Before site selection the planner should see the water table, besides the location of lakes, rivers, and the height of the ground from mean sea level. The character of soil and any dominant feature of the land should be carefully noted.

The first task of the geographer in an urban study is to determine exactly the characteristic of the site and situation of the settlement. The site embraces the precise features of the terrain on which the settlement began and over which it has spread. It includes such matters as relief, geology, water supply, the nature of the river that traverse the settlement, the areas liable to flood before embanking etc. It is usually taken to mean the physical conditions over a much wider area around the settlement. Nodality of the town should also be measured in this aspect on the basis of the functions of the settlements as a nodal centre. We need a measure of such centralized or nodal function in terms of relations of the settlement with the countryside settlements around it.

The development of towns in the early middle ages (before A.D. 1200) was favoured on valley and lowland sites. In the later middle ages the great majority of towns grew as local centres of trade and administration and nodality. Their sites indicate the need of defence rather than trade, and serve as castle town on strongly defensible natural sites[5] such as hill tops, river, spurs, etc.

The event of the railway in the middle of the nineteenth century added an entirely new element to the nodality factor and radically transformed the structure of the growing urban centres. The original conditions of the site were displaced by new means of communication, industry and commerce. But the great majority of towns are still small and have changed in population size many times within last five decades. The modern city has spread from its original nucleus along roads and in residential colonies to adjust with new demands of growing urban population with the study of physical conditions of site and situation which affected the beginning of the urban settlements, the geographer examines how, with the passage of time, the settlement utilizes, adapts itself to, and transforms these conditions in the process of its formation and expansion. Many urban studies especially in Britain, have traced the historical character of the adjustment of the settlement of the physical conditions of its site and situation and for that reason they do not give a clear picture of the present physiognomic and functional structure of the urban centre as an entity in space. The broad historical events decisive as the history of urbanism are common to western Europe. Towns in general grew little but changed much in aspect by the erection of new permanent and domestic buildings, gardens and avenues. New towns and extension of old ones were laid out on the geometric designs in accordance with the spirit of age as the same has been portrayed by Lewis Mumford.

FOOTNOTES

1. E. Ahmad, *Geographical Essays on India*, Kusum Prakashan, Patna, 1954.
2. V.N.P. Sinha, *Chhotanagpur Plateau: A Study in Settlement Geography*, K.B. Publications, New Delhi, 1976.
3. L.N. Varma, "Some Forms of the Town Sites in the Aravallis," *Indian Geographical Journal*, 1966, 4, pp. 52-63.
4. M.P. Thakore, "Sixteen Sites in Delhi," *Indian Geographer*, 1963, 8.
5. A.K. Tewari, "Mathura: A Study in Site and Evolution," *National Geographical Journal of India*, 1963, 9, pp. 48-59.

4

CLASSIFICATION OF TOWNS

THE STUDY of towns, as discussed earlier, constitutes a vital field of urban geography, and is based on a numerous distinct features, such as, age, size, population, geographic location, etc. taking into account such attributes, it can be seen that towns may vary from each other in terms of their characteristics. On this basis, various geographers have attempted to classify towns, which adds a new dimension in the study of urban development.

Though the towns have been classified in many ways, the most significant and meaningful classification has been made on the basis of the population size, the functional structure and the age.

Classification of Towns by Age

The relics of urban development came to-light after archaeological excavations. This reveals the earliest towns of the eastern world, e.g., Mohanjodaro, Harappa, Vaishali, Nalanda and Takshasila, Greek and Roman towns, medieval rennaissance Baroue followed by the industrial and modern towns. In the entire world at present the modern towns are thousands in number which developed into towns, cities, metropolis besides sub-urban area and conurbations. Lewis Mumford[1] had presented six fold classification of urban centres : Eopolis (small town), polis (major town), metropolis (large city), megalopolis (sectoral urban development in a country), tyranopolis (countrywide urban development), ecumonopolis (world urbanization) and nekropolis (as ghost city). Doxiades[2] has formulated dynopolis as the city of transport development. J.M. Houston[3] recognized three stages of urban growth in Europe :

The Nuclear Stage

This is represented by central area of a large town confined inside wall like the Boulevards of Paris, Vienna and Moscow.

The Formative Stage

During ninteenth century the extension of industrial revolution changed the transport network and trade. This favoured the spread of houses and factories outside the central area along lines of transport and communication.

The Modern Stage

This has developed the motor transport which favoured the growth of suburban towns in a big way. Griffith Taylor[4] while examining the urban settlements of Canada recognized four stages of urban growth which are as follows :

The Infantile Stage: The smaller town has been considered in this stage where the separation of commercial and residential areas could not be separated and the buildings and lanes are haphazardly distributed. Teghra in Begusarai district is an example of this stage.

The Juvenile Stage: In this stage the skyline of the houses rise and the factories are also established at places. Munger is an example of this stage.

The Mature Stage: This stage is marked by well separated residential and commercial area. The rise of new colonies on the fringe and vertical expansion is the rule. The urban centres of Ontario, Delhi, Patna, London, Dar'es Salaam and Cario come in this stage.

Senility: This stage is marked by cease of growth, decay of some areas and decline in economic development. In Britain, the industrial towns of Lancashire, Yorkshire and Durham are some of the examples. In Uttar Pradesh, Agra, Fatehpur Sikri, Mathura and Muzaffarnagar are towns of senile stage.

Mumford's Classification of Towns

As a physical entity, social phenomena, technological advancement, cultural ties and historical evolution Lewis Mumford considered the stages of growth of towns as follows.[5]

Eopolis: It is a small town of village origin-which are based on agriculture, mining and fishing.

Polis: It is a sort of market town with wholesale grain market and retail market. Some industries may also be found at this stage.

Metropolis: It is a large city with at least 10 lakhs of population. It has dominating position on neighbourhood towns and cities, along with number of industries and outlying residential colonies in its suburb. At this stage, the integration of culture, retail business and bankers are found along with the university level education.

Megalopolis: When several cities and metropolises coalese with each other forming a giant urban centre over a greater part of a nation than it is known as megalopolis. This is an urban area of material wealth, original art, varieties of business, industrial development and poly-nuclear city centres. New York, Boston, Philadelphia and Morrisville form a megalopolis on the eastern Atlantic sea coast of U.S.A. (Fig. 4.1).

Tyranopolis: In tyranopolis city countrywide urbanization predominates the scene. Here the display and expense become the measurement of culture, standard of living and the expansion of trade and commerce of both national and international levels. Great Britain may attain this stage where about 90 per cent people are living in urban areas.

Nekropolis: It is known as ghost city or dead city. This stage may attained due to war, famine and diseases and the decline of municipal services, decay of cultural institutions and the relict features of towns give deserted look like the ancient Babylon, Nineveh, Vaishali, Mohanjodaro and Harappa. This is the last stage of the decline of urbanization.

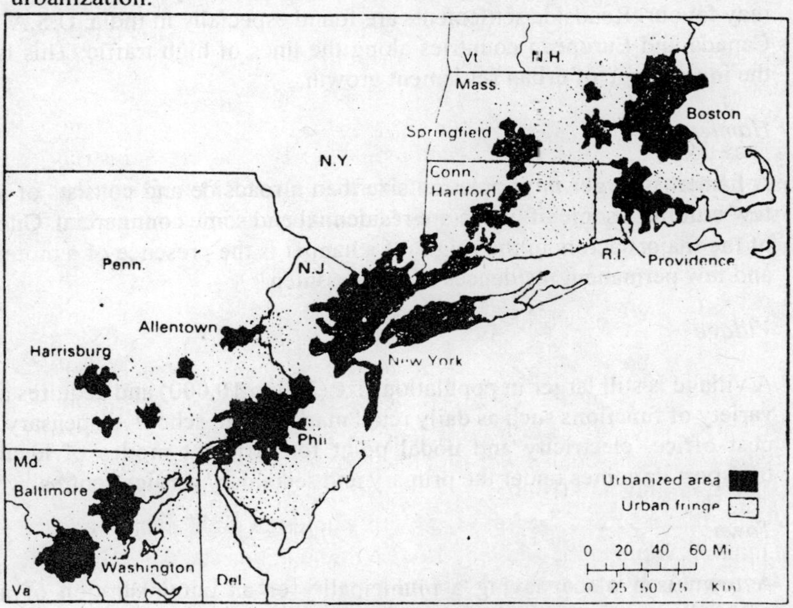

FIG. 4.1 : Megalopolis : Eastern Coast of U.S.A.

In contrast, we have the invisible city, the product of revolution in communication and electric transmission. This has helped even the remote villages to have the urban component of life and hence we avoid the city where the degeneration of urban environment by slums and squatters and heaps of garbage disfavour sound human living. Doxiades has favoured the term dynopolis a city not for man but meant for cars, aeroplanes, helicopters and rocket.

Cycle of Urban Development

The origin, growth and stage of an urban settlement favours its cyclic character. Whether it is a hamlet or a village, the availability of tertiary activities adds an urban character in it. Hence, the size of service centres ranges from roadside to ecumonopolis, a stage of world urbanization. Some of these terminologies are very loosely defined but some are very precise.

Roadside

A roadside is an isolated housing unit of gasoline station, petrol pump, betal shop or tea stall etc. situated along a highway. The owners of such units either reside in the same house or on the first story, as the situation may favour. Roadside settlements are found especially in India, U.S.A., Canada and European countries along the lines of high traffic. This is the initial stage of urban settlement growth.

Hamlet

A hamlet is larger in population size than a roadside and consists of a few buildings some of which are residential and some commercial. One of the major factors in the origin of a hamlet is the presence of a motel and few permanent residences of businessmen.

Village

A village is still larger in population size (150 to 10,000) and acquires a variety of functions such as daily retail market, high school, dispensary, post office, electricity and nodal point for different modes of local transport. It comes under the primary and secondary service centres.

Town

A town is a place having a municipality or an administration of a notified area committee. The population may range from 2,000 to 20,000 and it may have predominance of tertiary functions, e.g., shopping, transport, house renting and store house of various items along with a wholesale grain market. It may be a centre for Anchal Headquarter, telephone exchange, college and railway station as well.

City

In case an urban centre acquires a population of more than 1,00,000 than it is known as a city. This criteria has been adopted by the census of India. It acquires diversified functions such as administrative, business, education, transport, industrial and religious, etc. and different nodes of business area may be identified and it must be a railway junction or perform a university level service.

Conurbation

In a conurbation cities coalesce with each other due to expansion of industries and grow together economically. Politically it may be independent, viz., Mumbai and Thana constitute an urbanized cluster but they are independent from the point of view of administration.

TABLE 4.1 : Hierarchy of Urban Settlements

Sl. No.	*Level of Settlement*	*Example*	*Population (1991)*	*Characteristics*
1.	Roadside	Hirmani Gacchi Hat	20	Temporary businessmen
2.	Hamlet	Thahi	250	Satellite village of Majrahia
3.	Village	Baheri	15,000	Daily market
4.	Town	Rosra	35,000	Sub-divisional headquarter
5.	City	Bhagalpur	490,000	Divisional headquarter
6.	Metropolis	Patna	1,100,000	State capital of Bihar
7.	Conurbation	Bombay	12,600,000	Brick and mortar contiguity of cities
8.	Megalopolis	North-Eastern U.S.A.	90,000,000	Mid-Atlantic sea coast in North America
9.	Tyranopolis	Country/continental level	National level	Biggest urbanization continent or national level
10.	Ecumonopolis	World Urbanization	20,000,000,000	World urban community
11.	Nekropolis	Ghost City	Hrioshima (1945) and Basra (1991)	May develops in any stage of urban growth

Source : Prepared by Author, 1998.

Metropolis

According to the census of India an urban centre having 10,00,000 of populatior is known as metropolis. Hence, the term millionaire city is self-evident. The 1991 census of India showed that there were 25 metropolis in the country and about 150 in the entire world. It is also known as a cosmopolitan city.

Megalopolis

This term has been used for the most massive concentration of urbanized settlement on the Atlantic Sea Board of North America over a stretch of 900 kilometres from Boston in the North upto Florida in the South. It resulted obviously from the coalese of chain of metropolitan areas, each of which grew around a substantial urban nucleus. The super metropolitan character of this vast area and the greatest urban growth ever observed earned it a special name megalopolis given by Jean Gottmann.

Tyranopolis

When the whole country is urbanized than it may be called tyranopolis. Until now this situation of urban development has not aroused throughout the world but the increase of urban population of Great Britain, Germany and U.S.A. the spread of tyranopolis by 100 per cent urban development is not very far.

Ecumonopolis

This is the stage of world urbanization. Near about A.D. 2050 the world may pass through this last stage of urban development and people will get food from ocean which covers about 78 per cent area of the globe.[7]

Classification of Towns by Population

In terms of population, towns have been classified into: Small towns, major towns, city and metropolis, as shown in table below. These classifications are made on the basis of number of inhabitants in an urban centre and their density per square kilometre. Towns having a population below 50,000 are called small towns; and those whose population ranges from 50,000 to 99,999 are called major towns. A city has above 100,000 inhabitants, and a metropolis has a population of 1,000,000.

S.No.	*Class of Town*	*No. of Inhabitants*	*Example*
1.	Small town	Below 50,000	Meerut
2.	Major town	50,000—99,999	Patna
3.	City	100,000	Bangalore
4.	Metropolis	1,000,000	Delhi

Classification by Functions

The towns may be classified according to the predominant functions found in them. Such as, it may be a residential or dorminatory town, political or administrative town, transport or communications towns, commercial, industrial or a business mart besides mining, educational and religious towns. The details of functional classification of towns is given in the next chapter.

The towns may also be classified on the basis of environmental situations available, mode of transport, types of industrial establishments, kinds of administrative units, the size of population, areal extent and locational variability. Cultural variations in terms of festivals, dress materials and linguistic variations are other sorts of bases of classification.

FOOTNOTES

1. L. Mumford, *City Development*, Secker and Warburg, 1947.
2. C.A. Doxiades, "The Future of Human Settlements" in A. Tiselias and S. Nilson (eds.), *Place of Value in a World of Facts*, Stockholm Almquist and Wikselle, 1969, pp. 307-38.
3. J.M. Houston, *A Social Geography of Europe*, Duckworth, 1953.
4. G. Taylor, *Urban Geography*, London, Methuen, 1949.
5. L. Mumford, *The City in History*, Secker and Warburg, Penguin Books, 1966. .
6. J. Alexander, *Economic Geography*, New Jersey, 1963.,
7. F.S.A. Hudson, *Geography of Settlements*, McDonald & Evans Ltd., London, 1970.

5

FUNCTIONAL CLASSIFICATION OF TOWNS

THOUGH the towns have been classified on the basis of various features, viz., Age, population, geographical location, area, etc. the most significant classification has been made on the basis of functions which the towns perform. These functions are, of course, non-agricultural or tertiary in nature and lead to specialization of certain activities which differentiate towns from each other. For instance, some towns perform administrative functions, some name industrial functions, some are cultural centres and so on. Most of the towns are poly-functional in nature, i.e., they perform more than one function at the same time. But a town is generally known by a single dominant function it performs. For example, Calcutta is a port, a financial centre, an industrial centre and a cultural centre. Thus, on the basis of their dominant functions towns can be classified as follows :

FUNCTIONS TOWNS

Administrative Towns

They are mainly the capital city centres of provinces, states or the district headquarters, and deal mainly with the administrative organization. They are organized in certain public buildings, ministries and other government offices. They also house civil servants. Some administrative towns are: London, Washington D.C., Delhi, Mumbai, Lucknow, Patna, Chandigarh, etc. Geneva (Switzerland) is an international administrative town.

Commercial Towns

Commercial towns are those where retailing and which trading is the big function and they provide related facilities. They have business houses, banks, insurance companies and other financial organizations and offer many other commercial services. Examples of commercial towns are: London, New York, Frankfurt, Rosra and Muzaffarpur (in Bihar), Nagpur, Bhopal, Kanpur, etc.

Market Towns

Market towns are the places of collection and distribution of various goods, i.e., their main function is that of exchange. They are usually characterized by large business markets or *mandis* besides, a wide range of shops, stores, warehouses and godowns, supported by a good network of transport facilities.

Mining Towns

Exploitation of minerals from the mines is the main function of such towns. Thus, their location is largely governed by the availability of mineral resources. The mines can be of coal, limestone, iron ore, mica, bauxite, etc. Such mining towns are: Raniganj, Jharia, Bokaro, Beladila, etc. They have small settlements.

Industrial Towns

Industrial towns mainly perform secondary activities of transforming raw materials into manufactured goods. Such towns provide facilities for power, raw materials, labour, market and an efficient network of communication. These towns usually specialize in a single product, viz., textiles, iron and steel, machinery, chemicals. etc. for instance, Bokaro (steel), Ahmedabad (textiles), Bangalore (watches), etc.

Cultural Towns

They are the towns having cultural functions such as the inoculation of education, art galeries or religious buildings. These can be :

Educational: They are characterized by universities, college buildings, libraries and playgrounds. They also have shops which mainly catch to the needs of the students, such as bookshops, sports shop, etc. For example, Cambridge, Oxford, Heidelburg, etc.

Entertainment: Such towns are known for their theatres, art galleries and other cultural activities as that of film-making, e.g., Hollywood, Mumbai, Chennai, etc.

Religious: Religious towns can be centres of pilgrimage on the seats of religious leaders. They are characterized by religious monuments shops selling religious books, pictures, cassettes, candles, agarbattis, etc. These towns are visited by pilgrims from all over and often serve as centers of tourist attraction. Such towns are: Jerusalem, Mecca, Varanasi, Rishikesh, etc.

Resort Towns

Resort towns are those which provide recreational facilities for people and are located in favourable geographical surrounding and salubrious climatic conditions. Resort towns are characterized by hotels, guest houses, cinema halls, night clubs, amusement parks, shopping centres, etc. These towns may be: (i) coastal resorts providing sea-side recreation and facilities for water sports, such as, Miami, Brighton, Mumbai, etc.; (ii) Hill Resorts providing scenic beauty, cool climate and adventerous and thrilling sports, i.e., scating, trekking, etc. such as, Shimla, Mussooree, Nainital, Darjeeling, etc.; and (iii) Health resorts mainly based on health-giving waters, etc. These towns have usually a more neat and clean environment than other towns, and cater to the needs of medically unfit people.

Defensive Towns

Such towns have defensive functions. They may be naval dock or air bases and are characterized by barracks, cantonments, training facilities for the armed forces, special air fields, docks and harbours, e.g., Cochi, Chennai, Ambala, etc.

Residential Towns

These are mainly the modern towns which are usually located away from the congested cities and provide residential facilities for urban people. These are often well planned and have a neat and healthy environment, e.g., Chandigarh.

STUDIES IN FUNCTIONAL CLASSIFICATION

The concept of functional classification of town had been initiated in the early eighteenth century and today the concept has undergone tremendous changes so far as its definition and impact on urban planning is concerned. Before 1940 most of the classification techniques were based on empirical or subjective studies. In Britain the committee on the Health of Towns classified the towns cf Britain in 1840 in five groups:

1. The Metropolis;
2. Manufacturing towns;
3. Populous sea port towns;
4. Great watering Places; and
5. Country and other considerable inland towns not being the seats of particular manufactures.

Here the distinction is made between central place cities (5) and special activity cities (2, 3 and 4) and the variable size is indicated by the words 'populous', 'great' and 'considerable' small.

In 1921 M. Aurousseau in his paper "The Distribution of Population: A Constructive Problem," classified towns into the following six categories:

Class I—Administration	:	Capital cities, Revenue towns.
Class II—Defence	:	Fortress Towns, Garrison towns, Naval Bases.
Class III—Culture	:	University towns, Cathedral towns, Art centres, Pilgrimage centres, Religious centres.
Class IV—Production Towns	:	
Class V—Communication	:	
(a) *Collection*		Mining towns, Fishing towns, Forest towns and Depot towns.
(b) *Transfer*		Market towns, Fall Line towns, Break-o-Bulk towns, Bridgehead towns, Tidal Limit towns and Navigation head towns.
(c) *Distribution*		Export towns, Import towns, Supply towns.
Class VI—Recreation	:	Health Resorts, Tourist Resorts and Holiday Resorts.

In this classification he tried to distinguish the primary or basic, secondary or non-basic functions.

Mackenzi has classified the towns under four heads (i) first was the primary services community which linked the rural agricultural markets with the large economic centres as metropolis; (ii) second was commercial town. The type of commerce depend upon the linking primary service community; (iii) Third was the manufacturing centres which included heavy industrial centre as well as the cottage industry centre and the (iv) was recreation centre. The basic handicap was that Meckenzi did not taken into consideration several other functions besides these four.

Gist and Halbert classified the towns of other areas in the United States of America on more or less the same line followed by Aurousseau but he added one more type that is diversified towns.

Rosir lists of diversity of categories some cities may fall into more than one of these:

1. Cities of consumption (residential or resort places).
2. Cities of Production (mainly manufacturing).
3. Mixed activity cities (consumption and production).
4. Cities of storage and distribution (wholesale).
5. River and seaport cities.
6. Receiving and forwarding cities (Free ports where goods are transferred).
7. Cities of finance and credit.
8. Cities of working men and artisans.
9. Military cities (Malta, Belfast, Kiel).
10. Thermal or Bath cities (Rajgir and Allahabad).
11. Climatic cities (Nice, Annecy, San Remo).
12. Museum cities (Hyderabad).
13. University cities (Oxford).

Other special function cities are religious centres, pilgrimage cities, mining cities and oil cities. Among the unclassified types are listed twin cities and declining or dead cities. Among dead cities Rosir named Carcass one in France and Bruges in Belgium. One point made clear by Rosir is that classifying urban agglomerations by whatever method is arbitrary, but it is some how informing.

Weimer and Hoyt classified towns on the basis of sources of employment in a particular function. On this basis town have been classified into (a) commercial, (b) political, (c) industrial, (d) recreational and (e) health resort centres.

After this Robert B. Haul classified the cities of Japan under four heads:

1. Capital town;
2. Castle town;
3. Temple and shrine town; and
4. Commercial and modern industrial town.

It has been mostly classified on the basis of their growth, because Japan is a land of temple and castle. Hence, this classification could be said to be more effective in that particular country only.

In India in the early nineteenth century attempts were made to classify towns on similar objective levels. The largest city with function of a capital was termed as Rajdhani. The big agricultural town surrounding small rural marketing centre was termed as "Kheta." The commercial towns with relatively smaller population termed as "Pattana." A big commercial emporium near the mouth of a river or sea coast termed as "Dronamukha vivambra," and the forestress town was known as Durga, as Sasaram in India.

Categories

1. Manufacturing;
2. Retail;
3. Diversified;
4. Wholesale;
5. Transportation;
6. Mining;
7. University; and
8. Resort and retirement towns.

Criticism

In this classification :

1. The first drawback is that he followed different criteria for different functions.
2. Secondly, his classification does not distinguish the basic and non-basic classification of towns.
3. The third drawback is that he did not include administrative centres in the functional types.
4. Harris followed different methodology for the classification of cities of U.S.A. and former U.S.S.R., due to their difference in economic and cultural background.

In former U.S.S.R. he considered the towns as industrial on the basis of per capita value of sales while the transportation towns were those in which the percentage of gainful workers in transport was exceptionally high. The towns which are not specialized either in transport or industry were described as trading centres. This classification does not seems to be correct, because any town which is neither manufacturing nor transport in character need not necessarily be a trade centre.

Hart classified the cities of American South on the basis of percentage figures whom he termed as 'mineral quarta' and decay diffusion or 'mineral earners' minimum percentage in a particular occupation. Hart divided manufacturing centres on the basis of decile variation that is one-tenth workers of a particular town engaged in manufacturing termed as "manufacturing town" following the basic principle. But in most cases more and more people have been found in wholesale and recreational activities in comparison with the factory. He considered diversified towns under three heads manufacturing, retail trade and transportation.

Dutton Race classified the towns under two groups: (i) manufacturing and non-manufacturing; and (ii) trade and non-trade centres. The basis of classification was the volume of labour force in different sectors in a particular locality.

Pownell classified the cities of New England on the basis of positive and negative deviations. Naturally, if a particular town has got more percentage of people employed in a function than the average deviation supposed to be positive, if less than average then was negative. All towns coming under negative category was supposed to be specializing in residential function. He considered a group of cities of the same size. In that way, this method is more advance than Harris who did not consider population control behind the functions.

Nelson classified the American cities in the year 1950 in terms of 'deviations', but he did not consider the positive and negative deviations as scientific. Therefore, he found out the Standard Deviation from the average in a particular function of towns. The standard deviation considered for the purpose is as follows :

$$\sigma = \sqrt{\frac{d^2}{N}}$$

If diversified town has 20 per cent an average

Then Average + 1 *SD* = 20 + (1 × 15) = 36%
Average + 2 *SD* = 20 + (2 × 16) = 52%
Average + 3 *SD* = 20 + (3 × 16) = 68%

With the help of this method Nelson emphasized that this method is helpful for comparison between different functions of cities. Through this method he tried to categories those towns which fall below the average figure as diversified towns. Towns having average value in a particular function termed as 'average town'. Towns with a value of Average + 1 *SD* termed as *S.D.* one town, that with average + 2 *S.D.* as

S.D. 2 towns; and Average + 3 *S.D.* as *S.D.* 3 towns. This method is popular till today.

Criticisms

Firstly, this method is not suitable for the classification of various population size of towns and their relationship with a particular function. Secondly, the method is not suitable for inter-town comparison, but suitable for intra-town comparison. Third drawback is that the specialization of a town in terms of all the functions cannot be assessed by this method. Even then, this method is effective in planning a particular town, because it brings out personal association of towns.

A.E. Smiles had developed a method of gradation of town on the basis of services that is full-fledged town, which have at least five key functions such secondary school, cinema, insurance, retail business and transportation facility from the surrounding region. In case a town has all these functions that should be treated as the central town. According to this theory if in a particular town these five functions are not found than that is a sub-town.

In 1939 the political and economic Planning Authorities of U.K. had developed a method for classifying the cities of Great Britain. The classifying process is known as the 'quotient method'. In this method for each town quotient for a particular function has been found out in comparison with the local town quotient, and if the ratio is one then the former has got the average position in that function in comparison with the regional average. If the quotient value is more than one that signifies the increasing specialization of that function in that town. Later on each figure has been multiplied by 100, then four categories emerged at the interval of 40 to show the gradation of specialization.

John Webb classified the city of Minnesota USA with the help of 1950 census in a new way. He invented the method of weighing individual percentage and simple percentage ratio divided by the British P.E.P.; but this was not adequate in bringing out the specialization of a particular town. If a particular town gets 60 per cent of its population engaged in manufacturing and the regional average figure for manufacturing is 40, then the ratio between the local percentage and the mean percentage will be 3 : 2. Similarly in trade if the employed workers in the town is 3 and the corresponding mean figure is 2. Hence, the ratio again comes to 3 : 2. Naturally, 60 per cent of people employed in manufacturing cannot be given an equal status with 3 per cent of people in trade.

Therefore, he weighted the previous ratio by multiplying it with 60, and the later ratio by multiplying with 3.

The weighing may be as follows :

$$\text{Local Place Average of all Towns} = \frac{\text{Local\%}}{\text{Mean\%}} = \frac{10}{5} = 2 : \frac{5}{10} = 0.5$$

The balance should be maintained between the individual percentage and the final figures. After finding individual functional indices, he added all these figures to get a specialization index for each town. This specialization was discovered first of all by John Webb which was not a less stability in case of Nelson's Standard Deviation method. His method, however, does not as certain functional association of towns, nor does his classification demarcate diversified towns as demarcated by Nelson. In this classification all the towns came in the picture, however, less specialized they may be.

In India V.L.S. Prakasha Rao classified towns of Mysore State on the basis of regression method. In this method he correlated population size with the percentage of people in a particular function. As for example in Mysore he discovered that as the percentage of workers in industries decreased the town became more and more specialized. Simply speaking he correlated population size with the percentage of population in a particular function.

According to V.L.S. Prakasha Rao, Nelson's Standard Deviation method is useful only in case of homogeneous population. But in a state like Mysore where the population of town ranges from 10,000 to more than 200,000 the standard deviation method becomes fruitless.

The other Indian geographers who classified the Indian cities are Kashinath Singh and Amritlal of Ludhina. Kashi Nath classified the towns of Uttar Pradesh on the basis of Standard Deviation method. Amritlal classified the cities of India on the method applied by British P.E.P.

Mrs. V. Janaki of Baroda University classified the towns of Kerala mostly on the basis of their growth. She does not apply, however, any statistical technique. She classified the towns of Kerala State on five sub-heads: 1. Administrative, 2. Commercial, 3. Agricultural collecting and distributive centres, 4. Temple towns, and 5. Plantation towns.

Mattila and Thompson Method

The national proportion of workers employed in any occupation is obtained and the number expected to be so employed in a town of any size is calculated. If the actual employment when compared with the expected total shown a surplus, then this represents basic employment. This is the principle used in a clearest statement of this concept by J.M. Mattila and W.R. Thompson who proposes an index of surplus workers.

$$S = ei - \frac{et}{Et} Ei$$

where *ei* is the regional employment in the industry concerned; *et* is the total regional employment in all industries, and economic activity; *Ei* is the National employment in the industry concerned; and *Et* is the total National employment in all industries and Economic activity.

The Index can be used in classification of towns by the following procedure. The index is calculated for all occupational groups in the town and all positive values are totalled giving an aggregate of surplus workers for the town. The percentages of the aggregate formed by each occupational surplus are calculated and ranked. These rankings are then used as a basis for functional classification. An application of this index of surplus workers can be examined in provincial metropolis where it is used in the study of the Economic base in South East Lancashire.

They argued that the minimum percentage employed in any city provides the no-basic ratio since this must identify the smallest proportion for a viable city. This is an argument clearly open to serious challenge. Indeed the use of minima had been considered and rejected by Alexander on the ground that anomalous cases would become the decisive one's. His choice of the *K* point was specifically made to avoid that problem. To some extent Ullman with city size is taken into account. This non-basic ratio is subsequently used to classify cities not by comparing one employment sector with another but by measuring the extent to which a city departed from the various minima, thus providing an index of diversity (*D*) where :

$$D = i\frac{(Pi - Mi)^2}{Mi} = \frac{(ipi - iMi)^2}{iMi}$$

where *i* is each of the Employment group; *Pi* is the Percentage employed in each of *i* group; *Mi* are Minimum requirement for each group; and *i* is the Sum of all the groups.

The higher is the index the more specialized is the town, the lower it is the more diversified. It is apparent that this index demonstrates how statistical measures are increasingly employed and the towns differ from each other is the real basis of classification.

Classification of British Towns, Classes Identified by G.A. Moser and W. Scott (1961)

Mainly Resorts, Administrative and Commercial Towns

1. Mainly sea side resorts;
2. Mainly pass, professional and administrative centres; and
3. Mainly commercial centres with some industries.

Mainly Industrial Towns

4. Including most of the traditional railway centres;
5. Including many of the large ports as well as two black country towns;
6. Many textile centres in Yorkshire and Lancashire;
7. Including the industrial towns of the north-east sea board and mining towns of Wales; and
8. Including the more recent metal manufacturing towns.

Suberbs and Suburban Type Towns

9. Exclusively residential suburbs;
10. Mainly older mixed residential suburbs;
11. Mainly mixed residential suburbs;
12. Including light industry suberb, national defence centres and towns within the sphere of large conurbations;
13. Mainly old working class and industrial suburbs; and
14. Mainly newer industrial suburbs.

Not Allocated : London and Hyton

It is interesting to compare these results with the initial identification of urban functions which was suggested at the beginning.

1. Central place activities are directly comparable with the mainly administrative and commercial towns. But the resorts included would have to be replaced with the special activities.
2. Transport activities are not directly shown to produce a distinct set of towns. In this classification identification has been broadly included other towns which have no distinctive transport function.

3. Special activities include the remaining groups, although 'suburbs' have no distinctive functional connotation. They are parts of larger units which appear in the classification due to the use of local government units.

Hadden and Borgatta's Method

Another attempt regarding the classification of urban centres was made by Hadden and Borgatta in relation to American cities. Sixty-five variables were used and separate analyses for different city sizes were carried out.

From the Matrix sixteen factors were extracted and the first of these, socio-economic status was the same as that in the British study. The second, third and fourth factors, however, were linked to non-white population, age composition and education. Subsequently, the factor were used to select variables in order to construct a profile for each city by a simple decile diviation and this does not provide a classification as such.

M. Mukherjee's Method

M. Mukherjee in her study of functional classification of towns of Bihar had adopted a method to evaluate towns economic pattern in terms of surplus workers. Mukherjee suggested that if a locality possesses more workers in a particular function than the average for the region might indicate extra workers in that localities.

Her formula for finding out functional indices is :

$$\frac{ei - \dfrac{ef\,Ei}{Et}}{ef - \dfrac{Ei}{Et}} \times \left(\frac{ei}{ef} \times 100\right)$$

where *ei* is Local functional employment; *ef* is Local total; *Ei* is State functional employment; and *Et* is State total employment.

For finding out specialization indeed, all the *FI* (Functional Index) values are added and categories are grouped in arithmetic progression at an interval of 40.

A. Ramesh classified the towns of Tamilnadu on the basis of functions found in that town after considering Nelson's method.

P.S. Tiwari classified the towns of Madhya Pradesh based on the multi-variate analysis of urban functions. On the basis of functional attributes he classified urban centres into different groups and analyzed their characteristics.

On the basis of 1961 occupational structure figure V.N.P. Sinha classified the towns of Chhotanagpur Plateau with the help of following functional index :

$$\text{F.I.} = \left(\frac{Nw \times 10}{TW} \quad \frac{PW \times 100}{TN} \right)^2$$

where *Nw* is the number of workers in a particular function of towns; *TW* is the total workers in that particular urban function of the region; *PW* is the number of workers in all functions of a town; and *TN* is the total number of urban workers in a town.

The specialization indices have been calculated in the following table:

$$\text{Specialization Index} = \frac{FIA + FIM + FIH + FIM + FIC + FITC + FIT + FIO}{8\,(\text{No. of functions})}$$

where *FI-A, MI, H, M, C, Tc, T* and *O* represent functional indices value of agriculture, mining , household industry, manufacturing, construction, trade and commerce, transport and other functions respectively.

—

REFERENCES

Aurousseau, M., "Recent Contributions to Urban Geography," *Geographical Review*, 1924, Vol. XIV, pp. 444-45.

Berry, B.J.L., Ribbon Development in the Urban Development Pattern," *Annals of the Association of American Geographers*, 1959, Vol. 49, pp. 145-55.

Gest, N.P. and Helbert, L.A., *Urban Sociology*, New York, 1933, p. 18.

Hall, Robert B., "The Cities of Japan," *Annals of the Association of American Geographers*, 1934, Vol. XXIV, pp. 175-200.

Harris, C.D., "Functional Classification of Cities in the USA," *Geographical Review*, 1943, Vol. 33, pp. 86-99.

Hart, J.F., "Functions and Occupational Structure of Cities of the American South," *Annals of the Association of American Geographers*, 1955, Vol. 45, pp. 269-86.

Mukherjee, M.M., "Manufacturing Towns of Bihar," *Transactions of the Indian Council of Geographers*, Vol. 4, 1967, pp. 33-40.

Mukherjee, M.M., "Functional Association of Towns in Bihar, India," *21st International Geographical Congress*, 1968.

Meckenzie, R.D., *The Ecological Approach to the Study of the Human Community in the City* by R.E. Park, E.W. Burgess and R.D. Meckenzie (eds.), Chicago, 1925.

Nelson, H.J., "Some Characteristics of the Population of Cities in Similar Service Classifications," *Economic Geography*, 1957, Vol. 38, No. 2, pp. 95-108.

Pownell, L., "Functions of New Zealand Towns," *Annals of the Association of American Geographers*, 1953, Vol. 43, pp. 332-50.

Raffiullah, S.M., "A New Approach to Functional Classification of Towns," *The Geographer*, Vol. XII, 1965, pp. 40-48.

Ramesh, A., Functions and Functional Classification of Towns of Tamilnadu in Tamilnadu: A Study in Urban Settlement, Unpublished Ph.D. Thesis, Banaras Hindu University, 1965.

Rao, V.L.S. Prakasha, *Towns of Mysore State*, Indian Statistical Institute, Calcutta, 1964.

Smailes, A.E., "The Urban Hierarchy in England and Wales," *Geography*, 1944, Vol. 29, Part I, p. 41-51.

Smith, R.H.T., Methods and Purpose in Functional Town Classification, *Annals of the Association of American Geographers*, 1965, Vol. 55, p. 454.

Sinha, V.N.P., *Chotanagpur Plateau: A Study in Settlement Geography*, K.B. Publications, New Delhi, 1976.

Sinha, V.N.P., *Op. cit.*, 1976.

Webb, J.W., "Basic Concepts in the Analysis of Small Urban Centres of Minnesota, *Annals of the Association of American Geographers*, 1959, Vol. 49, pp. 55-72.

Weimer, A.M. and Hoyt, H., *Principle of Urban Real Estate*, New York, 1948, p. 107.

6

MORPHOLOGY OF TOWNS

URBAN SETTLEMENT is a part of the settlement fabric. It is treated as "an organic whole with a distinction in build and structure etc. Morphology evolved through various ecological processes."[1] The urban activities are expressed in physical forms. Thus, the study of form or morphology of urban settlement is very important.[2]

Morphology of a town is concerned with the ground build and skyline of the houses. The plan may be internal which concerns with the arrangement of streets and built space,[3] or external which concerns with shape and the bird's eye view of the street patterns developed in a settlement.

According to Cozen,[4] "A town plan comprises the geographical alignment of the urban built up area in its full morphological detail and diversity, bringing the plan into intimate relation with the aspect of building fabric and of landuse. A town plan can be defined, therefore, as the topographic arrangements of an urban built up area and all its man-made features." Brush says, "Interpretation of the morphology of Indian cities begins with mapping the existing layout of streets, the arrangement and characteristics of buildings, and the associated pattern of land use.[5]

Global Model of Urban Morphology

In I.G.C. abstract Merill R. Ridd and Dȯn H. Card (1992, p. 531) emphasized a Global Model of Urban Morphology from satellite remote sensing. Urban areas are expanding on all continents. Perhaps no other human activity alters the environmental character more dramatically than urbanization/industrialization. Not only is the immediate urban site altered, but areas nearby end for beyond the urban peace itself. In this area of global environmental concern, with extensive attempts to monitor and celebrate changes and processes, a model is needed to characterize urban landscape in terms that permit comparability across all continents and islands, and that allows for detection and characterization of change over time.

This chapter presents a vegetation—impervious surface-soil (V-I-S) model for characterizing urban and perurban landscapes. Here the V-I-S composition may be quantified and plotted for various parts of the city and nearby environments. In the core of the city, the central business district is typically composed almost completely of impervious surface. Outward from this core, not unlike Von Thunens' concentric rings, changing lend use and land values result in decreasing amounts of impervious surface and increasing amounts of vegetation as population density decreases. At the periphery, the transition is to more vegetation and/or more soil depending on the ambient environment, whether forest, farmland, or desert. Thus the model also serves to demonstrate and measure the environmental variation within the city, as well as the contrast between the city and its surroundings.

Employed through time, the V-I-S model serves as a measure of environmental impact, quantifying the change in environmental composition as the city grows. Satellite imagery facilitates continuous monitoring of the urban complex in systematic environmental terms, and permits the comparison of cities of all sizes around the world.

The model also serves as a valuable basis for various biophysical and engineering models. Runoff models benefit from V-I-S information. In addition, many human patterns have been seen to correlate with biophysical composition. For example, land values, family income, and demographic patterns are usually related to V-I-S composition.

In global terms, if we aspire to understand the linkage of urban/industrial process to other global processes of atmosphere, hydrosphere, lithosphere, and biosphere, we need to characterize and quantify urban environments more effectively.

External form of Town

Towns vary widely in their forms. They reflect the physical and cultural factors of the site and situation. Physical factors is not so important in determining the forms of town, as all towns are located along the bank of Ganga river. Consequently, the cultural factors, such as forts, old market place, roads, railway stations and administrative offices are more important in determining the form of the towns. Buxer, Dehri and Nasriganj have triangular shape. This is due to the attractive forces along the river and attractive sites along the main roads which connect the town with other places, Sasaram, Bhabhua, Dumraon and Bikramganj are more or less rectangular in shape (Fig. 6.1). Arrah has irregular shape. Jagdishpur has kidney shaped form. Elaborate planning both in the past and the present is an important factor in determining the form of the towns, but information as to planning in the past is non-existent and during modern times planned reconstruction of town, has been generally of hexagonal in character.

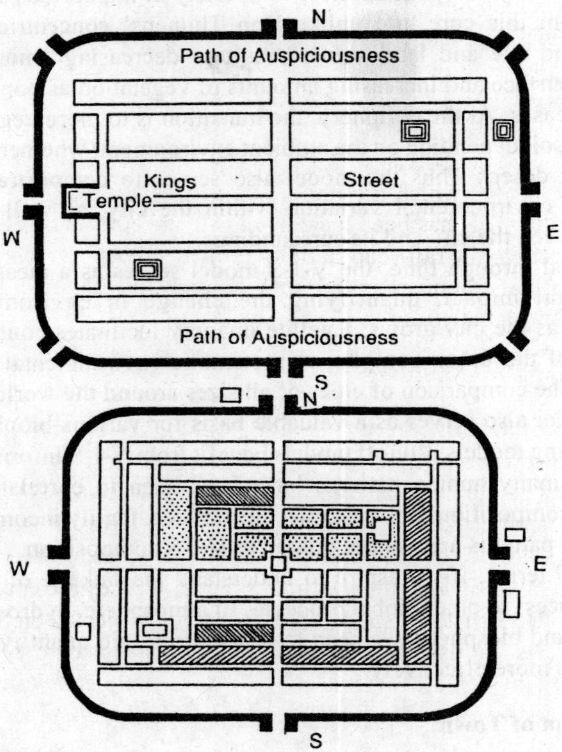

FIG. 6.1 : Aryan Settlement Plans

Functional Zones and Town Landscape

During the long period of evolution they have acquired irregular street pattern, complex aspect and function. They lack clearcut functional zones although a central zone containing a bazar and residence may be distinguished in the towns. Such central zones act as business core is known as *Chauk* which is the commercial core of the town. It comprises general bazar, the specialized bazars such a grains, vegetable and brassware market, as well as firms of native banks. The business sections are surrounded primarily by the residential zones with small markets and retails shops scattered in them. Industrial areas are not distinct although modern factories are located on the outskirts of the town. Industries such as cloth-making (tailoring), goldsmith and blacksmith shops are located in the bazar itself. Quite distinct are the civil lines and railway colonies adjacent to the town proper. All these are added during the British rule. Civil lines are mainly administrative and residential in function, containing government offices, court houses, posts and telegraph offices, public gardens, tanks, clubs and bungalow of officials clung along the pitched shady roads often in separate gardens. The railway colony in the towns which are important railway centres, is a distinct functional zone. This has been located outside the confines of the town proper, adjacent to the rail-road station, 'railway colony' which consists of railway offices and quarters for the railway employees as in Katihar, Samastipur and Jamalpur of the State of Bihar.

Street Patterns and Buildings

As the towns of the district are unplanned there is not any definite street pattern. The towns with pre-British origin which have grown at all during the last century shows a striking contrast between the indigenous part and the anglicized part. The indigenous part have irregular street pattern which are narrow and crooked. The main roads are rarely more than 10 to 15 metres wide, often without side walls. They are encroached upon by wooden structured shops whose roofs are projected on the road. The side streets are usually narrower and so crooked as to be almost impossible by the vehicle to pass. The ground is completely covered by some kind of structures. More than two storey buildings predominate, although lofty buildings or three, four or even six stories may exist along the main street. The common type of masonry construction are known as *Pucca* houses which are made up of brick, mortar and often covered with plaster and tile as roofing material. *Kutcha* houses with mudwall and tile roof or even thatch-roofs are found in the peripheries. There are some ferro-concrete structures belonging to the prosperous merchants and modern flats of the wealthy residents standing side by side with traditional structures but big houses even in the countryside appear not to have been reached by modern influence.[6]

In contrast to the indigenous parts of the town, civil lines, military cantonments and railway colony are planned and they have been constructed to serve definite purpose. They are characterized by rectangular layout of streets. The streets are broad, metalled and varied with shady trees. In civil lines all the buildings are brick-built, generally constructed in western style with some modification such as the addition of verandah and courtyard so as to suit the local climate. The density of population is low as compared to the old sections of the town with extensive lanes, shady avenues, flatish skyline looked sometimes like the spire of a church or high building. The civil lines present a distinctly modern appearance. The distinctive feature of a railway colony have perfectly uniform grid of streets and the monotonous row of brick dwellings graded and rented strictly according to the income status and occupational level of the employees.

Dehri

The township of Dehri and Dalmianagar presents striking morphological features. K.N. Singh writes "while Dehri provides a typical view of an essentially indigenous, although fast being modernized overgrown urban market with its richness and age long heritage of Indian life. Dalmianagar on the other hand with its chess board plan, large scale industrial landscape and fascinating history of migrants is an altogether different world." A simplified morpho-functional map of Dehri-Dalmianagar (Fig. 6.2) illustrates the morphological and functional aspects of the town.[7]

The main business section of the town is along the old Grand Trunk Road east of the western Sone canal. It is the oldest site which has been developing since the reign of Shershah. Its growth has been haphazard without any plan. On account of the transportation facilities provided by the National Highway its main function is commercial. As is common to the all indigenous part of the Indian towns it is highly populated and densely built. No open space is found. The streets are narrow and crooked. The houses are mostly single storied, mud-walled and tile-roofed. Few modern buildings have been built along the road which seem to be superimposition upon the old site. Most of the houses are used for commercial purposes, viz. shops and storages. However, residential houses are not altogether absent. In fact businessmen reside in this section of the town. New business centre has developed near the railway station. It is due to the attraction of the railway station and nearby industrial zone. Retail shops are also found in the residential

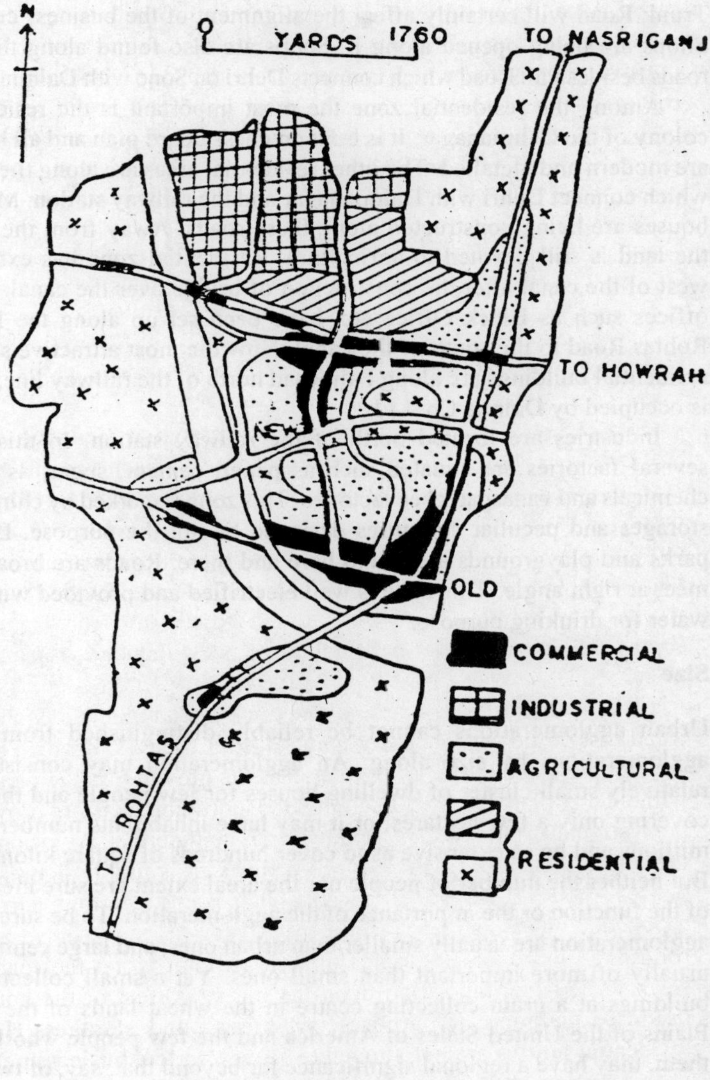

FIG. 6.2 : Landuse of Dehri-Dalmianagar

colony of Dalmianagar and to the west and south of the canal. It is due to the expansion of the built up area towards west and south due to the construction of bridge cover the canal. The construction of new Grand Trunk Road will certainly affect the alignment of the business centres. Shops are being opened along it. Shops are also found along the two roads besides Pali Road which connects Dehri on Sone with Dalmianagar.

Among the residential zone the most important is the residential colony of the Dalmianagar. It is built on chess board plan and all houses are modern and metalled. The other residential areas are along the roads which connect Dehri with Dalmianagar and the railway station. Modern houses are being constructed along these roads. Away from the roads the land is still devoted to agriculture. Residential zone has extended west of the canal after the construction of bridge over the canal. Some offices such as Block Office etc. have been set up along the Dehri-Rohtas Road to the south of the canal. Now the most attractive site for residential buildings are along Pali Road north of the railway line, but it is occupied by Dalmia Co. Ltd.

Industries are located north of the railway station. In this zone several factories are situated such as paper, cement, sugar, asbestos, chemicals and vanaspati ghee factories. This zone is marked by chimneys, storages and peculiar structures designed to suit the purpose. Lawns, parks and playgrounds are found here and there. Roads are broad and meet at right angle. This zone is well electrified and provided with soft water for drinking purpose.

Size

Urban agglomerations cannot be reliably distinguished from rural agglomerations by size along. An agglomeration may consist of a relatively small cluster of dwelling houses for few people and the land covering only a few hectares, or it may have inhabitants numbering in millions and be so extensive as to cover hundreds of square kilometres. But neither the number of people nor the areal extent are sure measures of the function or the importance of the agglomeration. To be sure, rural agglomeration are usually smaller than urban ones, and large centres are usually of more important than small ones. Yet a small collection of buildings at a grain collecting centre in the wheat lands of the Great Plains of the United States of America and the few people who live in them, may have a regional significance far beyond that, say, of twice as large as settlement in the more heavily populated north-east.

The size of population is often used arbitrarily, for convenience, as in the United States of America census in the past, to distinguish between rural and urban agglomerations. But such arbitrary distinction may lead to a misunderstanding of the significance or even the nature of the settlements of any area. Only broadly, as a contributing element to the whole problem of urban studies, the population size may also be considered.

Internal Patterns

The skeleton of any urban agglomeration is the important aspect of morphological analysis. Settlement pattern, and the scheme of its streets, which may be arranged either according to some geometrical plan or completely without regularity. Often the features of site conspire to favour certain lines of movement over others which determine the overall morphological configuration. Again, site may impose no control so that the pattern develops entirely in accordance with the desires or whims of the inhabitants.

In urban agglomerations that are old and had their origins in simple groupings of rural people, the patterns of streets may be quite irregular. The exact alignment of early houses and other buildings was not a matter of importance to the inhabitants. As the settlement grew large enough for actual formal streets to be laid out, they had to follow as best they could be the irregular lines between buildings. Many of the street patterns in lower New York or the heart of Paris are of this sort.

Sometimes pattern has been determined, directly or indirectly, by government decree. The urban settlements of Spanish American were laid out with a central plaza and streets parallel to its sides. Thus came a pattern of rectangular nature, with streets crossing one another at right angles. In the central and western United States of America the rectangular land survey previously described brought original focus upon roads crossing of one another at right angles. Generally, in cross roads agglomerations, additional streets developed logically along lines parallel to the roads forming the cross roads and thus at right angles to one another.

Increasing importance and size of an agglomeration create the need for speed and ease of access to its centre from the periphery. Thus came the development of the radial pattern whereby streets lead from points on the periphery directly to a common centre, much as he spokes of a wheel lead from the riers to the hub. Cross streets then tend to assure a weblike arrangement. The sketch map of streets in Washington D.C. is a complex combination of radial rectangular patterns. So is the relatively simpler skeleton of Indianapolis.

Beyond those three basic patterns-irregular, rectangular and radial—there are many others that have grown out of the vagaries of geographical factors or of real estate development. These are irregular in a sense but they are usually symmetrically so. Circular drives, curving and recurring lines and dead-end courts leading in from each of four sides of a large rectangular block are some of the familiar forms encountered in growing urban agglomerations in different parts of the world.

All the patterns may occur in combination as well as individually especially in large urban areas. As one proceeds uptown, the irregular pattern of lower New York becomes a well-crystallized rectangular pattern, with only one notable exception, broadway. The repeating radial patterns of Washington D.C. are imposed upon a rectangular base. Where peculiarities of site restrict full development of any one pattern. Local modifications are introduced as for example, the city of Syracuse New York, has a basically rectangular street pattern, but much of the city's site is characterized by drumlins.

Whatever single-pattern or combination of patterns exist, the character of the agglomeration is deeply affected. The alignment of houses and other buildings are crystallized by the position of the streets and thus the agglomeration acquires a fixed quality. Additions to it or subtractions from it, both measures of changes, must in some way be articulated with the existing pattern. Despite changes of detail, the agglomeration as a whole retains a distinctive appearance, determined basically by its pattern of streets.

Profile or Cross-Section

Something of the nature of any agglomeration is told by its profile, or skyline. In small town the buildings are often widely spaced. Individual houses rise above low sheds, and huge barracks may overshadow all. Rising near the centre of the settlement may be the thin spire of a church. The whole profile of a small urban settlement is irregular and jagged, although trees often raise their rounded corners to often the sharpness of angular roof lines. Large urban agglomerations, especially in the United States of America present quite a different profile. The outer edge may look like the profiles of small towns. As the centre is approached, the buildings become more massive and higher, whose profile gives almost an impression of the cross-section of a pyramid. The lines are straight or symmetrically rounded, man-made forms rather than natural ones.

In other parts of the world, this "piling up at the centre" is less characteristic than in the United States of America. For example, in most of the foreign urban agglomerations the tall buildings are largest indistinguishable from that of a small towns. Only by great length does it suggest large size and reveal the urban character of the settlement.

Functional Areas

Urban agglomerations perform various functions for the groups living in them and for the area that surrounds them. While these functions may occur singly or in combination, each one involves certain types of buildings; for example, the appearance of a retail store such as a grocery store, which carries on a commercial function, contrasts sharply with the appearance of a factory, which carries on the function of manufacturing. Similarly, a zone of retail stores or a shopping centre within an agglomeration creates a geographical setting quite different from the setting created by a group of factories. A residential zone is easily distinguished from the campus of a college.

Each function presents a distinctive form which tends to be localized within the agglomeration. An area that performs a special function is known as a *functional area.* Taken together function is known as a discernible pattern. They provide the remaining element of the morphology of agglomeration (Fig. 6.3).

It is common, in American, settlements at least, to refer to the down town section of an urban agglomeration. Usually, this means the section of the settlement that is given over to stores, public buildings, and the like. It is usually a part of the settlement definitely removed from individual residences, although large apartment buildings are often located in the down town section. Similarly, there is generally a factory section, on the part of Americans, the concept of functional area is thus clearly recognized.

Within older settled parts of the western world, distinctions between the specialized areas of an agglomeration are less clearly marked than in newer settlement. The history of the growth of an agglomeration often reveals why the various functions are performed in scattered rather than in collected zones. In European settlements, for example, residence is often maintained close to or in the same building with retail stores. The function of commerce may thus be so scattered as to leave no impression of a down town section. In original settlements, the same building may be craft workshop, the residence of the craft workers, and, at the same

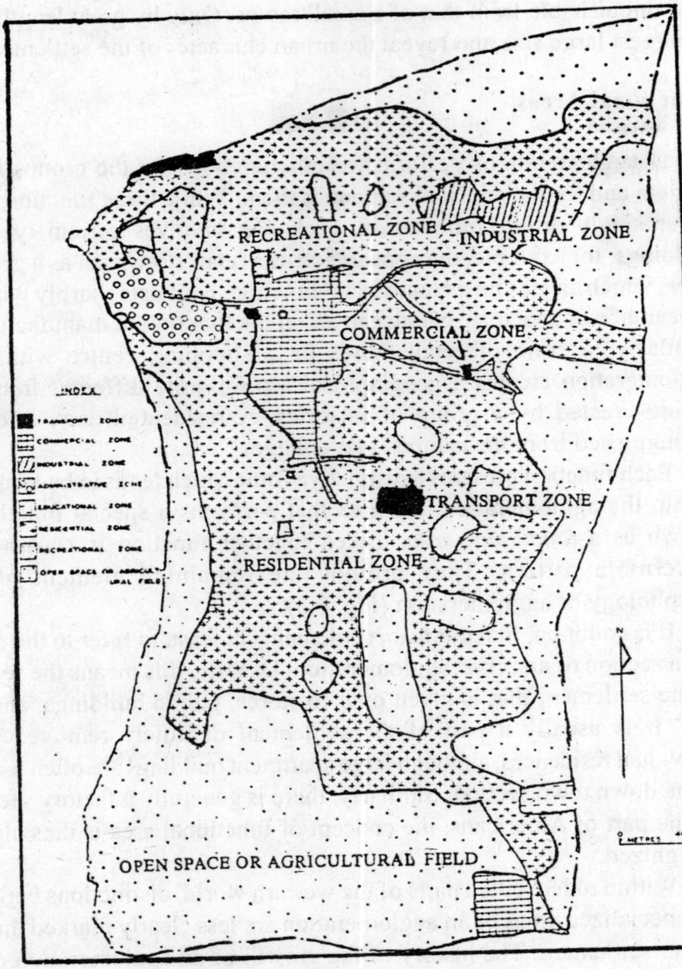

FIG. 6.3 : Functional Zones of the City of Munger

time, the home of persons engaged in intensive agriculture on a small plot of land nearby.

The arrangement of the functional areas results from the interplay of a wide variety of forces. At an early stage, the inter-section of two highways, which may have given rise to the settlement in the first place, may be an attractive site for a store. With the growth of a commercial nucleus, a distinct functional area develops a commercial core in this for instance.

The growth of any agglomeration brings with it continual, far-reaching adjustments of its functional areas. If a settlement has developed a commercial core surrounded by residences that in turn, are encompassed by a ring of manufacturing plants, further growth in population requires expansion of the residential area. Often the only possibility lies in jumping beyond the manufacturing ring, soon new residence areas appear on the fringe, sometimes close to the factories for the workers, and sometimes, for those who can afford to choose their own site, well away from the factory belt. Here and there small agglomerations may develop that are entirely separate physically from the original nucleus but are economically still a part of it.

As new residential areas spring up there is usually a decline in the size and quality of the old residential belt near the centre of the settlement. This decline is accompanied by outward spread of the commercial core and inward growth of the manufacturing belt. Old residential houses of the settlement become roomed houses, gradually deteriotating and eventually being removed to make way for expansion of the other functional areas. On such zones on the fringes of the commercial core, the name 'blighted area' or 'Slum' is frequently given in American cities. When the new scattered nuclei of commerce begin to appear irregularly in the outer residential districts, these are called the 'occasional corner stores' or 'neighbourhood shopping centres', which is the characteristic of growing American cities.

Modern transportation has made it possible for urban folk to have their dwellings at considerable distances from their work. Intangibles such as the desire for space and tangibles such as tax rates have both contributed to the migration of urban people away from the urban necleus, where the movement has actually occurred. A zone of interpretation of urban and rural settlement, is the *rural-urban fringe,* which has come into being with proper planning and with restrictions on an urban area's ultimate size. All these are matters of site and situation and morphology of settlements which could be controlled so as to produce the most economically sound solution to the problem of increasing urbanization.

Morphology of the Planned City of Jamshedpur
Regional Location

In order to know the regional location of Jamshedpur the study of its morphology is essential because it determines the patterns of future growth. Jamshedpur is situated between 86°9′ E to 86°16′30″ E longitudes and 22°44′ to 22°50′15″ N latitudes. It has reached a population of 6 lakhs in 1991.

Site of the City

The Industrial city of Jamshedpur is quite similar to any other city in Bihar. Its origin took place in the wilderness in 1907 and its amazing development at the site of old-Sakchi village. The city has grown on an area of 117 square kilometres bordered on the south by the residential colonies located south of the railway line, on the west by the Kharkai river close to its confluence with the Subarnarekha river, on the north by the Subarnarekha itself and on the east by a line of hills, e.g., Murakati, Chari, Dungri, Kalamati Pahar and others. The credit for the creation of this modern city belongs to the Tata Iron and Steel Company.

Planned Growth from 1907 to 1912

The area north of the railway line was chosen in 1907 as the site for the creation of Tata Iron and Steel Plant. The foundation for the blast furnace and Steel Plant was laid in 1909. This site has the facilities of Damodar Valley Coalfields in the north and the iron ore belts of Singhbhum in the south. The confluence area of Kharkai and Subarnarekha provides sufficient plain land even in a rugged terrain. The site is underlain the mica-schist and uranium furnishing a suitable foundation for the construction of a steel plant. The Subarnarekha provides an easy access to the port of Calcutta, whereas the Sanjaikoel gap is utilized by the main line of the Eastern Railway which connects Jamshedpur with the southern Peninsula and Mumbai. The planned industrial city of Jamshedpur is centrally located in terms of the availability of vital raw materials—iron ore, coal, dolomite and manganese. It is also well connected with the largest market of Calcutta in the eastern India. It gets connection of B.N. Railway in 1908. Mr. Sahlin of U.S.A. prepared the plan of Jamshedpur in February, 1908.

Form of the City

The form of the city of Jamshedpur has two aspect internal and external. The internal form largely depends on the road pattern, while the external form depends upon the built up area of the town in 1912. This was guided by two factors—(a) topography of the land, and (b) the limits of the TISCO acquired lands. The township has three main areas at that time : (i) the TISCO factory; (ii) Bistupur quarter; and (iii) North Town Bungalows. The city of Jamshedpur has extended in cardinal directions from this original nucleus of 1911 (Fig. 6.4).

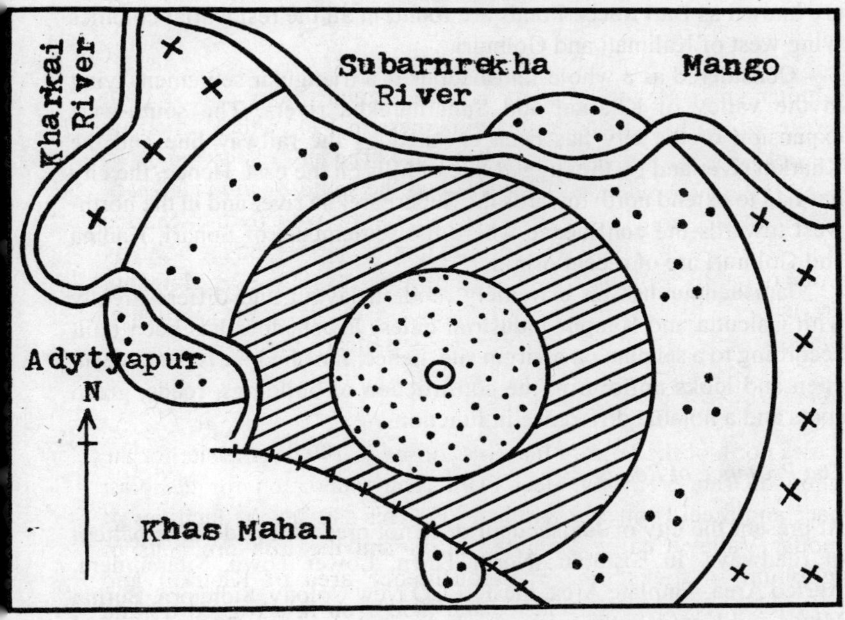

FIG. 6.4 : Morphology of the Planned City of Jamshedpur

Extension has not been found in the west and south-west of the old Bistupur area. This site has been partly influenced by the growth of the built-up area. The general landscape is partly undulating and the elevation of the surrounding country increases towards north and south. In the north Dalma hills rises upto a height of 100 metres from mean sea level. The general slope of the country decreases towards the east and south along the direction of flow of the Subarnarekha river.

Settlement Area of the City

In Jamshedpur the settlements are located either on the main ridge in Kalamati, Jogobera, Bistupur, Beldi and Mango areas. These are situated on the spurs of different heights. Except for the scattered house of the north-western Sakchi and Bera, houses are not found below the peak flood level of the Subarnarekha and the Kharkai.

Most of the residential areas tended to form linear patterns along the roads, but in some cases they are distributed wide apart in residential colonies such as in Jugsalai, Sakchi, Nildin, Beldi, Khutadi and Sonari. In most of the cases they were surrounded by agricultural lands which are known as Bari tracts. Ponds are found in all the residential colonies lying west of Kalimati and Golmuri.

Considered as a whole Jamshedpur is a triangular settlement lying in the valley of Kharkai and Subarnarekha rivers. The south-west expansion of the city has been checked by the railway line and the Kharkai river and by the rugged line of hills on the east. Hence, the city has had to extend north towards the Subarnarekha river and in the north-west towards the confluence where the settlements of Sonari, Kadma and Golmuri are of recent origin.

Jamshedpur has an extremely modern layout and differs greatly with Calcutta and Kanpur industrial cities. Jamshedpur has been built according to a set plan on a virgin site, hence, Jamshedpur is neat, clean, open and looks uniform in the construction of buildings, roads, green spots and a notable difference in functions.

The Presence of Roads

At present the city of Jamshedpur does not present any definite pattern of roadways. In Bistupur, North Town, Lower Town, Sitaramdera, Agrico Area, Tinplate Area, the TELCO New Colony, Sidhgora, Burma Mines and Kamda—the minor roads are rectangular in form; in Sonari north-west of the aerodrome and around Baridih T.B. Hospital the minor roads are like arch of a circle; in Baradwari some good residential houses have been built—here the minor roads are in cobweb pattern, quardrangular roads exist in the southern part; in Jugsalai the roads are totally assymetrical. The major streets are neither rectangular nor circular, but in some cases they are straight lines. On the whole, they are mostly irregular lines. Thus the present pattern of the roadways of the city of Jamshedpur can be termed as irregular in form and lacks a definite system.

Housing Pattern

The housing pattern of the present day Jamshedpur is free from typical slums as in Calcutta. This has happened due to the practical approach adopted by the two planners Temph and Keshingsburges. Both of them tried to accommodate the bulk of the workers in well maintained sanitary *busties* as the member of quarters built by the TISCO management was not sufficient to meet the housing demand. So, an effort has always been made to keep the *busties* neat and clean and healthy. But these *busties* have insanitary houses and they stand in contrast to the well built brick houses.

Adequate quarters must be built in the near future by the management of different factories and government to house all the employees of the city; but till it is done, the *busties* should maintain sanitary norms. *Busti* is essentially a rural mode of living. The workers of Jamshedpur, are mostly rural in origin and therefore, they do not feel incongenial in 'busti-life'. As long as rural character of *busties* are maintained it is not unhealthy.

Geographical Zones of Jamshedpur

The city of Jamshedpur has four broad zones : (i) inner zone, (ii) middle zone, (iii) outer zone, and (iv) suburban zone.

Inner Zone

The inner zone consists of the central business district, the original residential area of the township, the new extension in the west, such as the town administrative office, the health and education office, the main hospital of the TISCO etc. It is bounded by the boundary wall of the TISCO in the east, private road in the south and roughly the Kharkai link in the west and the straight road in the north. It includes whole of Bistupur and whole of north town, whose residential quarters are the oldest in the city. It includes the central business district of the city which is the hub of the city and the TISCO main office is also located there. The town administrative office, the Town Office, the health and education office, the main hospital, the united club, the beautiful TISCO Hotel, the Perin Memorial School, Regal Cinema house, the Women's College, etc. are all located in the inner zone. All these offices and institutions attract the daily movement of people from different corners of the city and it is these along with the central business district, which work as a centripetal force in the daily movement of people.

Middle Zone

The middle zone of Jamshedpur surrounds the inner zone, but the later does not occur in the geometrical centre of the former. The middle zone is more extensive in the east than in the west because the city has developed more towards the east than towards the west. It mostly includes within it whole of the main built-up area of the city of twenties. In contrast to the inner zone, the middle zone is not entirely '*Pucca* built' but a large number of *busties* of *Kutcha* house exist in this zone. It includes that part of Jugsalai which was built in the twenties. Unlike the inner zone, it is not yet fully built and still open unused spaces are brought under built-up areas. The factories which are non-existent in the inner zone, occur here. The middle zone contains the following factories : (a) TISCO, (b) the INDOXCO, (c) the Indian Cable Company, (d) the Tinplate Company, (g) The Agrico. The business area of Sakchi and the marketing area of Dhatkidih, Kadua, Agrico, Golmuri and Jugsalai fall within it. Because of the town planning, the density of population, and built-up areas do not decrease symetrically as one goes towards the outer zone. There are definite localities such as Sakchi, Kasidih, Mohullera, Dhatkidih, etc., which are densely populated, but they are all separated from each other. Though the residential area occupies quite a large proportion of the zone, yet the lands used for residential purposes are considerable.

Outer Zone

The outer zone surrounds the middle zone. Like the later it is intensive eastward. This zone is, in some respect rural in character while others are urban. Here large areas of open unbuilt space found. Unlike the inner zone and the middle zone, the outer zone has large tract of forest land in the north west at the confluence of the Subarnarekha, the Kherkai and that of the eastern margin of the township. The agricultural lands of Sonasi, Ulyam, Gyuriagora, Kadhua, Behtya and Bina are also included in it. The 'Busties' of east plants, Jogsalai, Bhetya, Ulyan, Kadua, Hume Pipe and Sonari are old settlement areas of the outer zone but the TELCO New colony, Sidhgora and the 'pucca'-built areas of Sonari have been built under definite planning. In this zone, the factories built in the twenties are found such as the TELCO, the ISWP, the TEMCO and the Hume Pipe Company. The Sullage disposal plant and the Dairy farms are situated in Bera in the north eastern extremity of the zone. The new and old colonies of the TELCO and the plant constitutes a separate township as such its recreational facilities and a small retail and vegetable market. The B.N.R. Colony and Tatanagar Railway Station are situated in south extremity of the city. They are also separate townships, but they have no marketing area within limits of the colony. The nearest one is at Jugsalai. Both the TISCO and B.N.R. Colony have well planned houses and roads. The magnificent jubilee Park along with the Baghakund lake and the Aerodrome lie in this zone spreading their wide open expanse.

The Subarnarekha Zone

The Subarnạrekha zone of the city is scattered and does not form a contiguous area like the other zones. Mango, Khasmahal and Adityapur are the three units of this zone. Now Mango has developed as a separate city along with a municipal boundary.

Build and Architecture

The buildings and their architecture in Jamshedpur are products of the twentieth century. Hence, the entire pattern is modern. Minarates and Domes, which are common in the medieval cities of India, are no where to be observed in Jamshedpur except in the Mosques; *Gurudwaras*, Temples and Churches. But even the fifty years life of the city demonstrates some significant changes in the house types. The two-roomed quarters of Bistupur, constructed in 1911, are either tiled entirely or the front *verandah* is tiled, but they are invariably devoid of descent appearance. On the other hand, newly built two-roomed quarters of Sidhgora and the TELCO new colony are not only built by cement and concrete but out verandah is descent.

Like other cities of India, Jamshedpur does not display colours on the buildings. There are large areas of company built quarters and bungalows which are one colour. It is mostly the light yellow which is the result of white washing. The Pipe line flates and houses are pinkish in colour as it is made up of tiles. The new bungalows have red coloured roofs and yellow coloured walls.

A decision about development, extension and modernization of Jamshedpur has been taken in 1988. Under this scheme the companies of hightake drilling, Tata Telecomm. Limited, Keltrone Telephone Instrument Limited and Tata Finance Limited have to expand 1000 crores of Rupees for modernization of Jamshedpur. In this way, we can conclude that the contributions of Tata is praiseworthy in the industrial development of Bihar. Only the planned development of its street pattern helps in the ushering of a clean environment and scientific development of this city.

FOOTNOTES

1. R.E. Dickinson, "The Morphology of the Medieval German Towns," *Geographical Review*, Vol. 35, No. 1, 1945, p. 75.
2. M.R.G. Conzen, "The Plan Analysis of an English City Centre in K. Norborg," (ed.) *Proceedings of the IGU Symposium in Urban Geography*, 1962, pp. 383-414.
3. J.E. Brush, "Spatial Pattern of Population in Indian Cities," *Geographical Review*, 1968, Vol. 58, pp. 362-91.
4. K.L. Dutta, *Morphology of Indian Cities*, Ph.D. Thesis Banaras Hindu University, Varanasi, 1968.
5. A.K. Dutt, *Jamshedpur and Its Umland : A Study in Urban Geography*, Ph.D. Thesis, Patna University, Patna, 1959.
6. C.S. Sargent Jr., "Towards a Dynamic Model of Urban Morphology," *Economic Geography*, Vol. 68, 1972, pp. 357-74.
7. R.A.G. Savigear, "A Technique of Morphological Mapping," *Annals of the Association of American Geographers*, Vol. 55, 1965, pp. 514-21.

7

THEORIES AND MODELS IN URBAN STUDIES

MOST of the models used in urban studies are either diagrammatic or mathematical. These models represent the landuse differentials found in different parts of the city or are based on increasing or decreasing population densities.

General Landuse Model

In 1969 Alice Coleman[1] suggested a model for landuse study in order to explain the second Landuse Survey of Great Britain. Through this model Coleman attempted to classify five categories of landuse, viz., farmscape, urbanscape, wildscape, urban fringe and marginal fringe (Fig. 7.1). In this way, it is felt that conventional landuse categories must be of a lower hierarchical order as the fundamental territories of landuse are generally found continuously up to the marginal fringe. The variations are due to the physical and socio-economic influences. The interacting zones are found in advancing, stable and retreating stages and it is hoped that these stages and the reasons for their formation are capable of wider application all over the world.

The characteristics of comprehensiveness seeks that all categories of land must be considered in the model or the related item should be there in some other form of integrated framework.

The integrative character does not require the formation of categories in isolation but it should include the association of variables dynamically in 'landuse territory'. The attributes of added insight reflect that the model should be prepared keeping in view the varying objectives, i.e., territorial, partial, sample, regional, principle, etc. Further, it is expected that the model should be problem oriented such as planning the area, classification of the area, and must have a high level of academic value. The idea of visual-spatial quality requires that the model should be

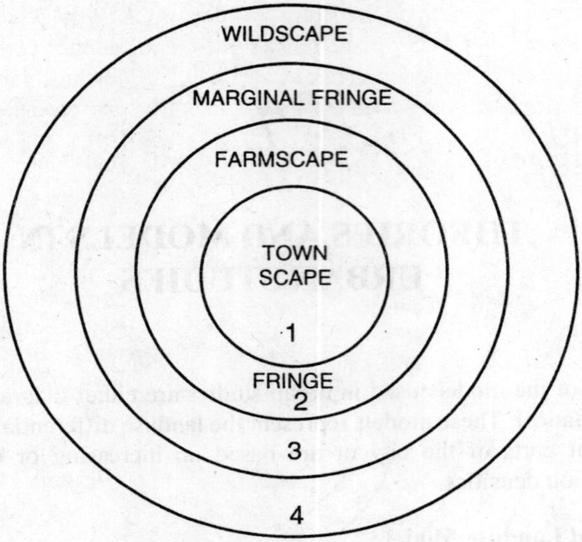

FIG. 7.1 : Coleman's Model of Urban Landuse (After Rawson and Sealy)

really geographical and one which should study the landscape territories and not just mathematical complexities. The model should be flexible enough so that the frequent change of landuse in the present situation could easily be included in the model and finally it should have an application to any part of the world. General Landuse Model of Coleman is applicable to any region having both urban and rural landscape and some forest land also. Since it was proposed to explain the landuse of a country like the U.K., it can't be applied to a city like Munger. However, the essential features of a model can be borrowed from this description in order to formulate an independent model.

ECOLOGICAL MODELS OF URBAN STRUCTURE

The theories of models in urban studies attempt to explain and describe the growth and development of a city and the evolution of functional zones within the city.

Concentric Zone Theory

This hypothetical pattern of urban growth was first propounded in 1923 by E.W. Burgess,[2] an urban sociologist, in his study of urban morphology of the city of Chicago. Through this model Burgess stated that the development of a city place outwards from its central commercial core in a series of concentric circles (Fig. 7.2). He thus, recognized five concentric zones expanding outward from the city core these are :

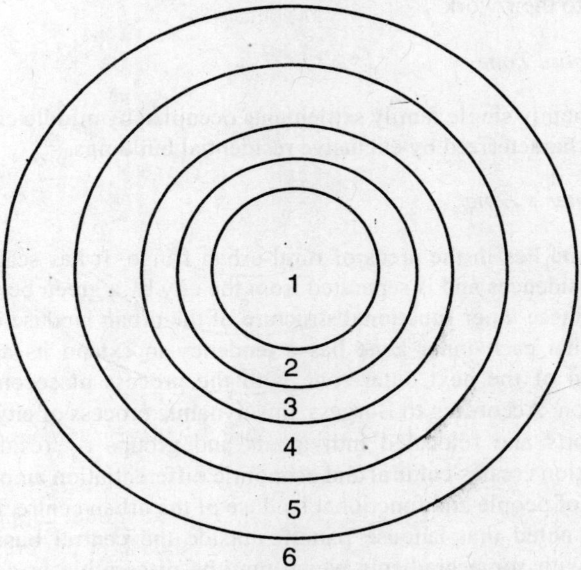

FIG. 7.2 : Urban Landuse in American Cities
1. Loop, 2. Factory Zone, 3. Zone in Transition, 4. Zone of Working Men's Home, 5. Residential Zone, 6. Commuter's Zone (After E.W. Burgess)

Central Business District

The Central Business District (CBD) is the central area of the city centre which is described as the commercial core. It has the downtown retail districts, stores, shops, office buildings, banks, clubs, hotels, theatres, museums and organizational headquarters. It has also the greatest concentration of tall commercial buildings and vehicular traffic. The CBD draws its business from all other encircling zones.

Zone in Transition

This zone encircles the CBD and supports business and light manufacturing. This is the zone of immigrants and has slums where poor people of economically weaker sections reside.

Zone of Working Men's Homes

This is the zone occupied mainly by industrial workers who have migrated out from the zone in transition, but have a desire of settling within easy access to their work.

Residential Zone

It has mainly single family settlements occupied by middle class groups and is characterized by exclusive residential buildings.

Commuter's Zone

This zone lies in the areas of rural-urban fringe. It has scattered high class residences and is separated from the city by a green belt.

In these inner-functional structure of the urban landuse it has been noted that each inner zone has a tendency to extend its area by the invasion of the next outer-zone with the process of second zone of transition. According to Burgess, this dynamic process of city structure, shift sorts and relocated individuals and groups by residences and occupation creates cultural and economic differentiation among various groups of people and functional landuse of the urban centre. It has been further noted that landuse pattern outside the central business zone varies with some gradients which may be discernible in delinquency rate, sex-ratio, percentage of foreign born individuals and poverty which to decrease outward from the central business district.

Sector Theory

The idea of sector or wedge like pattern of urban landuse was proposed by R.M. Hurd and Homer Hoyt[3] in 1939. This is the concept of axial development of landuse patterns along main transportation routes or lines of least resistance to form a star shapped city pattern (Fig. 7.3). It is certainly an improvement on the concentric zone idea since both distance and direction from the city centre are taken into consideration. The entire city is considered as a circle and various areas as sectors radiating out from the centre of that circle. In this way, a high rent residential area in the eastern quadrant of the city would tend to shift outward in the eastern quadrant only. In this way, if a low quality housing area is located in the southern quadrant, the same would tend to extend outward to that very margin of the city in the same sector. The migration of high class residential areas outward along established lines of trade is

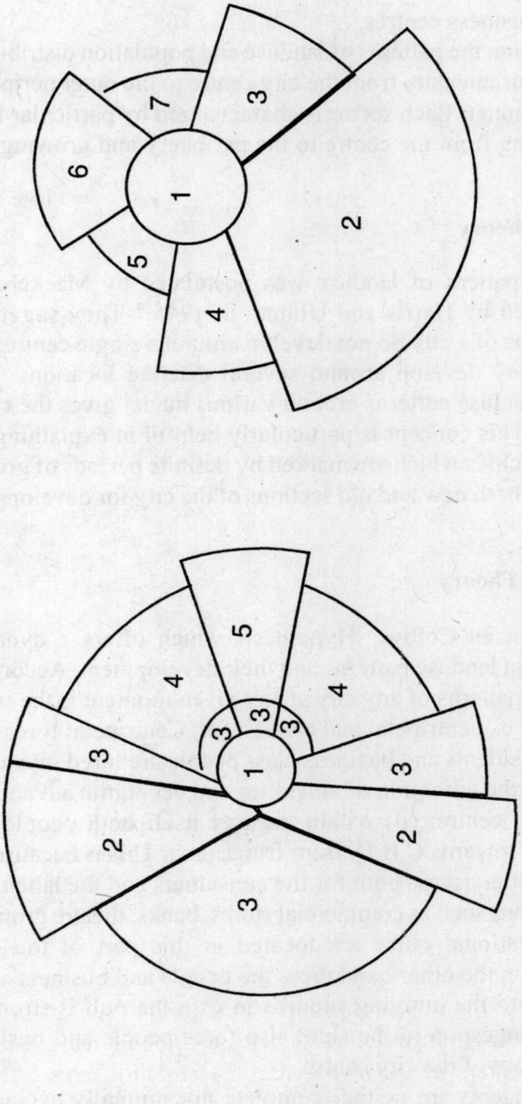

Fig. 7.3 : Sector Model of Hoyt

1. C.B.D., 2. Light Manufacturing Industry, 3. Low Class Dwellers, 4. Middle Class Dwellers, 5. High Class Residences

particularly pronounced on high ground towards the countryside, to the houses of community leaders along lines of fast transportation and towards existing business centres.

In such a situation the patterns of landuse and population distribution tend to differ in their intensity from the city centre to the outer periphery in a progressive manner. Each sector is characterized by particular form of landuse spreading from the centre to the periphery and growing in a sectoral fashion.

Multiple Nuclei Theory

This hypothetical pattern of landuse was postulated by Meckenzi in 1933 and developed by Harris and Ullman in 1945.[4] They suggested that landuse patterns of a city do not develop around a single centre. But on the contrary they develop around several discrete locations. This concentration of landuse patterns around various nuclei gives the city a cellular structure. This concept is particularly helpful in explaining the structure of Indian cities which are marked by definite periods of growth and often combine both new and old sections of the city for development (Fig. 7.4).

Colby's Dynamic Theory

This is also known an Colbys[5] Hypothesis which offers a dynamic explanation of urban landuse patterns and their development. According to this concept the patterns of any city at any given moment is the result of forces at work, i.e., centripetal and centrifugal. Centripetal forces are of two types : (i) residents and business class people are lured into a city from elsewhere by the attraction of amenities and economic advantages found in the urban centre; (ii) within the city itself both people and business are drawn towards C.B.D. from fringe area. This is because the C.B.D. provides better access both for the consumers and the labourers. Moreover, institutions such as commercial stores, banks, theatre grounds, libraries and recreational clubs are located in this part of the city. Centripetal forces on the other hand drive the people and business away from the C.B.D. into the outlying suburbs in case the pull is stronger. The problem of congestion in the slum also force people and business activities to move out of the city centre.

The above concepts are neither complete nor mutually exclusive. The concentric zone theory depends upon the growth and development of the city according to income and the circular development of city function are the result of spatio-temporal dimensions. In fact, this theory

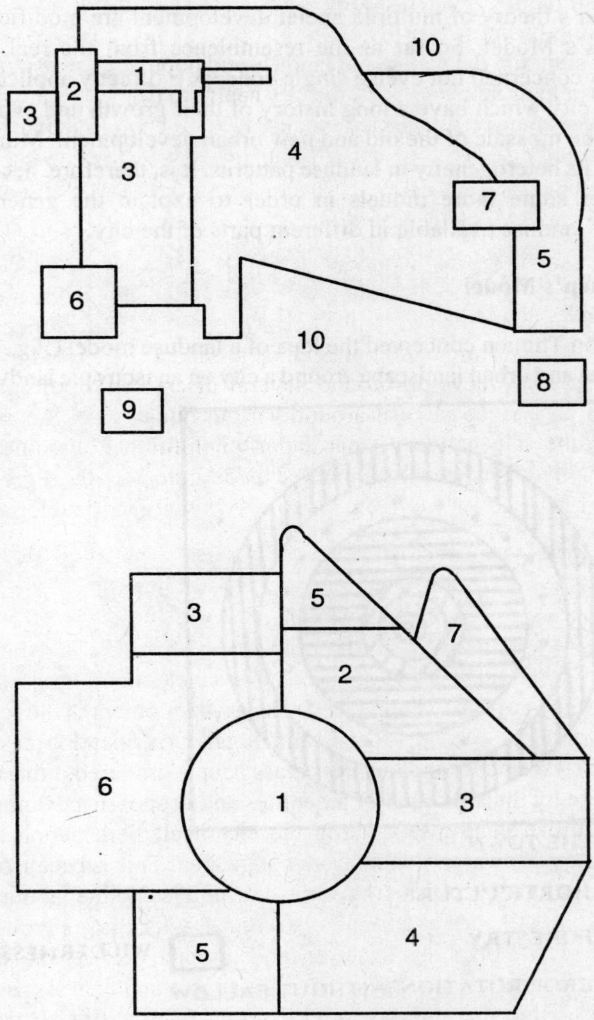

Fig. 7.4 : Multiple Nuclei Model of Harris and Ullman

1. C.B.D., 2. Light Manufacturing Industry, 3. Low Class Dwellers, 4. Middle Class Dwellers, 5. High Class Residences, 6. Heavy Manufacturing Industry, 7. Outer Commercial Belt, 8. Residential Suburb, 9. Industrial Suburb, 10. The Commuter's Zone.

was propounded to explain the landuse of a newly developing industrial city like Chicago. Hoyt's concept of sectoral development and Harris and Ullman's theory of multiple nuclei development are modifications of Burgess's Model. So far as the resemblence from the real world situation is concerned not even a single concept is directly applicable to the Indian city which have a long history of their growth and evolution and complex mossaic of the old and new urban development. Munger is typical for its heterogeneity in landuse patterns. It is, therefore, necessary to consider some more models in order to explain the generalized patterns of landuse available in different parts of the city.

Vön Thunen's Model

In 1826 Vön Thunen conceived the idea of a landuse model (Fig. 7.5) in case of rural and urban landscape around a city on an isotropic landscape.[6]

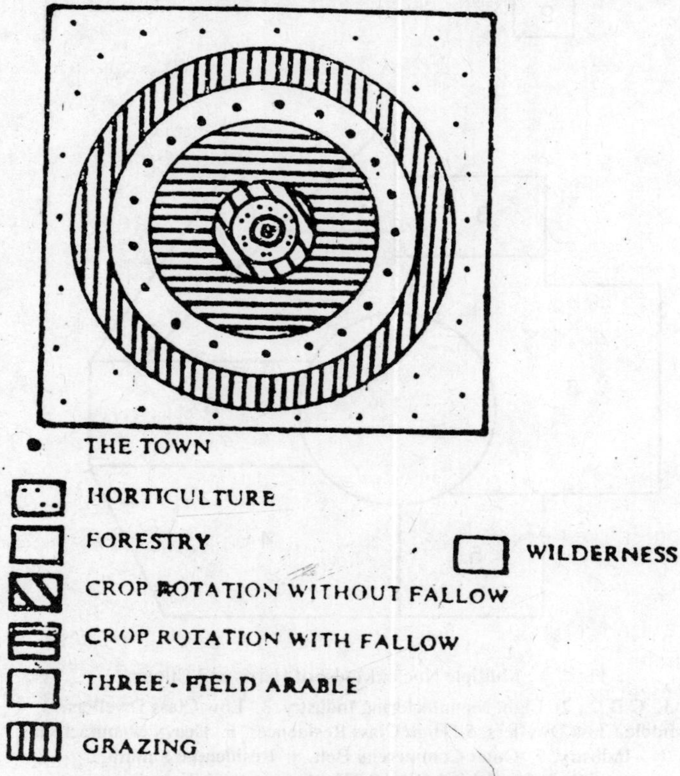

FIG. 7.5 : Vön Thunen's System: Original Model of Landuse Types

He applied his notion to the State of Tallow near Rostok in Germany. His idea is based on the consideration of economic rent which decreases from the centre of a city to its periphery. His system of landuse around a city which has not trade connections with any other nation, is ring shaped. Near the city he envisages rings of forest (for wood), crop rotation without fallow, horticulture and dairying and the like. It is quite clear that this theory of location explains the principles of agricultural landuse around a city rather than the landuse within the city.

Sinclair's Model

In 1967 Sinclair[7] had offered a ring type model (Fig. 7.6). The progression of the intensity of Sinclair's ring is directly proportional to the degree of urban influence in the form of high urban taxes, restrictive zoning, and nuisances stemming from near by urban areas. More recently, two English writers, Best and Gosson, have lend their support to this theory in their study of a trend towards extensiveness along the south-eastern margins of London. They believe that the shift is largely the result of: (i) increasing competition from areas more distant but with otherwise better production facilities, and (ii) the loss of casual labour to city jobs by rural-urban fringe farmers.

Sinclair's theoretical sequence of landuse around an expanding metropolitan area are as follows (Fig. 7.6).

1. Urban farming close to the city.
2. Vacant and grazing land.
3. Field crop and grazing land.
4. Dairying and field crop land.
5. Specialized food grain-livestock.

Anti-Polar Concept

According to the present distribution of landuse zones inside the city of Munger, R.B. Mandal[8] feels the need of a new model based on Indian conditions. In formulating this model—the increasing intensity of urban landuse from fringe to the city centre is regarded a reality. The overwhelming market oriented development facilities provided by the government have helped in the increasing intensity of landuse near the city centre. This shows that centripetal forces are at work from the hinterland to the city centre and we may also observe urban impact on rural areas as centrifugal forces. Decreasing intensity of landuse from city centre to the fringe area has been recorded in Munger. The central

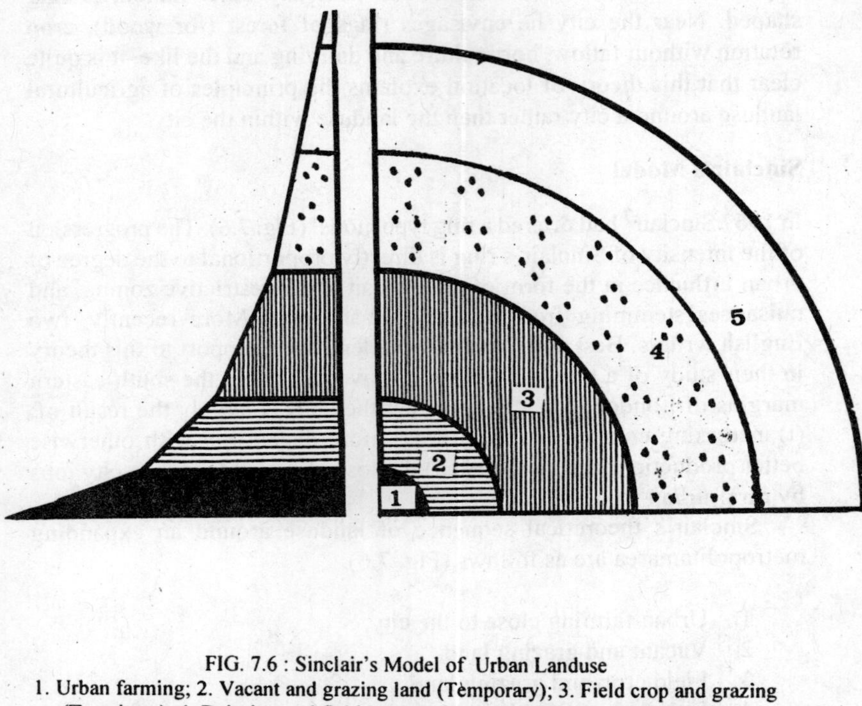

FIG. 7.6 : Sinclair's Model of Urban Landuse
1. Urban farming; 2. Vacant and grazing land (Temporary); 3. Field crop and grazing (Transitory); 4. Dairying and field crop; and 5. Specialised foodgrain livestock

part of the city is occupied by Central Business Area followed by light manufacturing, residential and zone of outlying business nodes. Lastly, the area of open space, agricultural fields, remnants of old villages, industrial centres, transportation nodes, religious centres and administrative centres are found (Fig. 7.7) while going away from the centre of the city. With the help of local relief, density and skyline of houses, land value and density of population the author supports anti-polar concept of urban landuse as a new model.

Model of Decision Analysis or Activity System

Chapin[9] has proposed a new dimension of urban landuse analysis. He considered several aspects of human behaviour as the product of particular

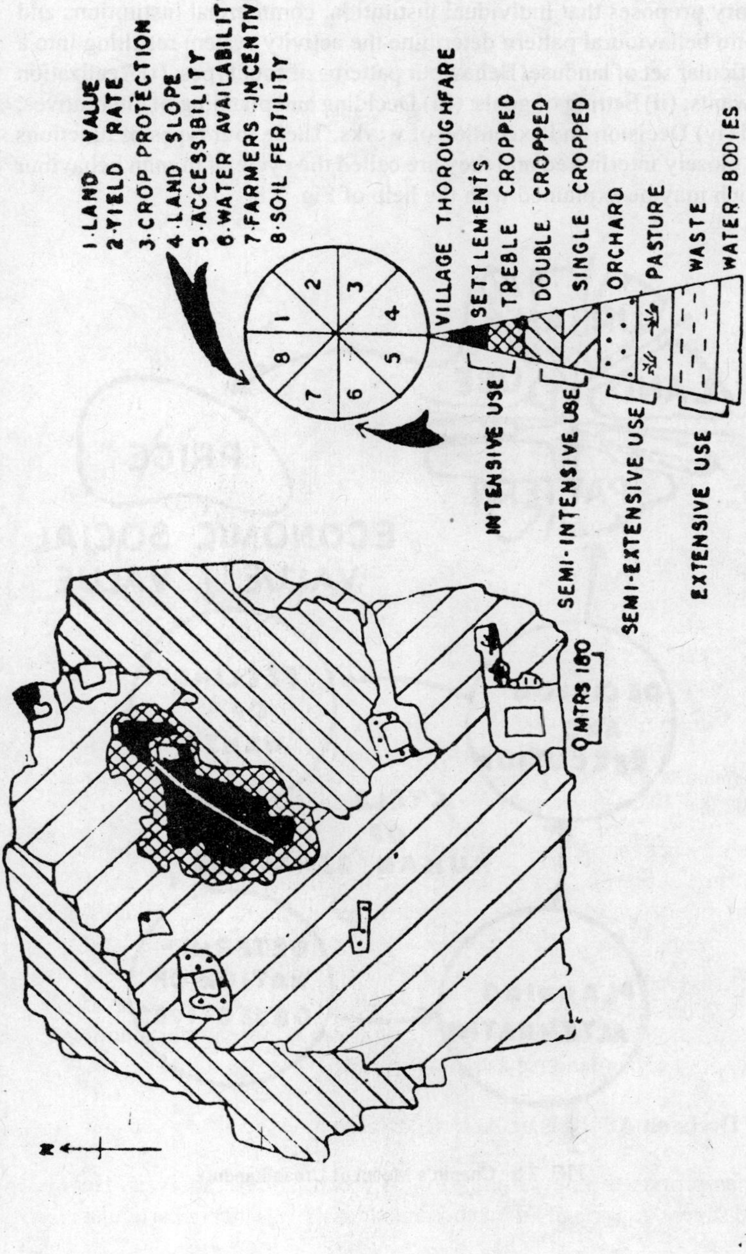

FIG. 7.7 : Anti-polar Conceptual model of Landuse

landuse. So his concept is known as "Theory of Activity System." The theory proposes that individual institution, commercial institutions and spatio behavioural pattern determine the activity system resulting into a particular set of landuse. Behaviour patterns of four types: (i) Realization of wants; (ii) Setting of goals; (iii) Deciding and planning of alternatives; and (iv) Decision and execution of works. These four types of functions are closely interlinked and they are called the cycle of human behaviour which may be explained with the help of Fig. 7.8.

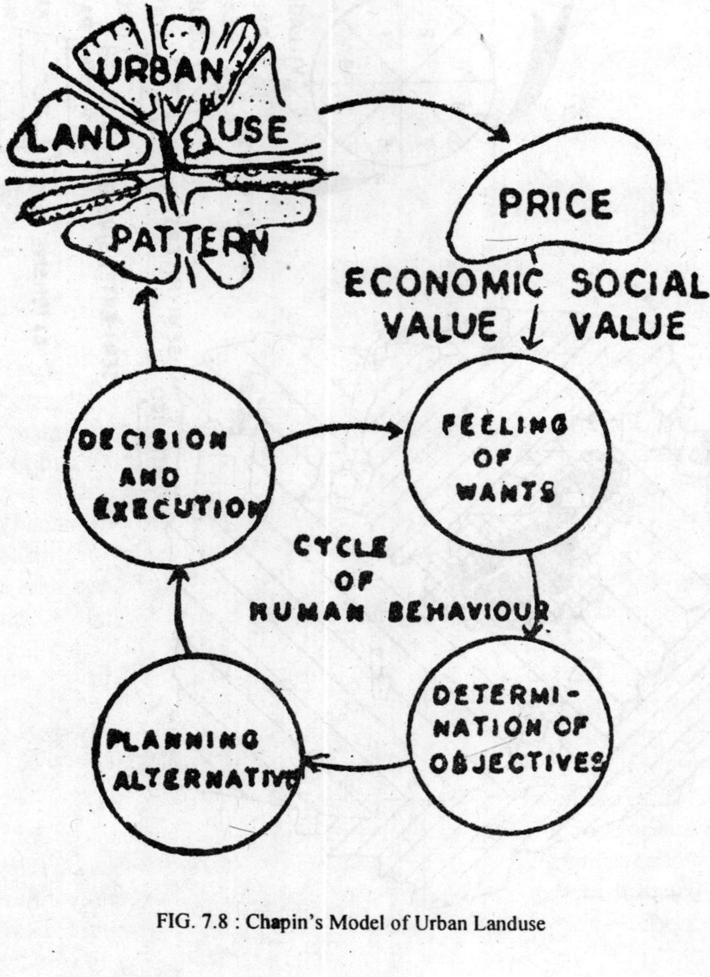

FIG. 7.8 : Chapin's Model of Urban Landuse

Gradient Models

The presentation of urban form by gradient analysis is an important dimension of describing and generalizing urban structure. It is related to zonal and sector theory, though it has a more pronounced and quantitatively rigorous empirical foundation. It is useful because any regular and logical sequences, or change with distance from the city centre; can be identified by line graph and represented mathematically. They can be employed in a broadly predictive way because the relationships observed in one city is often hold true for others.

Population Densities

Clark's work (1967)[10] on gradient analysis attaches special attention. He studied population density gradients of 36 cities, and using this statistics from 1801 to 1950, Clark was able to argue that, regardless of the particular landuse arrangement within a city, and the study of residential population density outside the central business area of a city declined as a negative exponential function of distance from the city centre (Fig. 7.9).

A logarithmic transformation of the curve reduces a straight line. Two important points should be made about Clark's generalization. First residential densities for the central areas of cities are extrapolated from the slopes derived from the outer residential areas. Since the central areas of some cities are occupied mainly by business and public buildings, the hypothetical density used in deriving the equation for the gradient is often very much low than the observed density. Instead of a solid curve the density profile resembles the profile of a volcanic cone. Secondly, the values used for densities outside the central business area are usually calculated by taking the mean of all administrative of census districts that lie at equal distances from the city centre. Since metropolitan areas have highly complex structures, with many subsidiary centres in addition to the main central areas, it is likely; that in most of the studies regular population-density gradients disguise important differences in urban form.

More recent work has confirmed that original empirical observations provide an economic rationale for declining residential densities. But, as the evidence improves, so that suggestions has arisen that Clark's relative exponential model is not a fully efficient and conceptually relevant description of urban density gradients. For example, Tanner (1961) and Sherratt (1960) have proposed that urban population densities decline exponentially as the square of distance while Newling (1969) has suggested an alternative hypothesis which not only regards densities within and beyond the central business district as a continuum, but also provides a dynamic framework within which the emergence of a density crater in the central business area can be demonstrated. Most recently Kostrubiec (1970) has outlined a three-part profile, based density statistics derived from London and Paris, against which the current stage of expansion of smaller urban agglomerations can be matched.

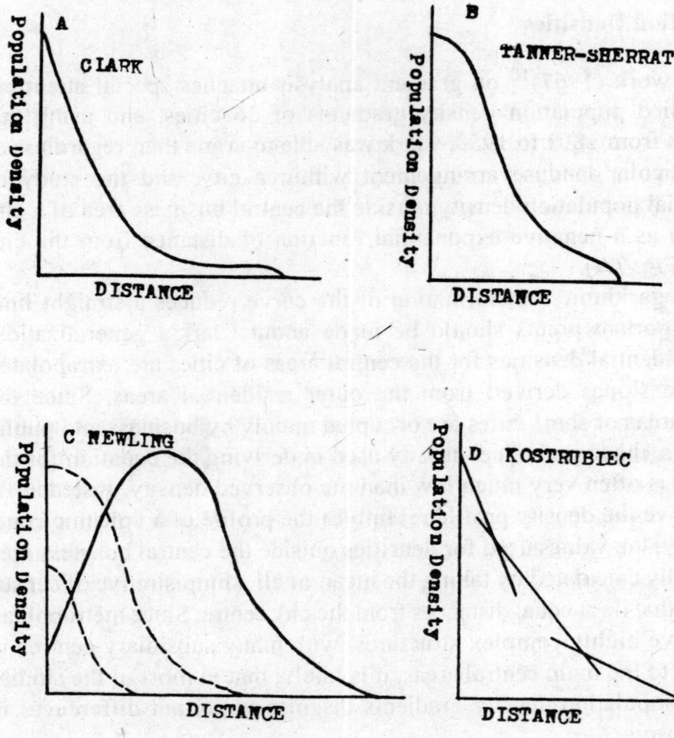

FIG. 7.9 : Model of Urban Residential Population Density Profiles

Most gradient analysis has been applied to western-style cities. But work by Berry, Simmons and Tennant (1963) has suggested that the population density of the non-western cities may also be described in a negative exponential gradient model. Present gradients are very similar, but unlike western cities where density gradients in the past were much steeper; the density gradients of non-western cities have remained relatively constant through time; partly due to their less flexible transport systems; which have encouraged more compact cities; partly due to the differing structure of society, which has led to a different distribution of rich and poor within a city. But it should be added that the differences noted by Berry are by no means universally agreed.

Growth Simulation Models

Most of the Urban Simulation Models are directed to replication and forecasting of population spatial patterns. Work has been undertaken within both deterministic and probabilistic frameworks and a short explanatory statement needs to be made about each of them.

Deterministic Models

Deterministic models make strict assumptions about human behaviour. Based on the above assumptions it becomes possible to predict with some degree of accuracy the future development of population in time and space. Bylund (1960)[11] using such a model, was able to analyze the process of urbanization in the central upland area of Sweden and established why in the early phase of colonization, settlers failed to bring the best quality land into use first; before cultivating poorer land. Adopting these assumptions and working from the initial settlement pattern, he was able to produce patterns which approximated closely, at each stage of development, to the actual distribution.

Though Deterministic Models of different kinds have been employed with considerable success; they do have a number of disadvantages. They tend to be inflexible; they require a fairly through understanding of the process of change in any given situation and there may be a few when there are large number of factors to be considered. The alternative, if these disadvantages are unacceptable, is the Probabilistic or Stochastic Model.

Probabilistic Models

Probabilistic models depend upon the assumption that, though the actions of individuals may be consciously determined, the aggregate effect of human decisions, within certain constraints may often be expressed randomly and concieved; for the purpose of chance variations analysis. By and large Probabilistic models are more flexible and often more realistic than Deterministic Models but they can also become Mathematically more difficult. This problem is normally overcome by the use of Montecarlo Methods that is by simulating growth and change with reference to random number sequences. The random processes are often made more realistic by adjusting their operation so that they conform with empirically-observed trends. This model has been quite often used by Morrill.[12] He began his work in Southern Sweden (e.g., 1968-1965); facing the growth in population from 1860 to 1960. Using Census Material and employing principles derived from central place, Industrial location, and transport location theory he simulated population growth and predicted future distributions in twenty year circles with considerable success. In assigning migrants, to take one element of his model as an example, he assumed that the propensity to move from a source area was a function of the size of population lies in that place and secondly of its character as age structure. The probability of migration between two areas was assumed to be closely related to their distance apart, the difference in their attractiveness and their previous migration history. Given these assumptions, the chances of migration taking place from one area to another could be assessed and the total number of migrants distributed among the destination areas by means of random number tables calculated. In other work Morrill simulated the urban growth on a city fringe. He considered the issue in a rendom fashion and visualized the inverse relationship to distance from the existing settlement and the neighbourhood quality.

Monte-carlo Simulation Methods are not without their difficulties. But provided that the assumptions which determine the operation of the random elements of the Model are valid, and that the Model is based on data that can be aggregated. They seem to provide a reasonably efficient means of representing the dynamics of urban land expansion. They are useful because they present a practical and realistic method of establishing the casual forces operating at present, because they enable prediction of the future development to be made with greater precision, and because they provide an effective tool for those engaged in applied studies, e.g., planners whose task is to make decisions about alternative courses of action to deal with current and future problems.

Urban Land Appraisal

Recently, several empirical methods have evolved to study the urban landuse patterns in towns and cities of the world. They discussed different types of urban landuse patterns especially when they agglomerated into city "quarters." The studies also had a dynamic element in the sense that they employed description and statistical material to create an impression that people in urban areas move from one place to another but the form and functions of buildings changed through time including basic nature of cities.

Urban Land Component

(a) *The Use of Central Business Area*

The commercial activities of the city centre have long attracted the attention of planners. The central business district has provided geography to study the nerve centre of the city (CBD). The city centre is the focus of highways within and beyond the city. It is highly desirable location for the commercial, social and cultural activities which depend upon accessibility. In smaller cities these activities are intermingled, but in large urban areas there is a decided tendency for separate sub-districts to be devoted to the government, administration, finance, retailing, theatres and other basic activities.

Since most central business district developed when traffic modes were different from those of today and when traffic flows created less intense demands upon the communication arteries. The streets are relatively narrow and railway lines are often inadequate to carry present loads. Traffic congestion results and is often severe in the large cities. To combat congestion in a limited number of traffic free precincts have been created, but a more typical response particularly to the motor car has been to build new car parking often in multi-story buildings, to construct new roads and to introduce traffic management schemes to facilitate traffic movement. Combined with the pressures from the central business district uses; the result has been to squeeze the less competitive uses; such as housing out of the inner city. Cities are, therefore, common lodging for permanent residential population in their central areas.

Wholesaling and Small-Scale Manufacturing Uses

Some cities are intimately associated with the central business district while in others occurs a zone in which wholesale activities, retail shops, workshop and manufacturing predominate. These are forms of economic enterprises which require direct access to extra city transportation and thus market location at a focus of routeways is desirable. Both uses derive support from the city but unlike the true central area activities they are not firmly drawn into the centre. Their ability to compete for city centre sites is rather lower and they tend to be found on the edge of the central business district, adjacent to the railway stations and main road arteries. They are set to reach the city besides regional and national markets. The two uses contribute substantially to the congestion inside the city. They not only attract labour into the centre but they also generate a considerable volume of heavy commercial traffic.

Quite often, wholesaling and small manufacturing business are able to occupy buildings not specifically constructed for such use. They often adjust in former houses meant for residential purposes around the centre of the city. These areas have not yet been subjected to the processes of urban renewal and hence rented buildings are available easily in low rent.

A New Model of Activity System

The author has tried his best to devise his own model of activity system for explaining the special variations of urban land utilization between the intensive landuse of central business area, i.e., Chowk Bazar and the extensive use of land of fringe area, i.e., Herudiara in the city of Munger. Besides there are several forces which interact in space as high or low density of population in central business area and fringe area respectively. Similar is the case for flow of urban oriented goods and services, infrastructure facilities and forces of migration. But inverse relationship has been established in case of individual participation, flow of land produce, consumers and the availability of pollution free environment. The areas of central business are characterized by intensive landuse whereas fringe area are land of extensive or sparse landuse. In between these two lie the zone of the dilemma of space search owing to limited income potential on the one hand, zone of functional interaction of medium intensity of landuse on the other.

The above forces act as carriers and barriers of urban land utilization. Actually they come under the human activity system as response to the environment. Thus the intensity of land utilization of a place is the end product of the demand of the land, supply of the land and spatial dimension of population concentration. This model helps us in analyzing the human activities in a particular ecological condition and the resulting pattern of urban land utilization.

Model of Urban Situations and Changing Character of Problems

From centre of the settlement towards fringe the housing structure of urban areas evolve in different stages. This leads to increasing land value and population from fringe of the urban centre towards the (Fig. 7.10) central core. On the other hand, the intensity of landuse decreases from core to the periphery. This favours increased unplanned economy towards the urban periphery while in the core problems increases due to over population and congestion of houses. This is in accordance with the distance decay function' from city core to the fringe area.

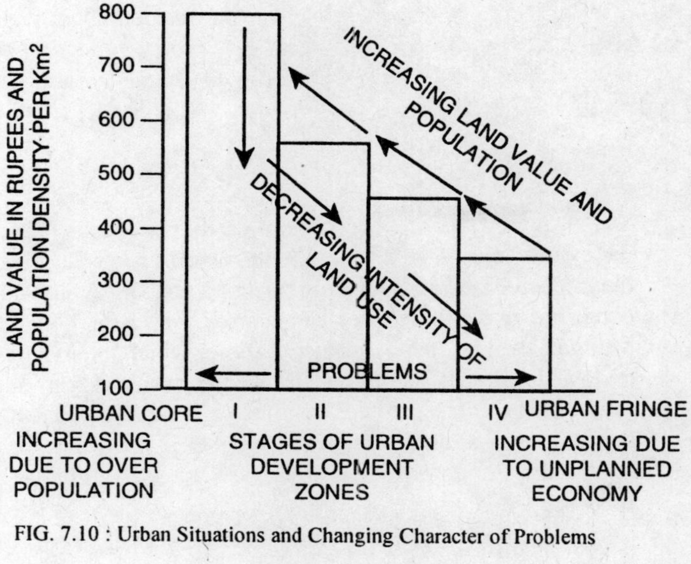

FIG. 7.10 : Urban Situations and Changing Character of Problems

FOOTNOTES

1. Alice Coleman, "A Geographical Model for Landuse Analysis," *Geography*, No. 242, Vol. 54, Part I, 1969, pp. 43-55.
2. E.W. Burgess and Bogue, D.J., (Ed.), *Contributions in Urban Sociology*, University of Chicago, 1964.

3. H. Hoyt, *The Structure and Growth of Residential Neighbourhoods in American Cities*, Washington, 1939.
4. C.D. Harris and Ullman, E.L., "The Nature of Cities," *Annals of the American Academy of Political and Social Sciences*, Vol. 242, 1945.
5. C.C. Colby, "Centrifugal and Centripetal Forces in Urban Geography," *Annals of the Association of Urban Geography*, Vol. XXII, 1933, pp. 1-20.
6. C. Von Thunens Clark, "Isolated State," *Oxford Economic Papers, New Series*, 19, 1967, pp. 370-77.
7. R. Senclair, "Von Thumen and Urba Sprawl," *Annals of the Associtión of Amrican Geographers*, Vol. 57, 1967, pp. 72-87.
8. R.B. Mandal, *Land Utilization: Theory and Practice*, Concept Publishing Company, New Delhi, 1990.
9. F.S.J. Chapin, *Urban Landuse Planning*, New York, 1957.
10. W.A.V. Clark, "The Spatial Structure of Retail Functions in a New Zealand City, *New Zealand Geographer*, 1967, No. 1 pp. 23-33.
11. E. Bylund, "Theoretical Considerations Regarding the Distribution of Settlement in Inner North Sweden," *Geografisca Annaler*, Vol. 42, 1960, p. 225.
12. R.L. Morril, "The Development of Spatial Distribution of Towns in Sweden: A Historical Predictive Approach," *Annals of the Association of American Geographers*, 1963, Vol. 53, pp. 1-14.

8

SYSTEMS OF URBAN PLACES AND CENTRAL BUSINESS DISTRICT

URBAN CENTRES having self-centred in a well defined boundary is known as "Closed System." The Aryan market centres like Swastika, Sarvato Vadra, Padmaka are examples of 'closed systems', where all functions are performed inside the settlement in different quarters.

The urban centres having wider linkages with its surroundings is known as "open system." In this system the flow of money, men and material comes to the city from distant places. Almost all urban centres of present day are examples of the "open system".

System denotes different tiers of functioning of different elements of urban environment. It may be related with the systems of physical, economic, social and political in nature. The urban system, particularly the land and property market, is subject to state intervention which facilitates and controls the distribution of resources and activities. At a subjective level, urban life is increasingly experienced as a series of encounters with complex public organizations and cultural heritage.

The systems of urban places means those who are involved in the production, supply, administration, management and regulation of public goods and services for collective consumption. Those goods and services range from land, roads, sewers, water, electricity, gas, etc., through the planning system, police and fire services, to health care, education, housing and social welfare.

Tiers of Urban Systems

Clearly these collective consumption facilities and resources are controlled by a variety of state bodies—local authorities, central government departments, nationalized industries, semi-autonomous national governmental organizations, such as Regional Health Authorities, Regional Water Authorities and so on. But the crucial point is that their role is distinctive and separate from private sector institutions. Their interventions are significant determinants of urban life-chances. However, some institutions operate universally throughout the country to carry out these functions, there are important territorial inequalities in service provision. There are widespread variations in the quality, quantity and type of urban public goods between different local authorities. Therefore, where you live determines not only your job and housing opportunities, but also the conditions of access to a whole range of services and facilities. The impact of the formal urban system is thus pervasive and substantial.

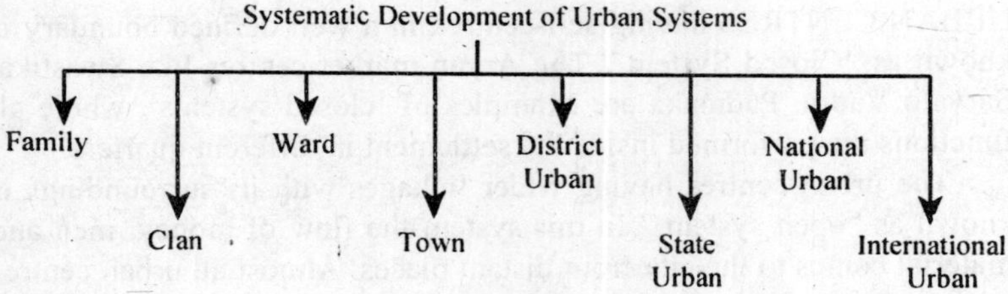

The urban development starts from the functioning of a family who involves in non-agricultural occupation. The family may be a part of the clan group. The association of different clan groups help in forming a ward. The group of different wards form a town. The towns and cities are again grouped at district level under the command of district magistrate. Again all the district level urban centres form the state level urban centres. The state level urban centres form national level urban centres, which are directly or indirectly related with the international level mega-cities where multinational companies are moving today for various kinds of trade under GATT agreement. The cities of Mumbai, Calcutta and New York come under this category.

Elements of the Urban System

A brief outline of the elements of urban system will show the range of organizations involved in urban affairs including the number of points at which people come into contact with different elements which shape the functioning of urban residents. These distinctive elements which described below have cumulative impact. This schematic outline is not a comprehensive description however, it is an approach to emphasize how the range and scale of these provisions contributed to the structure and process of urban life. For example, the interaction between the personal social services and participation in the labour market of women is not always recognized in conventional terms.

Systems of Trade and Commerce

Urban development is primarily 'driven' by private market forces. The state is only involved through the regulation and negative control of landuse by the planning system. Roads and other infrastructures are provided by the State Government.

Shopping system includes wholesale market, retail market, market nucleus in central business area, commercial ribbons along the road and the location of marginal business centres along the rural-urban fringe. At some places the businessmen use to come daily to sale consumable goods in which a small amount of capital is involved, i.e., vegetable shop, shoe mendors, cycle repairing, burnt earthen jar shop, fruit shop along with ready-made garments, *juice walas* and parched grain shops.

Ferrywalas often move daily in each and every *mohalla*'s of the city. Their movement is cyclic in character from town to town and *mohalla* to *mohalla*. They sale goods of smaller capital investment and get benefit at the shortest possible time. Their prices are not fixed and in order to get maximum profit they bargain.

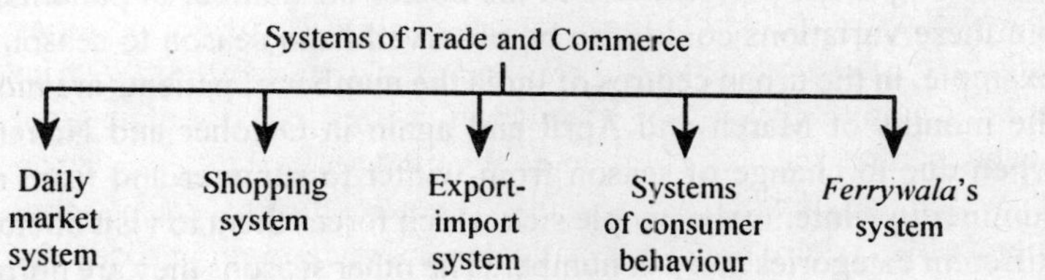

Systems of consumer behaviour denotes that for higher order goods consumers come from distant places but for lower order goods the consumers from shorter distance come to the market. In this system the consumers of different income groups, for different goods and temporal dimensions behave differently. For example, the consumers of betel come from local areas whereas for refrigerator they come from distant places.

So far as the export-import system is concerned different towns specialize in different items of surplus products. For example, in Muzaffarpur *Lichhi* is a surplus product, in Dhigha of Patna Maldah mango is a surplus product and from Madhubani *makhana* and painting items are exported like Jardaly mango from Bhagalpur, Silk products from Nathnagar, Mica from Kodarma and Giridih and *tilkut* from Gaya. Similarly according to their requirement all urban centres import kerosene oil, cloth, medicine, fertilizer and other industrial products from larger urban centres whose business is more commanding and the salable items are available in lower prices.

Systems of Urban Land Utilization

In urban areas the land has been utilized zone-wise according to the functions performed in those areas. The transportation node, health centre, commercial hub, administrative areas, postal and communication and central libraries are located in down town whereas the marginal services find their place in the fringe area of the city. This has been caused according to the price of the land. In central part of the city the land value is generally high whereas on the fringe area it is less and less.

Health Service System

The national health service provides a whole range of services in addition to general practice of clinics and hospitals. There is a wide variations in the systems of medicine from allopathic to Homeopathic, Ayurvedic, Unani, Hakim and Baidya including acquapuncture and neurologists. According to the performance of the doctor the number of patients vary but these variations could also be observed from season to season. For example, in the urban centres of India the number of patients are more in the months of March and April and again in October and November when due to change of season from winter to summer and from rainy summer to winter make people sick which forces them to visit doctors of different categories more in number. The other seasons they are normally low.

The system of hospitals range from primary health care to municipality level, sub-division level, district level, state capital level and national level.

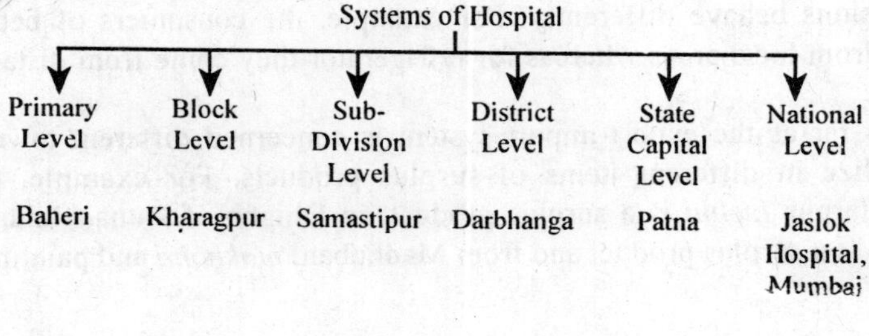

In spite of attempts to reorganize and to streamline the system, there are persistent regional and social class variations in ill health and in the amount and standards of medical care.

Community Care System

The personal social services run by municipalities are meant to provide support and a variety of services for people who cannot fully participate in economic and social life. Thus some of the old, the sick and the young are partially catered for, and the welfare state to that extent removes the burden of care from the women, allowing women in particular to participate in the labour market. With rising levels of unemployment, there has been a shift away from this kind of policy towards more 'care in the community' by the women.

Social Security System

The social security system is nationally organized but locally administered and makes cash payments to alleviate poverty caused by unemployment, disability or sickness. At the end of 1978, nearly 3 million people were receiving 'supplementary benefit'—59 per cent of these were pensioners and 20 per cent were unemployed. The system has been criticized because it stignatises claimants, discriminates against women and relies heavily upon discretionary decisions by officials in India.

Educational System

The educational system historically grew up to supply the needs of the labour market for trained workers, even though in practice it did not always seem to be doing that. Now with a contracting labour market, the state is more reluctant to support the system. Teachers training colleges and schools have been closed and this retrenchment may continue. Educational success does vary throughout the country and councils differ in the kind of school system provided, as well as in the investment of resources. For example, there are marked variations in expenditures on facilities such as career services.

Administrative System

It includes the system of wards, municipalities, corporations and urban development authorities. These are administered by police force, ward

Commissioners, Chairman and mayor at the corporation level. They control the regulation of street lightening, sanitation, control of sewage and water supply as well. The administrative machinery has a right to plan the place for bus-stand, railway station, rickshaw stand, water tank, stations of electricity grid, play grounds and botanical gardens, etc.

Transportation System

Transportation lines are considered as life line for the circulation of money, men and material inside the city. It commands road, rail, water and air transport as well. Among these road and rail have greater importance in transporting passenger and goods inside the city. In the urban centres main road and streets are the main arteries of movement. The means of transport are rickshaw, tempo, cycle, motor cycle, white line, green line and red line buses plying upto different places along with the taxies and private motor cars. In Calcutta tram service and underground rail are some of the important means of cheap transport services available to the public.

Jajmani System

In this system the social functionaries of cobler, oilmen, priest, Kumhar and barber do their job to their *jajman*. This practice is traditional and hereditory. In the urban centres of lower-order Jajmani system still prevails in India. In this system cobler makes shoe, oilmen gives oil, Priest help in worship, shradh and marriage ceremony; Kumhar provides earthen jar, barber dresses the hair and washermen washes clothes regularly as they have divided the families on their own that whom they have to serve.

Bank and Insurance System

There are different banks which help urban people in saving, running trade and commerce besides credit facilities. Insurance companies help in life insurance, crop insurance, building insurance, car insurance and others. Banks function as extension counter, branches, main office, lead bank and affiliated with reserve bank. Without bank it is not possible in urban areas to run different kinds of business by the companies. These days multinational companies cannot function without banking and insurance facilities as their business turnover is very high. The banks function at different levels. Sahara is a finance company which collects the savings of petty businessmen daily. Similar facility is available to the public in Canara Bank, Bank of India, Baroda Bank and the Punjab National Bank.

Marriage System

In urban areas marriage ceremony performed in different castes and different based of religion basis different clan basis, with dowry and without dowry besides urban to urban marriage and so on. Recently, inter-caste-love-marriage is also getting ground in urban areas but this type of marriage is transitory and its continuance is always questionable-marriage of a ten aged girl is sure and after marriage her migration from one urban centre to the other is sure. It is known as forced migration. In almost all castes priests and the barber help in the marriage under traditional Jajmani system.

Religio-Ritual System

This favours *Purdah* system by the Muslim women. Both the houses and women folk live in a secluded manner even in their small-hutments. Ritual favours maintenance of untouchability, performance of *shradh* after death, marriage ceremony by customary method, along with the dominance of *barnas* in the society. Aryan's have divided the whole Hindu society into Brahman, Rajput, Vaisyas and Sudras. These divisions are based on the occupational variations in the society. In this way, the whole society is divided into Hindu, Muslim, Sikh and Christian. Their temples of worship are different from one another whereas the God remains the same in belief but their statues are different.

Political System

In urban centres the political parties find their base due to multiplicity of views among masses and the place of concentration of population where any meeting and group discussion help in disseminating ideas of political parties upto far away places in rural areas. Nowadays, in India Congress (I), Bhartiya Janata Party, Jantadal, Communist Party and several others are trying their best to mobilize the people to gain favour. But recently the criminalization of politics have severally damaged the morality of people and disrupted the political ethics. Most of the political crimes in the country are occurred in towns as these are the nerve centre for any political decision.

Multiple Systems of Employment

Employment is completely controlled by the state agency known as employment exchange with local offices at district level centres. Since 1973, the manpower services commission has worked to improve the efficiency of local labour markets by providing assistance for people to select, train, obtain and retain suitable employment. With rising unemployment, these functions have become more politically sensitive. Throughout the country, local offices and 'Job Centres' attempt to match job vacancies with potential workers; provide advice and guidance; make training facilities, grants and transfer, etc. The former scheme provided the unemployed school leavers with some training and work experience, while the latter provided younger unemployed adults with temporary jobs. Although these measures are small in scale when measured against rising unemployment, they represent growing state of intervention in the labour market and the incorporation of employment as an element of social welfare.

Thus the formal urban system consists of a multiplicity of essentially bureaucratic state institutions whose activities influence the distribution of opportunities and resources in many aspects of people's lives. The administration of the welfare services is largely controlled by officials who often exercise discretionary powers. However, the exercise of these powers reflect political decisions by local authorities. In particular, attitudes the spending are politically structured and over long period of time authorities develop characteristic approaches to spending.

The formal urban system is characterized by complexity, heterogeneity and fragmentation. These features were identified as the causes of confusion, uncertainty and ambiguity in central/local relationships in recent government reports and have become the object of a major research programme. The implementation of policies is regarded as problematic because of difficult inter-governmental relations and variation caused by local autonomy. Moreover, the allocation of urban resources must be seen not only in the context of interlocking and overlapping state bureaucracies, but also in the context of interactions between the state and the private sectors.

Thus, the provision of the social wage and the management of the urban system by the local state are fundamental to the operation of the labour market and economy. Intervention in 'manpower planning', tightening controls over the social security system, moral panics over 'welfare scroungers', anxiety about the 'relevance' of education, and coercive aspects of social work, can all be interpreted as part of the state's overriding preoccupation with the need for a viable labour force and an efficient economy. Other local government and state welfare services (housing, health, transport, planning) have similar effects.

There is evidently a close connection between the formal urban system and the labour market, and since each operates as a determinant of life-chances it is essential that we analyze the bureaucracies involved. Given that 'social wage' allocation is not primarily governed by market criteria but by bureaucratically-defined criteria of need we should examine the role, influence and interests of those who manage the urban system.

Urban Managers and the Social Wage's

The recognition that inequalities in the urban system derive largely from variations in the allocation of scarce goods and services was the basis of Pahl's (1975) argument that certain groups of 'urban managers' are important and powerful decision-makers in the social and spatial structure. The emphasis was on the fact that these gatekeepers or managers determine the eligibility of clients, assessed their needs, and made allocation decisions which could reduce or increase existing social inequalities resulting from a household's position in the labour market. In other words, urban managers exercise some independent influence on life-chances, and, if this was so, it is important to examine their ideologies, action and degree of power.

This idea stimulated considerable discussion and criticism. The main difficulties with the concept of urban managers centred on its relevance for both public and private sectors, problems of identifying the manages empirically, and most important, assessing the extent of managers' influence in the context of structural constraints on their autonomy. These problems are substantial, but the question of the actual or potential mediating role played by state bureaucrats in the distribution of the social wage remains important.

Systems of Rurban Centres

The smaller urban settlements and the larger villages have been described variously as rurban settlements by R.L. Singh.[1] rurban community by Galpin,[2] urban villages by Dickinson[3] and subtowns by Smailes.[4] In India most of the smaller towns may be included in this category, which would also cover a large number of villages with population over 2,000.

The smaller towns have a distinctive feature of their own. "Most of the smaller centres are overgrown villages and still retain their rural character. They show a combination of economic, social and educational function.[5] While all towns have a city region around them, the small towns function as the economic and social capitals of their surrounding essentially agricultural areas.[6] By virtue of its intermediate location and the nature of evolution, the small town has certain advantages and offers unique opportunities in the general scheme of rural-urban and agricultural industrial continuum.[7] Thus, the study of smaller urban settlements is highly important "as they are urban islands in the vast sea of rurality[8] and thus represent the "basic connection between the dispersed agricultural population and the agglomerated urban population."[9]

The primary objective of this study is to examine the factors responsible for the spatial distribution of smaller urban settlements and examine their role in the systems of urban places. Such urban units are marked by the growth of certain significant urban functions. In fact, they are neither purely rural nor purely urban but a mixture of both. By virtue of certain urban functions they serve as service or social centres for the surrounding rural areas. They are the foci of a number of market villages. Such centres often have significant retail and wholesale marketing facilities, post and telegraph facilities, market, a dispensary, a veterinary hospital, rural professional services, a high school and community block development offices are other central services found in them. Sometimes, these centres are trade centre for the surrounding rural area, or, administrative centre due to their being the headquarters of the local administration. On account of their distinct commercial, cultural and administrative features such units are characterized by a greater number and variety of services, which are rarely present in a village. Many other functions are attracted to such centres by virtue of their being the seat of local administration.

Many services are found only rarely in rural areas, such as a cinema, hospital, library, etc. These are today available only in towns; as they are too costly to be operated in open country locations or in very small towns. Some of these services may be located on rural-urban fringe but this is basically a town location for all practical purposes. While the bus service and the metalled road have enormously lifted the horizons of rural families as far as the accessibility to these services are concerned, yet costs in time and money are still prohibitive. As a result, most rural families use less of these services than they would if they live in a town and belong to the same income group.

Special type of towns based on services are those involving governmental services of all kinds. Federal and state agencies of various kinds have their offices in rural towns often in the 'country-seat town'.[10] These include the Block Development Offices and the circle offices of the Government. The farmer will have occasion to visit such offices as part of his business operations, and to a greater extent the rural family will visit them for various aspects of family living. "Thus the small towns are quite often seen as the 'bridge' between the urban and the rural universe, as the kind of settlement which can offer the best, and none of the worst of both worlds."[11] Obviously a study of the systems of rurban centres would form as much a necessary part of rural settlement geography, as the study of the larger villages themselves.

The city level centres have distinct advantages over rurban centres in industrial development. But it is difficult, if not impossible to carry out a systematic review of the role of small towns. Although one might conceive of industrialization of the country mainly through large cities, undoubtedly rurban centres would play an important part in providing levers·of regional development as well as ancillary services, such as education and health facilities which are essential for economic growth.

In the system of rurban centres two sets of variables have deterministic influences on the proper functioning of the centre. The first set of variables includes those having predominance in rural areas and operate through larger villages controlling the satellite hamlets. The nearness of smaller villages to larger village or rurban centre attract a large number of visitors. The higher density of population provides the market for local produce, while a lack of population in the sphere of influence always restricts the development of any rurban centre. The second aspect in rural area is that of topographic condition of the region. Highlands with stony wastes prevent the development of rurban centres where as riverine productive plains become conducive to their development. Lowlands liable to be flooded obstruct the movement of vehicles. The third group is that of soil characteristics, economic situation, social and cultural factors each of which have impact upon the development of the centre.

On the urban side three tertiary functions are important for regulating the rurban system. The number of shops, types of shops, utility services and other tertiary services are the main activities of the rurban centre. The other group of factors are modes of transport. The centre may be situated on the roadside, railway side or have river side location while in some cases three variables may converge at a particular centre and give added impetus to all functions of the rurban system. Still another set of factors is wholesale market control. This regulates the capital investment and is related to the productivity of the hinterland as well as the incentive provided to the businessmen. They together determine the market turnover and in return determine the hierarchy of rurban centres.

When all these factors assemble at one place they give rise to a central place system. Here decision about the production, distribution and consumption of goods is taken through the system of exchange, the process of distribution, the supplies from producers and the demand of the consumers. Thus a centre is created where the social, economic and cultural functions are performed for the surrounding hinterland. When rurban centre ranks between the purely urban and the rural sector, it forms part of the rurban group. Further development of this system leads to the development of major trade and urban centres as well as sub-divisional and district administrative headquarters in India. Again these three help in the development and proper functioning of city level centres reaching ultimately to the regional capital (Fig. 8.1). This indicates that rurban centre is part and parcel of central place system envisaged by Christaller or Lösch.

Methodology

A Rurban centre is essentially a local focus of activities and is marked by commercial, industrial, administrative and cultural institutions. It is, however, difficult to differentiate on a functional basis between the large village and the smaller town, and a combined list of criteria has to be evolved. The measurement of functional indices, particularly in a developing region such as India, must be based on the variety and type of central functions other than those available in the bulk of villages. The central functions considered in this study have, therefore, been classified by giving adequate weight (Table 8.1).

Centres having at least three central functions given in Table 8.1 have been considered as rurban centre. Classification of rurban centre on the above basis from the list of villages is easy, but a problem arises in the selection of rurban centres from the list of towns. Both towns and cities have functions of a higher order as compared to the rural, which creates confusion. Thus for the selection of rurban centres from the list of towns, the following method has been devised.

$$RI = Pc \times \frac{tw \times 100}{TW}$$

where *RI* is the urban index; *Pc* is the total population of the centre; *tw* is the workers engaged in tertiary occupations in that centre; and *TW* is total number of workers in that centre.

After calculating the value of *RI* by the above method for all towns the mean was determined, and accepting the mean as the line of demarcation, all towns having an *RI* value less than mean were classified as rurban centres. Further a detailed list of villages selected as rurban centres on the basis of Table 8.1 was prepared. The two combined lists of rurban villages and rurban towns were given proper weight as per Table 8.1, to identify rurban centres in India. The Centrality Index values were put to statistical analysis for determining hierarchy and functional character.

TABLE 8.1 : Weighing and Classification of Rurban Centres for the Purpose of Obtaining Values of Rurban Index

Sl. No.	*Nodal Functions*	*Class I 15 points*	*Class II 10 points*	*Class III 5 points*
(1)	(2)	(3)	(4)	(5)
1.	Education	College	High School	—
2.	Medical	—	Hospital	—
3.	Communication	—	Telephone Exchange	—
4.	Finance	Bank	—	—
5.	Transport		Railways	Highways
6.	Administration	Sub-divisional office	Block Development Office	—
7.	Trade	—	Wholesale Market	—
8.	Recreation	Cinema	—	—
9.	% of total workers engaged in tertiary occupations	—	over 15%	25 to 50%
10.	Diversity Index	9 Nodal functions available	7 to 8 Nodal functions available	5 to 6 Nodal functions available

The concept of the rurban centre is useful because it serves to distinguish rural settlements with a bias towards public utility services, commercial services and non-agricultural population structure. Similarly, there are full-fledged villages which lack something in their basic urban character and yet cannot be termed as rural settlements. A text book on urban geography would be incomplete without identifying such hybrid settlements, which have a high potentiality of urban growth. By definition the rurban centres are partly rural and partly urban and have, therefore, to be viewed as part of the systems of urban centres.

Central Business District (CBD)

Central Business District is the commercial heart of the city where there is a maximum concentration of retail an service activities, highest land values and tallest buildings. It is the focus of maximum pedestrian and automobile traffic (Murphy and Vance, 1954).[12]

The CBD is the area where the residential population is the least and daytime population is high. It is also the area where the intensity of landuse is high and higher order functions are concentrated (E. Swaminathan and A. Raja Mohan).[13]

Central Business District (CBD) is described as the downtown area or the urban nucleus around which the city is structured. It is the city centre and is characterized by high concentration of tall buildings, working population and vehicular traffic. Other features of a CBD include most of the public buildings, many government and organizational offices, largest number of retail shops, headquarters of banks, hotels, clubs, cinema halls, etc. and has the highest land values as compared to other parts of the city. The CBD draws its business from the whole city and is thus the centre of its social and civic life. Some examples of CBD are: Cannaught Place (New Delhi), Tank Road (Mumbai), Chowringhee (Calcutta), M.G. Road (Bangalore), Patna Market and Hathua Market (Patna), etc. (Fig. 8.1)

The Central Business District is a definite region and in many cases it is the heart of the city. There are several detailed studies of the Central Business District of American cities. Methods have been evolved for delimiting the CBD on a rational basis and factors governing the shape of the CBD and typical features of landuse should be discussed. This has in turn stimulated studies of the CBD of towns in other countries of the world as well.

There is an urgent need to identify the CBD in Indian towns and cities in order to describe their main features. The Master Plan of these urban centres should take into account the problems and planning of CBD in comparison with other parts as in CBD the competition of land is high. Solution should also be searched out for haphazard development of lanes and houses insanitary conditions, traffic congestion besides shortage of houses and removal of slums and squatters. Most of the CBD cities of India suffer from these problems.

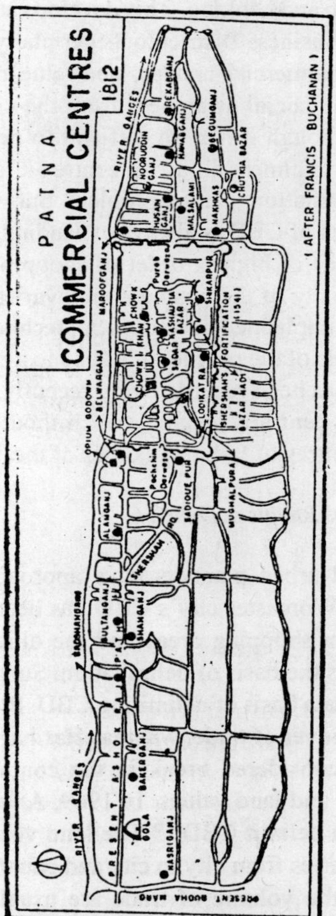

FIG. 8.1 : Patna : Commercial Centres

This shows the importance of the city centre as an urban nucleus around which rest of the city is structured. By virtue of its location and historical importance, CBD provides a wide range of services, e.g., retail, wholesale, traffic and banking. This develops an enormous traffic flow from the Central Business District to its periphery. This favours the declining intensity of commercial landuse, land value, population density, income distribution and social awareness from the city centre towards rural-urban fringe. Although it is a bit difficult to demarcate the CBD but in Indian situation in almost all urban centres it is known as Chauk Bazar area whose delimitation is not a problem, but with the expansion of city business the area of CBD is always expanding in nature.

The CBD is an area of highest order of shopping centre in a city. The location and intensity of the business area varies from city to city due to their planned or unplanned development, technological efficiency and socio-cultural levels of the people. In India most of the cities are old and their fast growth has been recorded very recently. In most cases the central business area is centrally located, which should be considered as CBD even in urban centres of India like cities of the Western World.

Delimitation of Central Business District

Both geographers and urban planners have approached the study on delimitation of CBD. Worcester city's CBD has been delimited on the basis of land values in shopping areas and the allied zone. Williams considered shop rent as the basis of delimitation. Sund and Isacsen used trade and block-wise data basis of delimiting CBD. Proud Foot (1937)[14] considered annual turnover of trade, whereas Harihar Singh (1968)[15] in case of Kanpur city considered break in the continuity of the main functions of the CBD, and land values. In 1969, Ananthapadmanabhan used street frontage to delimit CBD.[16] The land values may not serve the purpose and this varies from city to city and country to country. The shop rent index and the volume of trade are usually difficult to get because people won't furnish this data in their correct form. Murphy and Vance (1954)[17] used persons per unit of area within CBD supported by landuse survey. They found that people are not residing in CBD but here the central business intensity indeed is highest. On this basis they were able to achieve a more realistic delimitation of CBD's of nine American cities.

Central Business District of Shillong

In Shillong of Meghalaya, Bara Bazar is the western part of CBD. It is an old market centre, older than the town itself. Police Bazar in the eastern part of CBD was established in 1864. Other markets and shop in between the two bazars widened the area of CBD. The extension of CBD to the south and west was restricted due to the existence of the cantonment. In 1961 the commercial centre of Shillong was in the western part of Police Bazar on the Shillong-Gauhati road. In 1991 the core of CBD has shifted towards south due to coming up of new shops in other areas of the town. A large part of CBD is found within a radius of 1.5 kms and it is closely related with residential and industrial houses on either side.[18] (Fig. 8.2).

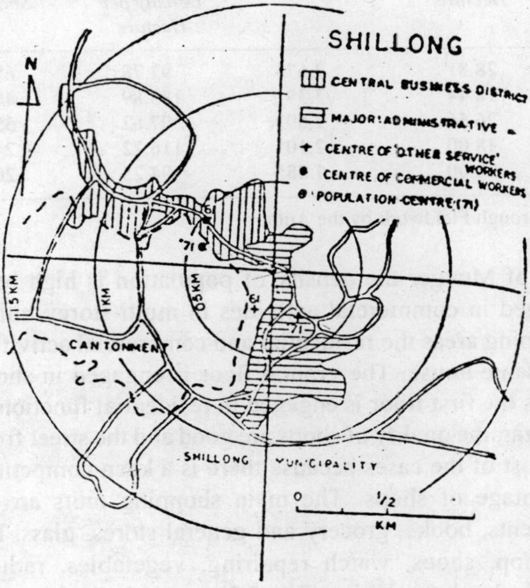

FIG. 8.2 : Shillong : Central Business District

CBD in the City of Munger

The Central Business District of Chauk Bazar lies very close to the Fort Gate and it is the main shopping centre of the City. The CBD in Munger spreads over Gandhi Chauk in the east and Fort Gate in the south. This is the core of the city where along cross roads most of the shops are located. It is one of the busiest market of the city where variety of shops are found within a small area. The present outline of Chauk Bazar provides planned development of roads on grid iron pattern. Such a morphological pattern made the Chauk Bazar more accessible from all directions. The shops are found in a mixed form, i.e., grocery, motels, Pan-Biri shop, fruit stall and others.[19]

The Chauk Bazar (CBD) of Munger covers an area of 112 hectares with a population of 12,663 in 1993. This comprise the commercial cum residential zone of Chauk Bazar, Sarwan Bazar, Bekapur and Azad Chauk besides the open shopping counter of Mirquasims Fort Gate.

TABLE 8.2 : Area, Population and Shop of CBD in Munger, 1993

Sectors	*Area in Hectare*	*Population*	*Population Density per Hectare*	*No. of Shops*
Chauk Bazar	28.81	2,673	92.78	65
Sarwan Bazar	18.44	3,391	183.89	40
Bekapur	26.56	2,593	97.62	55
Azad Chauk	18.00	2,101	116.72	24
Fort Gate	20.00	1,885	94.25	20

Source : Collected through Fieldwork by the Author.

In the CBD of Munger the density of population is high because people are engaged in commercial activities in multi-storey buildings and even in shopping areas the residential and commercial activities are going on in the same house. The ground floor is engaged in shopping activities whereas the first floor is engaged in residential function.

In Chauk Bazar, the quality of shops are good and the street frontage is narrower in most of the cases because there is a keen competition to acquire side frontage of shops. The main shopping units are cloth, readymade garments, books, grocery and general stores, glass, hotels, motels, fruits shop, shoes, watch repairing, vegetables, radio and television, utensils, books and wine (Fig. 8.3).

As the character of CBD is continuously changing, the internal landuse pattern will also change. The shop of lower order will give way to higher order shops. Similarly, the buildings also change their shape and size according to the changing demand of the society.

Conclusion

The Central Business District is of changing character in all respect. Its boundaries expand with certain fronts and retreat on other directions.

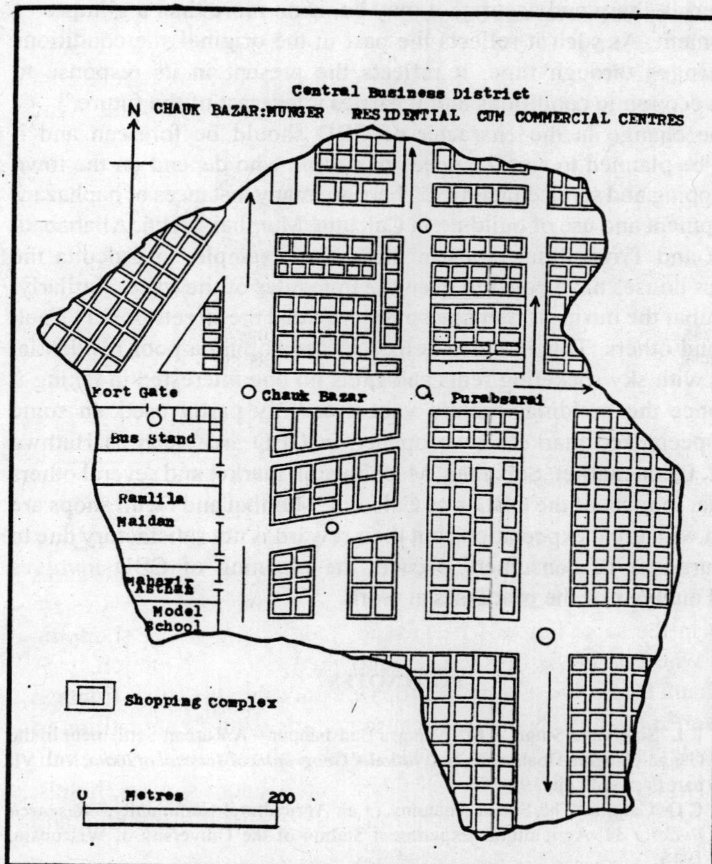

FIG. 8.3 : Central Business District—Chauk Bazar: Munger Residential-cum-Commercial Centres

The change covers the landuse and appearance of buildings. Due to lack of space vertical development of houses is typically found especially in areas of recently developed human settlements. In CBD the land values increase drastically and so is the house rent. "The picture of the CBD, however, is sharp and accurate it may be, is no more than a glimpse of the moment. As such it reflects the past in the original site conditions and changes through time, it reflects the present in its response to current economic conditions and it carries a forecast of the future."

The change in the character of CBD should be foreseen and it should be planned to suit the needs of people who depend on the town for shopping and service facilities. There are many instances of haphazard development and use of building in Calcutta, Mumbai, Delhi, Allahabad, Kandla and Trivandrum cities in India. For example, in Calcutta the business houses have covered even the tramsides of the tract. Similarly, in Mumbai the businessmen have over-crowded the streets of C.P. Tank Road and others. This has also helped in developing a poor residential houses with sky rocketing rents and finds no one interested in taking it and hence the buildings remain vacant or only partly used. In some cases specialized markets have opened in CBD area such as Hathwa market, Patna market, Sabzibag, Machhwatoli market and several others in Patna. In most of the CBD's of Calcutta, Mumbai and Delhi shops are opened with great expectations but their reward is not satisfactory due to poor turnover of consumers. Hence, the planning of CBD involves careful analysis of the processes at work.

FOOTNOTES

1. R.L. Singh and Singh, S.M., Mungra Badshahpur—A Rurban Settlement in the Ganga-Ghaghra Doab west, *The National Geographical Journal of India*, Vol. VI, part IV, 1960, pp. 199-206.
2. C.G. Galpin, "The Social Anatomy of an Agricultural Community," *Research Bulletin* 34, Agricultural Experiment Station of the University of Wisconsin, 1915.
3. R.E. Dickinson, "The Distributions and Functions of the Smaller Urban Settlements of East Anglia," *Geography*, No. 95, Vol. XVII, Part I, 1932, pp. 20-30.
4. A.E. Smailes, "The Urban Hierarchy in England and Wales," *Geography*, Vol. XXIX, Part-2, 1944, pp. 41-51.
5. R.L. Singh and Singh, S.M., *Op. cit*, p. 199.
6. K.N. Singh, "Changes in the Functional Structure of some Small Towns in Eastern Uttar Pradesh," *Indian Geographer*, May 1961, Vol. vi, No. 1, p. 21.
7. K.N. Singh, "Small Towns as Rural Development Centres in Uttar Pradesh," Abstract, Proceedings of Fourth Session, Indian Science Congress Association, p. 11.

8. Kingsley Davis, "Urban Research and Significance," *Urban Research Method*, Edited by Gibbs, J.P. New Delhi, 1966, p. xi.
9. Thomas N. Edwin, "Some Comments on the Functional Basis for Small Lows Towns," *Iowa Business Digest*, 1960, p. 10.
10. R.G. Golledge, Rushton, G. and Clark, W.A.V., "Some Spatial Characteristics of Dispersed Farm Population and their Implications for the Grouping of Central Place Functions," *Economic Geography*, Vol. 42, 1966, p. 288.
11. "Urban-Rural Differences in Southern Asia," Report on Regional Seminar, Delhi, 1962, London, 1964, p. 3.
12. R.E. Murphy and Vance, J.E., "Delimiting the CBD," *Economic Geography*, Vol. 30, 1951, pp. 189-222.
13. E. Swaminathan, "Coimbatore Metropolitan Area : A Study of the Hierarchy of Settlements," *Indian Geographical Journal*, Vol. 49, No. 2, 1974, pp. 78-86.
14. M.J.P. Proudfoot, "City's Retail Structure," *Economic Geography*, Vol. 30, 1937, pp. 425-28.
15. H.H. Singh, "The Business Structure of Million City—Kanpur," *Applied Geography*, N.G.S.I., 1969.
16. J.P. Singh, "Central Tendency in Population Distribution and Landuse: A Case Study of Shillong," *Indian Geographical Studies*, R.B. No. 2, March, 1974, pp. 58-61.
17. R.E. Murphy and J.E. Vance, "Internal Structure of CBD," *Economic Geography*, Vol. 31, 1955, p. 42.
18. H. Robinson, *Geography for Business Studies*, Macdonald Evans, 1965.
19. A.K. Dutt, "Shopping in Calcutta," *Town Planning Review*, January, 1967.

9

CENTRAL PLACE HIERARCHY

CENTRAL PLACE study is closely related with the town centred region. The theory of central place is related with the economically optimum location of services of different range and specialization both for the town as service provider and the countryside as the service receiver.[1] This study analyses all kinds of 'central functions' which serve the population of surrounding zone of influence of a town and its own population. Tn towns where central functions found are known as central places. Central functions cover retail business, medical services, administrative services, repair works, schools, colleges and industrial activities in which labourers are employed enbulk from the surrounding zone of influence of a town.[2]

"Central place is a place which performs services for the people living in its command area."[3] The Central Place Theory is concerned with the number, size, spacing and arrangement of central places. First of all Mark Jefferson[4] used the term 'central place' in 1931 for a settlement which is necessarily a focus of various activities such as economic and social for the surrounding hinterland. Walter Christaller[5] analyzed this in detail in 1933 in West Germany on the basis of telephone connections. The application of various theories and studies on central place has increased rapidly since Second World War (1945). On the basis of functional importance[6] and other parameters many other formulations have been suggested for determining the centrality,[7] and the central place hierarchy.[8]

Another method is to group the urban centres in terms of the 'scale of equipment'.[9] This measure either examines the overall facilities found within a settlement[10] or makes use of other functions. These methods could easily be applied to the urban centres to determine their respective importance and hierarchy.[11]

Hence, a more realistic method of market area definition is based on consumer behaviour data collected in the hinterland of a town.[12] Even complex market area could be classified on the basis of this data.

Overlapping areas, and the relationship between market area characteristics and the central place hierarchy could be determined (Fig. 9.1).

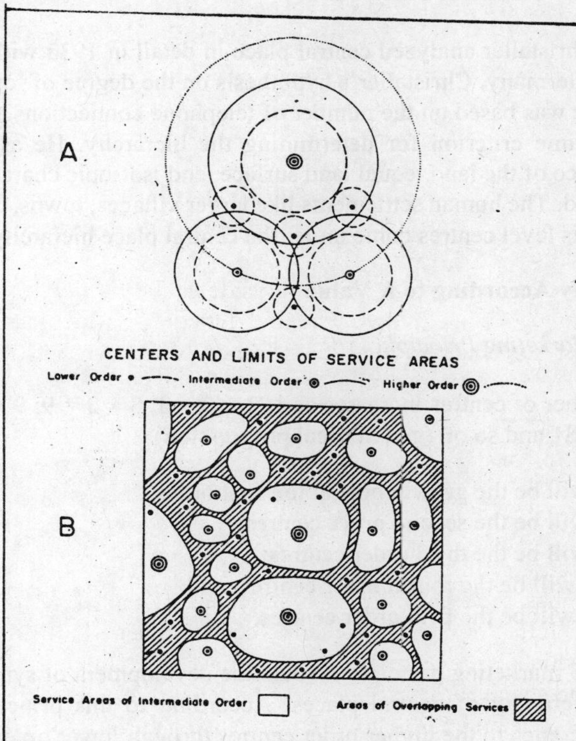

FIG. 9.1 : Empirical Pattern of Centres and Service Areas

Classical Central Place Theory

The work done by Christaller and Lösch regarding the spatial structure of economic development through nucleated settlements come under classical central place theory. Christaller pointed out that the number of retail establishments and total population give only the aggregate importance of a central place. According to Preston E. James[13] "the term nodality stands for the aggregate importance of a place as a provider of central goods and services both for its own residents and for the consumers. The term centrality will stand for the importance of a place as a provider of central goods and services in excess of the needs of its own residents." Christaller's each order of central places would be uniformly spaced in a hexagonal pattern.

Christaller's Central Place Theory

Walter Christaller analyzed central place in detail in 1933 with respect to West Germany. Christaller's hypothesis on the degree of 'centrality' of a place was based on the number of telephone connections at a place as the prime criterion for determining the hierarchy. He assumed a stable price of the land, equal land surface, and isotropic characteristics of the land. The human settlements like larger villages, towns, cities and metropolis level centres come under the central place hierarchy.

Hierarchy According to *K* Value

K = 3 : *Marketing Principle*

The number of centres increases as 1; 1 × 3 = 3; 3 × 3 = 9; 9 × 3 = 27; 27 × 3 = 81 and so on (geometrical progression).

1 will be the growth pole at the first order;
3 will be the second order centres;
9 will be the third order centres;
27 will be the fourth order centres; and
81 will be the fifth order centres.

K = 3 marketing principle favours the development of symmetrical nested hierarchy of central places. According to this principle rural produce comes to the higher order centres through lower order centres, and the goods produced in urban areas move through higher order centres to the lower order centres (Fig. 9.2).

K = 4 : Transport Principle

In this principle there are three hierarchical tiers :

1. Villages at the fourth order which serve towns and *vice versa.*
2. Towns at the third order which serve city level centres and vice versa.
3. Cities at the second order which serve primate city and *vice versa.*

4. **Primate city at the First order which serve national capital at the higher level and cities at the lower level.**

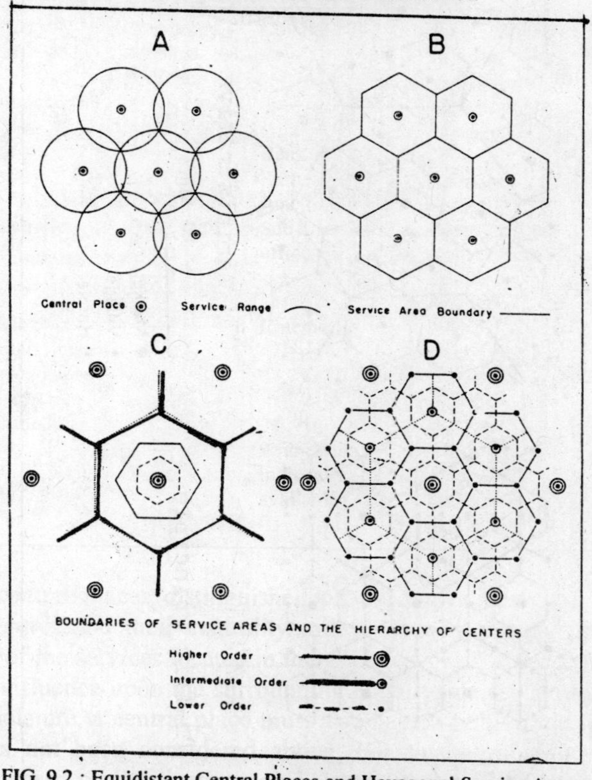

FIG. 9.2 : Equidistant Central Places and Hexagonal Service Areas

In transport principle the number of centres followed the geometrical progression as 1, 4, 16, 64,. . . and so on. At this level one big centre serves the 4 lower order centres (Fig. 9.3).

K = 7 : Administrative Principle

At this level one bigger central place serves the seven second order centres and so on as 1, 7, 49, 343, 2401. . . In areas of perfect transport development the $K = 7$ progression of central place comes true.

For determining the fixed K hierarchy of central place the following formula has been used :

$$Nt = K^t$$

where N is the number of dependent places, and t is the hierarchy tier.

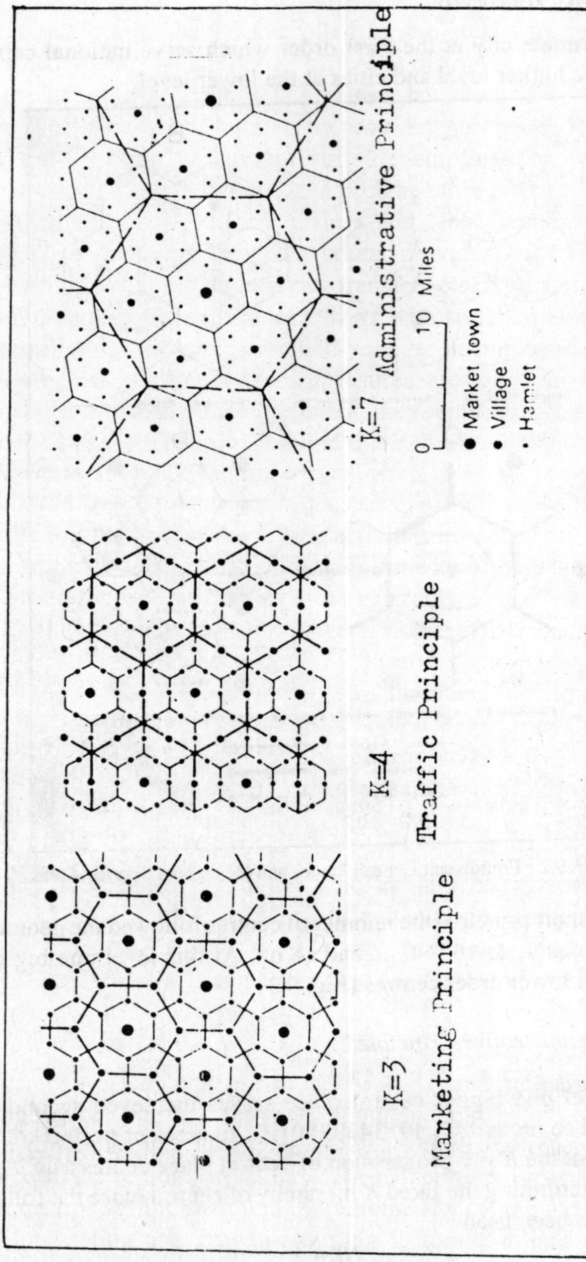

FIG. 9.3 : Tesselations of Triangles in $K = 3$, $K = 4$ and $K = 7$ Market Areas.

Variable K Hierarchy of Lösch

In 1954, Lösch used the same hexagonal lattices for his theoretical landscape, but he did not consider fixed K system ($K = 3$, $K = 4$, and $K = 7$) as Christaller. He developed a more sophisticated form of economic landscape by super imposing all the various hexagonal systems (Fig. 9.4). By super imposing all the lattices on one point, and rotating them he achieved six sectors with many relatively higher order central places as city rich sectors, and six sectors where relatively poor services are found, he considered them as city poor sectors (Fig. 9.4).

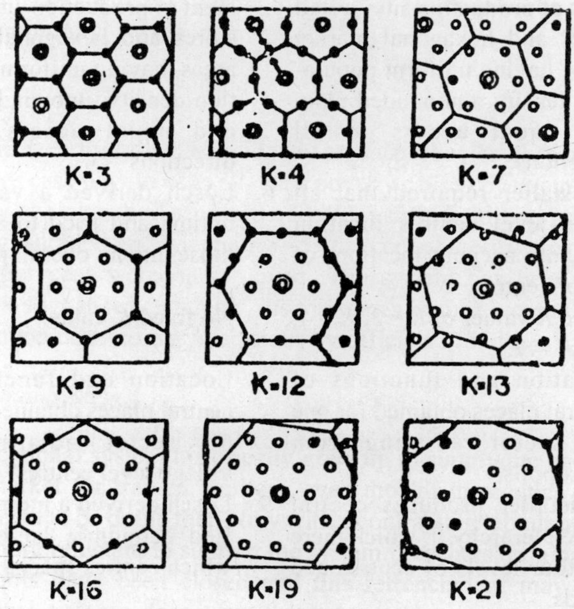

FIG. 9.4 : Increasing Number of *K* and Market Areas as suggested by A. Lösch

Thus the contributions of Lösch is mainly regarding the market areas, because his *K* values are variable. In making the central place hierarchy he considered the bottom top approach. Lösch also hypothesized that individual demand decreases as the distance of the consumer increases from the point where the goods are offered.

Conceptual Difference between Christaller and Lösch

The basic differences between Christaller and Lösch study of central places are as follows :

Christaller's Idea	*Lösch's Idea*
1. Region—South-West Germany	1. Iowa
2. Isotropic Plain Surface	2. Isotropic Plain Surface
3. Begins with City level Centres	3. Begins with village level least specialized centres.
4. Suggested Hierarchy from higher order goods to the lowest	4. Suggested hierarchy from village level goods upward.
5. Considered triangular arrangement of production sites, retail stores and hexagonal market areas, having uniform population density, unbounded plain and equal access in all directions.	5. Considered triangular arrangement of production sites, retail stores and hexagonal market areas, having uniform population density, unbounded plain and equal access in all directions.
6. Christaller required that all lower-level centres location take into account locations of larger centres.	6. Lösch derived a variety of optima and tried to assemble those into an overall pattern.
7. Fixed K values of $K = 3$, $K = 4$, and $K = 7$.	7. No fixed K value.
8. Location and functions of central places obtained for one 60^0 sector radiating from metropolis.	8. Location and functions of central places obtained for one 60^0 sector radiating from village level centres.
9. Christaller produces central place hierarchy in which there are four levels of centres and goods.	9. Lösch derived a more diversified economic landscape in which centres specialize.
10. Relevant for analyzing wholesale and retail business.	10. Provides a framework for analyzing the spatial distribution of market oriented manufacturing.
11. Christaller's agglomerative requirement is for travel cost minimization by rational consumers on multi-purpose trips.	11. Not necessary for hierarchy of Lösch.
12. In Germany smaller central places are larger in number in comparison with higher order centres.	12. Such a condition is not found in Iowa.

13.	All places in a particular tier are of the same size and have the same function.	13.	The same size central places need not have the same function.
14.	Number of central places be minimized.	14.	All excess profits be eliminated.
15.	All higher order centres contain functions of the smaller central places.	15.	Larger centres need not have the functions of lower order centres.
16.	Different hierarchical levels have differential supply of goods.	16.	One centre may be a supply centre for all goods.
17.	Model developed by the market areas of consumer behaviour pattern.	17.	Model developed by super imposition of fishnets (demand cone) of different goods threshold range.

Source : Complied by the author.

Philbrick's Nested Hierarchy Theory

A.K. Philbrick is an American geographer. His nested hierarchy theory is based on the following postulates.

Inter-connection between Different Occupations

The first postulate is that different occupations of men such as agriculture, cattle rearing, mining of minerals, manufacturing, industry and trade are inter-connected with each other. Some of these inter-connections are found in between uniform areas which are homogeneous in terms of occupance. Smaller homogeneous areas may be exemplified as a *mohalla* where people of same caste and income group resides in a town.

Origin of Nodal Areas

Second postulate favours the origin of nodal areas with the outer-connections of occupance units. In one nodal areas different uniform areas are found which are connected with the focal point.

The example of a nodal region is one town which is made of different *mohallas* having device uniform areas, e.g., agricultural farm, residential areas, and shops which are related with the town. In this way, human occupance has a focal character.

Nested Hierarchy of Nodal Organization

The third postulate is related with the occupance units arrangement functional areas a nested hierarchy. This hierarchy turns from uniform relationship to nodal organization. Hence, the human occupations may be of seven types in order of magnitude in which each of them composed of nodal and uniform areas:

1. A village farm.
2. Town as service centre.
3. Major towns as Purnea, Chhapra and Saharanpur.
4. The large city as Bhagalpur, Meerut, Ambala and Srinagar.
5. Metropolis as—Patna, Kanpur, Bangalore, Ahmedabad, Hyderabad and Jaipur.
6. I. *Nodal :* Large Metropolis and their commercial areas as Calcutta, Chennai and Delhi along with its trade areas.
   II. *Uniform Areas :* The neighbouring nodal areas came in this category.
7. I. *Nodal :* It includes Primate city such as greater Calcutta and Greater Mumbai.
   II. *Uniform Areas :* It includes neighbouring areas of a nodal region including zone of influence known as *holon.*

Modern Central Place Theory

Berry and Garrison[14] emphasized that retail establishment and consumer services should be the focus of central place study. They related the assumption of isotropic plain surface Christaller by introducing the concept of range of goods and demand cone. Dacey[15] analyses the spatial aspects of central places and uses the regional method to compare the observed pattern with three expected distributions: (a) hexagonal distribution; (b) random distribution; and (c) the clustered distribution.

Recently, it has been considered that the central place theory formulated by Christaller gives a picture of particular time period although the settlement evolved through time. Hence, Curry[16] emphasized the need to develop a dynamic central place throy and highlighted that the time factor in central place activities is the key to the understanding of geography.

From the Abstract to the Real World

Hexagonal service areas have been observed in Southern Germany by

Christaller (1933), in East Anglia, Great Britain (Berry, 1967) and the parts of China before the 1950's (Skinner, 1954).[17] However, it is only in old established and stable human landscapes where central places are predominantly found. Although both trade areas and hierarchies of central functions can be distinguished in the real world, they rarely display the regular patterns described above in the real world with respect to the theoretical landscape. The reason for this is that the real world is much more complex than the theoretical landscape. The distortions of the model in the real world are caused by :

(a) The uneven distribution of population;
(b) Variations in purchasing power and spending patterns;
(c) The unequal travelling facility to different directions;
(d) The failure of consumers to act as optimizers;
(e) The non-service function of many urban settlements; and
(f) Central places of different ages provide varying technological and organizational influence;

Maximum and Minimum Trade Areas

For each central function, be it a commodity sold or service provided, there is a 'threshold'. This refers to the number of customers necessary for the central function to be variable, i.e., for the shop-keeper to make sufficient profit, or for a Government service to serve sufficient people that the service providers regard their activity as worthwhile.

Suppose in planning a primary school there is a rule that the school should have at least 100 pupils to be worth the expenses of construction and the provision of text-book and teachers. The threshold of 100 pupils can be translated into a population threshold if the age structure of the surrounding population is known. For instance, 25 per cent of the total population might be between 7 to 14 years old, the primary school going age. This means that to obtain 100 potential pupils there needs to be a population of 400. Using the population density figure for the area, the threshold population of 400 can be converted into a threshold area which can be drawn as a circle. If the population density of the area is 10 persons per km^2 then the threshold area should be 40 km^2. Using the formula for the area of a circle (πr^2) the radius of the threshold area or minimum trade area is found to be just over 3.5 kms.

$$\text{Area of the circle} = \pi \text{ radius}^2$$
$$40 = 3.14 \times r^2$$

$$r^2 = 40/3.4$$
$$r^2 = 12.4$$
$$r = 3.52 \text{ kms.}$$

In a theoretical landscape the lattices will be circular but when packed to ensure that the parts of an area are served they will become hexagonal in shape. In case of relaxed assumptions and complexities of the real world allowed to enter into the model these hexagons may become distorted.

In sparsely populated areas where the purchasing power is low, in order to accumulate sufficient demand for a commodity to make profitable, a large supportive area would have to be served because the people in such a peripheral areas are beyond the range of services. It means in order to have a profit from the sale the minimum trade area exceeds maximum trade area.

In such a situation generally the street hawkers or the suppliers of goods and services move in their trade area. This increases the service area and fulfil the threshold demand. Throughout India such patterns have been now regularized in the form of periodic markets. Even then the hawkers are serving the area in the gap of two periodic markets especially in the off days. In this way, they move around in a cyclical way to different periodic markets in the area, and coming back to the same place once in a week. In the State of Bihar, India, the distance of one periodic market from the other is hardly 5 kilometres which could easily be covered on foot.

From rural township in Zambia, the radius of the area served is about 9 kms and for the market town approximately 22 kms. It means that the distance between villages and rural townships is about 5 kms between rural townships and market towns about 16 kms, and from village to market town not more than 22 kms.[18]

Functional Hierarchy of Central Places

Any study of central places has necessarily to include both the urban and the rural settlements as in any scheme of hierarchy, towns, perform more central functions than villages. However, the transitional areas consists of rurban centres, villages may sometimes be central places of a high order than the smaller towns. The purpose of determining the hierarchy of central places would be defeated if we confine our study to urban settlements only. In the following pages, therefore, both towns and villages have been examined together to determine their centrality.

Most of the settlements have acquired their urban status due to their achievement in socio-economic dimensions. They have grown to be the chief market centres of an area because bulk of the functions are retail shops, wholesale, warehouses, banks, garage and offices have developed to meet the requirements not merely of the towns own population but that of the surrounding area. Walter Christaller explained market areas in terms of the provision of services, constructed a theory with the help of smallest units of settlements which developed in areas of uniformity. He favoured that the most efficient system of clearly defining market areas should be based upon a hexagonal lattice. This geometrical form approaches close similarity of the circle, but a circular lattice suffers from the disadvantage of allowing unsupplied or overlapping areas.

As 'the chief functions of a town is to be centred at its rural surroundings' then the centre of the hexagon is the urban area where the central place find their location. Christaller clearly indicated that there would be mutual dependence between the central place and the supply region. In plain areas the equal spacing of settlement results in developing competition between them in the provision of services. The shops in small towns for example, usually stock a limited range of goods and thus small towns are lower-order central places with small market area, whereas bigger towns provide a wider range of services for a larger area and hence, they are higher order central places. In case higher order and lower-order centres are considered to be equally spaced then the hexagonal market area of higher-order is larger than that of the lower-order central places and the later areas and functions must be present in the former. Hence, in the model a hierarchy of central places is conceived, ranging from small hamlets to large towns.

In Christaller's hexagonal lattice, each hamlet is located at the meeting point of three uniform market areas and hence, with equal number of people having equal ease of movement and equal purchasing power one-third of the hamlets trade will be with each central village. Hence, the total hexagonal trade area is 2 around a village (one third of six hexagon) plus one of its own market. Christaller used the term '*K* value' to refer to the total number of settlements served by each central place. Hence, in this case $K = 3$. He also assumed that in one region the *K* value would be fixed ranging from hamlets, villages, small towns and so on.

In case of the establishment of third-order centres the number of units of trade of the central place is 27 and the calculations are as follows :

(a) Six centres are located at the vertices of the hexagon each share their trade between 3 centres and therefore, the central place *A* has $^6/_3 = 2$ units of trade.

(b) The other 12 peripheral settlements each share their trade between 2 centres and therefore *A*'s share is $^{12}/_2 = 6$.

(c) Inside the hexagon is 18 other centres whose entire trade is directed towards *A* and therefore, a further 18 units are added.

(d) The entire trade of *A* is conducted in that centre which provides 1 unit.

In this way, the total trade of *A* is found by summing up the above parts, e.g.,

$$2 + 6 + 18 + 1 = 27$$

As the hierarchy of the settlement is established so the trade of each central place develop as follows :

$$1, 3, 9, 27, 81, \ldots\ldots\ldots$$

i.e., $1, 3, 3^2, 3^3, 3^4, \ldots\ldots\ldots (K = 3)$.

This model of central place suggests that it is possible that the function of such places controls their size of the supplying area, besides it offers a partial explanation that why settlements are scattered in the region differ in size. $K = 3$ hierarchy of central places in southern Germany is as follows :

TABLE 9.1 : Christaller's Central Place Hierarchy in Southern Germany (*K = 3 Network*)

Class	*Average Radius of Supplying Area in km.*	*Typical Population Size*
Market Town	4.0	1,000
Central Township	6.9	2,000
Country Town	12.0	4,000
District City	20.7	10,000
State Capital	36.0	30,000
Provincial Capital	62.1	100,000
Regional Capital	108.0	500,000

Source : W. Christaller, *The Central Places in Southern Germany*, English translation by C.W. Baskin, 1966.

Centrality

Determination of the importance of central place is universal problem in the study of space economies. The only satisfactory model for its solution developed so far is the one offered by Christaller in 1938. His approach requires both in operational definition of his concept of 'centrality' and its system of measurement. Thus the areal distribution of central place function was established in this study by applying the following modification of Christaller's model to place with 2,500 or more inhabitants in 1966 and that appeared in the U.S. census of Business for 1963 and 1972.[19]

$$C = R + S - L\ MF$$

where C = centrality, R = total sales in retail establishment, S = total sales in selected service establishment, L = average percentage of medium family income for a central place spent on retail and service goods, M = medium family income in a central place, and F = the total number of families in central place. In this model R + provides an estimate of total retail and service activity in a town or its "Nodality," while $L\ MF$ provides an estimate of local consumption. The difference is an index of the volume of extra-local trade, or centrality. The results of this procedure were used to establish the magnitude of the central place function for places.

Criteria of Centrality

For declaring central places only the number of persons is not a satisfactory answer to determine the centrality. Christaller has considered telephone connections while studying the central places of Southern Germany as centrality index.

After considering telephone connections Christaller derives the degree of centrality of a place, that is, amount of service performed apart from those relating to the needs of its own inhabitants by means of the following formula :[20]

$$\text{Centrality of a place} = \text{Number of Telephone in a place} = \text{Number of inhabitants of a place} \times \frac{\text{Number of Telephone area served by a place}}{\text{Number of inhabitants and area served by a place}}$$

or

$$C = T_2 - E_2 \times \left(\frac{Ta}{Eq}\right)$$

or

$$C = \left(t - \frac{Pt}{P} \right)$$

where C is centrality, t is the number of telephone in a town, P is population of the town, T is number of telephone in the whole region and P is population of the whole region.

The method has been criticized in recent years as telephones are also used for non-central functions and most of the lower-order places have no connection at all.

The centrality of a centre can be measured after considering the goods and services for which the zone of influence depends on the centre. Bracey has applied this method numerically in Southern England. After considering the intensity of bus services. Green has measured the centrality of places in England and Wales. Godlund of Sweden has selected retail trade as an index of centrality and has given the formula:[21]

$$\text{Centrality} = \frac{St.100}{Pt}$$

where St is the number of persons employed in retail trade and Pt is population of the region.

Kashi Nath Singh has put forward a more feasible proposition in the shape of formula as:[22]

$$C = \frac{N.100}{P}$$

where C is centrality, N is the number of persons dependent upon commerce, and P is the total regional population in commerce.

In order to identify central places O.P. Singh (1969) covers up the functional as well as spatial gaps in the region. He distinguished two types of demographic indices. Relative Centrality Index (RCI) and Absolute Centrality Index (ACI) by the following formula:[23]

$$RCI = \frac{50(CB.RT + CT.RB)}{RT.RB}$$

where CB is basic population of centre (non-agricultural population); RB is basic population of the region; CT is commercial population of the region, and 50 is constant figure and $AC.I = \frac{CB}{5000}$; and ACI is calculated here by dividing the non-agricultural population of the centre by 5000.

This figure is taken here as a norm also to lowering the size of big centres for easy calculation.

M.S. Vishwanath while studying the growth pattern and hierarchy of urban centres in Mysore has used the following formula to measure the centrality index:[24]

$$C = \frac{P \times 100}{\text{Total State Population}}$$

where C is centrality, and P is the total regional population is commerce.

V.N.P. Sinha[25] while studying the hierarchy of urban settlements in Chotanagpur Plateau devised the following formula to measure the centrality of a place :

$$C = \frac{T \times 100}{P}$$

where C is centrality, T is the number of persons dependent upon tertiary services of a centre, and P is the total regional tertiary population.

According to Davis[26] before deriving the centrality value for a central place, one should find out the location coefficient of shops of the region with the help of the following formula :

$$LC = \frac{t}{T} \times 100$$

where LC is the location coefficient of shops in a central place, t is the number of shops in that centre, and T is the total number of shops in the entire region under study.

Multiplying the relevant location coefficient by the centrality index (derived from the weight given to the different functions performed by a centre) will give the final degree of centrality in terms of centrality value :

$$CU = CI \times LC$$

where CU is the centrality value, CI is the centrality index for services, and LC is the location coefficient for retail trade.

N.P. Saxena and R.C. Tyagi devised the following formula to measure the centrality index :[27]

$$C = A - \frac{P \times Ar}{Pr}$$

where C is centrality measure, A is total number of workers engaged in central functions at a centre, P is population of the centre, Ar is aggregate of workers engaged in central functions in the region; and Pr is population of a region.

In measuring the centrality index O.P. Singh[28] makes use of commercial population of the centre. With the following formula, he calculated relative and absolute centrality index :

$$RPI = \frac{PCI - \left(Pt - \frac{RC}{Rt}\right)}{Pc - \left(P + \frac{Rc}{Pt}\right)} \times 100$$

where *PCI* is commercial population of the centre; *Pt* is total population of the centre; *RC* is regional commercial population; *Rt* is total regional population; *Pc* is summation of *PcI*, PC_2, *P* and *Pt* is summation of *Pt*, Pt_2, Pt_n.

Absolute centrality has been calculated.

$$AC1 = \frac{Pb}{100000}$$

where *Pb* is basic population of the place.

The author has investigated the central place system of the Bihar Plain by means of functional indicators covering the categories of education, medical facilities, trade, administration, communication, transport and employment in tertiary services. Each category is represented by one type of central function and most of the types have been sub-divided according to their position in the functional hierarchy. A score has been allotted to each of the central functional indicator, irrespective of the number of occurrences (units) of that particular function as shown in the Table 9.2.

The following criteria has been adopted for selecting the central places :

1. Only those centres are listed where at least three of the above functional indicators were present irrespective of the functional units.
2. After the index has been calculated, weighing for functional diversification was added according to the scale given below :

No. of functional indicators present	Weighing for diversification
3 to 4	Zero
5 to 6	5
7 to 8	10
9 and over	15

TABLE 9.2 : Weighing and Classification of Central Functions

Functional indicators	*I* *20 points*	*II* *15 points*	*III* *10 points*	*IV* *5 points*
1. Education	University	College	High School	—
2. Medical	Medical College	—	Hospital	—
3. Communication	—	—	Telephone Exchange	—
4. Finance	—	Bank	—	—
5. Transport	Air Port	Railway Junction	Railway Station	Highway
6. Administration	Dist. Court	Sub-Divisional office	Block Development Office	—
7. Trade	—	—	Wholesale Market	—
8. Recreation	—	Cinema	—	
9. % of total workers engaged in tertiary occupation.	—	—	Over 50%	25 to 50%
10. Diversity Index	—	9 functional indicators available	7 to 8 functional indicators available	5 to 6 functional indicators available

The central places distinguished on the above basis are mostly service centres and their centrality index depends entirely upon the hierarchy of the services located in them. However, in order to have an adequate influence upon the surrounding countryside as also to justify its nomenclature, a central place must also have a retail trade function which has not been considered above. For this purpose a ranking technique based on localization coefficient of the retail trade units was applied.[29] This method is commonly used in industrial geography and gives a high value for centres having a large number of outlets for retail trade. The location coefficient is based on the number of shops located at a particular place in relation to the total number of shops in the area for centres of the same group, and is therefore, a measurement for centrality or degree of focality. The formulae used is :

$$LC = \frac{t}{T} \times 100$$

where LC is the location coefficient of shops in the town; t is number of shops in that centre; and T is total number of shops in the entire urban area or rural area of the Bihar Plain.

The total number of shops in the urban areas of the Bihar were found to be 71,073 and that in rural areas, 24,184. The location coefficient values for urban centres range from 13.04 for Patna to 0.02 for Lauthaha. In rural areas, the railway stations were among some of these centres as they provided a meeting place for the farmers. Enterprising businessmen took advantage of these facilities to open retail shops at such centres. This, in turn, led to the establishment of grain collecting centres, electric lines, telephones and other similar facilities. Combination of these elements led to the greater centrality for some of the places, while competition between the growing centres created the pattern of a central place hierarchy.

The smallest central place in the study area is a village with basic public utility services such as an allopathic doctor, middle school, post-office and a metalled road connection. Villages of this category are closely spaced and may or may not have a retail marketing centre. They do, however, have a weekly or bi-weekly market known as a *hat*.

Rural central places of a higher order generally provide services like a high school, telegraph office, dispensary, retail market, block development office and a bus stop or a railway station. Such centres are located on the main transport routes between towns and are, therefore, inter-spersed between towns and rurban centres.

Further development of these facilities and the addition of a telephone exchange, bank, cinema, wholesale market, etc. led to the evolution of a rurban centre. We have discussed earlier the characteristics of rurban centres in the Bihar Plain which include both smaller towns and larger villages. Villages which have grown into rurban centres are potential towns. A rise in their population coupled with functional segregation in land uses and increase in employment in tertiary services would make them eligible for classification as towns.

The evolution and spacing of rural central places largely depends upon economic prosperity and transport facilities. There is a large concentration of central places on either side of the river Ganges, but their number decreases as we proceed towards the Himalayas or towards Chotanagpur Plateau. Also, the density and grade of central places decrease while the relative distance between them increases. This pattern keeps up the economic prosperity of the two areas.

A Dynamic Theory of Central Place Location

"Location theory is now caught up in the movement towards a more dynamic economics, but the problem of dynamic formulation is great"

(Robertson, 1956).[30] The central place theory as proposed by Vön Thunnen (1828), Jefferson (1931) and Berry and Garrison (1958) are only possibilities that they had made for a static time period but no one studied the dynamic character of central places over the region. Since Second World War there is a flood of ideas over central place studies and system analysis for a static time period but it presents only one dimensional picture about central places as they are constantly changing in number, size, spacing, function, layout, consumer behaviour, market sharing problems, threshold relationship and *K*-values, etc. The purpose of this section is to analyze the dynamic character of central places and to put it into a theoretical framework for some generalized study. A new theory has been devised by the author to express the dynamic character of spatial organization of central places which is proposed here as "The Dynamic Theory of Central Place Location."

The theory is presented here with the help of transportation principle as formulated in Fig. 9.5.

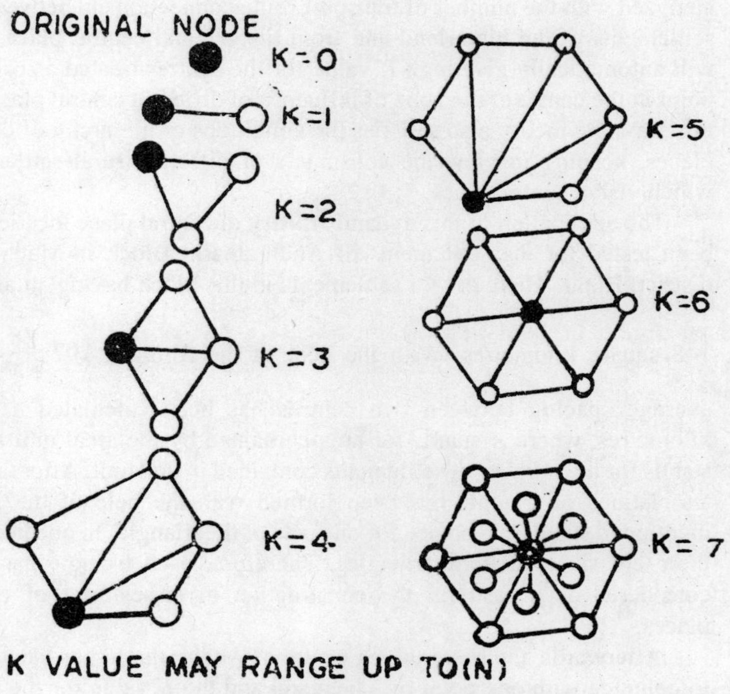

FIG. 9.5 : A Model of Transport Network Evolution

The beginning of a settlement by the first settler had taken place in ancient times in the forest environment when there were no transport route at all. At this stage, the position of K value for transport route may be regarded as $K = 0$. Later on with the beginning of second settlement in the neighbourhood might have given rise to one way connection of transport linkage. This may be denoted as $K = 1$. Further, in addition to the original point the growth of two centres provided two directional connections of transport routes as $K = 2$. Similarly, three centres gave three directional linkages as $K = 3$, four directions as $K = 4$, five directions as $K = 5$ and six directions in a hexagon as $K = 6$ for the densest packing of centres of the same order.

But with the increase in the number of centres in a triangle or in a hexagon, the K value will increase upto a very high figure. This will create problem areas for infrastructure planning of services in relation to the location of first ranking central place of the region. This theory also believes in the hierarchy of central places according to the increasing K value for a particular point in a equilateral triangle. This may be analyzed with the number of transport route connections in between the settlements of the hinterland and from the original central place. This will automatically give high K value for the centres treated as original point in the centre of the zone of influence of different central places. In this way, the theory also satisfies the conditions of hierarchy of central places, keeping in view the consumers of different rural settlements which visit a central place.

The application of this dynamic theory of central place location has been tested for the settlements of Andhratharhi Block in Madhubani district, Bihar. There are 46 settlements in this block having an area of 148 square kilometres. With the help of the formula $1.07\sqrt{\frac{A}{N}}$, the average spacing between two centres has been calculated as 1.91 kilometres, where A stands for area contained by the areal unit and N stands for the number of settlements contained in that unit. After this the tesselations of triangle has been formed with the help of the above mentioned standard distance for one side of the triangle. In one hexagon there are six equilateral triangles. Therefore, each triangle has been considered as a base for the construction of tesselations of central places.[31]

Afterwards, the first ranking centre of Andhratharhi has been made to coincide with one point of a hexagon and the K value for the actual distribution of settlements in different triangles and hexagons have been measured with reference to this original point. This gives the intensity of concentration of centres in a triangle including the problem areas for planning purposes and the siting conditions of settlements in a particular hexagon.

Around Andhratharhi there is a lack of settlements in that hexagon, where this central place stands. It is, suggested that for planning one should consider the original point of Andhratharhi in the centre of the hexagon which lies in the north-west of its present location, where K value is also high ($K = 4$). This is because only this particular location of Andhratharhi will be able to give efficient services to its zone of influence. Further, the above mentioned north-west hexagon lies in the centre of Andhratharhi block, where the favourable siting conditions also allow the concentration of settlement (Fig. 9.6). The intensity of locational problem in this block has been measured according to the number of K value from the point of view of rural settlement planning. Ten categories of areas have been demarcated on the basis of location of settlements and the need for the development of transport routes in relation to Andhratharhi.

To measure the distribution of discrete items as central place location over space Polar coordinate has been used in which Andhratharhi (0) is assumed as origin point and other centres as connecting centres of transport routes. To explain this in Fig. 9.6, *OA* is considered as initial line, *OP* as radius vector (r), O as a vectorial angle and r is represented by (r, O). This is because Polar coordinate deals with the discrete distributed items than regular distribution.

To draw the regular tesselations equilateral triangles have been used without considering any hierarchical level of settlements around Andhratharhi (Fig. 9.6). This has been considered as every urban settlement should earn normal profit from transport lines and with the nearby central place. In higher order centres hierarchical arrangement may also be observed in accordance with the service efficiency and area served by a particular centre.

It would appear that this theory proposes variable K values for varying number of settlements in relation to transport routes for urban community development. At the same time, this also embraces the spatial pattern and changing character of central places with the addition of newer settlements in relation to space and time.

The Contribution of Central Place Study in Regional Planning

1. It guides the planner in identifying the existing hinterlands of localized or mobile services.

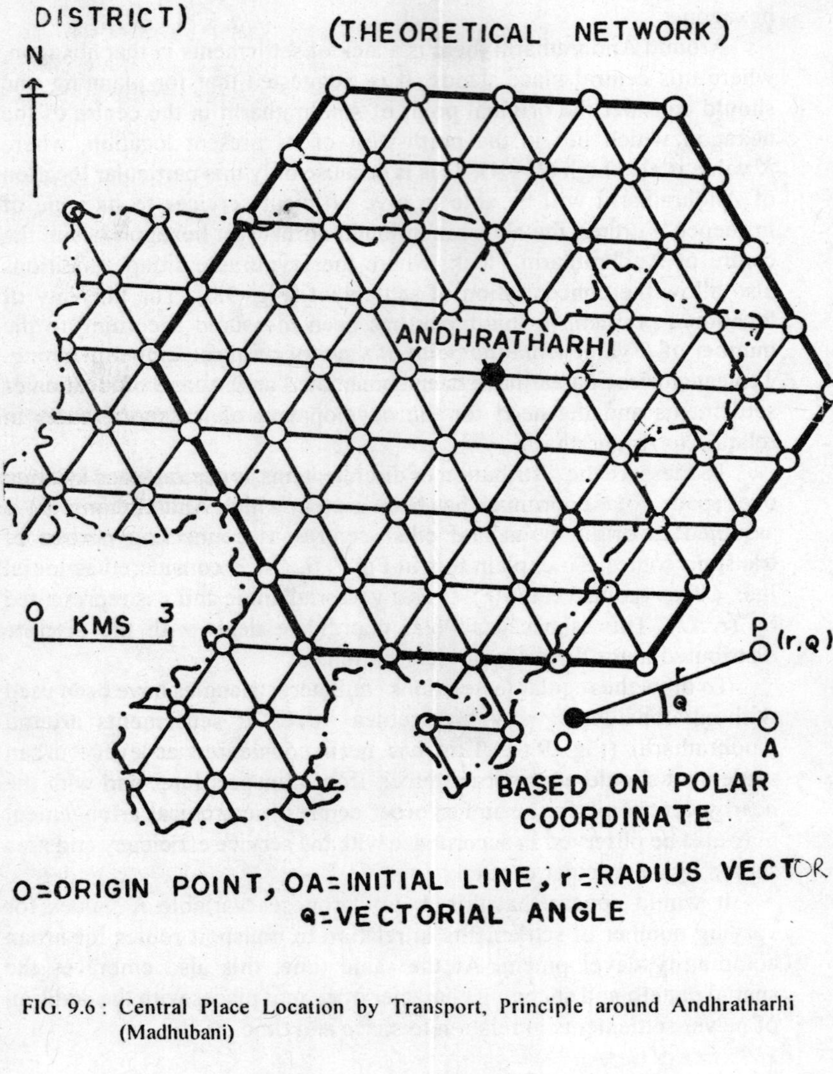

FIG. 9.6 : Central Place Location by Transport, Principle around Andhratharhi (Madhubani)

2. It helps in identifying the planner which areas and which people are out of reach of individual service centre.
3. It shows the planner which centres are stranger in terms of available services and so they could cater the needs of an increased local demand or for a demand in a larger hinterland.
4. It helps the regional planner to reorganize the central place in such a way that a combined hinterlands cover all the people in an area.
5. At last it requires from the planner to keep up-to-date information for each district level systems of central places on the following headlines :

   (a) Population distribution;
   (b) Locations of all government and private services;
   (c) Threshold values for individual services under different forms of management as for example mobile versus fixed businessmen, part-time versus full-time health worker; and
   (d) The real trade areas of the various services demarcated on the basis of consumer behaviour pattern as against the assumed one demarcated on the basis of gravity potential model.

The Value of Central Place Theory

1. It enables a comparative analysis of regional centres to be made;
2. A hierarchy of functions and settlements is devised;
3. It shows the interdependence of town and region;
4. The idea of competition between centres is stressed;
5. The intensity and spacing of centres suggests further investigation as to how these service centres have evolved and are changing their character at present time; and
6. On the basis of a theoretical structure it is possible to make a number of predictions about the pattern of future settlement location.

Criticisms of Central Place Theory

1. Christaller's central place theory is concerned with central place activity, and central functions for the hinterland. Other

aspects such as places of residence or industrial centres are not considered besides the sites of :

- — availability of mineral resources;
- — favourable climate;
- — strategic position;
- — growth of a port;
- — railway expansion; and
- — urban overspill requirements.

2. Central place theory may have played a significant role in the development of settlement in an agricultural area, but does not works well in industrial areas.
3. There may be planned delay in establishing full services in 'new towns' until they increased their population.
4. The theory does not work well in areas where the industries are expanding or retarding.

FOOTNOTES

1. L.M. Vanden Berg, "Central Place Theory and the Planning of Rural Service Centres in Africa," Zambia Geographical Association, *Occasional Study*, No. 11, 181, p. 1.
2. W.A.V. Clark, "Consumer Travel Patterns and the Concept of Range," *Annals of the Association of American Geographers*, Vol. 58, June, 1968, p. 386.
3. Om Prakash Singh, "Functions and Functional Classes of Central Places in Uttar Pradesh," *The National Geographical Journal of India*, Vol. XIV, 1968, pp. 1-2.
4. M. Jefferson, "The Distribution of Worlds City's Folks," *Geographical Review*, Vol. XXI, 1931, p. 453.
5. Walter Christaller, *Central Places in Southern Germany* (translated by C.W. Baskin); Prentice Hall, New Jersy, 1966.
6. G. Kuluczka, "Central Places and their Functional Areas in the Federal Republic of Germany," Paper read at the 21st International Geographical Congress, India, 1968 (Unpublished).
7. Scot Peter, "The Hierarchy of Central Places in Tasmania," *The Australian Geographer*, 1968, p. 135.
8. E. Ulman, "A Theory of Location for Cities," *American Journal of Sociology*, Vol. 6, 1940-41, p. 858.
9. H.E. Bracy, "Towns as Rural Service Centres," Transactions and Papers, *Institute of British Geographers*, Vol. 49, 1953, pp. 95-105.
10. H.E. Bracy, "English Central Places, Identification, Distribution and Functions," in *Land Symposium on Problems of Urban Geography*, 19th International Geographical, Congress, 1960.
11. A.E. Smailes, "The Urban Hierarchy in England and Wales," *Geography*, Vol. 29, 1944, pp. 51-55.

12. G. Rowley, "Central Places in Rural Wales," *Annals of the Association of American Geographers*, Vol. 61, No. 3, 1971, p. 537-50.
13. R.E. Preston, "The Centrality Models," *Yearbook Association of Pacific Coast Geographers*, Vol. 32, 1970, p. 62.
14. B.J.L. Berry and W.L. Garrison, "Recent Development of Central Place Theory," Papers and Proceedings of Regional Science Association, No. 4, 1958, p. 107.
15. M.F. Dacey, "Analysis of Central Place and Point Patterns by a Nearest Neighbour Method," *Lund Studies in Geography*, Series B, Human Geography, Vol. 24, 1962, p. 55-75.
16. L. Curry, "A Classical Approach to Central Place Dynamics," *Geographical Analysis*, Vol. 1, 1969, pp. 272-82.
17. G.W. Skinner, "Marketing and Social Structure in Rural China," *Journal of Asian Studies*, Vol. 24, 1954, pp. 343.
18. Doxiadis Associates, "Consultants on Development and Ekistics: Rural Townships in Zambia," General Report, Athens, 1967, p. 51.
19. R.E. Preston, *Op. cit.*
20. R.B. Mandal, "Classical Central Place Theory," in *Recent Trends and Concept in Geography*, Vol. II, Concept Publishing Company, 1980, p. 55.
21. S. Godlund, "The Functions and Growth of Bus Traffic within the Sphere of Urban Influence," *Lund Studies in Geography*, Series, B, Human Geography, No. 18, 1956, p. 12.
22. K.N. Singh, *Rural Markets and Rurban Centres in Uttar Pradesh*, Ph.D. Thesis, Banaras Hindu University, 1962, Varanasi, p. 187.
23. O.P. Singh, *A Study of Central Places in U.P.*, Ph.D. Thesis, B.H.U. Varanasi, 1969, pp. 101-109.
24. M.S. Vishwanath, "Growth Pattern and Hierarchy of 'Urban Centres in Mysore', *The Indian Geographical Journal*, Vol. XLVII, 1972, pp. 1-11.
25. V.N.P. Sinha, *Chotanagpur Plateau: A Study in Settlement Geography*, K.B. Publications, New Delhi, 1976, p. 175.
26. W.K.D. Daves, "Centrality and the Central Place Hierarchy," *Urban Studies*, Vol. 4, No. 1, 1967, p. 63.
27. N.P. Saxena and Tyagi, R.C., "Criteria for Determining Centrality in a Macro-Region," *The Geographical Observer*, Vol. II, March, 1975, pp. 53-64.
28. O.P. Singh, "Towards Determining Hierarchy of Service Centres : A Methodology for Central Place Studies," *National Geographical Journal of India*, Vol. XVII, Part 4, 1971, pp. 165-77.
29. W.K.D. Daves, "Centrality and the Central Place Hierarchy, *Urban Studies*, Vol. 4, 1967, p. 65.
30. C.J. Robertson, "The Loschian Landscape : A Review," *Scottish Geographical Magazine*, Vol. 72, No. 2, 1956, pp. 81-84.
31. M. Chisholm, *Rural Settlement and Landuse*, Hutchinson University Library, London, 1962, p. 12.

10

UMLAND OF TOWNS AND SPACING

The Concept

UMLAND commands the surrounding zone of influence of an urban centre. The umland is that portion of the land around an urban centre which is linked culturally with the town as a centre. A city cannot sustain the life of its people by itself but it should dependent upon the surrounding region in all respects. The dependence is not of one kind rather it is multi-directional in nature. On the other hand, the surrounding region is equally dependent upon the central city for all its woes. The interdependence of town and the countryside and their relationship have initiated the concept of umland. The surrounding area should not be attributed as a mere physical entity but acts as barrier in between town and the countryside. The word 'umland' has been derived from 'ome land'.

Keeping in view, the economic significance of umland, the Mughals have formed sub-divisions and districts. These are serving the political boundary and rent collecting area. These divisions were the basis for the collection of land rent even upto British regime and the present. Due to this disparity between the area and the extent it is quite difficult for the planners to come to any sound decision about the actual boundary of an umland. As for example, the professional and private service patron of cinema and threatre move even upto the interior village following the administrative boundary. But it is quite possible that they receive publication from these patron, do not fall under these territorial limits.

The word 'umland', with a geographical meaning, was first used by Andre Allix, a French geographer, in 1914.[1] He claimed to have coined it himself to express his concept of economic domain, meaning thereby the areas immediately around an interior city, more particularly a fair centre (1922). To him, 'umland' was comparable to 'hinterland' but he restricted the latter expression to ports only. The term was again used by the same author in 1922 in his article in the *Geographical Review*. Dodge (1932), an American Geographer, was perhaps the next to use the word 'umland' in 1932. Later in 1937, Whittlesey defined the Umland of Kano as the "immediately tributary area extending 50 or 60 kilometres on all sides of the city." Van Cleef also used the word independently in the same year as he divided the "economic domain into two parts, placing in the umland those communities that behave economically as they were located within the political limits of the primary centre, including the hinterland as those communities most, but not all, of whose economic activities are focussed on the primary centre." He also showed interest in establishing the German origin of the word 'umland' and traced its appearance in a German dictionary as early as 1883 attending Allix's claim of its coinage in 1914. Subsequently a number of English-speaking geographers have shown familiarity either with the word as such or with its substitutes like 'urban hinterland', 'sphere of influence', 'catchment area', 'urban-field', city-region', etc. However, R.L. Singh's work on Banaras (1955) happens to be the first to show a special preference for the expression 'umland'; and its definition contained therein, viz., 'umland' of a particular town or city is "the area in which the region and the city are culturally, economically and politically interrelated" has got due recognition in Stamp's glossary of geographical terms (1961).

In Indian Geography in 1955, the problem has been studied by several other scholars; although most of the researchers in the field have worked, more or less, on similar lines with similar understandings, a few have opted for expressions other than 'umpland'. Among those who have preferred to use 'umland' are Singh (1956), Singh (1961), Nath (1962), Mukerji (1962), Dutt (1963), Dwivedi (1964), Mathur (1968), and Krishan and Agarwal (1970), Mukerji has not been able to avoid the occurrence of words 'hinterland' and 'tributary area' as synonyms of 'umland'. But Janaki (1962) uses 'tributary area' and Singh (1964), and Mahadev and Jayasankar (1969) go in for 'urban-field' and 'city region' respectively. Alam (1965) finds that the service areas of Hyderabad-Secunderabad are not alike in their extent, and calls them areas of 'metropolitan dominance' and 'metropolitan association' (perhaps after Dickinson). Chatterji (1967) reserves the term 'hinterland' for Howrah, perhaps to denote its character of a shopping centre. Dikshit and Sawant (1968) use 'urban field' as parallel to 'hinterland' and have unhesitatingly associated it with an inland centre like Poona.

It may be stated that Singh (1956, p. 149), Singh (1961, p. 37), Nath (1962, p. 34), Mukerji (1962, p. 269), Singh (1964, p. 85), and Dwivedi (1964, p. 133) have shown the least deviation from R.L. Singh's view regarding the concept of umland. Dutt has a clear leaning towards Van Cleef when he states "a hinterland' is a wider economic sphere of a city, whereas an 'umland' is the inter-play of cultural, economic and political phenomenon, whereby the city and its surrounding region are functionally integrated to each other" (1963, p. 84). To Krishan and Agarwal, umland is "the contiguous tract which surrounds a service centre and is functionally integrated to it through certain specified criteria" (1970, p. 31). But on the contrary, contiguity is not essential for Chatterji who observes: "The hinterland of a city is composed of areas intimately linked with it. Most of these areas surround the city, others are located at some distance away from its immediate surroundings" (1967, p. 161). Explaining their concept of hinterland as a region, Dikshit and Sawant write: such a region is defined on the basis of nexus of relationships where the binding force of the region expresses itself in the circulation of goods, movement of people, flow of ideas and a considerable degree of interdependence between the core and the area around."

1. This study of umland could be highly objective in the regional plan with respect to public utility services, recreation and specialized retail service and the regional boundary is entirely confined within the city limits (Fig. 10.1)
2. The study of umland is helpful in the preparation of Master Plans for different metropolitan cities. In a Master Plan in case the town planner considers only the upliftment of Municipal boundary than the entire concept will change, because the municipal limits provide only compact form of Master Plans.
3. The concept of umland is also effective in the formation of town groups. First of all Robert Dickinson studied about the umlands of Leeds and Bradford. According to him these cities have got semi-common umland. The Leeds is more and more oriented towards administrative service and Bradford towards manufacturing. The industrial region under the influence of these two cities is known as Yorkshire region. From this region he excludes some towns located closer by the Shefield industrial umland. He considered West Riding under Western Yorkshire region. After demarcation Dickinson classified the umland into three regions : (i) Intensive region comprising residential area

of the city; (ii) the commercial region around the core of the city; and (iii) the surrounding hinterland around the town.

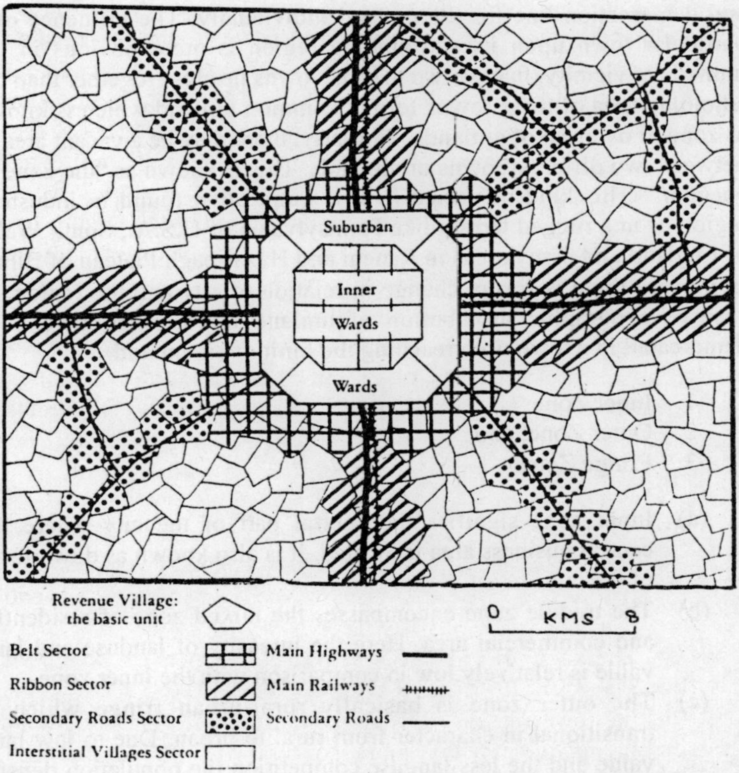

FIG. 10.1 : Schematic Representation of Indian City and its Hinterland

Recently, geographers have considered bus service index and newspaper circulation index to demarcate the area of umland. Definitely the area of newspaper circulation must be much more bigger than bus service index. In such a situation one has to more upto-date in the current changing situation of umland. For example, opening a new bus route is quite common in the 20th century because quite often this obliterates the real world situation. After several controversy the limit of 2.30 to half an hour walk through a vehicle has been considered as the umland of a town.

Nature of Umland

The nature of umland is difficult to decide especially in between two or more competitive cities lying close together. It is difficult to get access and demarcation line for the umland individually. The influence of a particular town upon its surrounding region is more obvious in the immediate vicinity. In case two to three towns lie close together than the adjoining area of these towns form a common umland which is known as zone of over-lap of umlands. Similarly, there may be a vacant area in between two different towns or umlands. This is known as "the zone of vacuum." This kind of vacant land is quite often found in industrial region or in a rugged terrain like Pennsylvania of U.S.A., South Wales of England and many areas in Ranchi and Hazaribagh Plateau of Bihar. In no case there is abrupt change in relation of city and the region. In case of geographic distribution of umland three zones have been demarcated in a town upto reaching the limit of an umland.

1. Inner Zone;
2. Outer Zone; and
3. Fringe Zone.

(a) Inner zone signifies the central part of the city where the central business area is located. It is also known as down town of the city.

(b) The middle zone encompasses the mixed zone of residential and commercial area. Here the intensity of landuse and land value is relatively low in comparison with the inner zone.

(c) The outer zone is basically rural-urban fringe which is transitional in character from rural to urban. Due to low land value and the less landuse competition the population density is low.

In India the most significant feature is that they do not coincide the administrative boundary of the political limit. In case of western world, especially England and other parts of Europe, the 1834 Act provides guideline that municipal boundary should coincide with the zone of influence and hence, the concept of umland has completely different bias in western world in comparison with India.

Defining Umland

Umland means land around a city. The concept was first of all formulated by Stanley Dodge in the year 1932. He identified the umland of the city of Howell in Central Michigan in 1937. Another geographer Darwest Whitsky found out the extent of umland for certain cities of England. In his study, he simply demarcated an area extending within a radius of 48 to 64 kilometres from the city. In the year 1938 Stiagendour identified the umlands of certain towns of England only demarcating the immediate areas around certain towns. He did not followed any geographic distance as an year mark for measurement. In 1950 H.L. Green has called it 'urban hinterland' and submitted a paper in Royal Geographical Society. He demarcated the extent of umland in some urban centres of England and Wales based on the distribution of bus services.

After 1950 most of the works done on this line, they are following the same methodology and same idea. It is yet for the geographers to give a new shape to the concept so as to be practically effective in future planning. So far as the nomenclature of the concept is concerned different geographers have advanced different views regarding the concept of umland. Robert Dickinson in his book *The City Region and Regionalism* has developed the concept as "city region." This has been criticized by geographers because if we speak a city region it means the region within the city. In case we speak State region—it means region within the State. It means the concept is grammatically not correct because when we say city region it means that city dominates over the region which is not a fact.

A.E. Smails forwarded the concept of 'urban field' for umland. The concept of urban field sometimes considered an agricultural sector within a town. In reality urban field means that field of the city which has been created by the town itself.

Another concept is sphere of location. It may be a geometric area to which any umland may conform at all. H.L. Green who has analyzed the service centres of umland in England and Wales used the word 'tributary area'. This has been highlighted by Stiagendour and is essentially linked with the concept of umland. He considers the entire accessible area from a town within the zone of umland.

Normally these measures are considered to measure the extent of umland.

(a) Vegetable and milk supply zone;
(b) The areas from where agricultural commodities are brought into the town;
(c) Wholesale zone;
(d) Retail zone;
(e) Zone of bus service;

(f) Zone of newspapers circulation;
(g) Professional service zone including medical, legal and educational services; and
(h) Miscellaneous service zone including business, recreation and personal services.

A brief discussion about these indices are as follows :

(a) Vegetable and milk supplies are very common indicies to measure the umland of an urban centre. Besides these, the supply of *goetha* to the urban centres in India is a peculiar feature to measure the spread of umland. R.E. Dickinson has also applied these measures to demarcate umland in England. But now-a-days the refrigeration system and the cold storage facilities obliterates the boundaries of umland to some extent.
(b) The area from where the agricultural commodities brought into the town covers wide area around the town especially under Indian condition. Naturally, the zone of influence derived from this index would be more wider in comparison with the previous one.
(c) Generally the wholesale of any items command a very big zone of influence upto far flung places. This is because the retail shopkeepers in large numbers are required to fulfil the aim of a wholesale dealer.
(d) The retail business area may not be very wide as we have seen in case of wholesale market area. It's threshold value is so small that it covers only daily walking distance area from the town. For example, the zone of medicine and cloth sales may be very wide in comparison with betal, bidi, cigarette and the business of a motel.
(e) The zone of bus service was strongly advocated by H.L. Green and under this head the points from where bus originates for an urban centre should be linked to form umland of an urban centre. In India and even in advanced countries this zone of influence is more theoretical because there are several areas which are completely devoid of bus services.
(f) The zone of newspapers circulation has been eliminated by the number of newspapers get published from a particular town and their intensity in respect of circulation. The circulation zone of newspapers gives a better picture in those countries where the literacy is high and the standard of living is such that

it commands constant cultural touch with the countryside. In this, we have a very big zone of circulation in respect of travel of newspapers especially for those which are published from capital cities like Patna, Delhi, Calcutta and others.

(g) The professional service zone is determined by three services—medical, legal advice and educational services. People from rural areas come to the town to avail the facilities of *X*-ray plant and medical treatment with specialized doctors, to study in technical institutes and universities besides they depend upon the pleaders of district and sub-divisional courts for legal advice in case of land and personal disputes. In case of India all these are very nice indicators for the demarcation of umland.

(h) Miscellaneous service zone including business, recreation and personal services are widely considered to demarcate the zone of umland. Rural areas are heavily depend upon the town for cultural and recreational activities especially cinema and theatre for which people from rural areas come to the town. In this regard, we may say that town depends on the countryside to sustain its business. Besides these, there are several other business performed in urban areas but they are meant to serve the people of surrounding rural areas only.

Delimitation of Umland

The boundary of umland should be such where one could reach within one hour walk from the town. A.E. Smails[2] studied the umland of different cities in England. He started investigation right from grocery shop to retail business of different items and knowing their influence around the surrounding area of a town. He completed the survey of England with the help of students and claimed that those maps are as reliable as ordinance Survey maps of England. In 1958 H.L. Green[3] started the delimitation of umland in England with the help of origin and destination points of bus services.

In any scheme of demarcating umland education and specialized services should not be ignored. On the basis of these services three zones inner-outer and fringe may be demarcated.

Criteria

As the city and its umland are 'culturally, economically and politically interrelated; the criteria used in delineation of umland are also generally chosen from these three fields. The six zones—vegetable, milk supply of grain and other agricultural products, bus service, newspapers circulation and administration used by R.L. Singh[4] in 1955 have been popular in all the later works, although a few new ones like the commuters zone, educational zone and hospital service zone are also adopted by a number of authors. Zones based on retail sales, bullock cart, bicycle, rikshaw, drug sale, cinema, police, banking fish, egg and poultry, livestock, etc. are less important having been only in casual use.

The vegetable supply zone has been so attractive to the Indian geographers dealing with the umland that it is common to most of the works, and has also been applied as a single criterion for measuring the regional relationships of at least two centres—Chandigarh (1963)[5] and Hyderabad (1961).[6] Though compact in nature, the vegetable zone is so small in areal extent that it hardly covers even limits of the primary umland. Similar is the case with milk supply zone also. The food grains and other agricultural products supply zone covers a wider area and has shown its merits by coinciding with the composite umland boundary in a few cases. Its growth is sometimes handicapped by imposition of Government controls on movement of foodgrains from one political unit to another as is the case with Jamshedpur[7] where the boundary betwèen States of Bihar and Orissa acts as a limit of the town's grain supply zone in east direction. The bus service zone has generally not yielded satisfactory results for two reasons: (i) The automobile transport is not yet so well developed in India as in the Western countries, and (ii) the zone is very irregular as it includes areas up to only 6-8 km. away from the bus routes. The newspaper circulation zone does not suit smaller centres as they hardly publish any local dailies. Administrative and hospital service zones are medium in size and may give useful results for certain towns. The commuter, education (upto Intermediate level), retail sale, bicycle, rikshaw, cinema and banking zones are indicators of deep relationship between the region and the service centre and thus may be helpful in demarcation of the primary umland. The drug sale and police zones indicate nothing new as they coincide with the hospital and administrative zones respectively. Fish, egg, poultry and livestock supply zones have been used only in the case of Howrah where they have not shown any promising results.

Distinction between City and its Umland

(i) Within a town there is a Peculiar distribution of shops and houses and divided into sections normally commercial, residential and others.

In commercial areas shops are the main features. Although people reside even in the commercial area but the total number of people living there permanently is very small. At some places especially on the marginal shops and residential houses are combined together. These features are completely absent in the rural areas.

(ii) The character of building is varied in case of the town. There may be one storied to multi-storied building. But in case of village all the houses are of a uniform character. In a town multi-storied buildings are found especially where, there is concentration of population and where there is no space for horizontal development of the houses. This leads to the construction of multi-storey buildings. Vertical expansion of houses in the centre of a town is the most noted feature.

(iii) There must be adequate public utility services within a town. A number of post office, banks and insurance companies are the features which are easily distinguishable by their presence in a town. Although every town and city enjoys these facilities but still there can be further differentiation among different types of towns. The present adequacy and efficiency of these services have great influence over the growth and importance of town.

A rapidly growing town whether industrial or administrative must have adequate number of units giving further facilities for the growth of population. In other words, the chances for further development of a town depends greatly upon such facilities.

(iv) Every town has a central place of activity which consists of shops, stores, wholesale and retail services, hotels, offices of various types and places of entertainment. These are usually located in the central portion of a city so that people might have shopping and service facilities. This central place in a town is well marked from other areas and is popularly known as Central Business District (CBD). This CBD is characterized by the concentration of transportation network made by the people in a town. Shops and stores may be located in other sections but they may not attract a large number of people in case they may not provide all sorts of facilities. But CBD is supposed to be the most developed area within a town which enjoys almost all the facilities of transport, communication, shopping and other services. The central location of CBD is mainly due to the attraction of people from all parts of town towards the centre. On the other hand, all these features are completely absent in the rural areas.

(v) The differences in the vertical dimension of buildings in different parts of the city is an important feature which never be true in a village. Within a town also the buildings are taller especially in CBD in comparison with fringe areas. It is chiefly because of increase in activities due to service facilities and lack of new lands to construct new houses.

Existing Gap and Needed Research

The research on umland of towns of India has completed a thirty-five years span but a number of gaps still exists. First, there is a lack of precision in the boundaries of umland. They are arbitrary, imaginary and vague. The responsibility for this lies not upon the researchers, but on the non-availability of necessary data. It is not possible to collect minute information on an individual level for many reasons. The problem may be more easily tackled by the Government, either by enlarging the scope of its census operations, or by making some other alternative arrangements for the data collection. Further, different geographers have taken different sets of criteria in their respective studies and this makes a comparison of their findings difficult or even impossible. Hence, a standardization of criteria is called for; and this, in its turn, will need similar data for all the centres. Whatever criteria have been used so far, the information is gathered mostly from the city. The collection of data from the rural areas by administering a questionnaire to the villagers is also equally essential. It may also be pointed out that the intensity of functional integration with the city is not evenly distributed throughout the umland. Settlements on one side of the city may be more intensely related than those on the other located at the same distance. Some villages are dependent on the city for all of their buying needs, there may be others whose needs are partially served by a smaller service centre in their neighbourhood.

The use of umland as planning regions will be of great significance. The hierarchy of umlands is also easily changeable into a hierarchy of planning regions. Further, the study of umland may be of immense help even in town-planning as no survey of a town is complete without a reference to its regional relations. Towns serve as growth poles and growth centres for the surrounding regions. Hence, the analyses of the growth-promoting and growth-retarding factors and their interactions are also to be taken into account fore long- and short-term planning as well.

Spacing of Cities in India

For the inter-spacing of cities it has been remarked by Watson that "geography is a discipline in distance."[8] According to Cole and King[9] "geography is concerned with the distribution of elements that occur on the earth's surface and with variations of distribution through time and space."

The inter-spacing of cities depend on the population size, areal expansion and functions of cities which are the focal themes for measuring the level of regional development. They also indicate the possibility of new cities growing up at places which have ample scope for raising the living standard of population, development of natural resources and increasing the potentialities of industry.

Spacing of urban settlements have a significant place in theoretical consideration of distribution, defined as the locational arrangement of central places with respect to one another. The theoretical basis of transformation of density into spacing is provided by an ultimate development of circle into hexagon to build spatial structure without leaving any empty space on the ground. In hexagonal arrangement of points, the spacing from one point to any of the six points is :

$$D = 1.0746^{d-1/2}$$

where D is the theoretical distance between points and d is the density of settlements per unit area (km^2).

Discrete spatial distributions occupy mainly specific locations so that members of the spatial set are separated by finite distances. The distribution of farmstead, shopping centres and cities exemplify these spatially discrete phenomena. Quite often such distributions are referred to as random, clustered, uniform or some other vague qualitative terms, but actually it needs objective quantitative means by which spatially discrete distributions can be characterized. It can be measured by nearest-neighbour of Clark and Evans (1954)[10] and standard distance (Bachi, 1962).[11]

This study may highlights the characteristics of distribution and measuring the interspacing of cities in India. This is because the higher order urban centres are the prime force of urbanization and the preferment of regional development. Further, it helps in evaluating areas of uniform, random and clustered patterns of urban concentration.[12] (Fig. 10.2).

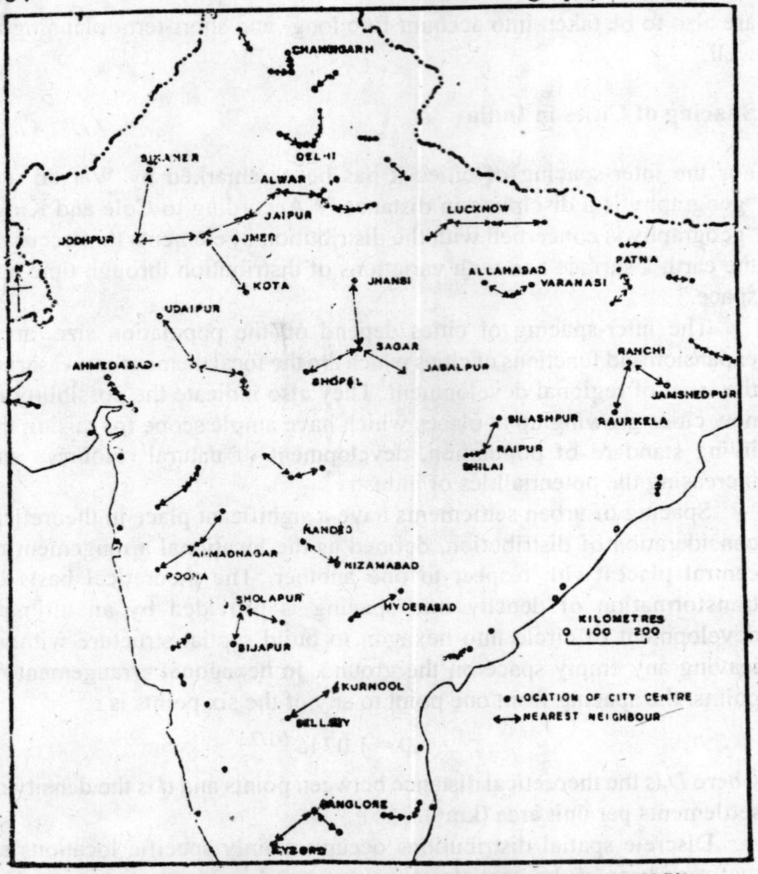

FIG. 10.2 : First Nearest-Neighbour of Cities in Central Part of India (1971)

In India, cities having more than 1,00,000 population are closely spaced near Calcutta, Chennai, Mumbai and in all over northern India. But wider spacing is observed in the forested and rugged terrain of north-eastern India and Peninsular-Vindhyan region. The developed transport facilities of North Indian Plains give easy and quick access to money, men and material together at one place for the distribution and exchange of goods and ideas which give impetus for the cities to be closely spaced. The productive riverine plains, port facility, capital regions of provinces, districts and sub-divisional headquarters are the urban centres of city level, e.g., the Ganges valley of North Indian Plain, Calcutta, Mumbai and Chennai Ports and in the nearby areas of the national capital of Delhi.

Cities in India are expanding in size and their inter-spacing is getting reduced day by day. This phenomenon has given rise to many new problems while the old ones continue to loom large for solution. A study of spacing of cities through nearest-neighbour analysis is thus significant both as an academic discipline and as a base for decision-making in administrative, industrial, social, commercial and religious matters.

The distance between two cities is directly related to the tertiary services and population increase under particular environmental condition. The multiplication and coordination of cities are two vital elements which give rise to closer spacing. Time factor also plays a dominant role because the spacing is narrowed down in course of time as natural phenomenon on with the multiplication of points of concentration as near Calcutta. The spatial interaction of a number of variables has a marked effect on the size and spacing of cities (Table 10.1) and is directly related to the development of commercial, administrative, industrial and transportational activities.

TABLE 10.1 : Spacing of Cities in India (1931-2001)

Census year	*No. of Cities*	*Average Spacing in kilometres*
1931	38	314.38
1941	53	266.20
1951	75	223.78
1961	107	187.35
1971	148	159.30
1981	189	140.96
1991	230	127.78
2001	271	117.72

Source : Calculated from Census of India, 1991.

According to the above Table 10.1 the inter-spacing of cities has undergone marked shrinkage from 314.38 kms. in 1931 to 149.30 kms. in 1991 and 117.72 kms. in 2001.

The cities around Delhi have a closer spacing of 29.10 kms. with an average population size of 18 lakhs. The widest spacing is 356.66 kms. in Jammu and Kashmir which have an average population size of about 3 lakhs in 1991. The largest number of cities are found in Uttar Pradesh having 22 in 1971. It has an average spacing of 124 kms. and the population size is about 3 lakhs.

The Pearsonian correlation coefficient for this comes to 0.43 and the regression value comes to $Y = 673.01 - 1.95\,X$ for the population size and spacing respectively. Such relationship clearly points towards the agglomeration of cities in greater part of the country around large metropolitan centres in recent decades in comparison with the beginning of the 20th century when most of the cities in India were British administrative strongholds.

Haggett[13] has noted that the spacing of urban places is largely governed by their size. Thomas (1962)[14] has also established the stability of population size and distance relationships among cities by employing a normal curve of error to determine whether cities were of similar populations at similar distances apart. Fewer number of metropolitan centres would be widely spaced, whereas small cities are found to be closely spaced. Both Christaller (1933)[15] and Lösch (1954)[16] have put forward evidence for discrete types of central places which clearly shows that this proposition may be valid in almost all parts of the globe.

The Concepts of Cluster, Uniform, Regular and Random Distribution

A close distribution of settlements is considered as concentration or clustered. In the distribution nucleation is a cluster as the enclosing area is large. In case of clustered distribution the *R* value of nearest-neighbour statistic tend to be below 1.00.

The equal distribution of point is called uniform when the enclosing area is moderate. At the stage of uniform (critical) distribution any change in spacing leads to either concentration or dispersion. A uniform distribution is an ideal which hinges on the concept of isotropic plain surface. At this stage Hagerstrand remarked that 'isotropic surface' is an elementary abstract geographical space that has no difference from place or in one direction to another.[17]

The regular pattern of distribution may be observed from cluster and uniform distribution in case the settlements are found on equal interval. In case the haphazard distribution of settlements are observed in space that is known as random distribution.

Model for Spatial Development of Cities

According to Charles and Stewart[18] the density of urban population, the ratio of secondary and tertiary labour force, market gardening and the standard of living in a city determine the spacing of cities in a particular region. In India, the spacing of cities are mainly dependent upon the following factors.[19]

(a) Religious centres and water bodies;
(b) Growth of administrative centres;
(c) Development of transport network;
(d) Growth of industrial centres; and
(e) The rise of trade and commerce.

The cities of Prehistoric Period were mostly religious centres located on the bank of rivers such as Allahabad (Prayag), Vaishali, Odantpuri, Champanagar, etc. In those days, the inter-spacing of cities were high and the population size were low (Fig. 10.3).

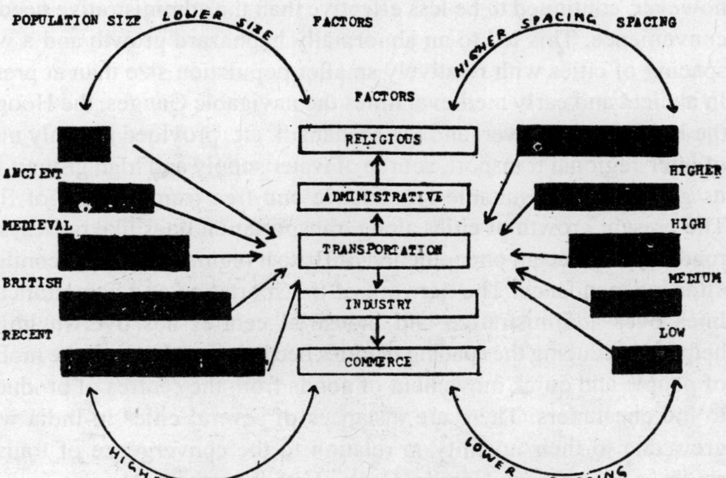

FIG. 10.3 : Model for Spacing of Urban Settlements

The growth of administrative centres is an important factor which attracts people of different trades at one place and provides incentive for raising a smaller town to the status of a city. This factor has been dominating in ancient times and is effective even today, with the establishment of new district and sub-divisional headquarters. The cities like Delhi, Calcutta, Mumbai, Ahmedabad, etc. these metropolitan towns were administrative centres of various rulers in the past and therefore, developed upto its present level of city where port facilities, industries and the model place for transport helped in the establishments of several city level centres in India. Administrative force leading to greater urbanization and regional development which are reflected in the emergence of newer administrative centres like Biharsharif, Ahmedabad, Chandigarh and others.

In ancient times the transport networks were in a very crude form, and at places practically non-existent. Wherever, they were developed, they were of low order. There was also lack of speedy vehicles for transportation of goods. It was only rivers which provided reasonable transportation facilities. Naturally, urban centres developed at those cities where rivers provided transportational facilities. This factor, however, continued to be less effective than the administrative need and convenience. This led to an abnormally haphazard growth and a wider spacing of cities with relatively smaller population size than at present. In ancient and early medieval times the navigable Ganges, the Hooghly, the Krishna, the Kaveri and the Mahanadi, etc. provided the only means of inter-regional transport, source of water supply and high ground from its general levels suitable for defence and free from ravages of flood. The present growth of cities along transportation lines like railways and roadways are recent phenomenon of British regime which has continued after independence. The network of transportation and communication lines over administrative and industrial centres has overwhelmingly helped in reducing the spacing of cities because they increase the mobility of people and quick movement of goods from the centres of production to the consumers. There are instances of several cities in India which grew due to their nodality in relation to the convergence of transport routes, e.g., Asansol, Kandla, Cochi, Visakhapatnam, etc.

After independence, the force of industrialization and development of mineral resources have attracted workers from the nearby region. This process of converging near mines and at newly established industrial centres has reduced the spacing of towns and increased the incentive for the development of large cities. In India there are about 25 cities which rose to this status due to the establishment of industries, as for example, Bhilai, Rourkela, Jamshedpur, Bokaro, Dhanbad, Kanpur, etc.

At present the rise of trade and commerce due to growth of population and increasing demand for consumer goods have highly helped in the rise of cities. As for example, Bikaner, Bareilly, Gauhati, Imphal,

Srinagar, Jammu, Biharsharif, etc. are centres of trade and commerce. This has highly helped in reducing the inter-spacing of cities and increasing the population size of cities in comparison with ancient, medieval and modern periods.

To sum up, the cities in Prehistoric period were few and far between having smaller population size. After the growth of administrative headquarters, roads, railways, sea ports and ferry ports, establishment of industries and increase of commercial centres give impetus for the growth of population size of cities and the relative reduction in spacing since Prehistoric up to the present.

Not only the above factors have profound impact on the growth of cities in India and their closer spacing but the variables like resources of the region, their historical evolution, agricultural prosperity, topography, soil condition, trends of population growth and others found in the surrounding zone of influence of the particular centre have multilateral effect on the size and spacing of cities. The above model can be explained with the help of practical differentiation in vector analysis. If a vector V (change in spacing) is a function of five scaler variables, religious site (*X*), administration (*Y*), transportation (*Z*), industrial (*K*) and commercial (*M*); and the partial differentiation is defined in scaler calculous, then the position vector which is the distance (*d*) of a point *P* in three dimensional space is expressed in terms of the cartesian co-ordinates of the point as follows :

$$r = \overline{X}_i + \overline{Y}_j + \overline{Z}_k + \overline{K}_m + \overline{M}_n$$

Then, $$\frac{r}{x} = i, \quad \frac{r}{y} = j, \quad \frac{r}{z} = k, \quad \frac{r}{k} = m, \quad \frac{r}{m} = n$$

In general if $V = V(X, Y, Z, K, M)$, then $\frac{r}{M}$ denotes the rate of change with respect to *X*. Similarly, changing of *Y, Z, K* and *M* also give the partial derivates $\frac{V}{Y}, \frac{Z}{Z}, \frac{V}{K}$ and $\frac{V}{m}$ respectively. If *X, Y, Z, K* and *M* change simultaneously by differential increments *dx, dy, dz, dk* and *dm* the total change or total differential of *V* will be :

$$dv = \frac{V}{X}X + \frac{V}{Y}Y + \frac{V}{Z}Z + \frac{V}{K}K + \frac{V}{M}m$$

Similary, $$dr = dx\overline{i} + dy\overline{j} + dz\overline{k} + dk\overline{m} = d\overline{M}n$$

Nearest Neighbour and Standard Distance Analysis

Nearest neighbour is a straight line measurement of the distance separating any phenomenon and its nearest-neighbour in space, and was originally developed by Clark and Evans, plant ecologists. With the help of this analysis it is possible to measure the departure from observed spatial distribution to a theoretical random distribution between clustered and dispersed pattern. This scaler distribution is known as *R* scale or nearest-neighbour scale and its value can be calculated with the help of formula.[20]

$$R = \frac{rA}{rE}$$

Thus, $rA = \frac{r}{N}$ where *r* is the distance from each point to its nearest-neighbour which is the actual straight line measurement, and

$$rE = \frac{1}{2\frac{N}{A}}$$

is the estimated mean distance. Hence, the assumption is that cities are ordinarily placed at the inter-section points of a rectangular grid and the best figure to use for inter-cities spacing is equal to he square root of the number of cities divided by the area of a province.

The ratio of observed mean distance (*rA*) to expected value (*rE*) is known as nearest-neighbour static (*R*). The *R* value ranges from zero (*O*) in agglomeration side to 1.00 towards random distribution, up to 2.1491 towards uniform pattern (Fig. 10.3). This signifies maximum dispersal, which is analogous to the hexagonal arrangement discussed by Christaller and others. In a random distribution a set of points in a region have the same chance of occurrence on any sub-area as any other point. Such specified area has the same chance of receiving a point as any other area of that size assuming that the placement of each point has not been influenced by any other point.[21]

Spatial Pattern of Urban Places according to Nearest Neighbour Technique

In North Indian Plains, the arrangement of urban places are very similar to the hexagonal arrangement of point[22] but elsewhere the arrangement is quite different. In Peninsular upland all cities are distant from their nearest-neighbour and hence, the term dispersed in applied to them. The cities lie in the river valleys of the Ganges, the Damodar, the Krishna, the Hooghly, etc. suggest linear clustering. Three types of spatial pattern of the basis of nearest neighbour distance have been identified with the help of *R* value for cities in India. They are :

Approaching Agglomerated Area

In India, West Bengal is the only state where cities are approaching towards agglomeration and more systematic distribution with decreasing entropy. In West Bengal the number of cities are 14 and the spacing of one city with other is just 16 kilometres.

Approaching Random Areas

The group comprises the State of Punjab, Orissa, Maharashtra, Madhya Pradesh and Gujarat according to the first nearest-neighbour *R* statistic (Table 10.2). In the *R* statistic of the second nearest-neighbour the above mentioned states come under more than random areas. This shows that the entropy is increasing towards second nearest neighbour.

More than Random Areas

There are five states in the *R* statistic for the first nearest neighbour. In the second nearest-neighbour group more than random areas are found in 12 States of India (Table 10.3). States lie under this group are Uttar Pradesh, Rajasthan, Assam and others. The increasing *R* value for the second neighbour shows the maximization of entropy and decreasing systematic arrangement of cities in India due to haphazard growth of cities.

Approaching Uniform Pattern Areas

According to Table 10.2 the areas approaching uniform distribution of cities are found in Kerala, Karnataka and Tamil Nadu. In these States the *R* value comes to 2.13, 1.93 and 1.89 respectively for the second nearest-neighbour whereas for the second nearest-neighbour the above mentioned States come under more than random areas. There are only 5 cities in Kerala, while it is 10 in Karnataka and 17 in Tamil Nadu in 1971. This shows that irregularity is increasing in the distribution of second nearest-neighbour centres.

In order to test the null hypothesis for the dispersion trend, chi-square test (χ^2) has been applied to analyze the relationship between

TABLE 10.2 : Nearest-Neighbour Statistics of Cities in India, 1971

States	*No. of cities*	*Density of cities per square kilometre*	*Mean Observed Distance of Ist. Nearest-neighbour in kms.*	*Mean Observed Distance of IInd Nearest-neighbour in kms.*	*Mean Expected Distance*	*R Value of Ist. Nearest-neighbour*	*R value of IInd Nearest-neighbour*
Andhra Pradesh	13	0.0000469	84	106	73.01	1.15	1.45
Assam	1	0.0000100	00	00	15.11	0.00	0.00
Bihar	9	0.0000517	63	98	69.54	0.90	1.40
Haryana	3	0.0000676	50	63	60.81	0.82	1.03
Gujarat	7	0.0000357	63	118	83.68	0.75	1.42
Jammu & Kashmir	2	0.0000089	150	260	167.00	0.89	1.55
Karnataka	10	0.0000521	94	134	69.27	1.35	1.93
Kerala	5	0.0001286	64	94	44.09	1.54	2.13
Madhya Pradesh	11	0.0000248	85	130	100.40	0.89	1.29
Maharashtra	15	0.0000487	64	105	71.65	0.89	1.46
Manipur	1	0.0000447	00	00	74.78	0.00	0.00
Orissa	4	0.0000256	83	152	98.82	0.84	1.54
Punjab	4	0.0000794	55	85	56.11	0.98	1.51
Rajasthan	7	0.0000204	141	173	110.70	1.27	1.56
Tamil Nadu	17	0.0001306	55	83	43.75	1.25	1.89
Tripura	1	0.0000954	00	00	51.19	0.00	0.00
Uttar Pradesh	22	0.0000747	63	93	57.85	1.08	1.60
West Bengal	14	0.0001593	16	26	39.61	0.40	0.66
Delhi	2	0.0013468	10	15	13.62	0.73	1.10

Source : Calculated by the Author.

observed and expected frequencies (Table 10.3). The results pointed out that at $N - 1$ degree of freedom with about 95 per cent probability level the value is coinciding, i.e., in Gujarat 5.11 is significant 95 per cent level for the first nearest-neighbour while for the second nearest-neighbour 14.07 is significant at 5 per cent level. This shows that up to a considerable level environmental factor is controlling the spatial distribution trend of cities in India. As such research hypothesis is accepted against null hypothesis.

TABLE 10.3 : Chi-square Statistic for Nearest-Neighbour Distance of Cities in India, 1971

Name of the State	*Ist. Nearest-neighbour* $\frac{(O-E)^2}{E}$	*IInd Nearest-neighbour* $\frac{(O-E)^2}{E}$	*Significance*	
			Ist. Neighbour	*IInd Neighbour*
Andhra Pradesh	1.65	14.90	No	Yes
Assam	0.00	0.00	—	—
Bihar	0.61	11.65	No	Yes
Gujarat	5.11	14.07	Yes	Yes
Haryana	1.92	0.07	Yes	Yes
Jammu & Kashmir	1.73	51.79	Yes	Yes
Karnataka	8.82	59.56	Yes	Yes
Kerala	8.99	56.94	Yes	Yes
Madhya Pradesh	2.36	8.72	No	Yes
Maharashtra	0.81	15.52	No	Yes
Manipur	0.00	0.00	—	—
Orissa	2.53	28.62	Yes	Yes
Punjab	0.02	14.87	Yes	Yes
Rajasthan	8.29	22.43	Yes	Yes
Tamil Nadu	2.89	35.21	No	Yes
Tripura	0.00	0.00	—	—
Uttar Pradesh	0.46	21.35	No	Yes
West Bengal	14.07	4.67	Yes	Yes
Delhi	0.96	0.14	Yes	Yes

Relationship between per 1000 sq.km^2 density of cities on *Y*-axis and spacing on *X*-axis shows that highest density of 1.35 is in Delhi while it is 0.16 in West Bengal, 0.13 in Tamil Nadu and Kerala, 0.10 in Tripura and in others it is in between 0.01 to 0.08. This shows that both urbanization and regional development are on the lowest ebb in India.

The upper and lower ranges of random matching for the first and second nearest-neighbour at 95 per cent probability level have been measured with the help of a formula :

$$= (2E \pm rE)\sqrt{rE}$$

where $rE = r$ expected

or $E = 0.26136\sqrt{nd}$,

where n stands for the total number of cities in the areal unit and d represents density of cities per km^2. The frequency curve of Rn value shows a normal curve which reveals that in first order nearest-neighbour statistic most of the area is found in the middle range of low regularity whereas for the second nearest-neighbour most of the States of India lie above the random line which shows the increasing level of irregularity and dispersed nature of the distribution of cities.

Standard Distance Analysis

It is a measure of the concentration and dispersion in geographical studies and it helps in identifying areas of concentration of points. Thus Bachi's standard distance is a measure of the dispersion of set of points in respect of two right angled axes.

It may be expressed as :

$$d_{ij} = |(x_i - x_i)^2 + (y_i - y_i)^2|^{1/2}$$

where x and y are two coordinates, and d_{ij} is the standard distance.

The above formula will give us distance between two points in Euclidian Space. For the n set of points n^2 of these distances can be calculated, so that the distance between the two points is itself included and every standard distance is measured twice, once for i and secondly for j.

For the set of n^2 distance, Bachi suggested the mean quadratic distance, as :

Then, the mean centre of the distribution have the coordination as :

Where d_{ic} is the distance from point i to the mean centre-

$$C = |(x_i - \bar{x})^2 + (y_i - \bar{y})^2|^{1/2}$$

then the standard distance is :[23]

Further "Bachi notes that the square of distance (d^2) for the entire area is equal to the sum of 'within regions' squared distance ($d^2{}_w$) and 'between regions' squared distance ($d^2{}_b$)."[24]

Then $d^2 = d^2\text{w} + d^2b$

where (x_{ij}, y_{ij}) are the coordinates of the ith point in the jth region.

$(\bar{x}_j, \bar{y}_j)$ is the regional mean centre.

$(\bar{x}, \bar{y})$ is the mean centre for the whole distribution.

$\left(d_b^2 / d^2\right)$ is the ratio, provides a crude measure of significance of the regional divisions. Ratio 1 is the maximum differentation. Where *O* shows the same mean centre as exists in the total pattern.

Standard distance may be defined as 'Mean Quadratic Distance', i.e., arithmetic average of square root of distance and can be measured with the help of latitudes and longitudes. In this study, the standard distance has been measured by taking the square root of the sum of the variances of longitudes and latitudes. The cities have been arranged according to classes of longitudes and latitudes in different States of India. Two frequency distributions were thus obtained for each state on which the deviations were calculated and summed up for finding the standard distance. The calculation for Tamil Nadu state is given in Table 10.4.

TABLE 10.4 : Calculation of Standard Distance for Tamil Nadu, 1971

Class Interval of Latitude in degree	*Frequency (F)*	*Deviation (D)*	*FD*	*FD²*
80°—81°	1	+1	1	1
79°—80°	6	0	0	0
78°—79°	6	−1	6	6
77°—78°	4	−2	8	17
Total	17			23

Squared Longitudinal Deviation of Cities = 23/17 = 1.35 (a)

Class Interval of Latitude in degree	*Frequency (F)*	*Deviation (D)*	*FD*	*FD²*
7°30′—8°	1	−2	2	4
8°—9°	2	−1	2	2
9°—10°	3	0	0	0
10°—11°	5	+1	5	5
11°—12°	3	+2	6	12
12°—12°45′	3	+3	9	27

Squared Latitudinal Deviation of Cities = 50/17 = 2.94 (b).

For mean latitude of Tamil Nadu

Radius of a parallel $= R\cos$

Circumference of a parallel $= R\cos \times 2$

1° of latitude $= R\cos \times 2/360°$

For 10° latitude, $1° = \dfrac{R\cos \times 10° \times 2}{360°}$

$$= \frac{3960 \times \cos 10° \times 2 \times 22}{360 \times 7}$$

$$= \frac{3960 \times .9849 \times 2 \times 22}{360 \times 7}$$

$= 64.12$ miles (102.59 kms.)

For Mean Longitude : $1° = 69.14$ miles

Standard Distance in Miles =

$$\sqrt{(a \times 1° \text{ mean longitude}) + (b \times 1° \text{ mean latitude})}$$

$$= \sqrt{(a \times 69.14) + (b \times 62.41)}$$

$$= \sqrt{(1.35 \times 69.14) + (2.94 \times 64.12)}$$

$$= \sqrt{93.33 + 188.51}$$

$$= \sqrt{281.84}$$

$= 16.76$ miles (26.81 kms.) (1971)

The difference between observed distance, distance according to Robinson and Barnes method and Standard distance have been exemplified. In this way, it clearly represents that there is a huge gap between observed distance, standard and distance observed by Robinson and Barnes method. Hence, the methodological treatment of spacing of cities in space is just a fraught and it is difficult to get a real picture in case methodological variability persists in geographical treatment of spacing.

Conclusion

This chapter represents the characteristic features of spatial distribution of cities which is estimated with the help of nearest-neighbour analysis, standard distance analysis and Robinson and Barnes method. There is a wide gap in the spacing of cities as these methods represent clearly. In this way, the characteristic features of spacing, concentration, dispersion, regularity, randomness and radial distribution of cities in India have been analyzed here in order to highlight the regional problems which put forward various concepts in regional development.

FOOTNOTES

1. G. Taylor, *Urban Geography*, London, 1949.
2. H.M. Mayer and Kohn, C.F., *Readings in Urban Geography*, Central Book Depot, Allahabad, 1967.
3. F.H.W. Green, "Urban Hinterland in England and Wales: An Analysis of Bus Services," *Geographical Journal*, 1950, Vol. 116, pp. 65-88.
4. R.L. Singh, "The Concept of Umland with Reference to Indian Cities: A Trend Report," in *A Survey of Research in Geography*, Popular Prakashan, Mumbai, 1972.
5. U. Singh, "Umland of Allahabad," *National Geographical Journal of India*, 1961, 7, pp. 37-51.
6. V. Nath, "The Concept of Umland and Its Delimitation," *Deccan Geographer*, 1962, 1, pp. 34-45.
7. A.K. Dutt, "Umland of Jamshedpur," *Geographical Review of India*, 1963, 26, pp. 34-98.
8. J.W. Watson, "Geography: A Discipline in Distance," *Scottish Geographical Magazine*, Vol. 71, 1955, pp. 1-13.
9. J.P. Cole and King, C.A.M., *Quantitative Geography*, Wiley, 1968.
10. P.J. Clark and Evans, F.C., "Distance to Nearest-neighbour as a Measure of Spatial Relationships in Populations," *Ecology*, 35, pp. 445-53.
11. R. Bachi, "Standard Distance Measures and Related Methods for Spatial Analysis," Papers and Proceedings of the Regional Science Association, 1962, pp. 83-132.
12. N.B.K. Reddy, *et al.*, "A Method for Delimiting and Measuring Concentrations and Dispersions, *The Karnataka University Journal*, 1970, Vol. XV, pp. 166-84.
13. P. Haggett, *Locational Analysis in Human Geography*, Edward Arnold Ltd., London, 1966, pp. 107-13.
14. E.N. Thomas, "The Stability of Distance-population Size Relationship for Iowa Towns from 1900 to 1950," *Lund Studies in Geography*, Sereis B, *Human Geography*, 1962, Vol. 2, pp. 13-30.
15. W. Christaller, *Central Place in Southern Germany,* translated by C.W. Baskin, Prentice Hall Inc, New Jersey, 1966.
16. A. Lösch, *The Economic of Location*, New Haven, Yale University Press, 1954.
17. J.D. Nystuden, "Identification of Some Fundamental Spatial Concepts," *Spatial Analysis* (Eds. Berry and Marbel), Prentice Hall, 1963, pp. 35-41.
18. T. Charles and Stewart, J., "The Size and Spacing of Cities," *Readings in Urban Geography* (Ed. Mayer and Kohn), Central Book Dept., Allahabad, 1969, pp. 245-50.
19. R.B. Mandal, "Spacing of Urban Place in Bihar, *Indian Geographical Studies*, R.B. No. 2, 1974, pp. 1-16.
20. M.H. Yeats, *An Introduction to Quantitative Analysis in Economic Geography*, McGraw Hill Book Company, London, 1968, p. 31.
21. L.J. King, "A Quantitative Expression of the Pattern of Urban Settlements in Selected Areas of the United States, *Analytical Human Geography* (ed.) Peter Ambrase, Longmans Green & Co. Ltd. London, 1969, pp. 89-102.
22. R. Bachi, "Standard Distance Measures and Related Methods for Spatial Analysis," *Papers Regional Science Association*, Vol. 10, pp. 83-102.
23. L.J. King, *Op. cit.*, 92-93.
24. G.S. Prasad, "Spatial Distribution of Towns in Bihar Plain," *Indian Geographical Studies*, R.B. No. 3, 1974, pp. 50-57.

11

URBAN ECOLOGY AND LANDUSE

'ECOLOGY' or 'Decology' is a branch of biology dealing with living organisms habitat, modes of life and relations to their surroundings. In this way, ecology is the study of habitat, mode of living and relations of habitants of a place to their surrounding environment. Hence, urban ecology include the habitat of urban people, their mode of life, relationship with the neighbouring areas which have a impact on the urban environment and the resident of the area.

Ecology of an area depends upon various factors namely climatic—which includes rainfall, temperature, humidity, and atmospheric pressure; edaphic which includes the effects of soil and topography of a place; and biotic which include effects of plants and any other items related to man and *vice-versa.* Nowadays with the clearing of forests, advancement of commercial agriculture, establishment of industries and over grazing of pasture land have created environmental stress especially near urban areas both for the sustenance of human life and animals. Forest cutting and insanitary condition in urban areas have adverse impact which resulted into the environmental pollution.

Nowadays, with the overwhelming increase of population, the burden on the available residential houses, cultivable land, roads, railways, water taps and industries located in the urban areas have surpassed all previous levels in terms of harnessing the resources and creating environmental hazards. Some of the important hazardous places are the Ganges river near Barauni and Mokameh, Patna, Bhagalpur, Buxar and Sahebganj besides Bermo, Chandrapura and Bokaro along the Damodar river and on the Sone near Dehri, on the Kharkai near Jamshedpur and others. In most of the urban areas the chocked drains, the predominance of service latrines, dump of garbage on roads, open drains and insanitary conditions are crux of the issue. Throughout Bihar, urban areas pose a deserted look on the environment which is serious for a sound socio-economic development of the country.

Urban Landuse

Urban landuse deals with the problems emerging in urban centres in the process of selecting and translating into action, the optimum utilization of limited land between shopping centres and residential areas.

The geographical distribution of landuse in urban areas is of prime interest to the geographer, urban sociologist and city planners. The differentiations in landuse pattern in urban areas vary due to internal differentiation in the location of commercial establishments, administrative offices, factories and residences. Due to rapid increase in population and size of the urban areas, the landuse changes. All these add a dynamic character to the urban landuse in comparison with the rural sector. Urban Geographers in India have also contributed to urban landuse study, e.g., Gist (1958), Rajgopalan (1960),[1] Chatterji (1965),[2] Alam (1965)[3] and Singh (1966).[4] This section deals with the basic principles of urban landuse study in relation to rural landuse.

An urban centre is a collection of houses of non-agricultural people. Here various uses of land are found on account of (a) the forces of attraction and integration, (b) the forces of dispersion and disintegration, and (c) the forces of spatial differentiation (Dickinson, 1964, 125).[5] The first two help in the vertical growth and horizontal expansion while the third results into segregation of buildings, persons and activities. A little change in social and economic conditions caused by natural, or political forces, brings a remarkable change in the landuse patterns. A plan through landuse study to eliminate some of the undesirable changes and to minimize the problems that have been already posing becomes imperative. In guiding the urban activities, the pattern of land utilization is of paramount importance, hence a clear-cut land policy becomes essential for guiding the use of the land, curving the misuses, preventing the abuses, regulating the non-uses and aiming at the reuses of land in the better interest of the inhabitants (Abrams, 1953, 34).[6] The study of existing landuse is essential before the start of planning urban areas.

Landuse planning in urban areas is a part of the larger process of city planning. It is basically concerned with the location, intensity and amount of land development required for various uses of space, functioning of the city, e.g., industry, wholesaling, business, housing, recreation, education, religious and cultural activities of the people (Chapin, 1957, XIV).[7] Landuse is determined by land economics, social values and public interests. In urban areas certain sites have greater value for certain specific purposes. This value is established by a web of relationships with other surrounding uses and human behaviour play a vital role in the evolution of landuse patterns, though such a socially rooted determinants are less understood and are frequently confused with the economic determinants. The urban landuse study is a new branch of geography and its importance has been aroused since the Second World War when the need for planning and replanning of urban places was universally felt (Dubey, 1969, 26-34).[8] Landuse mapping in planned urban centres are easier in comparison with the old and complex urban areas.

Landuse varies according to the differential demand and requirements of people, over a piece of land. For this fact, the landuse pattern, as well as the amount of land utilized for a particular purpose, and often the density of development are constantly under change, according to the changing needs of community.

Urban Landuse Classification

There is no agreement on uniform classification or urban landuses, adopted by individuals and organizations in different parts of the world. This is primarily due to the differential landuses in urban areas and difficulties in mapping various uses. It also varies due to differences in purpose for which they are mapped. There is no spatial relationship between landuses of the Indian urban areas and those of the western countries. The highly mixed landuse in Indian urban areas differs markedly from the usual segregation of landuses in Anglo-American cities (Breese, 1969, 58).[9] It appears to be the result of long history of growing compactness and continuing necessity of travelling between places of residence and that of employment.

There are four major categories of landuse in urban areas, e.g., agricultural, residential, commercial and industrial (Gallion and Eisur, 1965, 191).[10] Bartholomew (1955, 15)[11] distinguished two categories of urban landuse areas; developed and undeveloped (vacant). Then he divided the developed areas into (a) privately developed, such as residential, commercial, and industrial, and (b) publicly developed, such as streets, transport establishments, parks, playgrounds and public and semi-public property. In American cities, transport, communication, open public space and public buildings have been recognized as major landuse categories, besides others (Berry and Horton, 1970, 444-49).[12] Chapin (207) categorizes urban landuse into residential, retail business, transport and communication, industrial, wholesale and storage, public buildings and open spaces and vacant or non-urban use. There were six landuse categories in England in 1950's, e.g., residential, central areas (administrative and commercial) industrial, educational, open land, large establishments, hospitals and cantonments (Housing and Local Government, 1952, 37).[13] Collins (1965, 217-20)[14] has classified the categories of land in Britain as residential, open space, public buildings, and institutions, industry, commerce, statutory undertakings, vacant and derelict buildings while Conzen (1966, 56-78)[15] classified the urban landuse of Britain into the following categories :

1. Shops, business and professional premises with or without residence.
2. Premises with public or community functions.
3. Public transport premises.
4. Industrial premises including Builder's yards and warehouses.
5. Residential premises.
6. Open spaces.

In America, public buildings are public properties, while in Britain, public buildings are not essentially publicly owned. Even a privately owned house may be called a public building if it is used by the Government for various official purposes. A landuse map therefore, is not a map of land ownership or land occupancy (Jackson, 1968, 113).[16]

In India, landuse categories recognized by different scholars which belong to two different types, e.g., land planning and urban geographers. Town planners quite often categorises urban landuse as residential, commercial, industrial, transport, communication, public utilities, open space, agricultural, vacant land and waterbodies (Gowda, 1972, 86-87).[17] On the other hand, there are minor differences amongst the urban geographers and most of them classified the urban landuse into the following categories: residential, agricultural, open space, military land, parks and playgrounds, commercial, administrative, educational, industrial and burial grounds (Singh, 1964),[18] (Sinha 1970),[19] (Alam 1972,[20] Ram and Thakur 1972[21]).

The lack of uniformity in landuse categories is due to the absence of detailed data for various landuse categories, non-uniform distribution of land under different categories and difference in the character of individual cities and experience of the investigator. This difference is lesser in major categories while it is higher in minor categories such as religious, recreational, medical, etc.

In India, based on landuse classification used by different planning organizations and urban geographers J.P. Singh (1976)[22] prepared the following list of landuse categories for the city of Shillong Municipal area in North Eastern India.

Residential (R)	:	Boarding houses, clubs, welfare houses, residential quarters and hostels.
Commercial (C)	:	Hotels and eating places, stores, petrol pump and service stations, auto-repair, shops, transport depots, big commercial agencies, cinema and banks.
Industrial (I)	:	Factories, saw mills and auto-body builders and iron and steel fabricators.
Agricultural (Ag)	:	Vegetable and fruit gardens and farms producing food grains and cash crops.
Educational (E)	:	Schools, colleges, training centres, universities and libraries.
Medical (M)	:	Public dispensaries, hospitals, family planning centres, health and social centres and veterinary hospitals.
Administrative (A)	:	Government offices, post-offices, police stations and fire brigade.
Defence (D)	:	Cantonment, various defence camps, police lines and rifle range.
Forest (F)	:	Protected, reserved and other forests.
Places of workshops (W)	:	Churches, temples, mosques and Sunday schools.
Open space (OP)	:	Gardens, Parks, Race course, golf links, sports ground and vacant lands.
Grave yards (Gr)	:	Burial ground.

Ecological Zones of Urban Areas

Many scholars like Burgess, Ullman, Hurd, Hoyt, Feiry, McKenzie and Colby have presented their concept of urban landuse models to express functional zones of the city and to generalize the arrangement of landuse regions within a city. These models have been affected by the physical, economic, political and cultural factors of the urban areas in shaping the functional structure of cities over decades. These variables are interlinked with the process of feed-back mechanism in the repetitive functional zones in all urban centres. Four well known models of urban landuse developments are as follows :

Concepts of Urban Ecology

Nels Anderson's Concept

According to Anderson, "Human ecology is the study of man in his community life adjusting to the habitat."

Cuber's Concept

According to Cuber "ecology is the study of symbolic relationship and the resulting spatial patterning of human beings and human institutions in the community."

Concentric Zone Theory

A hypothetical model of urban functional growth was developed by E.W. Burgess,[23] a Chicago sociologist in 1926. This model suggests that any urban area tends to expand readily from its central commercial core to the outer periphery, so as to form a series of concentric circular zones which are as follows :

(*a*) *The Central Business District :* It has been considered as the focus of commercial, social, civic life and transportation. This is the commercial hub with its down town retail districts, stores, shops, office buildings, clubs, banks, hotels, theatres, museums and organizational headquarters. This zone of retail business is encircled by wholesaling.

(*b*) *The Zone in Transition :* It is known as the area of residential deterioration where business and light manufacturing encroach upon residential areas. Here slum prevails with their scattered pockets of poverty, degradation, diseases and under worlds of vice. In American, as well as in Indian cities it has been observed that recent immigrants inhabited in these regions due to lower house rent and easy access to the business centre, have created the zone of transition in urban areas.

(*c*) *The Zone of Independent Working Men's Houses :* In this zone mainly industrial workers live in American cities, who have a desire to live within easy access of their work.

(*d*) *The Zone of Better Residence* : This is known as a restricted area of high-class buildings, where mainly single family resides.

(*e*) *The Commuters Zone* : In the areas of rural urban fringe, the spotty development of high-class residences along lines of travel in known as commuters zone.

In these, the internal structure of urban landuse has been noted for each inner zone which tenders to extend its area by the invasion of the next outer zone with the process of second zone of transition. According to Burgess this dynamic process of city structure shifts, sorts and relocated individuals and groups by residences and occupation. This creates cultural and economic differentiation among various groups of people and landuse of the urban centre.

Sector Theory

The model of sector or wedge is called the Hurd-Hoy hypothesis after R.M. Hurd and Homer Hoyt (1939).[24] Axial development of urban areas and functional zones is proposed here along lines of transportation routes to form a star shaped city. In this theory along a particular axis of transportation, similar type of functional zone will follow. It is an improvement over concentric zone theory since both distance and direction from the city centre are taken into consideration. In this idea functional zone and population patterns tend to develop in the same manner, outward from the centre of the sectors and growing in a circular form especially towards the rural-urban fringe.

Multiple Nuclei Theory

This theory is propounded by Mckenzie (1933)[25] and developed by two geographers—Harris and Ullman (1945)[26] who considered that functional zones of a city has essentially cellular structure. In this idea, the functional zones are not built around a single centre but around severe discrete nuclei. This concept is particularly helpful in explaining the structure of Tanzanian and Indian cities which are marked by well defined periods of growth and often contain an old and a new colony.

The Colby Hypothesis

Charles Colby (1933)[27] hypothesis offers a dynamic explanation of urban development. In this concept the pattern of functional zones in any city at any given time are the result of the forces, i.e., centripetal and centrifugal. Centripetal forces are of two types :

(a) Residents and business communities rush towards the city attracted by the service amenities, employment opportunities and other economic advantages in an urban area.

(b) Within urban areas itself, both people and business on the fringe are drawn towards the CBD because the central point provides the least access to either customers or labourers or other institutions such as stores, theatres and libraries. On the other hand, centrifugal forces derive people and business out of the CBD into peripheral areas. This is due to the flight from congestion and longing desire for open space.

These theories of urban functional zones are not mutually exclusive. The concentric zone theory depends upon the concept of centripetal and centrifugal forces as envisaged by Colby, while the sector theory and Multiple Nuclei are modifications of the concentric zone theory. In Hoyt's concept, residential area expand outwards concentrically and Harris and Ullman's theory (Multiple Nuclei) is basically the same as that of Burgess, so far as Nucleus is concerned.

The socio-economic life of the city[28] is not affected by the same ecological concepts in all the towns. One can find government servants living mostly in areas where there are government offices and official buildings. In some cases the dominant factor is cantonments, e.g., in Meerut, Amballa, Chhattisgarh, etc. whereas in other places it may be air-force as in Delhi, Palam Air Port. Some places may be industrially centred—where schools, colleges, universities, courts, clubs, restaurants, theatres, hotels, wine centres and factories may be located. In industrial cities slums may be found because large number of labourers have to live in the city. It happens because the government is not able to give accommodation to all. On the other hand, even rich people live at a distance away from the din-and-bustle of the city life. They live in big palatial houses and grand apartments. The area around central business district are densely populated. The city also accommodate red-light areas, gambling places, wine den, toddy house, besides clubs, night schools, cinema houses and cremation ground. On the other hand, in small towns these vice of dens are found deep rooted. The labourers especially with who return to their homes after the whole day's work. In small towns and villages the social life is stable and controlled whereas in big cities the human life is full of tension, hurry and conflicts apart from great incidence of crime. There are strikes, ghearaos, bands' etc. but inspite of all these the cultural life in cities is rich and variegated in nature.

Factors Affecting Urban Ecology

The ecological elements like trade, transport, industries, cultural institutions and physical conditions of the urban centres are guiding force for deciding the patterns of urban ecology.

Some of them are as follows :

(a) Geographical conditions;
(b) Means of transport and communication;
(c) Business or industries; and
(d) Economic, social and cultural institutions.

We can analyze them as :

Geographical Conditions

The living condition of urban people is influenced by the topography of land and climatic conditions of the city. The people try to adjust themselves with the environment around them. Cities on the sea-coasts are densely populated whereas inside the continent the cities have relatively low density of population especially around mining and industrial centres. The hill stations such as Srinagar, Nainital, Darjeeling and Kumaon are silent in winter season whereas they become sufficiently busy during the summer season with incoming of tourists. In big cities such as Calcutta, Mumbai and London each and every food items are available throughout the year whereas this facility is not available in village environment.

In urban areas the living conditions of human being is also affected by geographical conditions. For example in Shillong people use wooden houses due to its cheapness and scarcity of burnt brick and cement as well. In Japan the wooden house is essential in order to have safety from earthquake which comes daily with the incoming of Tsunamis waves on the sea coast.

The morality in urban life has been relaxed a bit due to less social control. This favours in the indulgence of sex acts freely. In spite of these, the political, social and cultural life in big cities are varied in nature which makes the men dynamic and culturally advance.

Means of Transport and Communication

Transport and communication act as life line for urban ecological make up. The lack of transport favours low population density, trade and commerce on the lowest ebb. The outlook is narrow. The social life is slow, uneventful, traditional and orthodoxy dominate the society. On the other hand, the towns having better transport facilities are thickly populated, trade and commerce flourishes, social and cultural life is dynamic; the economic, artistic, religious and other activities are rich. The free movement, better transport connections and the facilities of higher education and employment makes the people prosperous and broader in outlook.

Business and Industry

These occupations add specific personality to the life of the city. The industrialization brings more money, men and material to the city, thereby creating the housing apartment of labourers along with the growth of slums and squatter settlements in certain parts of the city. Besides these big hotels, restaurants, parks, play-grounds, brothels and gambling dens are found along with juvenile delinquency. In this way, the industrial city is slightly different from the cities of trade and commerce.

Economic, Social and Cultural Institutions

Economic : Industrial, transport, commercial, agriculture, trade, commerce, storage, provision of tax, fishing, hunting and ranching.

Social : Political organizations, administrative make up, social customs, levels of education, migration of people and marriage.

Cultural : Sports, clubs, cinema, threatre, dress material, food habit, levels of crimes, religions places and institutions, health centres, water and electricity supplies and festivals.

These factors have definite impact on economic and cultural life of the communities. Besides these, the recent development of science and technology specially in the field of mass-media have immensely helped in the prosperity of city life and culture.

Structural Change in Urban Ecology

Understanding social differentiation and variability in cities has been and continues to be a major theme for those interested in urban systems. Though the identification of group-specific processes has become fashionable in the past decade, the necessary for measuring social dimensions and identifying spatial patterns of variation for the entire city or urban area remain important tasks in the research agenda.[29]

A body of theory exists about the ecological structure of American cities, but we know little about the nature of structural change. Whether we resort to theories of societal scale and functional differentiation, a macro systems approach, or to theories of residential location and household choice, a micro systems approach, our knowledge about ecological change is quite primitive. Clearly, uncovering and understanding structural change in the ecology of cities represents one of the more challenging research domains in urban systems research.

If the theoretical aspects of functional differentiation or societal scale possess substantive content, then one would expect increasing factorial separation across societies at one point in time and within a given society over time. The former aspect has received some attention via cross cultural studies, but relatively little attention has been devoted to tracking a given urban system over time in an ecological framework. Therefore, whether factorial separation increases over time and what is the substantive content of the resulting configurations are largely unresolved empirical issues.

Our approach involves defining a large and comprehensive, but uniform set of census data for Detroit for 1960, 1970 and 1980. These data sets are factor analysis for each time period, the derived structure are interpreted for each time period, and comparisons are made which indicate the nature and extent of structural change. It is via the comparisons of individual ecological structure that processes of social change are uncovered and verified, that concepts associated with theoretical aspects of societal scale are given empirical content, and the nature of urban system change is better understood.

Dalmianagar: A Case Study of Urban Landuse Pattern

The development of Dalmianagar as an urban centre in Rohtas district of Bihar is due to the connection of Grandchord line, Dehri-Rohtas light railway line and the development of number of industries, e.g., paper, sugar, cement, hydrogenated oil, heavy chemical, vulcanized fibre plant, asbestos, cement and soap.[30] In comparison with this, the landuse in rural sector is mainly devoted to agriculture, residential house site, etc.

In this study, the landuse of Dalmianagar is mainly divided into four classes, e.g., commercial, industrial, residential and agricultural. So far as the development of urban landuse is concerned it seems that there are two centres of business establishments location, i.e., one new market near railway junction and another old market located just on the bank of western Sone Canal, which suggests that Dalmianagar is a twin town. The industrial development has taken place on the northern side of railway junction. The development of commercial ribbons on the road side suggest that only the concept of sector theory is applicable in case of Dalmianagar.

FOOTNOTES

1. C. Rajgopalan, "Bombay : A Study in Urban Demography and Ecology," *Sociological Bulletin*, 1960, Vol. 9, pp. 19-38.
2. A.B. Chatterjee, "Ecological Structure of Calcutta Twin City," *The National Geographical Journal of India*, 1965, Vol. 11, pp. 59-62.
3. S.M. Alam, *Hyderabad-Secunderabad Twin Cities : A Study in Urban Geography*, Allied Publishers Pvt. Ltd., Mumbai, 1965.
4. U. Singh, "Changes in the Built-up Area of Kaval Towns in the Ganga Plain (India)," *The National Geographical Journal of India*, 1966, vol. 12, pp. 103-117.
5. R.E. Dickinsion, *City Region and Regionalism*, Routledge and Kegan Paul, London, 1964, p. 135.
6. C. Abram, Urban Land Problems and Policies : Housing and Town and Country Planning, *United Nations Bulletin*, No. 7, 1953, p. 34.
7. F.S. Chapin, *Urban Landuse Planning*, Harper and Brothers, New York, 1957, xiv.
8. K.K. Dubey, "Urban Landuse : A Critical Review," *Indian Geographical Journal*, 1969, Vol. 46, pp. 26-34.
9. G. Breese, *Urbanization in Newly Developing Countries*, Prentice Hall of India, New Delhi, 1969, p. 58.
10. A.B. Gallion and S. Eisus, *The Urban Pattern : City Planning and Design*, Affiliated East-West Press, New Delhi, 1965, p. 191.
11. H. Bartholomew, *Landuse in American Cities*, Harvard University Press, Cambridge, Mass, 1955, p. 15.
12. B.J.L. Berry and Horton, F.E., *Geographic Perspective on Urban System*, Englewood Cliff, Hall, 1970, pp. 444-49.
13. Ministry of Housing and Local Government, *The Density of Residential Areas*, London, Her Majesty's Stationary Office, 1952.
14. M.P. Collins, "Fieldwork in Urban Areas," *Frontiers in Geographical Teaching* (Eds. Chorley and Haggett), Methuen & Co. Ltd., London, 1965, pp. 217-20.
15. Conzen, 1966, pp. 56-78.
16. J.N. Jackson, *Op. Cit.*, 1968, p. 113.
17. K.S.R. Gowda, *Urban and Regional Planning*, University of Mysore, Mysore, 1972, pp. 86-87.
18. R.L. Singh, *Bangalore : An Urban Survey*, Tara Publications, Varanasi, 1964.
19. B.N. Sinha, *Sirsi—An Urban Study in Application of Research Models*, Karnataka University, Dharwar, 1970.
20. S.M. Alam, *Metropolitan Hyderabad and Its Regions*, Asia Publishing House, London, 1972.
21. L.N. Ram and Thakur, B., *Op. Cit.*, 1972-73.

22. J.P. Singh, "Patterns of Landuse in Urban Areas : A Case Study of Shillong," *Indian Geographical Studies*, Bulletin No. 6, 1976, pp. 25-32.
23. E.W. Burgess, "The Growth of the City," in R.E. Part E.W. Burgess and R.D. McKenzie, *The City*, University of Chicago Press, Chicago, 1925, pp. 47-62.
24. H. Hoyt, *The Structure and Growth of Residential Neighbourhoods in American Cities*, Federal Housing Administration, Washington D.C., 1939.
25. R.D. McKenzie, *The Metropolitan Community*, McGraw Hill, New York, 1933.
26. C.D. Harris and Ullman, E.L., "The Nature of Cities," *The Annals of American Academy of Political and Social Science*, 1945, Vol. 242, pp. 7-17.
27. C.C. Colby, "Centrifugal and Centripetal Forces in Urban Geography," *Annals of the Association of American Geography*, 1933, Vol. 23, pp. 1-20.
28. H.J. Nelson, "The Form and Structure of Cities : Urban Growth Patterns," *Journal of Geography*, 1969, Vol. 58, pp. 198-207.
29. E.D. Perle, "Structural Change in Urban Ecology," *IGC Abstract of Geography*, New York, 1992, p. 498.
30. R.P. Singh, "Dalmianagar : A Study of the Industrial Complex," *Indian Geographical Studies*, Bulletin No. 2, 1974, pp. 26-43.

12

SATELLITE AND DORMITORY TOWNS

THE SUBURB of an urban centre where due to locational advantage the residential, industrial and educational centres are developed are known as "satellite or dormitory towns." It has a benefit of providing clean environment and spacious ground for residential and industrial expansion. Sometimes satellite towns are regarded as reserve of labour pool, poultry farms, dairy farms, nursery centres and grain collecting centres from rural areas but located closer to the town (Fig. 12.1).

Dormitory towns may be tourist places, recreational centres, health resorts, bathing places,[1] administrative centres and religious places of historical or archaeological importance.

The satellite and dormitory towns are prolific especially in American cities. The American census since 1910 have identified 140 metropolitan districts having satellite towns. The study of satellite urban settlements has been considered as essential elements of urban hierarchy for efficient administration, municipal services and safety. The satellite urban centre is considered as essential especially in fertile plain besides saving time in movement from housing colonies to the place of work in the market centre, industries and government offices (Fig. 12.2).

Generally in Satellite and dormitory towns either rich or poor people resides because due to cheaper land it is easy to find a housing site in satellite town in relation to the congested city centre.[2] In this regard, the satellite town is considered as the reserve of industrial labourers or the paradise of the urban poor as well as the rich businessmen in disguise.

Dormitory or satellite urban settlement is a secondary settlement which maintains a closer link with the main urban centre of which it is a part. Satellite town may be created temporarily to solve a particular purpose or it may be built permanently. Sometimes satellite town presents a look of twin-town such as Dehri and Dalmianagar in Rohtas district of Bihar. They may be connected with over bridge as Mokameh and Barauni, Patna and Hajipur, Varanasi and Moghalsarai across the Ganges river.

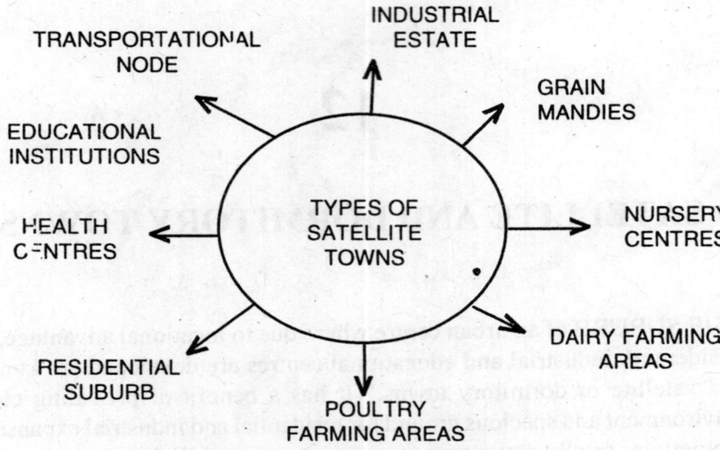

FIG. 12.1 : Types of Satellite Towns

The having off excess population into satellite and dormitory towns is not merely a process of historical interest, but this can be witnessed even today. This process is still a priority for solving the problems of over population and over concentration in metropolitan areas. These days most of the posh hotels of Delhi such as 'South City', the Heritage city' and 'the Lake View of Lucknow' and 'Devlok' of Mecruth are good examples of dormitory town. In some cases the process of de-urbanization may be the solution for the satellite and dormitory towns as this process could be observed in Australia, Neitherland and France.

Logic in Favour of Satellite and Dormitory Towns

Production and Distance Relations

The distance is an important element in relation to the movement of people and goods from the satellite towns to the metropolitan centres. Time spent in such a travelling could be put to other uses, which may provide increased production for the industrial goods in the neighbourhood areas (Fig. 12.3).

Satellite Towns in India

In the State of Bihar Lauthaha near Motihari, Jamalpur near Munger, Nathnagar near Bhagalpur, Danapur, Fatwah and Phulwarisharif near Patna, Sonepur near Hajipur, Laherisarai near Darbhanga, Mango near

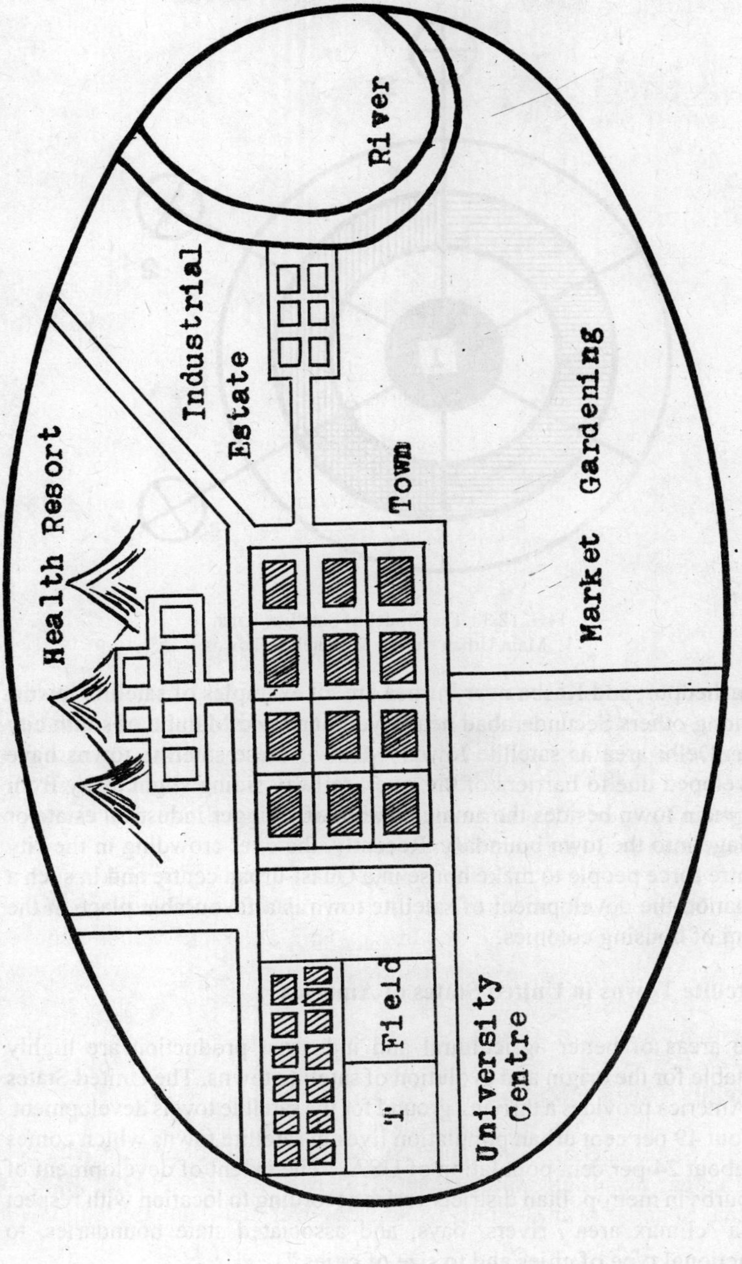

FIG. 12.2 : Model Plan of a Satellite Town

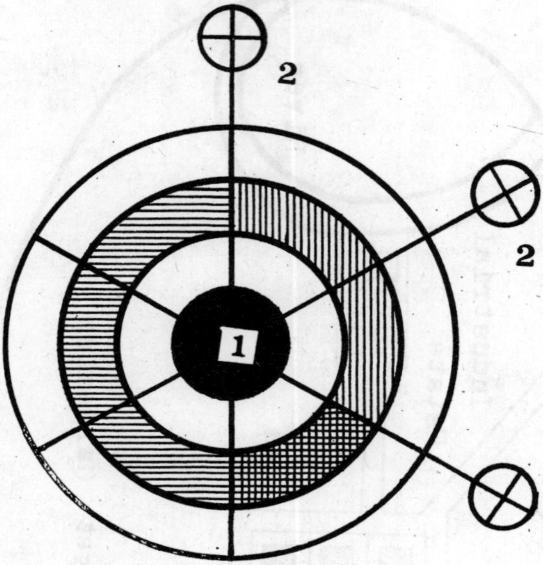

FIG. 12.3 : The Model of Satellite Town
1. Main Urban Centre, 2. Satellite Towns

Jamshedpur, and Kasba near Purnea are all examples of satellite towns. Among others Secunderabad near Hyderabad and Malubu of south city near Delhi area as satellite towns. Most of these satellite towns have developed due to barriers of the river, railway siding slight away from the main town besides the amalgamation of a bigger industrial estate or village into the town boundary. Recently, the over crowding in the city centre force people to make house in a Quasi-urban centre and in such a situation the development of satellite town is a favourable place in the form of housing colonies.

Satellite Towns in United States of America

The areas of better agricultural and industrial production are highly suitable for the origin and evolution of satellite towns. The United States of America provides a thriving ground for the satellite towns development. About 49 per cent urban population lives in satellite towns which comes to about 24 per cent population of USA. "The extent of development of suburbs in metropolitan districts varies according to location with respect to a 'climax area', rivers, bays, and associated state boundaries, to functional type of cities and to size of cities."

TABLE 12.1 : Percentage of People Living in Satellite Towns of USA

Population size group	*Satellite/Dormitory Percentage of Population*
10,00,000 Mere	50.8
5,00,000 to 10,00,000	65.1
3,00,000 to 5,00,000	55.2
2,00,000 to 3,00,000	33.2
1,50,000 to 2,00,000	51.0
1,00,000 to 1,50,000	42.2
50,000 to 1,00,000	38.8
Average of All sizes	49.2

New Towns Established as Satellite Towns to Metropolitan Cities

In order to decongest the metropolitan cities and diffuse the industrial and commercial activities located in a metropolitan city in the region surrounding the new towns have been established in the post-war period all over the world. In India such cities have been called satellite towns or ring towns and have been conceived as self-contained new towns. These are usually within 30-45 kms. from the main city and well connected with the same. The advantage of living within a commuting distance from the metropolitan city and availing all the facilities in the main city and at the same-time not being subjected to the congestion, traffic bottlenecks, high costs of housing and other services is an attractive proposition to those who live in suburban or satellite town.

The success of such new towns attracting effectively both migrant population as well as those who live in the metropolitan cities have been assessed in the last two decades. There have been some successful examples and also some failures. In the present context, where unlimited growth of metropolitan cities is causing serious concern, measures have to be taken to decentralize this growth over a larger region and decongesting the metropolitan city. Indian experience in regard to establishing such satellite towns will be valuable and they would also bring forth the environmental hazards and one may encounter in the establishment of new towns as satellites. Kalyani conceived as a satellite to Calcutta[3] which presents a case of both success and failure, success as a new town and failure as a satellite; even this failure as a satellite can be considered temporary as the necessary external inputs for making Kalyani a successful satellite has not materialized so far. With the completion of the road bridge across the river Hooghly at Kalyani, and the completion of Barrackpur Kalyani expressway a new wave of growth may be anticipated in Kalyani attracting industries and population to this satellite town.

Kalyani was established soon after the partition of India on the northern fringe of Calcutta Metropolitan District where an abandoned U.S. army base camp known as Rooswelt city had been located. The site enjoyed good climate being far away from the coast, was close to Kachrapara railway station, 50 kms. north of Calcutta connected by a road called today Barrackpur Trunk Road and also by boats plying on the Hooghly river. A number of institutions had been established at the site earlier and land which had been acquired during war extending over 4,000 ha was available.

Kalyani considered as a garden city was planned initially for a population of two lakhs. It was expected to attract some industries and activities from Calcutta where a large influx of refugee population was causing problems. Investments on water supply and drainage system were planned and were entrusted to the government departments. Subsequently in order to add to the attraction to Kalyani, a Kalyani University was also established. In spite of promotional measures undertaken by the Government of West Bengal, and the creation of the Kalyani Development Authority, growth of Kalyani proceeded very slowly and the few industries that came to be located were not the type to induce growth. The institutions that existed there earlier and even the newly established University did not offer sufficient attraction for industrial and commercial entrepreneurs and for a large population to move into the town. A major shortcoming was the very inadequate transport link between Calcutta and Kalyani although the distance itself was not much. The road link was not satisfactory and the rail system has no link across the river with the main line and to reach Kalyani one had to go down to Calcutta and come up again to Kalyani. The industries and industrial entrepreneurs would have welcomed to locate themselves at Kalyani had there been a rail link across the river at Bansberia connecting Kalyani to Bihar and other parts of India directly. The present rail connection was through Howrah, transhipment to Sealdah and from there by rail to Kachrapara and Kalyani. This obviously would not attract entrepreneurs and in fact in the 50s and 60s industries grew very rapidly on the west bank of Hooghly because of the rail facilities available and eastern bank of Hooghly only few industries are located.

A second reason which dampened the growth of Kalyani was that the large refugee population located in Calcutta did not have any capital and being mostly agriculturists. They depend upon manual work that could be found in the city of Calcutta and were engaged mostly in services. Kalyani did not offer scope for such employment. So *Bastis* and Slums[4] grow in leaps and bounds in Calcutta but very little population going towards Kalyani.

Hence, Kalyani grew not as a satellite of Calcutta but as moffusil town with a population of 40,000 in 1981 and 55,000 in 1991 instead of the 2 lakh population which it was originally planned for. The detailed examination of the growth of Kalyani which the Kalyani study has undertaken indicates the excellent infrastructure, institutional facilities and the low density residential area that is available at Kalyani with a large amount of unbuilt land available to entrepreneurs. The medical facilities at Kalyani is very sufficient with 2 big modern hospitals; so are educational facilities and vocational institutions. What it lacks today is the productive sector in terms of industries and commercial development providing employment and diversifying the economic base of the town. There are only 30 industries in Kalyani town today, some of them being Government undertakings and even these industries have posed some environmental problems. The impact of Kalyani on the surrounding region has not been significant as the town grew very slowly and became gradually merged into the general landscape of the area. The moderate urban character of Kalyani with its low density harmonizes with its surrounding agricultural land which is intensively cultivated with the help of borewells. Within the town, except for the industrial pollution emerging from the new chemical industry plants at Kalyani and at Kachrapara, no other pollution problem appears to exist.

Kalyani appears today poised for rapid growth if proper steps are taken for the purpose. The impending completion of the Kalyani bridge across the river Hooghli and the near completion of the Calcutta-Barrackpur express highway connecting Kalyani to Calcutta may provide a spring board of opportunities to rapid growth. It is perhaps at this juncture an assessment of the impact of its rapid growth on the surrounding area would help not only Kalyani but also lead to a programme of environmental protection and conservation all round.

The concept of Kalyani as a satellite town[5] to Calcutta was sound. The opportunities provided by availability of public land, basic infrastructure, a good location along the river was excellent. Had those elements that would have contributed to rapid growth of Kalyani been identified in the early 50s and the action taken to provide the rail and road links across the river, the picture of Kalyani and perhaps even

Calcutta would have been very different today. This gross neglect of opportunity contributed to the failure of Kalyani developing quickly and relieving Calcutta of all pressures. Today even if Kalyani should develop rapidly to a population of 2 lakhs or even 5 lakhs its impact on Calcutta would not be significant because the peripheries of Calcutta have grown tremendously in these two decades and Kalyani would not be able to compete with growth opportunities in the Calcutta Metropolitan area such as west Howrah, eastern Calcutta and areas to the south of salt lake. Calcutta seems to be growing southwards. Thus Kalyani presents an example of opportunity found and lost. The lessons to be learnt from the establishment of the township at Kalyani as a satellite could be summarized as :

(a) New Towns or satellite towns can certainly help to reduce pressures on the metropolitan cities and allow diffusion of activities and even function as sort of counter magnets to the city although at a lower key, provided the development of the new town. This has been conceived in a totally different manner and closely geared to the growth of the metropolitan city from the beginning.

(b) A new town as satellite requires to be developed faster than the rate at which the metropolitan city is growing and from the beginning, the new town should offer opportunities, basic facilities and amenities comparable to those obtained in the metropolitan city without the restrictions and difficulties imposed in the metropolitan cities. It is worth while to recall in this context that New Mumbai can never develop as a counter magnet to Greater Mumbai as it is being developed too slowly with too little facilities to be of much attraction let alone for industries, even for residence.

(c) The development concept of new town as a satellite has to keep in view not only the perspective for a future 15 or 20 years period but the growth needs of day-to-day type of activities which the city in its growing convulsions will experience. No day type of activities which the city in its growing convulsions will experience. No doubt there should be facilities for industrial growth such as developed land, transport and communications, power, credit institutions, markets, etc. But in addition there should also be facilities for employees to live, to recreate themselves and to be able to get their daily needs, to bring up their families, to educate their children and to get good medical assistance. Besides, the large construction force which is required in the building up of the new town has to be housed, and looked after in a permanent way, not in the camps or hutments leading to the development of slums and substandard development.

(d) The new town should provide for people who are migrating into it. They need rental accommodation of different types, they need jobs, and they need schooling and health facilities for the children. This should be available to the migrant from the beginning and will help the healthy growth of the new town. In this context, it is worth while to quote the example of Milton Keynes in England which advertises "For every worker coming into the Milton Keynes, there is a job and there is a house ready to occupy."

(e) In the conditions found in India and generally in the developing countries there is a large population casually employed or self-employed tending to the requirements of the regularly employed people. This service population partly formal and mostly informal needs to be provided for in the very beginning in terms of transport, etc. They should not become later an environmental liability living in slums, insanitation and catering to all sorts of people in unhygienic ways. It will also prevent unauthorized structures coming up on the road sides and at road inter-sections and on the peripheries creating a commercial chaos.

(f) The new town development concept would have to include the provisions for basic infrastructure and amenities not only for the regular population but for all the population and in advance so that the availability of these facilities become an attraction for the population in the metropolitan cities to move into the new town. The success of the English new towns have largely been dependent upon the better environment and better opportunities for work and greater choice of employment which the new towns offer to the population seeking to live in large cities.

(g) The availability of good communications system between the new town and the metropolitan city and other communities supported by a quick and efficient transport system amongst them substantially ensures the success of the satellite towns.

(h) The success of new town development very largely depends upon the mobilization of the private sector and its capital to participate in town development. Public investments can only be in the very initial period and mostly for infrastructure but should be quickly brought on a self-financing status and a balance has to be struck between private investments and public investments. Where this has succeeded as in the case of many new towns in England, the new town concept has also been successful.

The evaluation of the concept of new town development described as above and its implementation with continuous monitoring requires a dynamic management machinery, which has the necessary vision and at the same time can command the confidence among entrepreneurs and further demonstrate that it can deliver the goods. No new town in India founded as a satellite has been conceived on this basis nor implemented, and hence the failure both functionally as well as environmentally is to be ascribed not to the idea of new town as a satellite but to the manner of executing such a project, as in case of Kalyani.

FOOTNOTES

1. H. Thurston, "France Finds a New Holiday Coast," *Geographical Magazine*, Vol. 41, 1969.
2. R. Stead, "The New Villages in Malaya," *Geographical Magazine*, Vol. 27, 1955.
3. D. Miluer, "Calcutta : A City in Despair," *Geographical Magazine*, Vol. 41, 1968.
4. B. Gussman, "Shanty Towns in Southern Africa," *Geographical Magazine*, Vol. 25, 1952.
5. Ministry of Urban Development Report, 1988-89, New Delhi.

13

RANK-SIZE RULE

THE RANK-SIZE rule is one of the method of analyzing total settlement networks in a region. Hence, it is a tool for analyzing the national settlement system and help in the description and interpretation of the relationship between rank and population size of urban places.

The distribution of urban places according to their rank of population size, plotted on a double logarithmic graph paper (log *P*/log *R*) which gives a *S*-shaped curvilinear trend is known as the rank-size rule. But a straight line rank-size curve for settlements in many countries and regions of the world[1] suggests this as normal condition.[2] This empirical relationship was first presented by Auerback (1913)[3] in a study of German cities. Later on Lotka[4] observed that the urban concentration indicated by the cities of the United States fell in the same pattern.

The rank-size rule was first of all put scientifically forward by Zipf (1941)[5] as a theoretical model to express the relationship between observed and empirical regularity in the size of settlement hierarchy either urban or rural. This orderly development of urban places according to their rank and size is the basis of this relationship. However, the region in which this rule is applied must be a complete and self-sufficient region/country, i.e., not part of a larger region or greatly over-lapping with another region.[6]

Under this rule, the number of urban centres should continue to increase as their population size decreases.[7] Obviously one should expect fewer city level centres and more small towns than major urban centres. According to Haggett this relationship may not be universally true because small towns are not everywhere more numerous than major urban centres.[8] Probably, this relationship will result solely from the balance of local variability and the organizational needs for a system of urban centres under a given set of demand and supply conditions.[9]

The rank-size regularity as expressed by Zipf is as follows :

$$Pr = \frac{P_1}{rq}$$

where q is an exponent approximates to unity. This suggests that if the population of the largest city (P_1) is divided by the any city in the same region, the result will approximately be the population of the city (Pr) whose rank number is used as a divisor.

Further, the rank-size rule was generalized by introducing a constant as the exponent for the rank and the formula is as follows :[10]

$$Pr = ar^{-b}$$

$$\log Pr = \log a - b \log r$$

where rq is the rank of a city, Pr its population and a and b are constants. a can be interpreted as the expected population of the largest city and b as the slope of the linear relationship between $\log r$ and $\log Pr$.

According to Berry and Garrison (1958) rank-size regularity is a trivial geographical problem and it concerns with the systems of urban places. So far as the operating units of the system is concerned, urban transport workers must live near other urban dwellers, sugar mill workers near the factory and so on. There is a vast web of economic structural relations due to discontinuities in technological scale. The ordering of urban places will reflect that technical and social structure requires contiguity and thus affect the size of urban units by the lowering of number of people. At this juncture Von Foerster[11] has suggested a measure of entropy analysis which is very close to the rank-size rule and judging the primacy :

$$Rr = 1 - \frac{H_1}{Hr}$$

where Rr is the value of entropy or randomness, 1 is the constant, H_1 is the population of first ranking city and Hr is the population of the city in question.

In case of the primate city if the divisor is the population of the same, in such a situation the entropy shall be minimum, but Hr has no structural properties because it will always vary according to the urban centre in question, while H_1 measures all the structural relations occurring and requiring contiguity in the system. Clearly, if no structure exists, Hr and H_1 are equal and Rr is zero. On the other hand, if the system develops into a single tightly clustered web of activities around the primate city, H_1 would become zero and Rr would equal unity indicating complete order.

This empirical evidence has been questioned heavily by Rosing (1966)[12] and others who said that the rank-size rule is not very geographical because it is little concerned with the spatial analysis of urban places. While studying the empirical mathematical rules regarding the distributional equilibrium of population Stewart (1947)[13] has accepted the validity of rank-size regularity for the clusters of population. Similarly, the theory of agricultural location by Vön Thunen (1828),[14] Weber's theory of industrial location (1909),[15] Jefferson's Law of primate city (1939),[16] the Central Place Theory of Christaller (1966)[17] and Lösch (1954),[18] and Staller's allometric law of biology[19] are also lie in the same continuity and highly influenced by the development of this regularity.

Christaller and Lösch have put forward a functional size-class hierarchy of towns as a basis for the analysis of their spatial arrangement, functions and size. The basic assumptions are that a rural area supports a town which in turn serves the rural area, with small towns serving small areas and large towns serving large area. Most of the theoretical and experimental works in central place studies revolves around assumptions of a fixed number of constant *K* (Christaller) and non-fixed *K* (Lösch).[20] Berry and Garrison (1958)[21] have however, presented location theories of urban places as one of the several alternative schemes to the rank-size rule.

Factors Affecting Rank-Size Rule

The factors like administrative organization, transportation linkages, industrial development and commercial spread have played a multiple-role in changing the rank-size relationship of urban places in Tanzania. For example, the diamond mining and industrial centre of Mwadui exerts strong centripetal force attracting people to exploit natural resources. Similarly, all the ports such as Dar es Salaam, Tanga, Mtwara, Lindi and Zanzibar have strong forces of transport, commercial development, centre of break-o-bulk and favourable coastal cum administrative positions have highly helped these centres to obtain leading position in Tanzanian urban rank-size hierarchy.

The population change in a centre within a certain time interval changes the rank of a centre. The rank of any centre is dependent upon the hierarchical arrangement of the nodality value—which may be population size or an index based on different functions available at a place. The latter may depend upon the provision of administrative and transport facilities and the development of commerce and industry. The dissimilar and disproportionate development of these facilities causes variation in the centripetal forces involved in the growth of urban places. Further, with their greater force of attraction, the large centres have developed at greater distances, while the smaller towns and villages have a close proximity to each other.

This concept of close proximity has, however, changed today with the introduction of quick, mechanized transport, for example, people generally move from one village to another on foot which involves a certain amount of time. The smaller enjoys the facility of buses so that the inhabitants can also move from one centre to another more quickly. The bigger towns have train facility for travel between them. The development of airways has connected one city level centre to another. Hence, travel by the various above means though involving money, does not consume much more time than the movement from one village to another on foot. This shows the expansion in the zone of influence and spatial pattern of an urban centre based on transport facilities, which also control the rank and number of different order of urban places. The larger centres are more widely spaced and are, therefore, fewer in number while the small towns are larger in number and more closely spaced. This regularity has proved that as the nodality/centrality of places increases, the number of urban places decreased according to the rank-size rule (Fig. 13.1).

The Law of Primate City

A number of studies describe primacy of cities and levels of development of different countries. While investigating the urban systems of various countries, Berry (1961)[22] hypothesized that 'primacy is a feature of underdeveloped countries and a normal rank-size distribution is one of the developed world'. El-Shakhs (1974)[23] has also analyzed the primacy of urban systems for 75 countries of the world, relating primacy and development. It has been found that there is a curvilinear relationship between primacy and development. Even in Tanzania, it has been found that in the primary stages of urban development, primacy tends to be comparatively low, in the intermediate stage it attains its maximum development and in the last stage of development the level of primacy again reduces. This finding has been supported by the historic analysis of development of the urban systems of USA and Great Britain. According to Hagett (1979),[24] the countries with rank-size of urban systems tend to be large, have a long history of urbanization and are

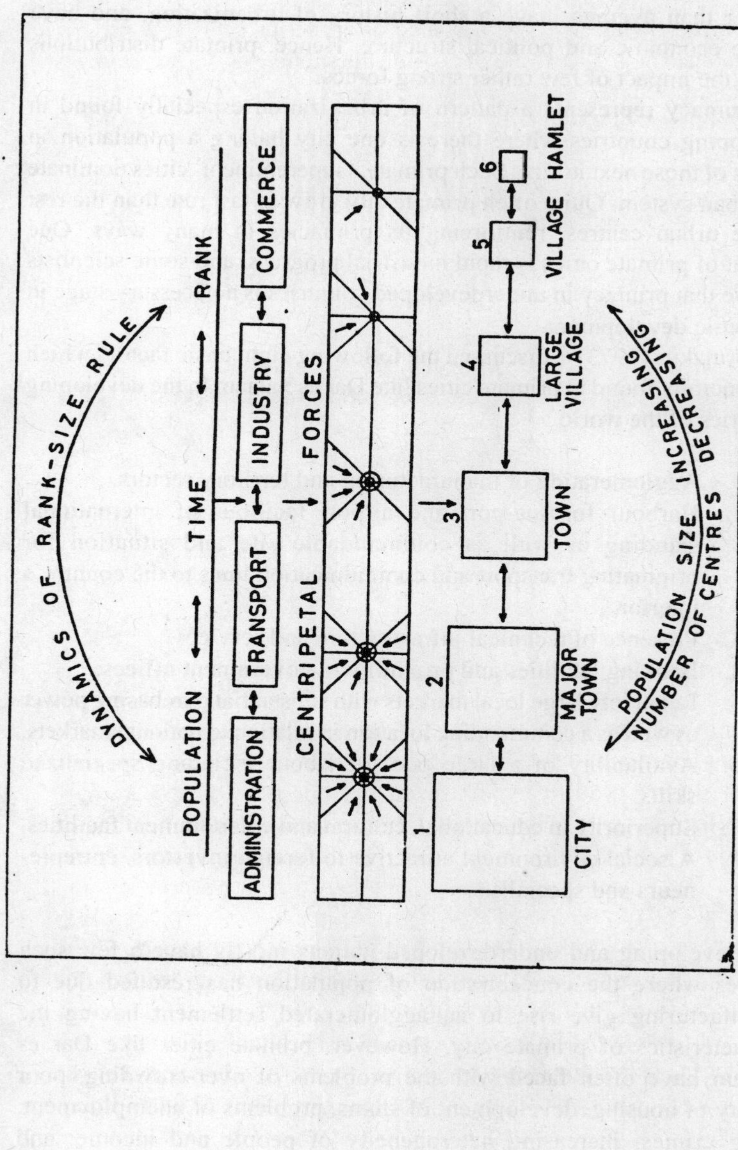

FIG. 13.1 : Model of Rank-Size Rule

economically and politically complex, "whereas primate distributions appear to be the product of urbanization processes in countries that are smaller than average, have a short history of urbanization, and have simple economic and political structure. Hence, primate distributions typify the impact of few rather strong forces."

Primacy represents a pattern of urbanization especially found in developing countries where there is one city having a population in excess of those next in size. Such primate, 'supereminent' cities dominate the urban system. Quite often primate city grow at fast rate than the rest of the urban centres, reinforcing its primacies in many ways. One benefit of primate cities is rapid industrial progress, and some scientists believe that primacy in underdeveloped countries is a necessary stage in economic development.

Brutzkus (1973)[25] discussed the following eight basic factors which are generally found in primate cities like Dar es Salaam in the developing countries of the world :

1. Agglomeration of manufacturing and territory sectors.
2. Harbour for sea-port and airport facilities of international standing as well as commandable site and situation for originating transport and communication lines to the country's interior.
3. Presence of technical infrastructure and services.
4. Banking facilities and proximity to government offices.
5. Relatively large local markets with substantial purchasing power as well as a commanding location in relation to national markets.
6. Availability of a large pool of labour including specialized skills.
7. Superiority in educational, cultural and entertainment facilities.
8. A social environment attractive to foreign investors, entrepreneurs and specialists.

Developing and underdeveloped nations mostly have a few such centres where the concentration of population has resulted due to manufacturing give rise to an agglomerated settlement having the characteristics of primate city. However, primate cities like Dar es Salaam have often faced with the problems of over-crowding, poor quality of housing, development of slums, problems of unemployment, rising crimes, increasing heterogeneity of people and income, and disparities of development. The primate city of Dar es Salaam on the edge of Tanzania is causing severe complications in areas of its hinterland where the distance from one place to another is already a factor impeding integration, and hence the rest of the country is out of touch in terms of development from the primate city of Dar es Salaam.

Like other primate cities it has been found that the percentage increase of population in Dar es Salaam is very high (Table 13.1).

TABLE 13.1 : Dar es Salaam : Percentage Increase of Population (1867-1991)

Year	*Total Population*	*% Increase*	*% Growth Rate*
1867	900	—	—
1887	5,000	455.55	22.77
1894	10,000	100.00	14.28
1900	20,000	100.00	16.66
1913	22,500	13.00	1.00
1921	24,600	9.33	1.16
1931	34,300	39.43	3.94
1948	69,277	101.98	6.00
1957	128,742	86.00	9.55
1967	272,821	111.91	11.19
1978	769,447	182.04	16.54
1983	1,120,000	46.85	9.37
1991	1,350,567	20.00	2.50

Source : Calculated by the author from *Atlas of Tanzania*, Dar es Salaam, 1968; The United Republic of Tanzania. 1978, Population Census, Vol. II, 1981, and 1991, Population from the speech of J. Nyerere.

Table 13.1 shows that the rate of population growth in the primate city of Dar es Salaam was high in 1867-1887, very low in 1900-1915 and 1921, but again high (16.54%) in 1967-1978. 1900-1913, and 1913-1921, were the decades when population growth were either negative or very low throughout the world due to war, famine, diseases and drought, but since then the population has increased slowly and perceptibly up to the present due to natural increase, improved health facilities and excessive migration from rural to urban areas. In this way, Dar es Salaam is leading the destiny of Tanzania among the East African cities in all dimensions as an outstanding primate city.

Dynamic Radius of Rank-Size Rule

The ever changing conditions of rank-size of urban places in Tanzania both in spatio-temporal dimensions could be seen in the above Table 13.1. In 1948 the first ranking town Dar es Salaam had a population of

69,227 and the last 25th ranking Songea had a population of 612 only. These two first and last ranking towns show a gap of 68,615 persons whereas in 1978 the first ranking Dar es Salaam has a population of 749,445 against the last ranking 98th urban centre Manda having a population of 527. This has a total difference of 768, 918 persons. It indicates that the structural gap between the primate city of Dar es Salaam and the last ranking town of Manda has increased in 1978 in comparison with 1948 situations in terms of population concentration, economic development and transportation linkages due to excessive pull of the above attributes in the environs of Dar es Salaam in respect to other parts of Tanzania. But the change in spatio-temporal dimensions and increasing number of urban centres, is a clearcut sign of progress and prosperity in future if this growth has been channelized in a systematic, scientific and planned way, instead of the present haphazard and lopsided development.

The rank-size curve of urban centres for 1978 census is *S* shaped which is the sign of primacy and a more rational distribution of urban size class of towns in Tanzania in comparison with 1948 census. In the curve of 1978 several steps are found. Harris (1970)[26] in his work on the cities of the former Soviet Union, suggested that "steps in a rank-size curve may indicate separate sub-systems of cities or relatively independent regions with little functional integration." But in case of Tanzania poor transportation linkages, functional independence due to rough terrain conditions and socio-economic backgrounds are some of the major causes of the availability of different steps on rank-size curve.

Rationality of Rank-size Rule

The relationship between actual population (1978) and estimated population of urban centres in relation to the first ranking city of Dar es Salaam in Tanzania, it has been found that second, third, fourth and 91st and above ranking urban centres have very low actual population in comparison with the Primate city of Dar es Salaam in terms of rank-size rule.

Urban centres having a rank of 7th to 90th have also lesser number of actual population according to rank-size rule, but here the discrepancy is low in comparison with Tanga, Zanzibar, Mwanza besides 91st and above ranking small towns (Table 13.1).

The placing of all the points for actual and estimated population of urban centres above line of actual rank-size distribution show a sign of underdeveloped economy (Fig. 13.2). In case of a developed economy all these points should either run along the line or below in case of over developed. This represents the paucity of service facilities, poor transportation linkages with the area of umland, unbalanced regional development and the availability of meagre income and low productive people in most of the urban centres of Tanzania in relation to the Primate city of Dar es Salaam.

Size Class Distribution of Urban Centres

The analysis of rank-size rule of urban places involves the problem of hierarchical ordering, the distributional pattern of the centres of various sizes and the causes of such distribution (Fig. 13.2).

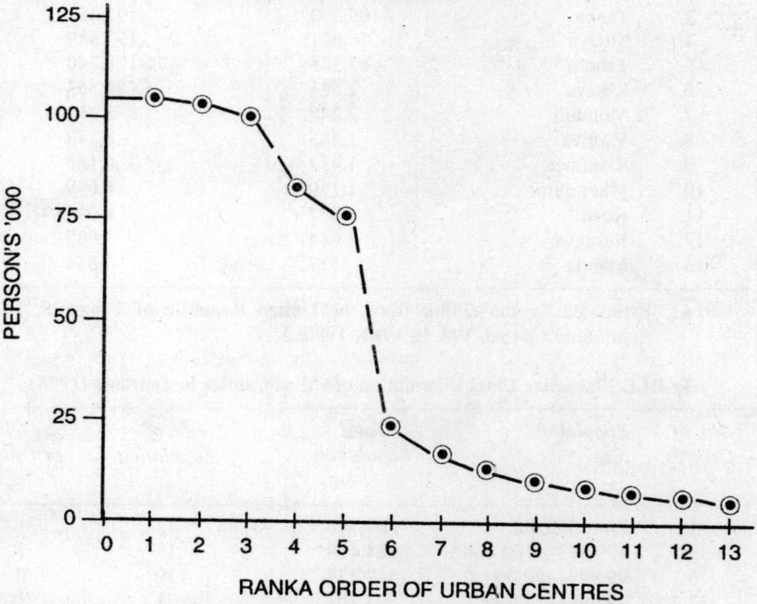

FIG. 13.2 : Rank-Size Rule of Urban Centres in Tanzania

So far as the size class is concerned Dar es Salaam accounts for 32 per cent of the country's urban population. In general, the urban pattern in Tanzania is most diffused. Four Class I cities namely Dar es Salaam, Mwanza, Tanga and Zanzibar have 45.35 per cent of the country's urban population and the balance is found in dispersed urban places of medium and smaller sizes. In Tanzania 63 per cent urban places are of smaller population size, i.e., below 10,000 persons in 1978 census (Table 13.2).

Note : 100,000 or more persons—Class I
50,000 to 99,999 persons—Class II
10,000 to 49,999 persons—Class III
Below 10,000 persons—Class IV

TABLE 13.2 : Margin of Actual and Expected Population in Relation to Primate City as some Major and Smaller Urban Places, 1978

Bank order	*Urban Centre*	*Actual Population (O)*	*Expected Population (E)*
1.	Mwanza	110,553	384,722
2.	Zanzibar	110,506	256,481
3.	Tanga	103,399	192,361
4.	Mbeya	76,601	153,889
5.	Tabora	67,388	128,240
6.	Kibaya	2,621	84,553
7.	Monduli	2,248	8,363
8.	Vwawa	1,585	8,273
9.	Kisarawe	1,434	8,185
10.	Mkokotoni	1,170	8,099
11.	Koani	832	8,015
12.	Karagwe	644	7,932
13.	Manda	527	7,851

Source : Estimated by the author from the United Republic of Tanzania, 1978, *Population Census*, Vol. II, 1981, Table 2.

TABLE 13.3 : Size Class Distribution of Urban Centres in Tanzania (1978)

No. of Centres	*Population Size*	*Total Population of Urban*	*% of Population*	*Size Class of Towns*
1	Over 500,000	769,445	32	I
3	499,999—100,000*	324,458	13	I
6	99,999—50,000	419,279	19	II
27	49,999—10,000	533,701	21	III
61	Below 10,000	365,917	15	IV
	Total	2,412,800	100%	

Source : Calculated by the author from 1978 *Population Census of the United Republic of Tanzania*, Vol. II, 1981.

Note : * More than 100,000 persons—City
50,000 to 99,999 persons—Major Town
10,000 to 49,999 persons—Intermediate Town
Below 10,000 persons—Small Town.

In Tanzania urban places serve as the nodes of transportation, regional development and inducement of administrative efficiency since colonial era of Dutch, Arab and German rules (Fig. 13.2). Although, the evolution of town sites took place a thousand years ago,[27] but increased size is truly the sign of recent urban development. Before the 20th century, most of the towns were confined to the eastern sea-shore only, but in colonial times towns have been developed in interior areas also, for example, Tabora, Mpwapwa and Ujiji. The last one has been the outcome of Arab traders in East Africa. These towns were established in interior areas during colonial rule to spread the Islam religion and the dissemination of Kiswahili language.

Rank-size on the Basis of Metropolis in India

1981 census has recorded 12 metropolitan cities in India. They are Calcutta (91.65), Mumbai (82.27), Nagpur (12.97), Lucknow (10.06) and Jaipur (10.04), besides seven others (Table 13.4). Fig. 13.3 shows that in the metropolises of India the rank-size rule cannot be applied as the metropolis of Mumbai and Calcutta both act as primate cities. Both of these are industrial, commercial and transport centres according to rank-size rule as the population of Calcutta 91.66 lakh than the second rank metropolis Mumbai should have a population of about 46 lakhs but in reality it has a population of 82.27 lakh. This shows that they are the growth poles of separate physiographic units and their economic, political, physical and social developments are quite different from each other.

TABLE 13.4 : Rank-size Rule of Metropolis in India, 1981

	Metropolis	*Population*	*Rank*	*Estimated Population from Calcutta*	*Estimated Population from Jaipur*
1.	Calcutta	91.66	I	91.66	120.48
2.	Mumbai	82.02	II	45.83	110.44
3.	Delhi	52.27	III	30.85	100.40
4.	Chennai	42.76	IV	22.91	90.36
5.	Bangalore	29.13	V	18.33	80.32
6.	Hyderabad	25.65	VI	15.27	70.28
7.	Ahmedabad	25.15	VII	13.09	60.24
8.	Kanpur	16.85	VIII	11.45	50.20
9.	Poona	16.85	IX	10.18	40.16
10.	Nagpur	12.98	X	9.16	30.12
11.	Lucknow	10.06	XI	8.33	20.08
12.	Jaipur	10.04	XII	7.63	10.04

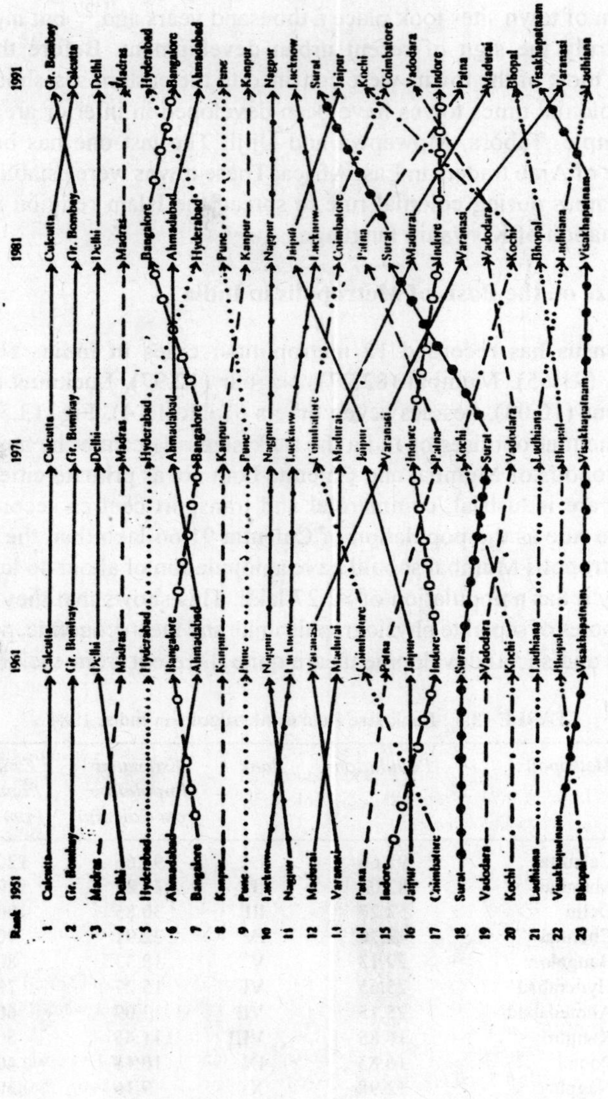

FIG. 13.3 : Metropolitan Cities of the 1991 Census Ranked by Population 1951-1991

The estimated population of different rank metropolis from Calcutta gives lower population in comparison with actual population. This gives us an ideas that the growth of Calcutta is slightly lower due to its bad siting condition in comparison with other metropolis of India. The result is gain ratified by the estimated population from the lowest rank metropolis of Jaipur. This shows that most of the metropolis development is far short in comparison with their actual population.

Ultimately it can be said that the rank-size rule does not fully operate in India and Tanzania. Berry and Garrison (1958) remarked that rank-size rule is certainly not the answer to all city-size problems.[28] The reasons are several including inadequate development of trade, transportation lines, industries, establishment of port, administrative centres besides landscape features as carriers and barriers of urban development. The rule would, however be applicable if the urban centres develop on isotropic plain surface without artificial incentives to chosen centres, or limitations to the growth of other centres.

FOOTNOTES

1. F.W. Boal and Johnson, D.B., "The Rank-Size Curve : A Diagnostic Tool," *The Professional Geographer*, Vol. XVII, No. 5, 1965, p. 106.
2. B.J.L. Berry and Garrison, W.L., "Alternative Explanations of Urban Rank-Size Relationship," *Annals of the Association of American Geographers*, Vol. 48, 1958, p. 83.
3. F. Auerback, "Das Gests Der Bevolkungs Konzedtration," *Petermanns Geographic che Mittilungen*, Vol. 59, Part-I, pp. 74-76.
4. J. Clarks, *Population Geography*, Pergaman Press, Oxford, 1966, p. 52.
5. G.K. Zipf, *National Unity and Disunity*, Principia Press, Blookington, 1941.
6. G.T.J. Steward, "The Size and Spacing of Cities," *Readings in Urban Geography*, (ed. H.M. Mayer and Clyde F. Kohn), Allahabad, 1967, p. 241.
7. R.B. Mandal, *Introduction to Rural Settlements*, Concept Publishing Company, New Delhi, 1979, p. 125.
8. P. Haggett, *Locational Analysis in Human Geography*, Edward Arnold, London, 1965, p. 106.
9. B.J.L. Berry, *Geography of Market Centres and Retail Distribution*, Prentice Hall, 1967, p. 78.
10. R. Henkel, "The Urban Systems of Kenya, Tanzania, Zambia and Zimbabwe," Paper presented to South Africa Conference of the Commonwealth Geographical Bureau, Lusaka, June, 1982.
11. H. Vön Foerster, "On Self-Organizing System and their Environments in Self-Organizing System," (ed. M.O. Yo Vitts and S. Gameron), Pergaman Press, New York, 1960.
12. K.E. Rosing, "A Rejection of the Zipf Model (Rank-Size Rule) in Relation to City Size," *The Professional Geographer*, Vol. 18, 1966, pp. 75-82.

13. J. Stewart, "Empirical Mathematical Rules Concerning the Distributional Equilibrium of Population," *Geographical Review*, Vol. 37, 1947, pp. 461-85.
14. J.H. Vön Thunen, Der Isoleirte Staat, Tr. C.M. Wartenberg and (Ed. P. Hall), Pergamon Press, New York, 1966, quoted by B. Thakur, *Urban Settlements in Eastern India*, Concept Publishing Company, New Delhi, 1980.
15. A. Weber, *Theory of the Location of Industries*, University of Chicago Press, 1929.
16. M. Jefferson, "The Law of Primate City," *Geographical Review*, 1939, pp. 226-32.
17. W. Christaller, *Central Places in Southern Germany*, translated by C.W. Baskin, Englwood Cliffs, Prentice Hall, 1966.
18. A. Lösch, *The Economics of Location*, translated by W. Stolper, New Haven, Yale University Press, 1954.
19. B.J.L. Berry, "Cities System within Systems of Cities," Papers of the *Regional Science Association*, Vol. 13, 1964, pp. 147-64.
20. L.J. King, *Statistical Analysis in Geography*, Englwood Cliffs, Prentice Hall, 1959, p. 130.
21. B.J.L. Berry and Garrison, W.L., *Op. Cit.*, 1958.
22. B.J.L. Berry, "City Size Distributions and Economic Development," *Economic Development and Cultural Change*, Vol. 9, 1961, pp. 573-587.
23. S. El-Shakhs, "Development, Primacy and Systems of Cities," *Journal of Developing Areas*, Vol. 7, 1972, pp. 11-36.
24. P. Haggett, *Geography : A Modern Synthesis*, 3rd Edition, New York, Harper and Row, 1979.
25. E. Brutzkus, "Centralized Versus Decentralized Pattern of Urbanization in Developing Countries," Tijdschrift Voor Econi. In *Soc. Geografie*, Vol. 64, Vol. 1, 1973, pp. 11-23.
26. C.D. Harris, *Cities of the Soviet Union : Studies in Their Functions, Size, Density and Growth*, Chicago, Rand Mcnally, 1970.
27. R. Coupland, *East Africa and Invadors*, Oxford, 1965.
28. N.B.K. Reddy, "A Comparative Study of the Urban Rank-Size Relationship in the Krishna and Godavari deltas and South Indian States," *The National Geographical Journal of India*, Vol. XV, Part II, 1969, pp. 63-90.

14

DEVELOPMENT OF CONURBATIONS

CONURBATION is a specific type of geographical region. It has occupied a dominant position in the patterns of urban development. Conurbation has recently attracted the attention of geographers and hence there is a utter lack of literature on this topic. The growth of conurbation has attained a significant form due to the industrial and technological development. Both the physical and population growth have taken place to such an extent that urban centres coalesce with each other like brick and mortar contiguity, and thereby forming a conurbation.[1]

The aim of this chapter is to assess the viability of the concept of conurbation in context of some of the metropolitan cities of India. At such centres due to the location of several industries the expansion of cities boundary and the coalesce of one urban centre with the other is a slow but continuous process of urbanization and regional development. How far the evolution of conurbation helped in regional economic development as Fawcett has observed in cities of U.K., is to be examined.

This chapter throws light on the concept of conurbation by definition, characteristic features and evolving a criteria of identification. A brief, discussion about Calcutta conurbation has been examined along with assessing its impact on economic and regional development.

The Concept of Conurbation

The word conurbation originated from two words—'Continuous' and 'urban area'. According to R.L. Dwivedi "conurbation means continuous urban development over a considerable area." First of all the word conurbation has been used in 1915 by P. Geddes. He explained the continuous urban area of more than two urban centres which may have a separate territorial units. Conurbation highlights the characteristics of neighbouring towns as a city region, agglomerated town, constellation of towns, millionaire cities, besides urban agglomeration, urban aggregate or complex, urbanized area, metropolitan area and aggregates of local authority areas. According to C.B. Fawcett,[2] "A conurbation is an area occupied by a continuous series of dwellings, factories and other buildings including, harbours, docks, urban parks and playing fields etc. which are not separated from each other by the rural land; though in many cases in this country such an area includes exclusion of rural land which is still in agricultural pursuits. "R.E. Dickinson has called it 'urban tract' followed by Jean Gottamman who called it "extended city" or "Super Metropolitan Region."[3] According to J.C. Saoyne "conurbation is an area of urban development where a number of separate towns have grown into each other and become linked by such factors as common industrial or business interest for a common centre of shopping and education." Conurbation is known as 'Stadt region' in Germany, and 'Standard Metropolitan Area' in U.S.A. and 'Standard Urban Area' in India.

This type of metropolitan expansion occurring in the industrially developed countries have favoured the extreme cases of urbanization. Conurbation is an asymptotic form of growth of certain sections of the town. Nowadays, these are the nucleus of regional, political, and economic development in India.

The term 'conurbation' applied to a group of cities that have coalesced and grow together economically while politically remain independent, as in the Rural wabs where Dortmund Essen-Duisburg and their suburbs constitute as urbanized cluster of over 5,000,000 persons. Conurbation is also known as an 'urban nuclei' which has been more strictly defined as a continuously built-up areas in which there is no apparent gap between several towns which have coalesced. Conurbation considers the 'brick and mortar continuity' of the urban centres.

In any region when the clusters of urban nuclei in many areas reached toward one another, and numerous cities converged into larger masses in the form of megalopolitan concentrations that is known as 'conurbation'. The conurbation consists of one or more central cities with their contiguous urbanized areas.

The term 'conurbation' has been first used in the earlier part of 20th century by 'Geddes' to refer to 'city-regions' or 'town-aggregates'.[4] According to Geddes the urban built-up area embracing several continuous administrative districts was called conurbation such as Greater London, Paris, Tokyo, New York, Liverpool, Greater Mumbai, Calcutta and Shanghai.

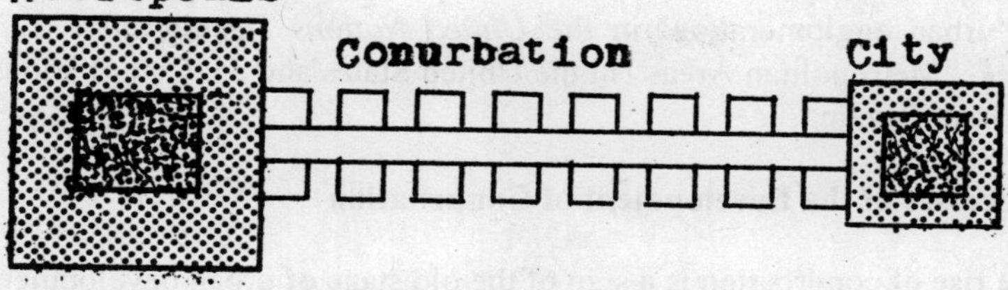

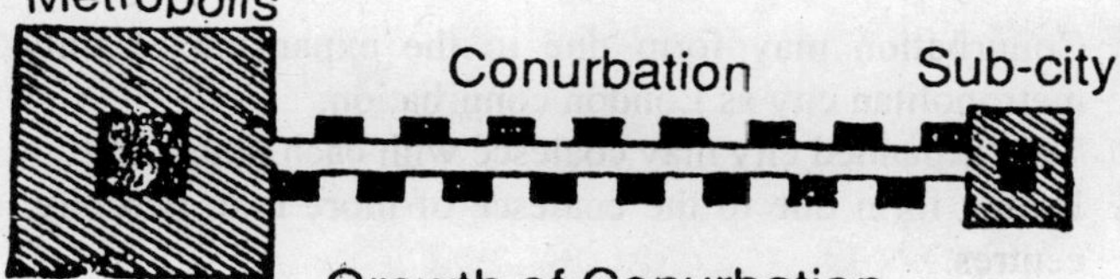

FIG. 14.1 : Development and Growth of Conurbation
'Conurbation means continuous urban development over a considerble area'—R.L. Dwivedi

According to Fawcett "A conurbation is an area occupied by a continuous series of dwellings, factories, harbour, docks, urban parks and playing fields, etc., which are not separated from each other by rural land though in many cases in India such an urban area includes enclosures of rural land which is still in agricultural occupations. Further, Fawcett studied about conurbations of Britain based on 1921 and 1931 censuses respectively and published a book in 1935.[5] In 1951 British census defined conurbation as "aggregates of local authority areas." In 1971 census in India the term "urban agglomeration" is used in order to indicate the town group but except few metropolitan centres of India other 'urban agglomerations' are not liable to be called 'conurbations'.

Flourishing cities and towns quickly expand beyond their formal limits, which seldom move far enough to keep pace with the growth of natural community. A recent historical development associated with industrialization has been the rapid massing of population in a few giant urban centres or super cities, which have engulfed the whole neighbouring towns and over-run local government boundaries as in case of Delhi metropolitan expansion in India. This type of urban complex described as 'urban agglomeration' in the *United Nations Demographic Year Book*, "Metropolitan Areas" in the United States and as 'Conurbations' in Great Britain.

Processes of the Development of Conurbation

The rise of conurbation is a sign of the old stage of urban development. The urban centres may have a poor contact with the neighbouring towns in the initial and youthful stage of its growth and development but in the old stage due to the development of transport, industry and trade-conurbation may be formed due to the following reasons :

(a) Conurbation may form due to the expansion of only one metropolitan city as London conurbation;
(b) Two expanded city may coalesce with each other; and
(c) It may form due to the coalesce of more than two city level centres.

Conurbations in Other Parts of the Globe

In the north-eastern seaboard of United States of America right from Washington in the south to Boston in the north, there developed a well marked conurbational trends which leads to the development of 'megalopolis' over a distance of 960 sq.km. No other continuous chains of metropolitan areas or conurbations, shaping now in other parts of the world, is indeed comparable in size as yet to the American 'megalopolis'.[4] The conurbation of New York is located on either side of the Hudson river in the States of New York, New Jersey and Connecticut and presents a good example from the United States of America. St. Paul and Minneapolis developed into a binuclear conurbation. Although they are politically and geographically separated but there is a wider contact zone, especially in sectors of transport, industry and urban development. Similar form of binuclear conurbation has been formed by Canbera and Queanbeyan of Australia although the 11 kilometres long link constitutes arterial life line with two way flow between the two.[5]

In Canada—Waterloo and Kitchner conurbation is also an example of twin cities where one cannot easily recognize that there are two separate civic administrations.[6] The other conurbation is formed by

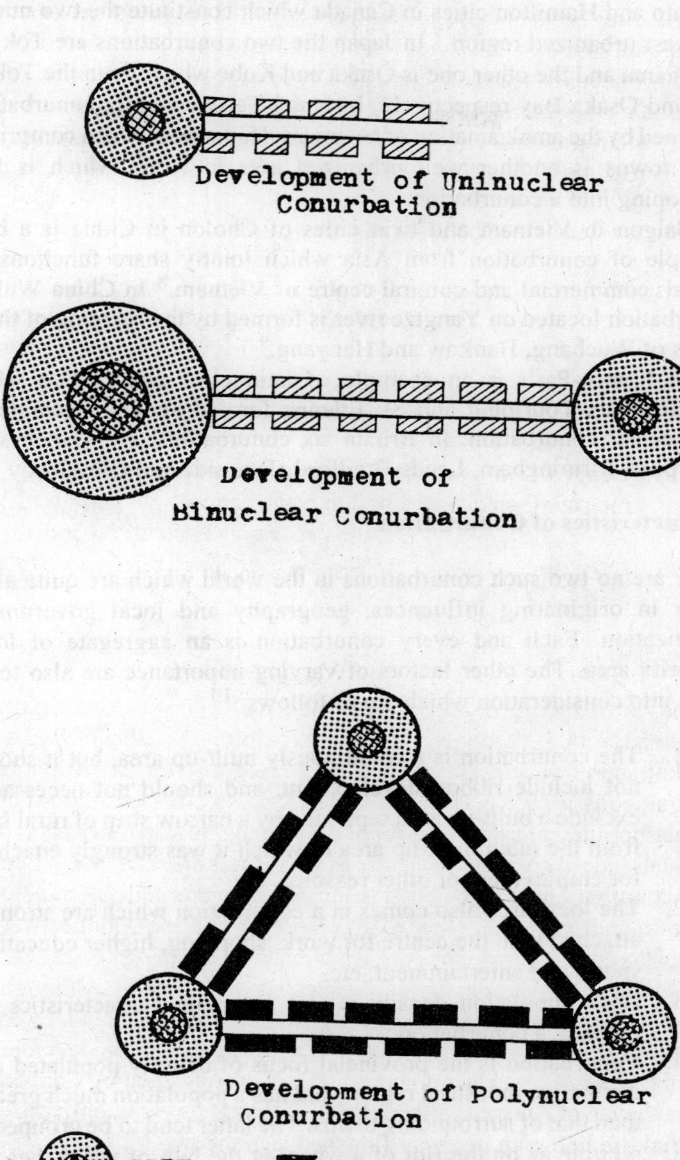

FIG. 14.2 : Theoretical Aspects of the Development of Conurbation

Toronto and Hamilton cities in Canada which constitute the two nuclei of a vast urbanized region.[7] In Japan the two conurbations are Tokyo-Yakohama and the other one is Osaka and Kobe which lie in the Tokyo Bay and Osaka Bay respectively. In Israel-Tel-Aviv-Jaffa conurbation is formed by the amalgamation of ten towns. Haifa-Acre which comprised eight towns is another well urbanized area in Israel which is fast developing into a conurbation.

Saigon in Vietnam and twin cities of Cholon in China is a best example of conurbation from Asia which jointly share functions of capital, commercial and cultural centre of Vietnam.[8] In China Wuhan conurbation located on Yangtze river is formed by the merging of three towns of Wuchang, Hankow and Hanyang.[9]

In France-Paris is an example of uninuclear conurbation while Lille-Roubaix-Tourloing and St. Etienne formed another example of polynuclear conurbation. In Britain six conurbations are Manchester, Liverpool, Birmingham, Leeds-Bradford, Tyneside and Glasgow.

Characteristics of Conurbation

There are no two such conurbations in the world which are quite alike either in originating influences, geography and local government organization. Each and every conurbation is an aggregate of local authority area. The other factors of varying importance are also to be taken into consideration which are as follows :[10]

1. The conurbation is a continuously built-up area, but it should not include ribbon development, and should not necessarily exclude a built-up area separated by a narrow strip of rural land from the main built-up area to which it was strongly attached, for employment or other reasons.
2. The local area also comes in a conurbation which are strongly attached with the centre for work, shopping, higher education, sports and entertainment, etc.
3. High population density is also the main characteristics for forming a conurbation.
4. Conurbation is the provincial focus of densely populated and distinctive industrial region, and has a population much greater than that of surrounding towns. The latter tend to be grouped in a circle as on the rim of a wheel at the hub of which lies the metropolitan centre.
5. Due to location, site and marketing organization each conurbation effected the world wide distribution of the goods

of basic industries of its surrounding area; and the collection and distribution of vital supplies. A metropolis, however, is not necessarily an independent but adequately dependent upon transport facilities as is clearly available in case of the surrounding area of Ahmedabad, Kanpur and Calcutta.

6. Each conurbation possesses a great number and variety of miscellaneous industries which owe their origin to the demands of regional market and their localization in the conurbation to its reserves of labour and excellent transport facilities. These are the regional secondary industries, due to the distribution of their products throughout the surrounding area and its business transactions are by no means restricted to its own industrial production.
7. Conurbations are endowed with a varying degree of financial individuality, as they have served in the past as regional centres of banking and insurance.
8. By virtue of cheap and frequent travelling facilities conurbation serves as an outstanding shopping centre for the surrounding hinterland. The maximum extent of the shopping area is roughly coincided with the area which can be easily travelled within two hours from the shopping centre of the conurbations.

Methodology

Conurbations in India has been determined here after taking the composite index of density of population per square kilometre (*DP*), per cent of built-up area (*BA*), per cent variation of population (*VP*), number of towns in the urban agglomerations (*UA*) and the number of working factories (*WF*) in some of the principal cities of India.

Thus the conurbation can be determined on the basis of the attainment of at least average number of value. This can be determined with the help of a formula :

$$C = \frac{\sum_{i=1-n}^{n} C_{DP} + BA + VP + UA + WF + M_{DP+BA+VP+UA+WF+\cdots+n}}{\sum N}$$

where C is conurbation; C, M are Calcutta, Chennai, etc.; and N is number of cities taken into the study.

In this way, the city above average value of composite index considered here as conurbation. Moreover, such cities are fulfilling all the characteristics of a conurbation as recognized by Fawcett, Geddes and others.

In case of eight principal cities in India total composite index for the five factors as mentioned above comes to 74,093 and the average value is just 9,260. Hence, only Calcutta, Mumbai and Chennai comes under the category of conurbation in India. The total composite index for Delhi, Kanpur, Bangalore, Hyderabad and Ahmedabad are very low. hence these centres are merely metropolitan expansion and the same never bear the requisite traits of a conurbation.

So far as per square kilometre density of population is concerned the highest density of 30,276 persons is recorded in Calcutta conurbation and the declining trend of the same is found in Chennai and Mumbai which have a density of 19,293 and 9,901 respectively. In percentage of built-up area the highest per cent of land is engaged in Bangalore (95) but it has also better position in Calcutta, Mumbai and Chennai which have also very high built up area come to about 79, 96 and 90 per cent respectively.

The variations in the land utilization of built-up space is presented by residential, commercial, industrial, public utility, parks and railway property. In this regard it is clearly discernible that Calcutta, Mumbai and Chennai have 53 per cent, 34 per cent and 45 per cent of built-up land used for residential purposes. The land used by industrial establishments and also moderately high, e.g., 14.5 per cent in Mumbai, 9.22 per cent in Calcutta and 7.9 per cent in Chennai.

Above mentioned conurbations in India have not shown a higher growth rate as the recent trends is achieved by Delhi Metropolitan centre of 1,170 per cent growth in between 1931 to 1981 census. The biggest conurbation of Calcutta has a per cent variation of 268 in the same period.

Table 14.1 shows the decadeal growth of urban population in principal cities of India covering a period of 50 years from 1931 to 1981. Before 1951 census the growth of urban population may be regarded as slow but continuous growth of 6.47 per cent in Chennai in 1931, to 52. 53 per cent in 1951 in Calcutta. The census period beginning from 1951 to estimated 1981 show higher percentage growth of population, which is mainly due to the heavy rural-urban migration of industrial labourers and greater dependence on industrial activity due to mounting unemployment problem both in the rural and urban sectors of India. In the later stage of rapid growth Chennai shows a lower growth of 17.29 per cent where as Calcutta is heading 74.95 per cent growth in 1971-81.

In 1971 census a special category of urban agglomeration has been formed which help us in determining the zone of influence of different cities in India. In this sense, Calcutta conurbation encompasses 74 towns, whereas Chennai 58 but due to island location the conurbation of Mumbai has no satellite urban centre just in the nearby areas (Table 14.1).

TABLE 14.1 : Growth of Population in Indian Cities 1931-91

(*in lakhs*)

City	*1931*	*1941*	*1951*	*1961*	*1971*	*1981*	*1991*
Calcutta	24.85	40.54	52.53	65.75	70.31	91.60	108
Mumbai	13.03	16.95	28.39	41.52	59.70	82.00	129
Chennai	6.47	7.77	14.16	17.29	31.69	42.80	54
Delhi	4.47	6.96	14.37	23.44	36.47	52.30	92
Bangalore	3.11	4.11	77.86	12.07	16.53	29.10	27
Ahmedabad	3.14	6.07	8.28	12.06	17.41	25.70	29
Kanpur	2.44	4.87	7.08	9.71	12.75	16.90	20

In industrial development the highest point of 1220 is achieved by Calcutta, 1065 by Mumbai whereas in Chennai just 251 factories are found to be engaged in production. Delhi has better position in relation to industrial development but in other sectors Delhi lag behind Chennai in order to form a conurbation.

FOOTNOTES

1. P. Geddes, *Cities in Evolution*, revised edition, 1949.
2. C.B. Fawcett, "Distribution of the Urban Population in Great Britain 1931," *Geographical Journal*, Vol. 79, 1932, pp. 100-112.
3. J. Gottamman, "Megalopolis or Urbanization of the North-Eastern Sea Board; in *Readings in Urban Geography*, (ed. Mayer and Kohn, Chicago, 1959, pp. 46-56.
4. C.B. Fawcett, *Op. cit.*
5. R.B. Mandal, "The Development of Conurbations in India," *Urbanization and Regional Development* (Ed. R.B. Mandal and G.L. Peters), Concept Publishing Co., New Delhi, 1982.
6. J.P. Gibbs, *Urban Research Methods*, Von Nostrand East-West Press, Mumbai, 1966, p. 52.
7. R.L. Dwivedi, "The Concept of Conurbation : A Review," *The National Geographical Journal of India*, Vol. XV, Part-I, March, 1969, pp. 45-55.
8. J.P. Gibbs, *Urban Research Methods*, Von Nostrand East-West Press, Mumbai, 1966, p. 52.
9. U. Singh, "The Calcutta Conurbation," *National Geographical Journal of India*, Vol. Vi, Part-2, June, 1958, p. 95.
10. A.B. Chatterjee, "Calcutta Metropolitan District Changing Structure of some Towns," *Applied Geography* (Ed. R.L. Singh), National Geographical Society of India, Varanasi, 1968, pp. 74-78.

15

MUMBAI METROPOLITAN REGION

MUMBAI is the premier port city of western coast of India. It is a capital of the State of Maharashtra. It is situated on Salsette Island in the costal plain of Konkan. Mumbai is approachable through Thalghat and Bhorghat of Sahyadri mountain. It is also known as the Gateway of India.

Mumbai is also known as Mumbay in the name of Goddess of Fishermen. The administration of Mumbai handed over to the Sultan of Gujarat in 1368 from Hindu chiefs. It passed on the hands of Portuguese in 1534 and to British in 1661.[1]

In southern side lies the backwaters followed by Malabar Hill on the north-west which has a height of 60 metres from Mean Sea Level. The Marine Drive of Mumbai is very famous sea-beach in the world followed by multilane highway along the sea coast and multi-story buildings on the land. The fort area contains the railway junction of Victoria Terminus, the Gate Way of India besides banking facility and commercial establishments. In southern part the military base of the State of Maharashtra lies at Kolaba. Parel and Majagaon dock which are the important industrial sectors of Mumbai.

Mumbai and its Region

Mumbai has a deep rooted influence of economic and social in nature on its hinterland. Its hinterland includes rural, semi-urban and urban centres. The workers from all these areas commute daily to the conurbation. On western railway from Bassein and Berur on Central Railway from Dombivli and Kalya—the commuters come daily in huge number to the metropolis. This has increased the population of Dombivli more than five times within the last decade.[2]

Mumbai receives 65,000 tonnes of fishes daily from Otan, Bassein, Arnala, Satpathi, Uran and other sub-urban settlements—which have substantial fishing ground on the sea shore. Mumbai also produces a substantial amount of salt in its hinterland including total requirements of vegetables and milk. Are and Berli are dairy centres which supply milk to Mumbai.

Mumbai Metropolitan Region has an extension of Tansa River Basin in the north, Patal Ganga River in the south, vast expanse of ocean in the west and Bhatheran Mountains in the east. It includes the Tehsils of Thana, Bassein, Bhivandi, Ulhasnagar and Kalyan in Thana district; and the Tehsils of Uran, Kharagpur, Karjat and Ponvel in Raisarh district.[3]

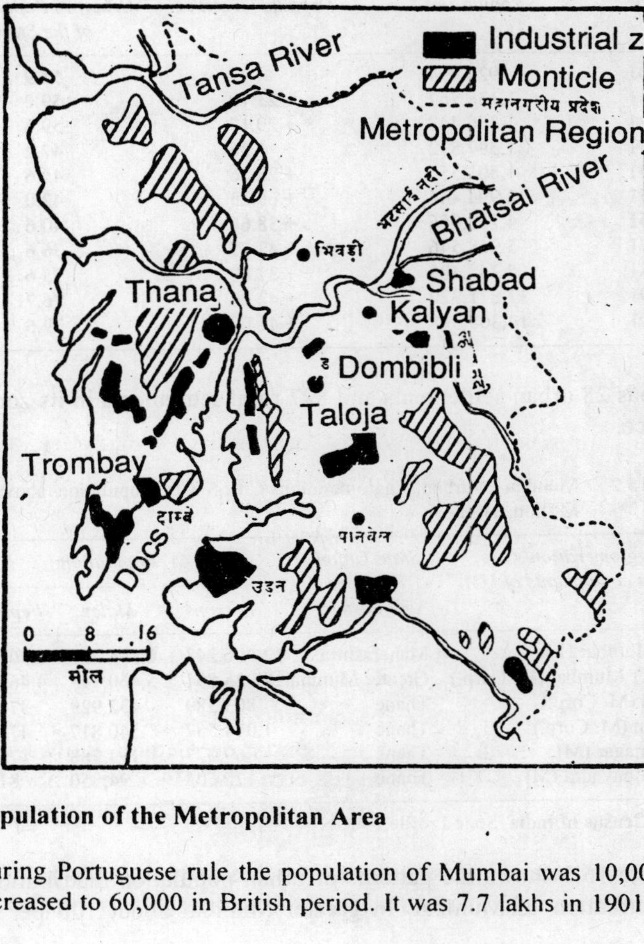

Population of the Metropolitan Area

During Portuguese rule the population of Mumbai was 10,000 and it increased to 60,000 in British period. It was 7.7 lakhs in 1901 and 128 lakhs in 1991, and has an area of 3361 km^2. It is a cosmopolitan and dynamic city with multi-religious and multi-lingual culture. Mumbai has institutions of regional, national and international level. The population growth was 2.37 per cent annually in 1911 which gone up to 7.67 per cent in 1991. During 1901-91 the growth of population in Mumbai is tremendous 1000 times which may be termed as population explosion in the real sense. In the year 2001 it may cross the 1025 times limit (Table 15.1).

TABLE 15.1 : Growth of Population in Greater Mumbai, 1901-1991

Year	*Population*	*Decadal Growth*	*Urban % Population of the State*
1901	927,994	—	54.9
1911	1,138,757	+ 23.79	59.0
1921	1,380,448	+ 20.17	59.6
1931	1,397,812	+ 1.26	47.2
1941	1,801,556	+ 28.36	45.6
1951	2,994,444	+ 66.23	42.0
1961	4,152,056	+ 38.66	40.6
1971	5,968,546	+ 43.75	36.6
1981	8,243,405	+ 37.80	75.6
1991	12,871,720	+ 42.00	76.7
2001	14,305,398	+ 41.00	79.6

It has 28 urban settlements and 897 rural settlements in its zone of influence.

TABLE 15.2 : Mumbai : Urban Agglomerations/Cities with Population above one Million, 1991

Urban Agglomeration/ City with main components of U.A.	*State/Union Territory and District*	*Population*		
		Persons	*Males*	*Females*
Greater Mumbai U.A.	Maharashtra	12,596,243	6,891,222	5,705,021
1. Greater Mumbai (M. Corp.)	Greater Mumbai	9,925,891	5,460,145	4,465,746
2. Thana (M. Corp.)	Thane	803,389	432,929	370,460
3. Kalyan (M. Corp.)	Thane	1,014,557	540,817	473,740
4. Ulhasnagar (M)	Thane	369,077	193,788	175,289
5. Mira Bhayadra (M)	Thane	175,605	94,550	81,055

Source : Census of India, State Profile,1991.

Table 15.3 shows the patterns of urban population distribution in the conurbation of Mumbai. In greater Mumbai about 100 per cent population is urban while in others rural population is also found. For example, in Karjat about 93 per cent population is rural and in Panvel only 20 per cent population is urban. In Ulhasnagar although spontaneous settlements have grown fast but here about 85 per cent population is urban followed by 75 per cent in Thana in 1991.

TABLE 15.3 : Percentage of Population in the Metropolitan Region, 1991

Taluka	*Urban % (of total)*
Thana	72.87
Bassiene	28.73
Bhibandi	57.55
Ulhasnagar	84.76
Kalyan	69.49
Khalapur	32.73
Karjat	16.54
Uran	24.23
Panvel	19.98
Greater Mumbai	100.00

Industrial Development

The industrial development of Mumbai includes cotton textile, engineering, wool, leather, chemical, petroleum, drugs, food, film and several other light industries. The location of fertilizer and atomic power installations speak of the bright future.

The heavy chemical industries of Ambarnath, match box and explosive factories of Badlapur, hydro-electricity, paper and chemical industries of Khopli, the mixed industries of Taloja and Rasabani have very wider impact on Mumbai and its hinterland.

Spread of Retail Function in Mumbai

With explosive population growth since independence and consequent urban sprawl, most Indian cities developed integuments that are characteristically Indian. One such feature of spontaneous development is informal pavement shopping and hawking. These shops deal in a wide-range of goods and cater to a variety of tastes particularly of the lower and middle income strata of the society. The spatial development also reveals a close affinity to the nodal junction alongwith the arterial roadways and points where the communting population converge and disperse on their way to work or home.[4]

Mumbai, the Prime urban centre in India, is definitely the commercial and financial capital of the nation. These factors led to Mumbai's rapid rise to primacy. The number of city grew as urban functions enforced one another, and this led to further accretion of more economic attributes. One important commercial dimension added to the commercial function was informal retailing, which at present contributes roughly 50 per cent of the retailing facilities to city dwellers. This new dimension creates intra-urban spatial differentiation and diversification to meet the dictate's of the urban and metropolitan environment.

This unique feature of the Indian metropolis with special reference to Mumbai has not been systematically studied by geographers either in terms of its spatial arrangement, its structure and service functions rendered or its relation to the hinterlands it serve.

The development of a subsidiary informal retail zone in the northern part of the city, in fact in the geographical centre of the island of Dadar, was sequal to the development of the early suburb. This response is possibly the first indication of the impact of the spread of population on the dormitory suburb beyond the industrial area, and the friction of distance being felt on the procurement of daily needs. The whole informal retail function at Dadar Junction stands on the convergence and transfer of a commuting population which caters to the lower and middle income population groups of the northern part of the city as well as the suburban population because of greater proximity. This is unlike planned Western cities where retailing is organized in a central shopping area with large enterprise department and chain stores.

However, in the process of retail suburbanization along the Central and Western Railway, there has been the formation of an informal retail zone along the commuting suburban rail heads that are located further away from the commercial core, and distinctly different from each other in character and composition. This clearly reflects the socio-economic status of the consuming population. Thus the growth of informal retail function is unplanned, sporadic and spontaneous under the pressure of growing urban population, and function and meet the growing needs for such retail services.

The investigation is mainly based on the primary data collected through interviews, fieldwork and sample studies. Discrete quantitative techniques and cartographic methods are applied for the analysis of data and mapping.

Mumbai Metropolitan Region and its Planning

In order to solve the problems of Mumbai and its environment the Government of Maharashtra has established "Mumbai Metropolitan

Region Planning Board" in 1971. This has helped in solving the problems of Mumbai and its fringe area development.

The metropolitan plan of Mumbai aims at removing the inequalities of services and ushering an era of the equal distribution of service facilities which may provide true urban scenario. The plan included :

1. Decentralization of economic and urban functions of the city;
2. Reducing the distances and number of commuters;
3. Establishment of residential and industrial centres at favourable site in the metropolis;
4. Reducing differences between the warring social groups;
5. Construction of new port;
6. A new Mumbai Township should be created near the harbour;
7. Change in functions of Greater Mumbai;
8. Development of other urban centres of the region;
9. Preparation of a definite policy of industrial location; and
10. Increasing the inter-city and intra-city transport facilities.

The plan of Mumbai has been prepared to accommodate 10,000,000 of population. Trans-Thana Chauk area will accommodate a new city. Kalyan, Kolset-Balkhum, Mandaur and Apte-Khurare have been going to be developed as industrial complexes. A bridge has been constructed between railway and road besides the transport network has been modified both inside and outside the city.

In order to lessen the population pressure from the old Mumbai—a new Mumbai has been going to be established in the narrow bay of Thana especially on the eastern coast of Trans river. From this new site one road goes to Kalba-Belapur and the other area is going to be developed as residential. The commercial belt will be developed in between Thana-Belapur and narrow bay. Residential areas have been divided into sectors. Each sector has been endowed with community facilities, green belt and open places. Nahebaseva has been developed as part.

Problems of Mumbai Conurbation

The metropolis if Mumbai is facing several problems, e.g., increasing pressure of population, extreme shortage of houses, lack of transport facilities, lack of service facilities, etc. The population of Mumbai was 82 lakh in 1981 and rose to 128 lakh in 1991. Most of the problems are faced due to this spectacular growth of population. The growth is more due to migration and less due to natural increase. This has created problems of unemployment, lack of space for industrial expansion besides the dearth of electricity and drinking water. About 20 per cent population is living in spontaneous settlements and one lakh forty-two thousand live on roadside hovels. All these have given birth to a variety of diseases, Juvenile delinquency and problems of insanitary hazardous environment, etc. Some specific problems are :

(1) A huge crowd of houses;
(2) Deterioration in the urban living conditions;
(3) Spread of work places along the road side only;
(4) Some areas have industrial concentration while others are completely devoid of it;
(5) Influx of migrants to a huge number;
(6) Increase of air pollution;
(7) Huge crowd on railway junction and aerodrome;
(8) Dearth of electricity and water supplies besides increasing municipal taxes for these;
(9) Lack of open spaces and recreational places;
(10) Increase of land prices continuously;
(11) On green spots on the side of roads the growth of squatters are problematic;
(12) Haphazard growth of industries on non-municipal land;
(13) Lack of cooperation and friendliness among families forced them to live in a secluded manner; and
(14) There is a huge difference in the social make up.

FOOTNOTES

1. S. Arunachalam, "Bombay City and Stages of Development," *Bombay Geographical Magazine*, Vol. 3, 1955, pp. 34-39.
2. C.D. Deshpande, "Cities and Towns of Bombay Province," Aspects of Urban Geography," *Indian Geographical Journal*, 1941, p. 16.
3. D. Singh, "Bombay : Study in Historical Geography," *National Geographical Journal of India*, 1960, Vol. 6, pp. 19-29.
4. D. Mukherji ad Rohine Karani, "Spread of the Informal Retail Function in Bombay," *IGC Abstract of Geography*, New York, 1992, p. 438.

16

CALCUTTA CONURBATION

CALCUTTA as an urban monster, is a city of hell, the worst siting condition amongst the cities of the world[1] and it is an urban area of misery, violence, dilapidation and juvenile delinquency to say a little. Even then Calcutta is a dream, which demands love as it is the heaven of labourers, pick pocketers, business cities and the scholars of repute. Calcutta is a fascinating city of millions of people, who have adopted a mosaic of languages, customs, variety of cultures and vibrating sensitiveness which automatically deserves love.

Calcutta as the capital of West Bengal, financial stronghold of India, nerve centre of regional economy, higher education and port location, it has been regarded as the most beautiful city of the East. The growth and decline of Calcutta conurbation may be explained by the political factors (the shifting of the capital to New Delhi in 1912), historical factors (the partition of Bengal in 1947 when millions of refugees came) and geographical factors (the agricultural development in the hinterland). Still the conurbation of Calcutta has supremacy over all other cities in Eastern India since times immemorial and it will remain.

TABLE 16.1 : Calcutta : Urban Agglomerations/Cities with Population above one Million, 1991

Urban Agglomeration/ City with main components of U.A.	*State/Union Territory and District*	*Population*		
		Persons	*Males*	*Females*
(1)	(2)	(3)	(4)	(5)
Calcutta U.A.	West Bengal	11,021,918	6,022,489	4,999,429
1. Calcutta (M.C.)	Calcutta	4,399,819	2,445,328	1,954,491
2. Haora (M.C.)	Haora	950,435	528,396	422,039
3. Bally (M)	Haora	184,474	106,341	78,133
4. Ulberia (M)	Haora	155,172	82,043	73,129

(*Contd.*)

TABLE 16.1 (*Contd.*)

(1)	(2)	(3)	(4)	(5)
5. Bhatpara (M)	North Twenty Four Parganas	304,952	176,104	128,848
6. Panhati (M)	North Twenty Four Parganas	275,990	146,338	130,905
7. Kamarhati (M)	North Twenty Four Parganas	266,889	146,338	120,551
8. South Dum Dum (M)	North Twenty Four Parganas	232,811	122,164	110,647
9. Baranagar (M)	North Twenty Four Parganas	224,821	120,134	104,687
10. North Dum Dum (M)	North Twenty Four Parganas	149,965	77,073	72,892
11. Barrackpur (M)	North Twenty Four Parganas	133,265	70,136	63,129
12. Naihati (M)	North Twenty Four Parganas	132,701	73,457	59,244
13. Titagarh (M)	North Twenty Four Parganas	114,085	67,121	46,964
14. Halisahar (M)	North Twenty Four Parganas	114,028	63,039	50,989
15. Barasat (M)	North Twenty Four Parganas	102,660	52,922	49,738
16. North Barrackpur (M)	North Twenty Four Parganas	100,606	53,166	47,440
17. Kanchrapara (M)	North Twenty Four Parganas	100,194	51,938	48,256
18. Bidhan Nagar (NA)	North Twenty Four Pargana	100,048	51,202	48,846
19. Khardaha (M)	North Twenty Four Pargana	88,358	47,343	41,015
20. Garulia (M)	North Twenty Four Parganas	80,918	43,128	37,790
21. New Barrackpur (M)	North Twenty Four Parganas	63,795	32,563	31,232
22. Dum Dum (M)	North Twenty Four Parganas	40,961	23,054	17,907
23. Barrackpur Cantonment	North Twenty Four Parganas	19,931	10,812	9,119
24. Budge Budge (M)	South Twenty Four Parganas	72,951	41,583	31,368
25. Hugli-Chinsurah (M)	Hugli	151,806	77,902	73,904
26. Serampore (M)	Hugli	137,028	75,398	61,630
27. Chandan Nagar (M.C.)	Hugli	120,378	63,388	56,990
28. Rishram (M)	Hugli	102,815	59,245	43,570
29. Uttarpara Kotrung (M)	Hugli	101,268	53,500	47,768
30. Champdani (M)	Hugli	101,067	57,709	43,358

(*Contd.*)

TABLE 16.1 (*Contd.*)

(1)	(2)	(3)	(4)	(5)
31. Bansberia (M)	Hugli	93,520	50,765	42,755
32. Baidyabati (M)	Hugli	90,081	47,288	42,793
33. Bhadreswar (M)	Hugli	72,474	40,633	31,841
34. Konnagar (M)	Hugli	62,200	33,237	28,963
35. Kalyani (N.A.)	Nadia	55,579	28,843	26,736
36. Gayespur (N.A.)	Nadia	52,158	26,598	25,565

Source : Census of India, State Profile, 1991.

The image of Calcutta emerged as the locus of life, place of palaces, mechanized life and a city of furious creative energy. Dr. Sinha described Calcutta as "to the poor and middle class citizens of Calcutta, the poverty and violence ridden city has gained the image of a massive uncoordinated human aggregate which has reached an irreversible State of autonomy."[2]

Genesis of the Metropolis

The city of Calcutta started its career about 290 years ago. It is situated 92 kilometres upstreams from the Bay of Bengal on the eastern bank of the Hooghly river. The existing Calcutta Corporation Area has an area of 104.0 sq.km. with more than 100 wards and 12 million population. The conurbation of Calcutta originated out of the three villages of Sutanuti, Kalikata and Gobindapur. In these villages the British had first of all settled and subsequently added new villages to the growing port-city. The Calcutta urban agglomeration is a post-independence addition thereby demarcating much wider territories to indicate a conurbation for the sake of urban planning (Fig. 16.1).

The Growth of Suburbs

The suburban towns of Calcutta has developed into Calcutta Urban Agglomeration. Chinsurah, Hooghly, Chandernagore and Serampore were thriving markets patrovized by the Portuguese, the Dutch and the French.

Transportation linkages make effective integration of suburbs and the metropolis into an organic whole. Good transportation helps in increasing the population of suburban residential areas and leads to all round development of the suburbs through opening of new streets,

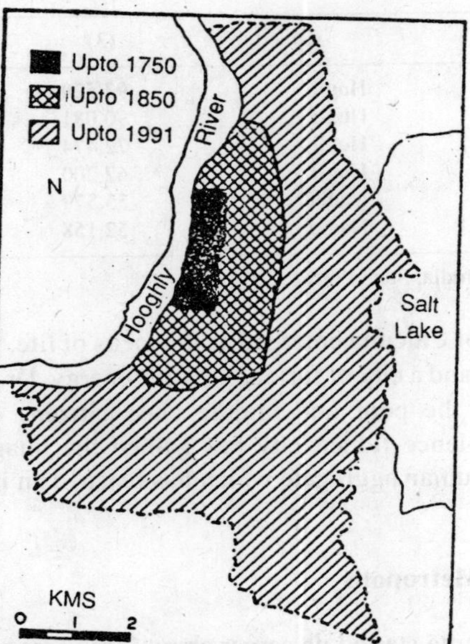

FIG. 16.1 : Expansion of Calcutta Conurbation

better men of civic service schedules and so on. A suburban area is thus essentially dependent on the central city for its growth (Fig. 16.2). In

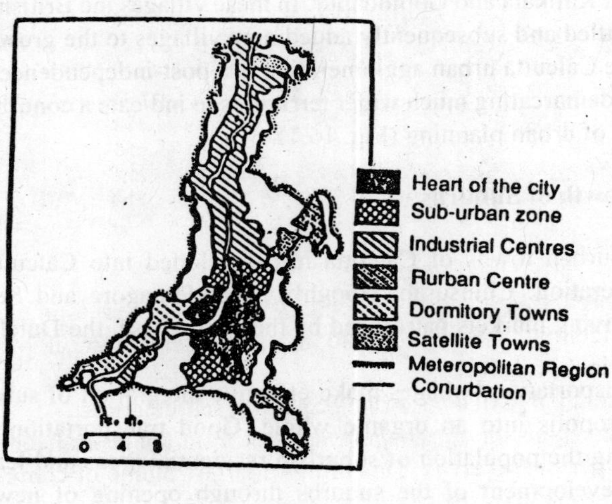

FIG. 16.2 : Calcutta Metropolitan Region

case of Calcutta a massive geographical barrier in the form of the Bhagirathi-Hooghly river has further complicated the problem.[3]

Growth of Population

The 1951 census first of all reported about Calcutta conurbation. It encompasses industrial area on the Hooghly side which is quite distinct in terms of demography, economic condition, industrial development and geographical unity. It covers an area of 425 km^2. Out of this, the core cities of Calcutta and Howrah occupied 92 km^2 and 26 km^2 respectively. In 1991 census per km^2 the density of population is 59,000 persons which was just 10,085 persons in 1941. Within last four decades (1951-91) the Calcutta conurbation has expanded tremendously in spatial, demographic and socio-economic dimensions (Table 16.2). In 1991 census the conurbation of Calcutta has a population of 10,800,000. Its position is second after Mumbai in India. This has been achieved through the following processes :[4]

TABLE 16.2 : Population of Calcutta and Greater Calcutta (1901-1991)

Year	*Calcutta*		*Greater Calcutta*	
	Population	*% Growth*	*Population*	*% Growth*
1901	933,754	—	1,488,323	—
1991	1,016,445	8.86	1,718,426	15.46
1921	1,053,334	3.63	1,850,650	7.69
1931	1,221,210	15.94	2,105,708	13.78
1941	2,167,485	79.49	3,577,789	67.91
1951	2,698,494	24.50	4,588,910	28.26
1961	2,927,289	8.48	5,736,697	25.01
1971	3,148,746	7.57	7,031,382	22.57
1981	3,291,655	4.54	9,165,650	30.35
1991	4,388,262	8.68	10,916,272	12.84
2001	3,875,453	8.32	12,555,275	10.75

Source : Census of India, West Bengal.

(a) Accretion of Urban settlement, industrial development and residential suburbs due to massive concentration of non-agricultural works;

(b) Municipal take over of neighbouring settled areas;

(c) The expansion of transport network from the central core upto the peripheral areas. This brings a large volume of commuters to the conurbation everyday; and

(d) The rapid growth of population density, heavy influx of migrant, industrial labourers and displaced persons from Bangladesh have worsened the situation enormously.

In 1961 census the conurbation of Calcutta encompasses 67 urban centres located in the surrounding area. This has engulfed Tollygunge and several peripheral towns (81) which are closely linked the Calcutta conurbation. The Calcutta Metropolitan District (CMD) encompasses an area of 1040 km^2 including the area of 2 corporations, 33 municipalities and 42 non-municipal urban areas. CMD extends from Baruipur Budge-Budge in the south upto Bansberia-Kalyani in the north. This also includes 475 rurban centres around the Calcutta Metropolitan District besides one per cent of area, 17 per cent of population, and 75 per cent of urban population of the State of West Bengal. The population load of CMD is more than 12 million. The population density per km^2 varies from 59,000 in core area to 1500 in rural-urban fringe. The outer boundary of CMD has been decided not only by the density of population, heavy built up areas but also by the incidence of non-agricultural population found in the area, suitability of future expansion of settlement and potentiality of future urban expansion.

Industrial Activities

In the conurbation of Calcutta the industrial structure includes the manufacturing activities like jute, textile, hosiery, apparel, leather, basic metal, metal products, engineering goods, machinery, transport equipments, chemicals, paper, printing, food, rice mills, pulse mills and backeries. In more than 20 towns manufacturing activities dominate the scene at significant level. These industrial centres are Titagarh, Naihati, Halisahar, Uttarpara, Bally, Champdani, Kotrung, Bhadreshwar, Uluberia, Bauria, Kamarhaf and Budge-Budge. All these are Jute mills of very high production and well managed. In Halisahar about 75 per cent workers are engaged in industrial activities. In other urban centres industrial activities predominate the scene at significant level associated with heavy transport, e.g., Howrah is a railway terminus, Kanchanpara a railway workshop town, Chinsura is a district headquarter, Barrackpur is a sub-divisional town and Ishapure is a gun factory town. In this way, manufacturing activities is dominant in the industrial landscape along the Hooghly river in Calcutta. Service activity is significant in Kalyani, Barrackpore, Baruipur, Nabapally, Jadavpur, Bansdroni and Purba Putiary.[5]

The Morphology of Calcutta

About mid-16th century Sutanuti extended from Chitpur to the old Mint House. Thus the area between old Mint House and Chandpal Ghat could be said to belong to Calcutta proper. Gobindapur occupied the site of the new Fort William, including the Race course. Sutanuti was the principal seat of commerce and industry while Calcutta and Gobindapur are the places of residence.[6]

North of the Bowbazar street extending from the Dalhousie square to Sealdah in an east-west direction, lay the native Indian town. The area between Bowbazar street and Dharmatala street was a mixed residential area dominated by the portuguese, the Greek and the Armenians. European resides in South of Dharmatala street.

The Imperial Gazetteer describes the morphology of Calcutta as "surrounded by the huge Maidan, the Fort William stood at the heart of the city. North of the Fort European shops and offices of business houses are located. The European residences were situated east of the Maidan. Alipur housed the palatial buildings of Lt. Governor. North of the European business centre of Lal Dighi Burabazar was located. Three roads run from the native town from north to south and six roads link the Hooghly river bank with the eastern most parts of the city.[7]

The Hinterland of Calcutta Conurbation

The zone of influence or hinterland of Calcutta is a very large ensemble of Indian regions of at least 780,000 km^2 and ranges upto Jabalpur and Lucknow in the west. According to 1991 census this hinterland has a population of about 300 million comprising most parts of Northern India. The Gangetic Valley in the west is a plain land, the plateau of Deccan in the south is a rough terrian while the Valley of Brahmaputra and the plateau of Meghalaya in the east presents a mixture of plain and mountain topography both.

Conclusion

In order to solve the problem the planned expansion of Calcutta conurbation is quite essential and in no case illegal construction of houses should be allowed. The lowland areas should be slowly filled with the urban refuse or garbage. The CMDA should monitor the development of roads and the supply of clean drinking water. The industrial estates should be established in specialized areas to guide pollution control.

FOONOTES

1. U. Singh, "The Calcutta Conurbation," *The National Geographical Journal of India*, Vol. VI, Part-2, June 1958, p. 95.
2. S. Sinha (ed.), *Cultural Profiles of Calcutta*, The Indian Anthropological Society, Calcutta, 1972, p. 283.
3. A.B. Chatterjee, "Calcutta Metropolitan District Changing Structure of Some Towns," *Applied Geography* (Ed. R.L. Singh), The National Geographical Society of India, Varanasi, 1968, pp. 74-78.
4. E. Ahmad, "Industrial Regions and Centres of India," Proceedings of the International Geographical Seminar, Aligarh, 1956, pp. 365-76.
5. R.B. Mandal, "The Development of Conurbations in India," *Urbanization and Regional Development*, Concept Publishing Company, New Delhi, 1982.
6. K. Bagchi, "The Howrah Conurbation : A Study in Urban Sprawl," *Geographical Review of India*, vol. XXVIII, 1966, pp. 16-31.
7. R. Murphy, "The City in the Swamp : Aspects of the Site and Early Growth of Calcutta," *Geographical Journal*, Vol. 130, 1964, pp. 241-55.

17

DELHI METROPOLITAN REGION

Importance

DELHI is the political, social, economic and cultural capital of India. It symbolizes ancient values, civilizations and present aspirations of not only Indians but foreigners too. Its ancient Fort and Cyclopean walls of Purana Quila tells the stories of abandonment, plundering and renovation by the succeeding rulers. Thus the saying that 'he who rules Delhi rules India' is seems to be true. Delhi is the capital of the Republic of India (1947 onwards) and lies on the Yamuna river where low sandstone ridge separates the Punjab Plain from the Ganga Plain. As Delhi is well served by railways and airlines from all parts of the country, hence it is easily accessible.[1] At present Delhi comprises the old city and the wonderful modern city of New Delhi which is the actual capital of India along with its palatial government offices and allied buildings.

Regional Location

Delhi has a northerly location in the Indian sub-continent. The site of Delhi over Aravalli Ridge-300 metres high from mean sea level is commanding amidst Peninsula on the south, Thar desert on the west, mountains of the Himalayas on the north and the great Indian plains in the east. Situated on the bank of Yamuna river the air routes, roads, railways all converge at Delhi since millennium. Thus the nodality of Delhi is inherent in its regional location which acts as the nerve centre of routes converging from all directions. In the selection of a site for Delhi, the consideration of defence, the high natural levee of the Yamuna river and the convergence point for different modes of transport play a vital role.[2]

Delhi extends from 28°53' N latitude and 76°50′ to 77°22′ E longitude. Due to its nodality and regional location Delhi acts as the Gateway of India.

TABLE 17.1 : Delhi : Area and Population

Area	*Population*
Delhi U A	8,427,083
Delhi M C	7,174,755
New Delhi	294,149
Sultanpur Majra	111,253
Palam	98,968
Bhalawa Jahangirpur	94,888
Delhi Cant	94,326
Nasirpur	81,254
Nangloijat	75,944
Gokalpur	49,052
Bakarpur	47,387
Bindapur	36,148
Tigri	33,207
Deoli	33,112

Total No. of Areas : 26

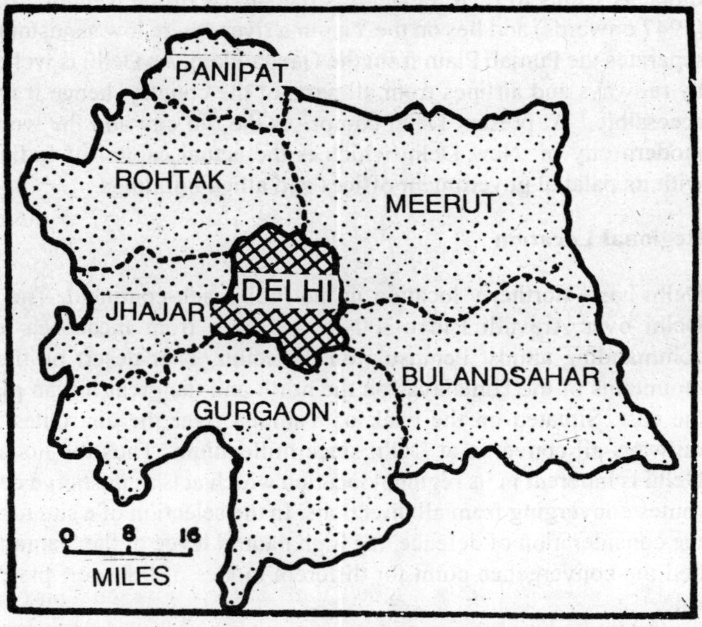

FIG. 17.1 : Demarcated Boundary of National Capital Region

History of Metropolitan Growth

About 10th century B.C. the city of Indraprastha was established by Pandavas on Western bank of the Yamuna river. It was rebuilt by Anangpal I in A.D. 1052 as Lal Kot (Delhi) south-west of Indraprastha. Prithvi Raj Chauhan enlarged the old Delhi upto Lal Kot in A.D. 1170 to 1192 and renamed it as Qila Raj Pithora. Allauddin Khilji made a new fortress town to the north of Pithora and the old city renamed as Siri. Siri was beautiful city having lofty and splendid buildings. It came to be known as Jahanpanah in A.D. 1327; Firozabad in A.D. 1351; Mubarakabad in A.D. 1433; Din Panah in A.D. 1533; Shergarh in A.D. 1555; Red Fort was built by Shahjahan in A.D. 1648. The site of New Delhi was planned by Edwyn Lutyens and Baker (1911) on south-west of Shahjahan Panahbad (Old Delhi) after considering the administrative personality and nodality of Delhi.

This clarified that the site of Delhi had changed many times according to the whim of rulers because the earlier site on ridges suffer from improper water supply and the western bank of Yamuna floods the area annually. Nearly four embankments have been made to protect the area from flooding.[3]

The growth of population in Delhi Metropolitan region could been seen in the following Table 17.2.

TABLE 17.2 : Delhi : Growth of Population

Years	*Population*	*Growth*	*Sex Ratio*
1901	214,11	—	816
1911	237,344	11.13	740
1921	304,420	27.95	672
1931	447,442	46.98	669
1941	695,686	55.48	677
1951	1,437,134	106.58	754
1961	2,359,408	64.17	780
1971	3,647,023	54.57	800
1981	5,713,581	56.66	808
1991	84,207,083	58.00	815
2001	105,000,000	65.00	825

The Table 17.2 clearly shows that Delhi is now the Third largest metropolis in India after Mumbai and Calcutta. It has experienced a decadal growth of 50 per cent in 1951 and 57 per cent in 1981 in comparison with 30.8 per cent in Calcutta, 38.2 per cent in Mumbai and

35.3 per cent in Chennai. Although the natural increase is not too much, that immigrants influx is so large that it contributes substantially leading to a density of about 5000 persons per km^2. This favours the congestion of houses and heavy pressure on the limited resource available to the metropolis. As a primate metropolis Delhi has a commanding position in India in terms of banking, wholesale business, industrial and transport development, besides higher education. It acts as the seat of learning, administration, recreation, publishing of books and foreign relations. Thus due to indigenous values, culture and superlative in many respects Delhi acts as the most modern capital city in the World.[4]

The growth of Delhi Metropolitan Region is associated with the increase of industries and diversification of economic base. From 1950 to 1965 the number of industrial units increased from 8160 to 19038. Due to lack of comprehensive plan industries are getting located wherever vacant space is found. This has created overcrowding of traffic and growth of slums. Delhi Master Plan prepared by Delhi Improvement Trust in 1941 includes industrial centres along Najafgarh and Kalkaji roads and Delhi-Mathura rail road. Most of the industries like Delhi Cloth Mills, Birla Cotton and Weaving Mills and Ganesh Flour Mills are located in this area. According to the Master Plan enforced in 1952 about 1.6 of the total urban area was under industries but it increased to about 5 per cent in 1981 and 7 per cent in 1991. This intense urban activity has favoured the growth of urban sprawls beyond Delhi triangle upto Hindan and National by-pass on other side of the Yamuna river (Fig. 17.2)

Regions of Delhi Metropolitan Area

Here three levels of regional set-up have been found are :

(a) The metropolitan region of Delhi is closely linked with the urban core of Ghaziabad and Loni in Uttar Pradesh, Faridabad, Gurgaon and Bahadurgarh in Haryana and Narela urban communities of Delhi.

(b) Delhi acts as a 'National Capital Region' which comprises several urban settlements such as Rohtak, Sonepat, Meerut and Bulandsahar.

(c) The third level covers the urban core of Delhi itself which is distinct in terms of commercial, administrative and transportational linkages in comparison with the surrounding countryside.

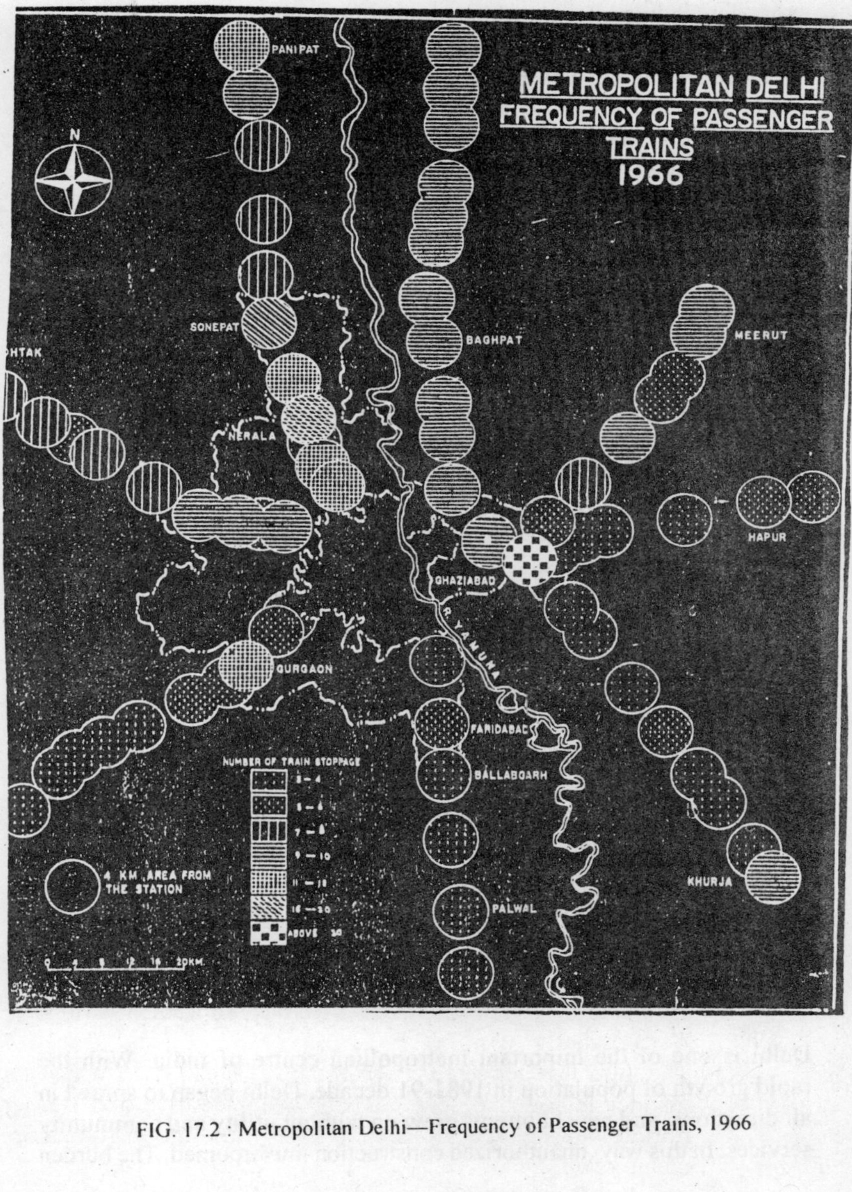

FIG. 17.2 : Metropolitan Delhi—Frequency of Passenger Trains, 1966

The Master Plan considered Delhi as National Capital, regional capital and commercial centre of a very vast region. Thus the Master Plan of Improvement Trust considered the functional aspect of this expanding industrial giant which brought in its fold quite a large number of towns, e.g., Ghaziabad, Bahadurgarh, Gurgaon, Sonepat and Faridabad. All these are coming closer to Delhi every year by accommodating more and more industries and commercial activities. They act as ring towns and urban regional planning one has to develop them in an integrated and coordinated manner with Delhi so that their planned expansion will not only help them to achieve a better economic and social base. They also releases pressure from Delhi especially for space and employment on both sides of the Yamuna river the lowlying marshy areas are creating barriers for urban expansion in Delhi.

Plan of Delhi Development Authority

From the planning point of view Delhi Development Authority has divided the whole area into eight planning regions and it has been further divided into 136 development zones in order to accelerate the planning process. The plan of Delhi Development Authority includes 42.9 per cent land for residence, 8.8 per cent for commerce, industry and government offices, 23.7 per cent for recreation, 8 per cent for public and semi-public facilities, 10 per cent for transportation lines and 6.3 per cent in military cantonment with the provision of 1.6 kilometres roadside under social forestry programme the concept of 'garden city' comes into reality. The community facilities like schools, colleges, hospitals, health centres, shopping complexes and cultural institutions have been planned along with slum clearance, controlled densities of urban population, the regular supply of drinking water and electricity. For smooth plying of traffic the width of road has been kept 60 to 90 metres in metropolitan, regional, ring and arterials roads. Even then the expanse of metropolitan Delhi is so vast that transport problem is always faced in metropolitan Delhi for proper movement from one place to another.

Conclusion

Delhi is one of the important metropolitan centre of India. With the rapid growth of population in 1981-91 decade, Delhi began to sprawl in all directions, and new hutments grew up without utility and community services. In this way, unauthorized construction mushroomed. The burden of undoing and substantially improved conditions presents financial and administrative conundrum. For example, the cost of providing tenements to rehouse about 40,000 squatter families in Delhi is estimated to be about Rs. 50 million, yet Rs. 100 million is needed to clear just a fraction of the city's slums and to rehabilitate these slum dwellers in rurban villages.

FOOTNOTES

1. A. Mitra, *Delhi : Capital City*, Thompson Press, New Delhi, 1970.
2. M.P. Thakore, *Aspects of Urban Geography of New Delhi*, Ph.D. Thesis, London University, 1962.
3. S.S. Bhatia, "Historical Geography of Delhi," *Indian Geographer*, 1956, pp. 17-43.
4. M.P. Thakore, "Sixteen Sites in Delhi," *Indian Geographer*, 1968.

18

WORLD URBANIZATION

THE GROWTH of population in urban areas is known as 'urbanization' whereas the decrease of population is known as 'de-urbanization'. The gonedays have seen the birth of thousands of towns and cities, many of which have largely expanded during the last 50 years. The growth is almost explosive. At present urban population explosion is found in each and every country, leading to the formation of fabroblast and sacroma settlements.[1]

The economic development of society and the progressive division of social labour have of necessity led to the formation and expansion to towns and cities, with ever increasing migration of people to perform a wide variety of functions. Urbanization throughout the world is not reflected in a steady direct linear increase in the proportion of town dwellers in each given society in temporal dimensions. Towns and cities have taken shape as centres of administrative and cultural significance through the centuries.

Urbanization depends on the socio-economic system of the society and the rate of its development. Each nation which has founded towns and cities imparts to them their own special appearance and their own unique content. With time this content alters, reflecting the changes in the development of the productive forces. While towns are necessary as strong points of power and military force they express the preparatory stage of urbanization. This process is characterized by the growth in labour productive in the rural and urban economies, class stratification of society, and the growth in the country's population.

The urbanization as a social phenomenon seem to have percolated even in countries which were rural and the greater acceleration in growth rates have been observed for towns and cities which are old and new. It is closely associated with the spread of industry, commerce, manufacturing and the improvement of transport and industries. The main factors contributing to the urban development are as follows :

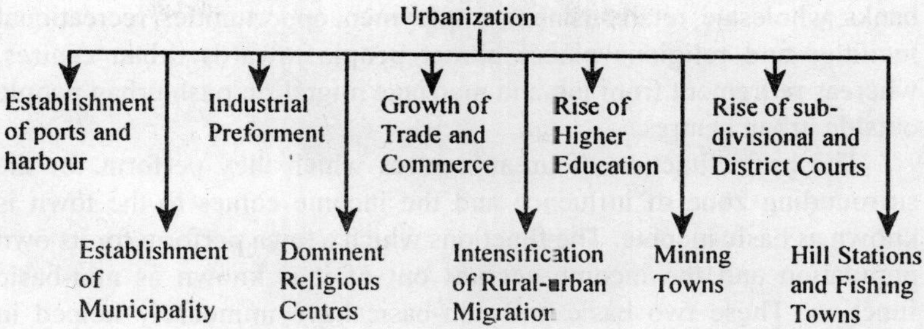

Urban growth is infinenced by physical, socio-economic, political, demographic, cultural and technological factors. It is therefore, essential to study the impact of these factors and as to accelerate the tempo of urban development.

Urbanization is a dynamic socio-economic forces which involves urban element that suffer considerable temporal and spatial variations. In case the process is viewed with a definite regional background, urbanization influences the surrounding rural-urban fringe, which could really be seen in rural-urban linkages. Hence, urbanization is equipped with the potentialities of development of non-agricultural sector to fulfil the economic requirements or urban manufactured goods in the hinterland. Nowadays, the term 'urban' has developed much wider connotation than ever before. In the context of regional development, reference may be to the various economic and social problems which are closely linked with the process of urbanization and creates a strange behaviour by which one could find that there is more impersonality and higher specialization.

The World Urbanization has four directions :

1. Coastal location of very many cities.
2. Peripheral location of small towns, industrial suburb and urban sprawls.
3. Complex building structure in developed countries but simple structure in developing countries.
4. Urban centres are the emblem of polarization of population and economic development.

Pull and Push Factors

Changes in the level of urbanization has been caused by pull and push factors. Due to pull factor the urban amenities like higher education, banks, wholesale, retail business, employment opportunities, recreational facilities and religions places attract people towards urban centres, whereas retirement from job and marriage migration push urban people outside urban centres.

The basic function of urban centres which they perform for the surrounding zone of influence and the income comes to the town is known as basic income. The functions which a town perform for its own population and the income accrues out of it is known as non-basic function. These two basic and non-basic have immensely helped in changing the levels of urbanization in most parts of the globe.

Sectoral Changes

So far as the sectoral changes of urbanization is concerned the industrial sector grows fast in comparison with the agriculture sector of sub-urban zone. It means with the industrial establishment the population supporting capacity of the urban centres improves a lot in comparison with the agricultural economy.

Degree of Urbanization

The degree of urbanization refers to the percentage of total population living in urban areas. The formula for computing degree of urbanization is as follows :

$$L = \frac{U}{T} \times 100$$

where L is degree of urbanization, U is urban population, and T is total population.

In accordance with the international comparison, this index is further defined in two ways :

(a) Instead of total urban population living in cities with 100,000 and more inhabitants is considered as urban population.
(b) In place of total urban population, the population living in urban places at least 20,000 inhabitants is considered as urban population.

Rate of Urbanization

It is simple arithmetic growth rate of the degree of urbanization, between two or more census figures.

$$\text{Rate of Urbanization} = \frac{L_2 - L_1}{L_1} \times 100$$

where L_1 and L_2 are the degree of urbanization at two point of time.

There are two ways of measuring the urban growth. One is the instantaneous method which ascertains population in all urban categories in each and every census, tracing the changes in each class regardless of the cities that make it up. Other is the continuous method which begins with particular centre and traces the subsequent extension of these groups.

Tempo of Urbanization

In 1991 about 45 per cent population of the world is living in urban areas. The level of urbanization is 84 to 85 per cent in Australia and New Zealand; 77 per cent in Japan; 72 to 75 per cent in North America, South America, Latin America and Europe; 66 per cent in Central America and former U.S.S.R.; 34 per cent in Africa and Asia; 32 per cent in Pakistan and just 26 per cent in India.

In South America the percentage of urban population in Argentina is 83 per cent, Chile 81 per cent, Uruguay 84 per cent, Venezuela 84 per cent, Brazil 68 per cent and Cuba 66 per cent. In Asia Singapore is 100 per cent urbanized, Kuwait 89 per cent, Hongkong 90 per cent, Yemen 11 per cent, Bhutan 4 per cent, Nepal 6 per cent, Myanmar 28 per cent, Sri Lanka 27 per cent and Bangladesh 15 per cent urbanized in 1991.

In 1991 the density of population in bigger urban centres vary from 67,00 per km^2 in Calcutta to 57,000 in Ottawa, 55,000 in New York, 52,00 in Montreal and 49,000 in Moscow.

Trends of Urbanization

The trends of urbanization differ substantially from country to country. In the United States of America in A.D. 1800 only 5 per cent people were residing in urban areas. This has gone upto 70 per cent in 1980 and it will be 85 per cent by A.D. 2000. So far as the logistic curve of urbanization is concerned, it is typically of *S* shape for Costa Rica, Tanzania and Switzerland.

The average annual population increase for 34 countries comes to 4.5 per cent especially for the continents of Asia, Africa and Latin America. In the entire 19th century the continent of Europe experienced an annual growth of only 2 per cent. Most of the urbanization of

19th century was migration based, whereas today it is in the form of Shantytowns,[2] development of colonies and flow of industrial workers to towns and cities. In Switzerland, Tanzania and India about 80 per cent urban growth is due to rural-urban migration. In the year 1900 the four leading cities of the world are London, Paris, New York and Shanghai. In the year 1920 only 24 cities have touched the million mark but in 1991 it is about 150 in number.

TABLE 18.1 : Millionaire Cities, 1870-1991

Date	*Number of Cities*
1870	7
1900	20
1920	30
1939	57
1951	95
1964	140
1979	213
1986	257
1991	300

Density of Urban Population

Cities are always very crowded places in the world. In Calcutta about 69,000 persons reside in one square kilometre and the corresponding figure for New York is 55,000, Montreal 52,000 and Moscow 49,000. The overall density is 25 for United States of America, 3 for Canada and only 11 persons for the Common Wealth of Soviet Union.

Levels of Urbanization

In 1991 about 40 per cent of the world's population is urban, and about 25 per cent urban population live in the centres of 20,000 persons or more. About 10 per cent people live in cities of 100,000 persons and more than 5 per cent people live in million cities. In case the 1981-91 growth rate of urban population is 3.5 per cent per annum continues by the year A.D. 2000 about 50 per cent world's population would be urban and by the year 2050 the entire world will be urbanized. In such a situation the people will get food from ocean and our mighty forests. In most of the industrialized countries, e.g., Britain, Germany, Australia, Argentina, California, Canada, New Zealand and the Common Wealth of Soviet Union, there will be a decreasing demand for farm labour but an increase in demand for industrial workers, and skilled people in transport, trade and commerce, administration and in educational institutions have revolutionized the scene. In most of the above mentioned countries the percentage of urban population ranges between 60 to 90 per cent. In older settled countries the urban people live in medium and small sized towns besides market towns, small industrial towns along with a few large cities.

In countries of peasant economy, e.g., in Tanzania, Kenya, Uganda, Sudan, India, Pakistan and China the percentage of urban population has increased fast but the pace of urbanization is still very slow between 5 to 20 per cent. In developing countries of the world the urban population is growing fast in comparison with total population.[3] In between 1900 and 1990 the urban population grew by 300 per cent. This increase is spectacular in Tropical Countries where due to the provision of health facilities the mortality rate has declined but on the other hand, this has enhanced the rural-urban migration thereby giving rise to shanty towns.[4]

In economically advance countries the growth of urbanization is not very high, because they had already covered 70 to 90 per cent during the past decades and further there is little scope for development on the current technology.

Urban Regions

The world could be divided into two regions from the point of levels of urbanization :[5]

Highly Urbanized Regions

In some countries of the world the level of urbanization has reached at a very high stage except few towns, cities and regions. Such areas are Australia, New Zealand, the East Coast of Brazil, China, Japan, Venezuela, Canada and the eastern parts of U.S.A. and Argentina. In all these areas the towns having coastal location are bigger in population size and the inland towns are of smaller size. The four fully urbanized regions of the world are as follows :

1. North-Western Europe—London, Paris.
2. North-Eastern U.S.A. and Canada—New York, Philadelphia.
3. Eastern Coast of Japan and Australia—Tokyo, Sydney.
4. Easter Coast of Asia—Shanghai, Calcutta and Manila.

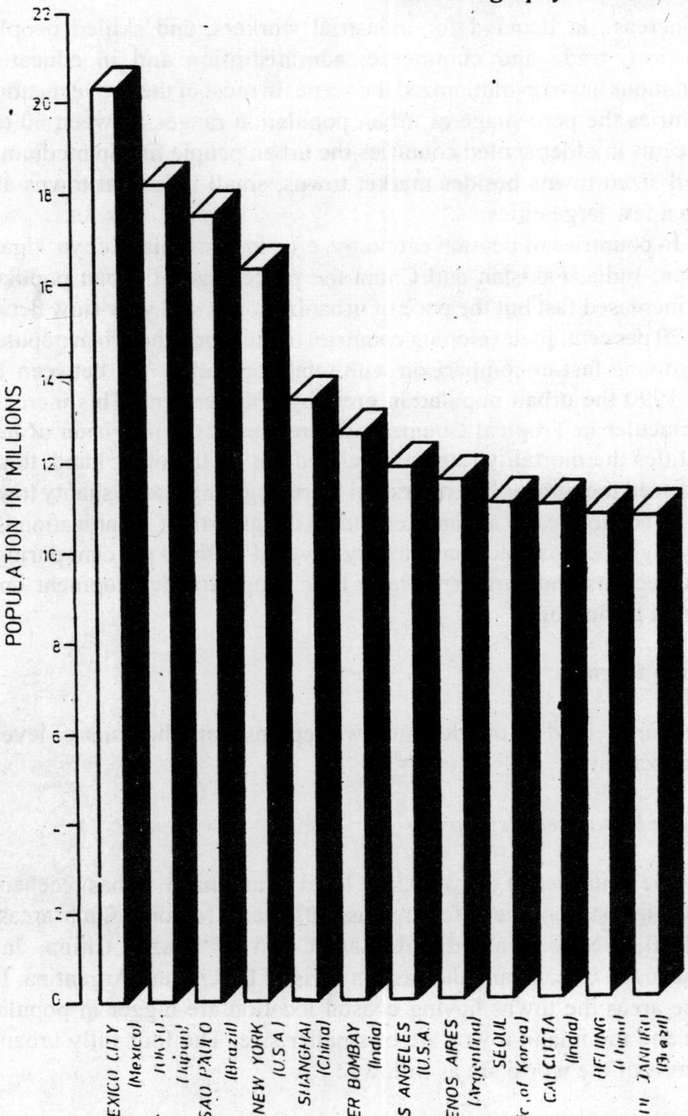

FIG. 18.1 : World's Largest Urban Agglomerations (10 Millions or More Inhabitants in 1990)

Note : **Population figures for Indian Cities are given as per the 1991 Census.**

In these regions industrialization and urbanization dominate the scene along with port location.

Less Urbanized Regions

Most parts of Asia, Africa,[6] South-America and Siberia have very low levels of urbanization. In these areas less than 10 per cent population are urban. The causes are low levels of transport development, lesser industrialization and the predominance of farm economy and forest economy. We may say that urban centres dominate the scene and even in city level centres the facilities are very poor and spontaneous settlement dominate the scene.[7]

The problems created by urbanization and urban centres are :

(a) Urban development threatens the expansion of world's agricultural sector;
(b) It exacerbates the water supply and aggravated the situation of waste disposal;
(c) It produces almost incurable traffic problems;
(d) It provides obvious targets for aerial attack;
(e) It has favoured the decline of traditional culture;
(f) It has brought refinement in criminal activities;
(g) Urban machine civilization have damaged the rural cottage industries and seriously affected the morality of people; and
(h) Urbanization favoured the transport of pollution to the nearby rural areas in terms of waste disposal and emitting smoke.

FOOTNOTES

1. G.L. Peters, "Third World Urbanization : Towards the Year 2000," *Urbanization and Regional Development* (Ed. R.B. Mandal and G.L. Peters), Concept Publishing Co., New Delhi, 1982, pp. 21-39.
2. G.L. Beier, "Can Third World Cities Cope I," *Population Bulletin*, Vol. 31, No. 4, December, 1976, p. 3.
3. J.M. Hamilton, "Time Bomb in the City," *Agenda*, Vol. 2, No. 5, June, 1979, p. 3.
4. C. Stadel, "Colombia : An Essays on World Urbanization," (Ed. by Ronald Jones), George Philip and Sons Ltd., London, 1975, p. 258.
5. Davis Kingsley and Hartz, H., *The Pattern of World Urbanization*, New York, 1966.
6. A.L. Mabogunge, "Urbanization Problems in Africa," In *Urbanization, National Development and Regional Planning in Africa*, Praeger Publishers, 1974, p. 23.
7. M. Jefferson, "Distribution of the Worlds' City Folk : A Study in Comparative Civilization," *Geographical Review*, Vol. 21, 1931, pp. 441-65.

19

THIRD WORLD URBAN DEVELOPMENT

TODAY the urbanization is a worldwide phenomenon. Every nation, every city and probably every individual in the world is directly or indirectly involved in the continuing process of urbanization which is transforming the world rapidly. By urbanization generally we refer to the process whereby land and inhabitants become urban oriented rather than engaged in primary occupations. In other words, urbanization refers to the process by which a section of people change from the way of agricultural operations in rural living to an urban sub-agricultural life with predominance in secondary and tertiary functions.

The purpose of this chapter is to discuss urbanization as Third World experience and as the area is vast so the discussion will not confine to any specific country, but it will be generalizing some characteristics, observations and general urban development pattern in the Third World countries. The area of study comprises the continents of Latin America, Asia and Africa (Fig. 19.1).

Historically most of the first ancient towns (e.g., 6th to 4th centuries B.C. towns and cities in Asia) were found in areas which now come under Third World countries, but surprisingly, today those areas are the least urbanized in the world. Smailes (1966)[1] pointed out, those urban groups of Asia were restricted by geographical conditions that made them possible; besides the low level of science and technology which always leads to industrial development and larger commercial activities, limited the scope of their urban development and hence were outpaced by Western Europe and other developed countries until when external influence was abound. Despite the existence of some pre-colonial cities in Third World countries, most contemporary cities reflect their origin as part of a colonial economic and administrative structure. The impact of colonial rule is apparent in the urban systems of West and East Africas' as well as in the internal spatial structure of individual city. In almost all Third World cities owing to the lack of coherence in the urban

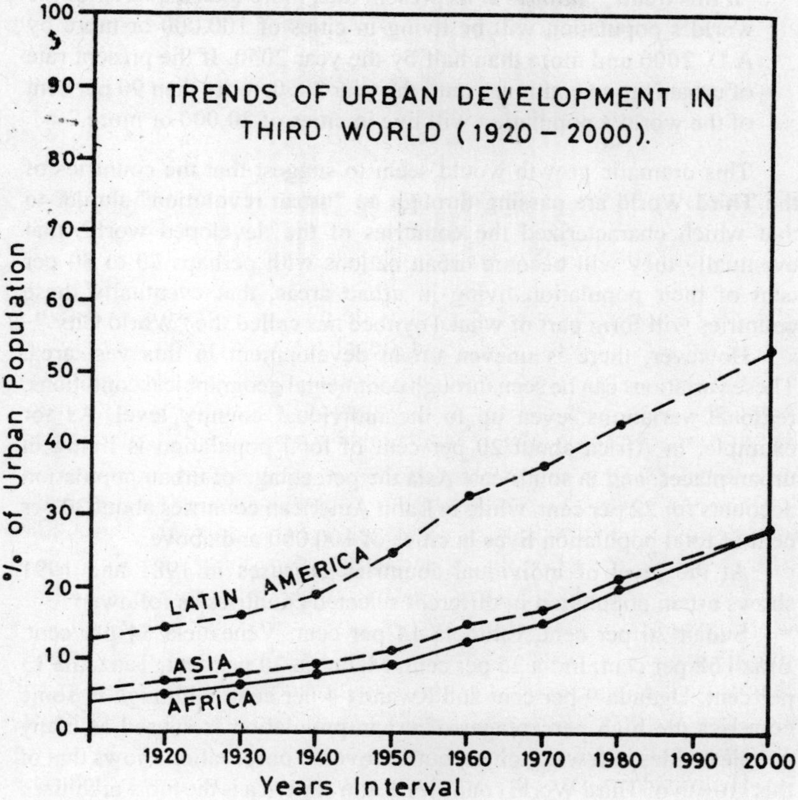

FIG. 19.1 : Trends of Urban Development in Third World (1920-2000)

system, growth generated in one part of the system may not be effectively transmitted to other parts of the system. It seems that the generative effects of urban growth, the "spread effects" of economic development may not have as widespread impact as might be hoped. The cities of the Third World are growing rapidly than the growth of urban population in their comparative period of European growth.

As for example, Jones (1975) tells us that urbanization was more remarkable in developed countries in the 19th century, while for the Third World countries it is in the 20th century whereby cities are growing at annual rate of 5 to 8 per cent and urban population doubling in every 10 to 15 years.[2] While writing about the trend to urban growth in the past 180 years, P. Mack (1982)[3] said :

"If this trend continues at its present rate, more than a quarter of the world's population will be living in cities of 100,000 or more by A.D. 2000 and more than half by the year 2050. If the present rate of urbanization continues until the year 2050, more than 90 per cent of the world's population will live in cities of 20,000 or more."

This dramatic growth would seem to suggest that the countries of the Third World are passing through an "urban revolution" similar to that which characterized the countries of the developed world, that eventually they will become urban nations with perhaps 70 to 80 per cent of their population living in urban areas, that eventually these countries will form part of what Toynbee has called the "World City."

However, there is uneven urban development in this vast area. These variations can be seen through continental geographical conditions, regional variations, even up to the individual country level. As for example, in Africa about 20 per cent of total population is living in urban places, and in south-east Asia the percentage of urban population accounts for 22 per cent, while in Latin American countries about 30 per cent of total population lives in cities of 100,000 and above.

At the level of individual countries censuses in 1981 and 1991 shows urban population in different selected countries as follows :

Sudan 26 per cent, Ethiopia 14 per cent, Venezuela 84 per cent, Brazil 68 per cent, India 25 per cent, Malaysia 30 per cent, Tanzania 13 per cent, Uganda 9 per cent and Rwanda 4 per cent. Although in some courtries the high percentages of urban population is caused as many people resides in few big cities, but the overall observation shows that of this growth of Third World countries, Latin America is the most urbanized followed by Asia and Africa which are the least urbanized continents. Therefore, to be realistic in our discussion it is better to analyze each continent separately in order to get an overall experience of urbanization in the Third World countries.

Latin American Experience

The idea of city was introduced by European colonizers especially the Spaniards and Portuguese of which the pace of development was slow and gradual and came to be remarkable in the 20th century. Latin America on the first instance experienced urban settlements in sea ports followed by commercial and manufacturing towns as the socio-economic development and exploitation of European colonizers outraged. Later on, as transport and communication improved internally urban places were opened up at nodal points, but mostly on the places where economic exploitation was possible. Putting aside historical background of urban places in Latin America, nowadays, the continent on the whole is considerably more urbanized than the average of the world's regions. It is much more urbanized than either Africa or Asia and some how more urbanized than southern Europe, and in industrialization and economic development it is considered less advanced than Southern Europe. In general about the outlook of world urbanization, it is said that only North America, North-Western Europe and Australia are very far ahead of Latin America. Some of the countries with high percentage of urban population are Argentina, Chile, Uruguay, Venezuela, Brazil and Cuba, as shown in Table 19.1.

TABLE 19.1 : Urban Development in Latin America

Country	*% of Total Population*		*% Average Annual Growth rate*		*% of Population in Cities over 500,000*	*% of Population in Larger Cities*		*Number of Cities over 50,000 Persons*
	1960	*1980*	*1960-1970*	*1970-1981*	*1960*	*1981*	*1981*	*1981*
Haiti	16	28	4.0	4.7	0	56	56	1
Bolivia	24	45	3.9	6.9	0	44	44	1
Honduras	23	36	5.4	5.5	0	0	33	0
El Salvador	38	41	3.2	3.4	0	0	22	0
Nicaragua	41	54	4.0	5.0	0	47	47	1
Cuba	55	66	2.9	1.9	38	32	38	1
Guatemala	33	39	3.8	3.9	41	36	36	1
Peru	46	66	5.3	3.5	38	44	39	2
Equador	34	45	4.4	4.6	0	51	29	2
Jamaica	34	42	2.4	2.5	0	66	66	1
Dominica Rep.	30	52	5.6	5.3	0	54	54	1
Colombia	48	64	5.2	2.6	28	51	26	4
Costa Rica	37	44	4.2	3.6	0	64	64	1
Paraguay	36	40	2.9	3.3	0	44	44	1
Panama	41	55	4.4	3.6	0	66	66	1
Brazil	46	68	4.7	3.9	35	52	15	14
Mexico	51	67	4.7	4.2	36	48	32	7
Argentina	74	83	2.0	2.0	54	60	45	5
Chile	68	81	3.1	2.4	38	44	44	1
Uruguay	80	84	1.3	0.6	56	52	52	1
Venezuela	67	84	4.7	4.2	26	44	26	4
Trinida	22	22	1.8	1.4	—	—	—	—

Source : World Development Report, 1983, pp. 190-91.

Urbanization in the majority of Latin American countries is of megalocephalic type. That is, concentrated to a large extent in the principal city, without any sub-towns in the nearby area. Out of 22 countries in Latin America, 16 countries have more than half of the population concentrated in megalocephalic city. On the other extreme includes Jamaica, Guatemala and Haiti with more than 80 per cent of their urban population concentrated in the area of the capital city. The mushrooming growth of cities is an conspicuous feature of recent population trends in Latin America. Estimates of 1950-1960 decade show that the population in the localities of 20,000 or more increased by 5 per cent in comparison with total population growth rate of 2.7 per cent annually. It appears that rate of urban population growth in Latin America was nearly twice that of the total population. The increase was mainly due to increased amount of rural-urban migration. But the contemporary urbanization as also added by the increased number of qualified localities as urban places and the expansion of other boundaries of major cities plus the development in the number of industries.

The rapid growth of Latin American cities, particularly the large ones, since the Second World War, together with the high cost of providing public services has similarly led to the prescription that cities other than the metropolitan areas should be stimulated. The focus of the Cornell Conference on 'The Role of the City in the Modernization of Latin America's was on the positive aspects of the city in the development process rather than on the 'usual urban problems'. The aim to identify those components of urbanization that can be accelerated without unduly aggravating the normal problems associated with rapid urban growth (Bayer, 1967).[4]

While increased urbanization was general throughout the world during the 1980's and 1990's, Latin America surpassed all other regions in the rate of growth of its urban population. But on the other hand, this rapid urbanization in Latin America as it is in all developing countries does not correspond to economic development and industrialization. There is low levels of development both in agricultural and non-agricultural sectors of the economy. Also there is uneven development between the urban and rural poor. As most of the Latin American cities are following the trend of West European experience where the problem of rural-urban dichotomy exists. Other problems facing them are congestion and unemployment which are the result of mass migration whereas unplanned squatters are found in many cities in Latin America.

Asian Experience

Asia has some of the ancient towns in the world but urbanization is still at its early stage. Ancient towns which dates to the 4th millennium B.C. were Mohenjodaro and Harappa in the valley of Indus. Early manifestations of urban life, however, were especially notable in the alluvial plains near the head of the Persian Gulf. Here were Susa in Elem and the cities of Summeru (Ur. Lagash, Erech and others) situated among the former distributaries of the Euphrates. Further inland near the site of latter Babylon was Kish.

These ancient Asian cities were predominantly a political and cultural phenomenon rather than economic. As history continued to shape these cities, later on the capitals and provincial administrative centres of the great Asian Empires in China, India, Japan and South East Asia functioned and were policy symbols of authority, legitimacy and power, creators and moulders of literature, culture, seats of the dominant religious ideologies and being the thrones of the great tradition. And in such a situation commercial and industrial functions were decidedly secondary and almost incidental.

Some of the Asian cities are of West European type, which derived their built structure directly from western experience due to the effect of expanded Europeans who moved eastwards in search of trade. The new centres were constructed at the time of Asian Empires. Generally, Europeans chose sites which were commercially profitable. Many ports were built of which their physical and morphological structures revealed their western and colonial origin. In urban areas, the skyline of houses come to be dominated by taller buildings in almost purely European style and since the population of all of them remained Asian, in such a situation at least the city centre looks western. The principal business of these urban centres was overseas trade and they were mostly port cities. But as the port cities also needed to maximize access to their commercial hinterland so some of them arose even near the navigable rivers.

Recent urbanization of 20th Century shows the remarkable rapid urbanization in Asia especially after 1945. Putting aside Japan which is now among the highly industrialized few countries of the world, the rate of population increase in Hong Kong since 1945 tops the list and also represents the highest urban population density in Asia. At this period, one must underscore the varying degrees which existed in Asian urbanization since 1945, which shows that it was lower in Yemen (11%), Bhutan (4%), Nepal (6%), Myanmar (28%), Sri Lanka (27%) and Bangladesh (12%) as shown in Table 19.2. These are the reflections of low level of economic development compared to their counterparts of Japan 78 per cent, Kuwait 89 per cent, Singapore 100 per cent and Hong Kong 90 per cent. In Asia, the large city population growth since 1945 has grown by almost 450 per cent in comparison with 160 per cent in Europe and North America. This shows that the growth of urban population in Asia in the recent decades are very high compared to other parts of the Globe. In fact, it is the larger cities which have contributed to such a sizeable proportion of increase in urban population even than the bulk of the Asian population is rural. For example, India which

TABLE 19.2 : Urban Development in Asia, 1981

Countries	*% of Population*		*Average Annual Growth Rate %*	*% of Urban Population in largest city*
	1960	*1981*	*1970-81*	*1981*
Bhutan	2	4	4.4	—
Bangladesh	5	12	6.5	30
Nepal	3	6	5.0	27
Afghanistan	8	16	5.8	17
Myanmar	19	28	3.9	23
India	18	22	4.2	6
Vietnam	15	19	3.3	21
Sri Lanka	18	27	3.6	16
Pakistan	22	29	4.3	21
China	18	21	4.5	—
Yemen	3	11	8.2	25
Indonesia	15	21	4.0	23
Thailand	13	15	3.4	69
Philippines	30	37	3.7	30
Papua New Guinea	3	19	8.2	25
Mongolia	36	51	4.0	52
Korea D.R.	40	60	4.3	12
Korea R.	28	56	4.6	41
Iran	34	51	5.0	28
Iraq	43	72	5.3	55
Malaysia	25	30	3.3	27
Hong Kong	89	90	2.5	100
Israel	77	89	3.1	25
Singapore	100	100	1.5	100
Saudi Arabia	30	68	7.4	18
Kuwait	72	89	7.5	30
United Arab Republic	40	73	16.6	—
Japan	62	79	2.0	22

Source : World Development Report, 1984.

follows China with largest population in the world, has an urban population of 26 per cent (1991) and more than 74 per cent rural is the characteristic feature of South East Asian urbanization. The tendency for urban hierarchy of many south east Asian countries is to be dominated by 'One great metropolis, the primate city' which dominates the urban situation. In fact, the slow growth rate of some of the large cities such as Bangkok and Manila is almost certainly due to the out migration of population from the crowded city core. Thus Manila city grew by 209.3 per cent between 1947-1990, and in the same period Quizon city on its boundaries grew by a marked increase of 268.5 per cent. The slowing down of the growth rate of the large cities in some cases simply indicates the growth of sub-urbanization.

According to United Nations estimate, the urban future of Asia by A.D. 2000 will be 993 million people. It means such a massive increase does not correspond with the economic development, and hence more urbanization will be hazardous to most of the urban population in Asia. In Asia the number of metropolitan cities were 19 in 1950, 49 in 1970 and 86 in 1990. In 1991 China had 38 metropolitan cities followed by 27 in India.

Because in 1990 the urban population in Asia was 465 million where in terms of physical facilities most of the cities were described as in a state of crisis. Clearly, it is in Asia that the core of most Third World urbanization problems lie.

African Experience

The continent of Africa has some of the ancient cities in the world such as those found in the Valley of Nile which is known as Aphroditopolis near Cairo, Hierakempolis near Aswan, but urbanization in Africa and even in North Africa is still at its early stage. Contemporary rapid urbanization in Africa was mainly due to the external influence especially during the colonial era and after. In Africa long distance trade was primarily responsible for the growth of towns during historic periods. Trading centres were established along the coast as early as the 1st century A.D. Later on in West Africa towns expanded in interior areas following the lines of Trans-Saharan route such as Timbuktu, Gao Jenne, Kano, Walata, etc. were established. All these urban centres have greater impact in African ancient scene, but the dimension has increased with the encoming of Europeans and establishment of industries (Table 19.3).

TABLE 19.3 : Percentage Share of Manufacturing in West African Capital Cities, 1990

Capital Cities	*Percentage Share*
Dakar (Senegal)	87
Bathurst (Gambia)	100
Conakry (Guinea)	50
Freetown (Sierra-Leone)	75
Monrovia (Liberia)	100
Abidjan (Ivory Coast)	63
Accra (China)	30
Cotonou (Dahomey)	17
Lagos (Nigeria)	35

The concentration of colonial administrative functions in the capital city, and the fact that these were often the principal ports for their respective countries, contributed to their growth as the principal commercial and industrial centres as well. According to McGee (1967) most of the colonial cities of the Third World owe their dualistic structure to the instructions of Western capitalism in the form of traditional culture.

Hence, the recent location, size and distribution of urban centres in Africa is almost entirely the product of colonialism. The key urban foci of this imposed system were the major ports and administrative headquarters, and their location reflect the strategic and exploitation objectives of colonial power. The major sea ports emerged as the base for colonial penetration into the continent and as the centres for export of commercial products from the interior. The colonial urban systems thus, contained a single 'core' or primate city, a specialized secondary centres and an array of small administrative and commercial centres acting as control and transmission points. Transport and communication lines emanated spoke-like of the wheel of a bicycle from the primate cities to tap controllable benefits from the hinterland, and connect this skeletal network to an efficient funnelling outward the resources available from the land.

Africa is considered as the least urbanized continent in comparison with other Third World countries, but in the decade 1981-1991 urbanization has taken place rapidly. On the whole, Northern Africa is more urbanized in comparison with south west, East Africa and Central Africa. In South Africa urbanization is relatively equal to that of North Africa. The percentage of cities of 100,000 and more inhabitants in 1960's was the lowest in the world with 9 per cent and the number of such cities was about 100.[5] According to 1981 figure the high urbanized countries are Libya (54%), South Africa (50%), Ivory Coast (41%) and

Congo (46%). The least urbanized countries are Rwanda (4%), Burundi (2%), Malawi (10%), Uganda (9%) and Upper Volta (11%) (Table 19.4).

TABLE 19.4 : Urban Development in Africa, 1981

Regions	*% of Total*		*% of Average Annual Growth Rate*	*% of Urban Population in largest city*
	1960	*1981*	*1970-81*	*1981*
Chad	7	19	6.5	39
Ethiopia	6	14	5.5	37
Mali	11	19	4.6	24
Malawi	4	10	7.0	19
Zaire	16	36	7.5	28
Uganda	5	9	3.4	52
Burundi	2	2	2.7	—
Upper Volta	5	11	6.0	41
Rwanda	2	4	6.4	—
Somalia	17	31	5.4	34
Tanzania	5	13	8.6	50
Guinea	10	20	6.0	80
Benin	10	15	4.1	63
Sierra-Leone	13	22	4.4	47
Madagascar	11	19	5.2	36
Niger	6	13	7.2	31
Mozambique	4	9	8.2	83
Sudan	10	26	7.1	31
Togo	10	21	6.6	60
Ghana	23	37	5.0	35
Kenya	7	15	7.3	57
Senegal	23	34	3.7	65
Mauritania	3	24	8.1	39
South Africa	47	50	3.1	13
Liberia	20	44	5.6	12
Tunisia	36	53	4.0	30
Lesotho	2	12	16.1	—
Algeria	30	44	5.6	12
Zambia	23	44	6.5	35
Libya	23	54	8.1	64
Egypt	38	44	2.9	39
Angola	10	22	5.8	64
Ivory Coast	19	41	8.3	34
Morocco	29	41	4.6	26
Nigeria	13	21	4.8	17
Zimbabwe	13	24	6.3	50
Cameroon	14	36	7.4	21
Congo	30	46	4.4	56

Source : World Development Report, 1984.

This indicates that, though Africa is contemporarily experiencing rapid urban growth, but this continent is still least urbanized in the world.

Some problems facing urbanization in Africa are pointed out by Berry (1976)[6] that the increasing primacy of the former colonial capitals, the weak development of middle sized towns, the concentration of infrastructure facilities in certains regions and urban residential sectors, the explosive growth of squatter settlements and unemployment, the pervasive migration of rural to urban areas lead people to a considerable discontent among the urban inhabitants.

Rate of Urban Growth

The rate of urban population increase in recent years for the world as a whole has been calculated at 3.2 per cent, but in the developing world this figure is over 5 per cent and the annual rate of growth for larger cities is even higher. Lusaka for example grew at an annual rate of almost 12 per cent between 1963 and 1998. By 1999 it has been expected that 25 cities in Latin America will have passed the million mark and four of them—Buenos Aires, Sao-Paulo, Rio de Janeiro and Bogota will each have more than 5 million inhabitants. In Asia, Calcutta already notoriously over crowded and with an estimated 600,000 entirely homeless and is expected to more than double its population to reach 15 million by 1995. This shows heavy migration from the countryside instead of natural growth. This alarming growth rate also shows 'Pull' of urban centres and 'Push' from the countryside. Paradoxically, the rate of urbanization and the current size of the major cities throughout the developed world seem to be more expressions of lack of economic development. Hoselitz for example, has drawn attention to the marked lack of coincidence between rates of urbanization and those of industrialization in India and this is more or less due to other parts of the world also.

Economic Growth and Patterns of Urbanization

The stages of economic growth by Rostow's assumes an 'urban revolution' of the same dimension in developing countries which occurred in the developed societies. This model is known as the historical model of urban economic growth.

The pattern now appears to be that urbanization is proceeding at a more rapid pace than the expansion of manufacturing, employment, which resulted into a direct shift out of agriculture into urban oriented services. Thus an understanding of the economic structure of the society and the economic processes which are offering is central to the analysis of the process of urbanization.

Due to difference in the current level of industrial development among under developed countries, there are distinct variations in their urbanization experience which are best taken into account through a typology. The basic criteria of differentiation are consequent on the current level of industrialization and urbanization (Table 19.5).

TABLE 19.5 : Comparison between Levels of Urbanization and Percentages of Active Population Employed in Manufacturing Industries

Continent	*Growth Rate of Urban Population A*	*% of Active Population in manufacturing Industry B*	*Relationship of A to B*
Africa (1990)	23.4	7.0	+ 91
Latin America (1990)	42.8	14.5	+ 126
Asia (1990)	23.7	9.0	+ 52

Source : Paul Bairoch : The Economic Development of the Third World since 1900.

The first category of countries comprises those in Latin America such as Argentina, Brazil, Chile, Colombia, Mexico and Venezuela, which have attained relatively a high degree of industrialization and the process of urbanization has been going on for over a century. At least 60 per cent of their population is now urban and only a small proportion remains in the rural areas. In some cases, however, urban growth has been concentrated in a single primate city as in Buenos-Aires of Argentina, Santiago of Chile and Lima of Peru.

The second category is represented by semi-industrialized countries whose urbanization experience is more recent. Most of these countries are to be found in Asia and North Africa and include such countries as Algeria, UAR, Korea, Malaysia and the Philippines. These countries have become increasingly urban with the growth of manufacturing activities and about 30 to 50 per cent of their population live in urban places. The urban development trends in Africa and Asia are more or less the same (Fig. 19.1).

The third group of countries is found in Africa south of the Sahara where below 25 per cent of the population is urban. These remain predominantly rural even though rural-urban income differentials have stimulated massive rural-urban migration.

TABLE 19.6 : Levels of Urbanization in Underdeveloped Regions (% of Total Population) in Town above 20,000 People in Third World Countries, 1920-2000

Year	*Africa*	*Latin America*	*Asia*	*Total*
1920	4.8	14.4	5.6	6.7
1930	5.9	16.8	6.6	7.8
1940	7.2	19.3	8.6	9.7
1950	9.7	25.1	11.6	12.9
1960	13.4	32.8	14.5	16.7
1970	16.0	37.8	17.1	19.7
1980	20.0	43.1	20.6	24.2
1990	25.0	46.0	25.3	27.5
2000	28.3	53.6	28.1	32.6

Source : United Nations Growth of the World's Urban and Rural Population, 1920-2000, New York, 1969.

Migration and Urbanization

The Third World today is the scene of the greatest movement of people in history, and though the waves of political refugees from war torn areas usually make the headlines, it is a quiter and spontaneous movement of people from the countryside to the cities which constitutes the mainstream. Hence, all over the developing world and especially in Latin America and Africa, there has begun a flight from the land, a tidal wave of country folk who now threaten to overwhelm the struggling cities and possibly even to bring about their social and economic disintegration.

Migration, therefore, put great strain on the city life to cope with rapidly growing numbers. In broad perspective the shift of people from rural to urban areas mainly reflects the process of industrialization and the changes it brings in the demand for labour. Some conditions in rural areas such as unequal land distribution, landlessness, agricultural mechanization, natural calamities, in the past forced labour migrations—have strongly influenced population movements in many countries. But by and large, people move to towns and cities for higher income and better job opportunities.

For one family, these attractions can be considerable. A move from the rural northeast of Brazil to Rio-de-Janeiro, for example, can roughly treble the income of an unskilled worker; the family income of a manual labour in Sao-Paulo is almost five times that of a farm labourer in the northeast. The higher costs of urban living may narrow rural-urban wage differentials in real terms, but urban dwellers also generally have much better access to basic public services, e.g., in rural areas of sub-Saharan Africa only 10 per cent of population have access to a safe water supply compared to 66 per cent of the urban population. Hence, the majority of the Third World countries are characterized by rapid urbanization. Much of this growth is greatly aided by migration from rural areas. In Cairo in 1991: "More than one-third of the permanent residents have been born outside the city." In Asia similar patterns also occur. In a survey conducted in Jakarta in 1998, over 75 per cent of the household heads were born outside the city.

In Latin America rural-urban migration contributes a much greater component of population increase of the cities. The population branch of the Bureau of Social Affairs, United Nations Reports that "in most of the countries under consideration . . . in the last interval between censuses, internal migration contributed half and in some instances even as much as two-third of the increase of the urban population."[7] During the last intercensal period the range varies from as much as 71 per cent of the urban population growth in Venezuela to 26 per cent in Cuba. Similar patterns have been reported in Africa, where the number of rural immigrants in towns was very high, especially in the industrial centres of the central and southern Africa. In Asia, Jakarta had over 75 per cent of the household heads born outside the city in 1991. In India, where the rural population is relatively more village oriented, the rate of net immigration into the cities between 1981-91 was equivalent to 20 per cent of the 1941 urban population.[8]

In Latin America rural-urban migration contributes a much greater component of the population increase of cities. During the last intercensal period the range varied from as much as 71 per cent of the urban population growth in Venezuela to 26 per cent in Cuba.[9] Similar patterns have also been reported in Tanzania, Africa where the rural to urban migrants is very high. The situation is more alarming especially in the industrial centres of central and southern Africa.

It has been observed that the Ecuadorean provinces of Quito and Guayaquil had the highest rates of immigration in Ecuador between 1950 and 1991 while receiving on an average over 50 per cent of public sector investment between 1955 and 1991.[10] Similarly heavy emphasis on the construction industry by the Colombian authorities during 1983, especially in Bogota has reduced unemployment of 11.3 per cent of the labour force in 1982 to 6.3 per cent at the end of 1983 even due to influx of people from rural areas. The experience in Brazil, particularly that of the Sao-Paulo region strongly suggests the importance of factors other than public policy in channelling migratory flows, and consequently on the formulation of national policies for urban growth, and it is not accidental that the Sao-Paulo state accounted for a full one fourth of the increase in the number of cities for Brazil as a whole over the pericd 1950-1990. The improved road access under the Sao-Paulo action plan is said to have fostered the cities growth rather than spurring deconcentration.

Except some problems, migrants are considered as assets to the urban economy. Evidence from Brazil, Colombia, Kenya, Korea, India and Malaysia suggests migrants increases the urban income. A World Bank study of Bogota, Colombia found that migrants earned more than non-migrants at all educational levels.

Migration and Aspects of Urban Deprivation

In the summer (April) 1971 India's largest city, Calcutta was fallen in crisis by the encoming of 5 million refugees in camps from Bangladesh. In the real sense there were no shelters and they huddled in the wilderness. Every form of shelter—even trees, bushes and overhanging rocks—had been occupied early in the exodus. Taki, a small town on the border, had 8,000 people as against its normal population of 3,000. Eventually most of the Taki's refugees would join the March to Calcutta. Cholera was already established among them and would add an extra dimension of horror to Calcutta's troubles. And even before the influx, almost 2 million of Calcutta's population were living either in cramped and squalid huts or literally on the streets. Though gross in degree, these glimpses of migration and the problem of urban shelter in India are by no means unique in kind in the Third World, but it is a common practice.

Food and Fuel Situation

By the year 2000, about 50 per cent of the developing world will be living in cities and hence Isabel (1981) stressess[11] the urgent need to develop urban ecosystems to provide food and fuel for the cities.

In Taipli, Taiwan, for example, the urban fringe used to supply 70 per cent of the total vegetable for the city. The loss of agricultural holdings on the fringe, however, reduced this amount to 30 per cent in 1974. These trends are most directly affecting the urban poor in many of the largest cities in the developing world. Examples from Cambodia and Somalia provide strong evidence of the difficulties associated with such a task in the face of a food shortage.

In the first half of twentieth century most urban areas around the world, produced a significant amount of food and other items required by local residents. Production was not limited to the urban fringe but included substantial yields in market gardens within the cities themselves. The Parsian 'marais' or market gardens of 100 years ago serve as an excellent example of this type of productive urban agro-ecosystem.

The cities of Addis Ababa, Ethiopia, and Lae, Papua New Guinea are examples of urban settlements which have become more self-reliant in fuel needs by developing urban forestry programmes. Upto 1984, the urban forestry of Addis Ababa reportedly had grown to 250 square kilometres covered with eucalyptus wood, which is used as fuelwood, lumber and fence posts. On the other hand, Lae's urban forestry scheme, begun in 1977, was designed to produce both fuelwood and tree crops and to arrest the erosion of hillsides around the city caused by rampant cutting of trees.

In the urban areas of China and especially other areas of Asia are promoting the recycling of human, animal, and vegetative wastes to create fuel. In the urban communes of China, small biogas units are used to produce fuel for the cooking and lighting needs of groups of upto 25 families. Other large-scale biogas plants such as one at Maya farms on the outskirts of Manila and the other one in Lae are constructed from industrial use.

Regarding food self-sufficiency for urban areas China offers the best example of large-scale urban food production in any country. At least 85 per cent of the vegetables consumed by urban residents in China are produced within the urban municipal boundary. The megalopolises of Shanghai (population 15 million) and Peking (population 12 million, 1991) are 100 per cent self-reliant in vegetable production.

Besides these, some of the other large cities in Asia such as Hong-Kong and Singapore, show impressive levels of food self-sufficiency. Hong Kong (population 6 million) with only 10 per cent of its total area used for farming, produced about 42 per cent of the fresh vegetables, 15 per cent of the pigs and 68 per cent of the live chickens consumed in 1991. Altogether 14,000 hectares of irrigated land in the Rimac Valley were lost to the growing urbanized region. In UAR more than 10 per cent of the nation's most productive land has been lost to urban expansion since 1965.

Due to this uncontrolled urban land market, the chaotic pattern of urban growth is evident in virtually all major Third World cities (Table 19.7). Metropolitan centres with more than a million people living in illegal developments are becoming increasingly common. This chaotic condition is found in Mexcio city, Caracas, Manila, Rio de Janeiro, Lagos, Calcutta, Cairo to name only a few as an example. In Bogota, a city with more than four million inhabitants, an estimate in the late 1970's and 1990's suggested that more than half the entire population live in areas which have been illegally occupied. In such a situation while the richer areas enjoy good infrastructure facilities and services, the illegal developments generally receive little or none.[13]

TABLE 19.7 : Uncontrolled Settlements in Selected Development Countries

Country	*City*	*Year*	*No. of Settlement in '000*	*% of City Population*
Tanzania	Dar es Salaam	1969	98	36
Senegal	Dakar	1969	150	30
Pakistan	Karachi	1964	752	33
Turkey	Ankara	1970	750	60
Colombia	Buenaventura	1964	88	80
Mexico	Mexico City	1966	1500	46
Peru	Lima	1969	1000	36
Venezuela	Maracaibo	1966	280	50

Source : UN General Assembly, Housing Building and Planning, Problems and Priorities in Human Settlements, 1970.

To combat the squatter problem, in 1965 Brazil set-up a National Housing Bank. Since 1965 it has provided man for over 750,000 dwellings, but the *favelas* increase. In case of Nigeria, it has been estimated that urban housing needs by 1985 has been about 200,000 units per annum. It seems that up to 1985 the set targets cannot be met and the housing and slum problems will further deteriorate.

To the city migrant, particularly of Africa and south east Asia, the shanty town is a reception area which has resemblances to the village. In some countries of Latin America the renovation of shanty town is progressing on self-help basis, especially for sanitation, water supply, lighting, education and dispensary services. Unless the measures are taken to reduce the flow of migrants from the countryside and to rehabilitate, those who are already in shanty slum, the urban life in a number of such cities may collapse.

About Caracas Charles wrote over a decade ago, "Caracas, Venezuela . . . is so crowded with squatters, traffic and people that unless other cities are built or at least expanded, the continued inflow of people and automobiles will throttle it."[14] Today, the situation is much worse. John Friedman has described the serious lack of regional balance and the dominance of Caracas in the life of the country and has suggested policies of regional development to counteract this.[15]

Problems of Urban Development

In urban areas of most Third World countries houses has been built without official permission as illegal settlements. This has increased the urban population manifold and hence growth from natural population increase has been swelled by the migration of people from smaller towns and rural areas. In these countries, the regional and national centres had a colonial origin and mainly grew with the export of primary agricultural produce and as colonial administrative centres. These centres generally developed as major urban centres after independence. Most centres have trebled their population since 1950, while the population of metropolitan cities like Bogota, Mexico, Manila, Jakarta, Sao-Paulo and Khartoum have more than quadrupled and those of many sub-Saharan African nations increased more than seven times. In such a situation the traditional approaches such as Master Plans, building codes and zoning norms have provided to be of little use in the face of such rapidly growing agglomerations where hundreds of thousands of people are added to the cities population even in urban slums.

We know that most cities of Latin America or the Third World are terminally ill, if not dead. We know that Lima, Buenosaires, Calcutta and Mumbai are beset with an awesome array of problems. Why then we are talking about urban development? Why should we be interested in the growth of cities? When all of us know that the problems faced by cities have doomed them. There will be about 6 billion people on earth by the year 2000 and as many as one-half of them may be urban. Unfortunately, however, in terms of individual Third World cities what Carrol (1966, p. 585) wrote of Manila in 1966 all too often typifies the contemporary urban situation.

"In the past twelve months, Manila has had a water shortage in which some 70 per cent of the Metropolitan area was without regular service . . . it has had a garbage crisis . . . there has been a school crisis, though a minor one this year . . . electric service went through a bad period some months ago . . . mail is in a continual state of crisis and in general it seems better to give up trying to use the telephone . . . police and fire protection are unreliable . . . the constantly increasing burden of traffic and the condition of the roads discouraged one from venturing beyond walking distance." In other words, social organization in the Manila area has not been able to maintain these services in the face of population increase and normal wear and tear on facilities, and at times it appears we are returning to a village type of existence. Yet in spontaneous settlements the shiftly made housing areas erected by urban squatters, are growing at even higher rates. For example, such rates are now reaching 12 per cent annually in several countries notably Mexico, Turkey, the Philippines and Peru. Lastly Rashmi Mayur highlighted the problems of Third World cities in the following words.[16]

> Runway population increases are at the root of the Third World urban crisis. These centres collect the over flow resulting from rural saturation and the exhaustion of resources to be shared there. Mexico city, Jakarta, Lagos, Mumbai are all moving in the same direction. Mexico city is already one large among polluted car parking lot.
>
> India is experiencing an annual 4.2 per cent growth in its cities. At the end of 20th century, 353 million people will live in its cities which will be the largest single urban population on the earth. Yet how is a country like India with limited resources to cope with such an normality?
>
> These cities hold the villages to ransom, steal their water and are parasitic on their resources. For example in Nagpur, middlemen from the city buy almost the entire orange crop. The villager pays sky-rocketing prices for whatever oranges remain in the region."

Our insatiable elities—rich, corrupt and manipulative are making profit from these large agglomerations. In their hands, our cities exploit, they do not civilize. We have paranoid urbanization rather than the city as liberator. Urbanites have an exploitable explosive power. Elites manipulate and frustrated city dwellers for political purposes. The media reinforce this.

These cities are centres of a false survival, paralyzing the rural environs. Thirty per cent of Third World urban populations are immigrant slum dwellers and are in a real sense refugees—though no international conference will be called on their behalf.

Solutions must involve controlling overall runaway numbers; use of space-age communications technologies to get knowledge and jobs into villages and stem the internal refugee tide in search of amenities; creative urban use of waste to produce agro-cities capable of meeting some of their food needs, new forms of habitat management as a government priority and as a national programme decentralization and subsequent grass-roots participation in decisions and policies; and reform of culturally alien, recently adopted lifestyles which our economies cannot support if the poor are ever to cross the poverty threshold in almost all Third World countries.

Hence, as we are living in an urban society, we cannot allow our cities to die or faced with the problem of deterioration. We should build and rebuild cities, because cities have much to offer us when the quality of life within them is high and much to trouble us when it is not.[17]

There is not enough sound housing for the poor in cities; there is too much crime, and too little high quality public education. Health care is not enough, besides, there is too often air pollution, and too little affordable public transportation. During the decades of 1960 to 2001 most countries of the world were so involved in fighting the war on poverty, nurturing the urban society and building model cities that we believed that all that was lacking was the will to solve our urban problems and reshape cities.

The population increase in Latin American and Afro-Asian countries by the end of 20th century will create enormous planning problems by five hundred million people, almost treble the present number. This growth will involve an acceleration in the expansion of existing urban centres to unmanageable levels unless steps are taken now to ensure that there are plans for the proper distribution of employment throughout each country. Such plans should include the establishment of growth points providing the economic basis for a more equitable distribution of social and other services. Furthermore to encourage stability of growth on a properly distributed basis, the gap between the income of the rural and urban populations must be reduced.

At the same time, Government should avoid the temptation of allowing all investors, whether they be public or private from home or abroad, to establish their interprises in the capital cities. They are in need of positive guidance and encouragement of investors towards the establishment of new industries and other facilities at the new growth points and the governments should assist in the building of these new centres by providing basic services so that they are attractive to investors. It may be desirable to give priority to a limited number of these growth points in the first instance so that efforts are not spread too thinly over the country as a whole and there are prospects of demonstrating the success of this policy at a reasonably early stage.

At the same time, it will be necessary to recognize that in the existing urban areas the self-employed in the informal employment sectors make a valuable contribution in the present circumstances. In differentiating between populations living in rural and urban areas the United Nations figure of 20,000 as the minimum size for an urban area is considered too high for Africa and, in making comparisons, it is desirable that figures should also be given a minimum size of 2,000. Too high a proportion of new development is going to capital cities as compared with rest of the country. For instance, 7 per cent of Kenya's population which is living in Nairobi receives 50 per cent of the country's registered employment. These are benefits for the capital city, but it encourages the rural-urban migration.

In many cities there are urgent problems of housing, ... transportation, and uncontrolled development, which in totality are lacking conditions and an ugly environment. Some of the difficulties arise out of lack of understanding between the politicians and the planners. Often there has been an opportunistic approach with commercial development encouraged at any price, even at excessive densities.

Commercial exploitation of land is increasing. In central area land values heights may endanger more modest developments designed to conform with previous height's control. The plot ratio method of control is encouraging to reduce height's as bulk of buildings leading to a deterioration in the appearance of the city and its skyline, as well as to traffic congestion.

The aim of development control should be to ensure a comprehensively designed urban environment with human scale. There is also developing a lack of consistency in the use of materials. It is desirable that every planning department should re-examine the effectiveness of its planing control, in the light of experience of the past five years.

FOOTNOTES

1. A.E. Smailes, *The Geography of Towns*, 5th edition, Hutchinson University Library, London, 1966.
2. R. Jones (ed.), *Eassys on World Urbanization*, George Philip and Sons Limited, London, 1975.
3. P. Mack, "The Trend Towards Urban Growth in the Past 180 Years," *North-Western Alumnie News*, November, 1981.
4. G.H. Bayer (ed.), *The Urban Explosion in Latin America: A Continent in the Process of Modernization*, Cornell University Press, New York, 1967.
5. G.W. Breese, *Urbanization in Newly Developing Countries*, Englewood Cliffs, Prentice Hall, 1966.
6. B.J.L. Berry, *Urbanization and Counter Urbanization*, Vol. II, Sage Publications, London, 1976, p. 259.

7. T.G. McGee, *The Urbanization Process in the Third World*, G. Bell and Sons Ltd., London, 1971, pp. 97-120.
8. J. Bogue Donald and Zachariah, K.C., "Urbanization and Migration in India" in Roy Turner (ed.) *India's Urban Future*, Berkeley, 1967, p. 29.
9. Philip M. Hauser (ed.), *Urbanization in Latin America*, New York, 1961, p. 31.
10. J.A.S. Ternent, "Urban Concentration and Dispersal: Urban Policies in Latin America" (ed., A. Gilbert) *Development Planning and Spatial Structure*, John Wiley & Sons, Chichester, 1978, pp. 169-95.
11. Isabale Wodi, "Fertile Cities, *Development Forum*, September, 1981, p. 7.
12. Alan B. Mountjoy, "Urbanization in he Third World," *The Third World Problems and Perspectives*, The McMillan Press Ltd., Hong Kong, 1981, pp. 102-11.
13. S. Blitzer, Hardoy, J. and Satterthwaite, D., "Plotting Out the Urban Future," *Development Forum*, March 1981, p. 3.
14. Abrams Charles, *Housing in the Modern World*, Faber and Faber, London, 1966, p. 137.
15. John Friedman, *Regional Development Policy: A Case Study of Venezuela*, M.I.T. Press, Cambridge, 1966.
16. Rashmi Mayur, "Mayur on Third World Cities," *Development Forum*, October, 1984, p. 16.
17. W.A. Sampson, "Is There Help for our Aling Cities," *North Western Alumni News*," January, 1982, p. 6.

20

TRENDS OF URBANIZATION IN INDIA

URBANIZATION is the most powerful socio-economic component of modernity. As the West has modern urban civilization than the east has ancient traditional civilization. The current is regarded as metropolitan civilization based on industrial and multi-national commercial development. The massive influx of population from rural to urban areas have created the problems of slums and squatters throughout the urban world.

According to G. Trewartha the level of urbanization is defined as the proportion of urban population to total population residing in urban places. The shift of population from village to city and the process of transformation of villages into city is called urbanization.

In India urbanization has come to occupy an important place in economic development of different regions. Urban development is a direct and immediate concern to 26 per cent of population who live in towns and cities.

History of Urban Growth

It is often said 'India lives in her huts and cottages of olden days' but it is also true that India has a tradition of urban living and town planning which goes back to 3000 B.C. The cities of Indus Valley Civilization, Mohenjodaro and Harrapa, which flourished 3000 B.C. were large and well planned (Fig. 20.1). The Indus Valley people had attained high standard of town planning and architectural style. Hence, in India urban tradition continued through centuries and during the ancient period of our history when there were many large well planned and beautiful cities in different parts of the country. Patliputra of Chandragupta Maurya, Ujjain of the Gupta's Kannauj, Banaras and Mathura were some of the great cities found in Northern India. In the South, cities were established in the Medieval period by the Chalukya's, the Rashtrakuta's, the Chola's, the Hoysala's and others. However, the urban tradition of the south could seem to continue through Vijayanagar which flourished during the 15th and 16th centuries to modern Mysore and Bangalore.

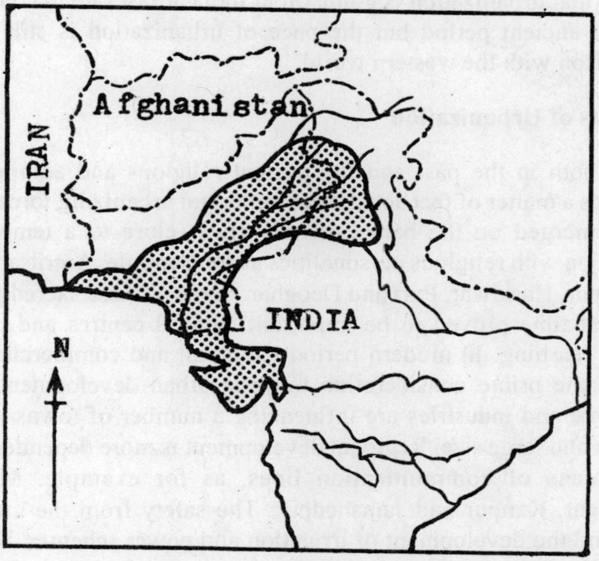

FIG. 20.1 : Ancient Urbanized Area of India and Pakistan

The Muslims and later the British made their own distinctive contributions to the evolution of cities in India. Many of our cities especially those in the North Indian plains bear the imprint of the Muslim influence. The metropolitan port cities like Calcutta, Mumbai and Chennai are the outstanding examples of British contribution. The cities like Delhi and Bangalore represent a syntheses of different traditions both indigenous and foreign including Hindu, Muslim and British. Thus the cities that arose in Pre-British India were few in number and archaic in type. They did not rest primarily upon industrial and commercial development but rather on political and religious functions. The Muslim rulers frequently shifted the seat of Government, and when they did so the capital city suddenly dwindled to dust and a new city arose around the next administrative headquarters. Cities also arose at strategic points where military required a permanent camp, such as bridge heads, mountain passes and cross roads. Still other cities arose at religious sites where pilgrimages were made and temples were built. In this way, for a long time, commerce and industry followed the cities located for non-economic reasons. Modern city has come to predominate just recently, which are based on industry, trade and natural resources. This new type indicate that urbanization is going on in India at fast rate in comparison with the ancient period but the pace of urbanization is still slow in comparison with the western world.

Elements of Urbanization

Hence, both in the past and the present religious and administrative centres as a matter of fact acted as an important urbanizing force. Sacred towns emerged on the banks of holy river, close to a temple or in association with religious personalities as for example, Amritsar, Ajmer, Gorakhpur, Haridwar, Puri and Deoghar. Some of these sacred places in course of time proved to be important cultural centres and principal seats of learning. In modern period industrial and commercial factors become the prime consideration towards urban development. Trade, commerce and industries are influencing a number of towns of small, medium and large size. Regional development is more dependent on the easy means of communication lines, as for example, Mirzapur, Modinagar, Kanpur and Jamshedpur. The safety from the menace of rivers and the development of irrigation and power schemes of various 'plans', further led to the growth of varied types of urban centres in the region. A number of new mill towns in the Western drier States of U.P., Haryana and Punjab on the one hand and on the other a number of frontier guard towns or check posts with industrial-cum-commercial activities have grown up recently along the border zone of North India. Towns also grew along the railway lines, which are especially noted for their busy railway activities. In southern India most of the cities have been developed as port, e.g., Mumbai, Ernaculam, Chennai, Vishakhapatnam, Cochi, Tutikorin and Marmogoa, etc.

In North Indian plains most of the cities have been evolved till the middle of the 19th century. When rivers were the only main arteries of movement and the settlement clung to their banks. In the last one hundred years the development of towns have been greatly influenced by the advantage of rail transport. The recent development of city centre takes the shape of star like formation pushing out the built up area along the main channel of movement. The urban expansion on the cost of productive agricultural land is continuously going on since the beginning of human civilization. This sort of growth is clearly visible in Calcutta, Kanpur, Allahabad and Delhi of northern India. These are the largest centres of trade and commerce, manufacturing and almost all of them are poly-nuclear origin. Calcutta urban agglomeration has 74 satellite urban centres, Kanpur has 6, whereas Allahabad and Delhi both have several smaller points of nucleation.

So far as the urbanization from demographic point of view is concerned, i.e., the proportion of population live in towns and cities, India is in the midst of gigantic urban increase. The growth of urbanization is an index of national development as it is most intimately related with industrialization and commercial development in the region.

Rural-Urban Ratio

In India 74 per cent population is rural and 26 per cent is urban. From 1921 to 1991 there has been a slow but steady growth of urbanization which is shown here in Table 20.1.

TABLE 20.1 : Percentage of Rural and Urban Population in India (1921-2001)

Year	*1921*	*1931*	*1941*	*1951*	*1961*	*1971*	*1981*	*1991*	*2001*
Rural	88.8	88.8	86.1	82.7	82.0	80.1	78.0	74.0	70.0
Urban	11.2	12.0	13.9	17.3	18.0	19.9	22.0	26.0	30.0

Urban centres in India, however, has grown at a fast rate, than the provision of infrastructure facilities mainly because of the low living standards of bulk of population who migrated to the urban places in search of livelihood.

The 'Urban-Rural Growth Differential' (URGD) was expected to go up, but this has declined during the last decade. Besides, the rate of growth of rural population has actually gone up during the last decade. The annual growth rate of rural population which had declined from 1.96 per cent in 1961-71 to 1.78 per cent in 1971-81 went up to 1.80 per cent in 1981-91 decade. Coupled with the sharp decline in the rate of growth of urban population, the increase in the rate of growth of rural population is one of the astonishing result of 1991 census. The 'Urban-Rural Growth Differential' was 1.25, 2.05 and 1.29 per cent respectively.

Growth of Urbanization

The growth of urban centres was slow from the year 1881 until 1941. The increase was from 9 to 13.9 per cent of the total population. Regarding the rapid growth of population since 1931 census, Yeat's Commissioner of the 1941 census wrote that "Another and much more potent reason than is usually realized, is the fact that city life has began really to appeal to the ordinary middle class or lower middle class Indian, because for the first time accommodation within his means and to his taste has become available. The huge blocks of flats which in less than a decade have completely altered the face of Mumbai and parts of Calcutta, with their amenities of running water, electric light and the city features of the tram, the bus, the cinema, etc. have meant that every year an increase in the number of persons could be seen who seek to pass their retirement or leisure in a city instead of their former home. The education question is also a powerful tool as the best education is not available in a country like India where even if the middle class attach much value to education. This is a powerful influence."[1] During 1941 census the urban population grew by about 40 per cent and by 1951 it comprised 17 per cent of the total population. The census data show that while the total population grew by 21.5 per·cent during the 1950's, urban population grew by 26 per cent. In between 1951 to 1961 the urban population increased from 17 to 18 per cent of the total population. This clearly shows that the tempo of urbanization had really slowed down, due to the changed definition as in 1961 more than 800 places were deurbanized in case the census definition of urban had not changed, the urban population would have increased by 48 per cent in 1991. During the period 1961-71 over all population grew by 7 per cent but the urban population grew by 37.9 per cent. The 1991 census data shows that India is urbanizing now in terms of population growth on overwhelming rate.

Comparing the decennial growth rates of urban population over last two decades, we find that States which were less urbanized in 1971 had a very high growth rate during 1971-81. In this decade, Orissa had a decadal urban growth of 68 per cent followed by Uttar Pradesh 61 per cent, Haryana 60 per cent, Rajasthan 59 per cent, Madhya Pradesh 56 per cent and Bihar 55 per cent. On the other hand, the States at high levels of urbanization are : Maharashtra, Tamil Nadu, West Bengal, Kerala and Gujarat. The Punjab had relatively much lesser growth rates of urban population. During 1981-91, however, the picture is somewhat confusing. The States of Tamil Nadu, West Bengal, Karnataka, Punjab, Gujarat and Maharashtra are more urbanized but the pace of urbanization is relatively low. In the States like Bihar, Orissa, Uttar Pradesh and Rajasthan the less urbanization is found due to agricultural economy (Fig. 20.2).

Table 20.2 indicates the increase of urban population from 15.05 to 212.86 millions in between 1872 to 1991 in India. From 1971 to 1991 the growth of urban population is almost double within 20 years and still India lags behind many well urbanized countries of the World.

TABLE 20.2 : Percentage Growth of Urban Population in India (1872-1991)

Census Year	*Urban Population in Millions*	*Decadal Growth in %*
1872	15.05	32.89
1881	20.00	10.85
1891	22.17	16.60
1901	25.85	0.35
1911	25.44	8.27
1921	28.08	19.12
1931	33.45	31.97
1941	44.15	41.42
1951	62.44	26.41
1961	78.93	38.22
1971	109.09	38.17
1981	156.18	23.34
1991	212.86	25.72

Source : Census of India, 1901, 1961 and 1971, 1981 and 1991.

Patterns of Urbanization

In India the total number of towns were 4,689 in 1991 census. The level of urbanization is 25.72 per cent whereas it was 23.34 per cent in 1981. Annual growth rate is just 0.24 per cent which is rather very slow. The highest urbanized State is Maharashtra 39 per cent and the least urbanized is Assam 11 per cent only. In Maharashtra the industrial development, transport linkage and business activities are relatively more developed in comparison with the counterpart of Assam and hence the States of Orissa, Bihar and Assam is least urbanized.

The level of urbanization in India shows that it reached the stage of Industrial Revolution of Western Europe of A.D. 1917. Although urbanization is a process of population growth in urban areas either through natural growth or immigration. In India, the highest urbanization has been found in the metropolitan areas of Delhi besides Chandigarh, Lakshadweep, Pondicherry and Daman & Diu. In these areas about 90 to 45 per cent people lives in urban areas. In West Bengal about 27 per cent people lives in urban areas followed by 34 per cent in Tamil Nadu, 29 per cent in Punjab, 26 per cent in Kerala, 38 per cent in Maharashtra, 34 per cent in Gujarat, 25 per cent in Haryana and 27 per cent in Andhra Pradesh. In India on an average 25.72 per cent people lives in urban areas. This clearly signifies that the majority of population in India still lives in rural areas (Table 20.3).

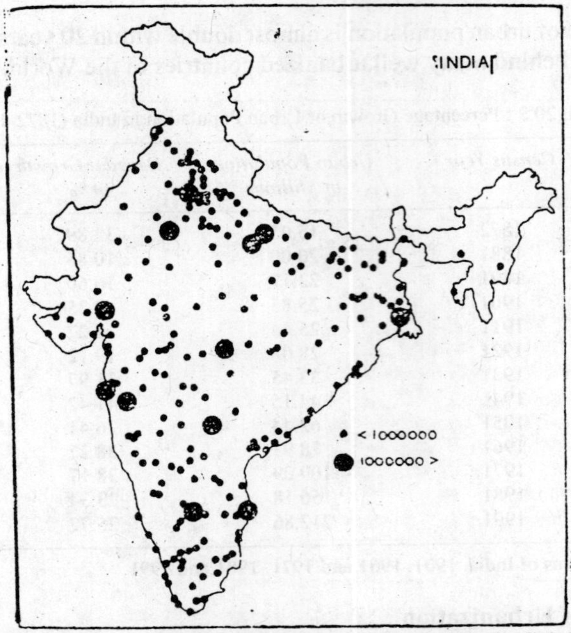

FIG. 20.2 : The Analysis of more than one lakh Population of India's Town

TABLE 20.3 : Urban Area, Number of Towns Population and Household in India, 1991

	States/Union Territory	*Area (in sq.km.) Urban*	*Towns Total Number*	*Urban Population*	*Number of Households*
(1)	(2)	(3)	(4)	(5)	(6)
	India	**64,613@**	**4,689**	**217,611,012**	**40,418,141**
1.	Andhra Pradesh	5,171	264	17,887,128	3,610,493
2.	Arunachal Pradesh	N.A.@	10	110,628	25,317
3.	Assam	828	93	2,487,795	480,219
4.	Bihar	3,744	271	11,353,012	1,836,794
5.	Goa	385	31	479,752	98,781
6.	Gujarat	5,137	264	14,246,061	2,688,348
7.	Haryana	967	94	4,054,744	732,335
8.	Himachal Pradesh	270	58	449,196	107,573
9.	Jammu & Kashmir	587@	74	1,839,400	N.A.
10.	Karnataka	4,270	306	13,907,788	2,591,441
11.	Kerala	3,365	197	7,680,294	1,411,033
12.	Madhya Pradesh	7,908	465	15,338,837	2,769,571
13.	Maharashtra	6,228	336	30,541,586	6,084,994
14.	Manipur	145	31	505,645	80,899
15.	Meghalaya	154	12	330,047	61,703
16.	Mizoram	493	22	317,946	57,295

(*Contd.*)

TABLE 20.3 (*Contd.*)

(1)	(2)	(3)	(4)	(5)	(6)
17.	Nagaland	147	9	208,223	42,287
18.	Orissa	2,544	124	4,234,983	831,226
19.	Punjab	1,441	120	5,993,225	1,069,570
20.	Rajasthan	4,864	222	10,067,113	1,715,858
21.	Sikkim	N.A.@	8	37,006	7,116
22.	Tamil Nadu	6,176	469	19,077,592	4,108,915
23.	Tripura	147	18	421,721	85,870
24.	Uttar Pradesh	5,603	753	27,605,915	4,353,385
25.	West Bengal	3,078	382	18,707,601	3,604,895
	Union Territory				
1.	Andaman & Nicobar Islands	14	1	74,955	16,439
2.	Chandigarh	78	5	575,829	128,306
3.	Dadra & Nagar Haveli	7	1	11,725	2,471
4.	Daman & Diu	23	2	47,543	9,351
5.	Delhi	685	32	8,471,625	1,699,618
6.	Lakshadweep	13	4	29,114	4,553
7.	Pondicherry	141	11	516,985	101,481

Source : Census of India, 1991, State Profile 1991, India.

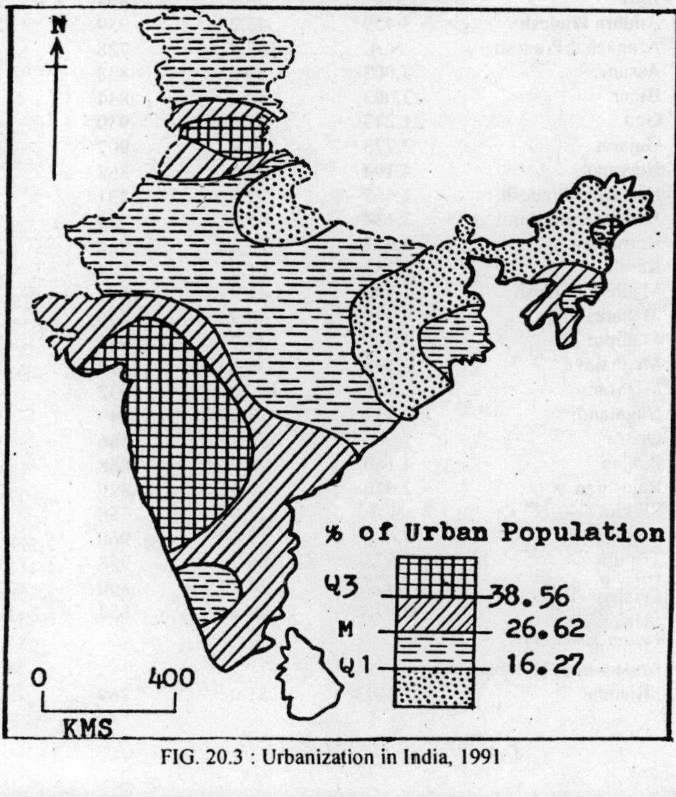

FIG. 20.3 : Urbanization in India, 1991

The low level of urbanization 12 per cent found in Arunachal Pradesh, 11 per cent in Assam, 13 in Bihar, 9 per cent in Himachal Pradesh, 13 per cent in Orissa, 9 per cent in Sikkim, 15 per cent in Tripura and 8 per cent in Dadra & Nagar Haveli. The causes of low-levels of urbanization are poor regional economic development, lack of good transport linkage, besides lesser development of industrialization. In this way, these states could be said as backward in the achievement of normal levels of civilization and culture (Table 20.4).

TABLE 20.4 : Density, Percent Decadal Variation, Sex Ratio and Percentage of Urban Population, 1991

	State/Union Territory	*Population Density, 1991 Urban*	*Percent Decadal Variation, 1981-91 Urban*	*Sex Ratio, 1991 (No. of Females per 1000 Males) Urban*	*Percentage of Urban Population 1991*
(1)	(2)	(3)	(4)	(5)	(6)
	India	**3,370***	**36.5**	**894**	**25.7**
1.	Andhra Pradesh	3,459	43.2	959	26.9
2.	Arunachal Pradesh	N.A.	167.0	728	12.8
3.	Assam	3,003	39.6	838	11.1
4.	Bihar	3,003	30.2	844	13.1
5.	Goa	1,247	48.6	930	41.0
6.	Gujarat	2,773	34.4	907	34.5
7.	Haryana	4,194	43.4	868	24.6
8.	Himachal Pradesh	1,665	37.8	831	8.7
9.	Jammu & Kashmir	3,134	45.9	930	23.8
10.	Karnataka	3,257	29.6	930	30.9
11.	Kerala	2,283	61.0	1,034	26.4
12.	Madhya Pradesh	1,940	44.9	893	23.2
13.	Maharashtra	4,940	38.9	875	38.7
14.	Manipur	3,479	34.7	975	27.5
15.	Meghalaya	2,146	36.8	910	18.6
16.	Mizoram	645	16.10	932	46.1
17.	Nagaland	1,414	73.2	749	17.2
18.	Orissa	1,665	36.2	866	13.4
19.	Punjab	4,160	28.9	868	29.5
20.	Rajasthan	2,070	39.6	879	22.9
21.	Sikkim	N.A.	27.6	750	9.1
22.	Tamil Nadu	3,089	19.6	960	34.2
23.	Tripura	2,873	86.7	958	15.3
24.	Uttar Pradesh	4,927	38.7	860	19.8
25.	West Bengal	6,079	29.5	858	27.5
	Union Territory				
1.	Andaman & Nicobar Islands	5,301	51.0	769	26.7

(*Contd.*)

TABLE 20.4 (*Contd.*)

(1)	(2)	(3)	(4)	(5)	(6)
2.	Chandigarh	7,382	36.2	810	89.7
3.	Dadra & Nagar Haveli	1,763	69.6	817	8.5
4.	Daman & Diu	2,035	63.8	1,024	46.8
5.	Delhi	12,361	46.9	830	89.9
6.	Lakshadweep	2,189	56.3	930	56.3
7.	Pondicherry	3,656	63.6	985	64.0

Source : Census of India, 1991, State Profile 1991, India.

The level of urbanization in India has gone up from 23.34 per cent in 1981 to 25.72 per cent in 1991 which recorded an increase of 2.38 per cent over the hole decade. This has recorded an annual increase of 0.24 per cent only. Whereas during the decade 1971-81 the average annual rate of increase in the level of urbanization was about 0.34 per cent which has higher than recorded in 1981-91 decade. The pace of increase is demographically known as 'tempo of urbanization' which has considerably slowed down over the last ten years as the economic growth takes off the level of urbanization being at a low level.[2] This should normally increase fast every decade and it is only than the level rich persons are sufficiently high for which the slow-down takes place.

A fall in the rural population growth rate is of fundamental importance for increasing the level of urbanization and for the implied increase in the level of economic development. In the 15 major States, which account for 97 per cent of the country's urban population, there has been fast increase in the annual growth rate of rural population during the last decade.[3]

When compared to the rest of the world, the level of urbanization in India is very low. The world as a whole had a level of 45 per cent in 1990, the highest level being 84-85 per cent in Australia and New Zealand followed by 77 per cent in Japan, between 72 to 75 per cent in North America, South America, Latin America and Europe, 66 per cent in Central America and former USSR, and 34 per cent in Africa and Asia and even Pakistan had a urbanization level at 32 per cent.

It has been estimated that by A.D. 2000 half of the world urban population will reside in India and most of the bigger cities, e.g., Mumbai, Calcutta and Shanghai will reach a mark of 1.5 crore persons.

From the economic point of view, the higher the level of urbanization, the higher is the stage of economic development which has been typified by a higher level of industrialization and development of modern sectors of commerce, transport and services. Economic development is further typified by a shift in the structure of work-force from agricultural to non-agricultural activities which, in turn, means a shift in population from the rural to urban areas, either by rural-urban migration or the development of rural areas themselves into industrial and commercial areas. Economic development is also characterized by a rise in agricultural productivity which also leads to a shift of the surplus population from agricultural to non-agricultural activities as has happened in the Punjab and Western Uttar Pradesh.

By definition, the Indian census treats those areas as urban which are either statutory towns or which have got a (a) minimum population of 5,000 persons coupled with; (b) three-fourth of its male work-force being engaged in non-agricultural economic activities, and (c) a density

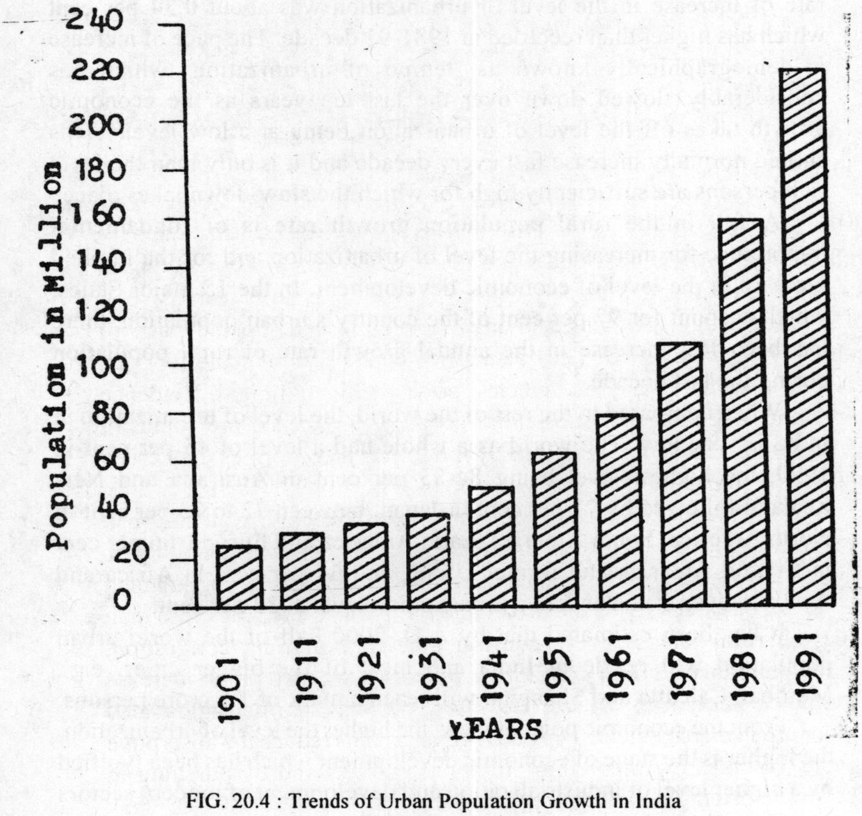

FIG. 20.4 : Trends of Urban Population Growth in India

of at least 400 persons per square kilometre. That is, even those areas which are classified as rural administratively would be treated as rurban in census if they fulfil the above three criteria. Such urban areas are called "census towns." In 1991, there were 1,693 census towns out of a total of 4,689 towns in the country.

The level of urbanization differs very sharply from State to State. Among the fifteen major States of India, four States had levels about 30 per cent, e.g., Maharashtra 39 per cent, Gujarat and Tamil Nadu 34 per cent and Karnataka 31 per cent, four States had levels between 25 and 30 per cent, Punjab 30 per cent, West Bengal and Andhra Pradesh 27 per cent and Kerala 26 per cent, four States were between 20 to 25 per cent, e.g., Haryana 25 per cent, Madhya Pradesh and Rajasthan 23 per cent and Uttar Pradesh 20 per cent while three States had levels below 15 per cent, e.g., Orissa and Bihar 13 per cent and Assam 11 per cent.[4]

The 1991 census data shows that India is urbanizing now in terms of population growth on overwhelming rate. India was farther behind in urbanization in 1991 than U.S.A. (66%) and Japan's (77%). These figures highlight the slow and steady progress of India's urbanization.

The Costs-benefit of Urbanization

Urban growth gives rise to various types of economies. Industries benefit from concentration of suppliers and consumers, which allow saving in communications and transport costs.[5] Large cities also provide big, diffentiated labour markets and may help to accelerate the pace of technological innovation. They also allow economies of scale for such services as water supply and electric power to be exploited. Evidence from India suggests that substantial economies of scale are found in cities of upto 150,000 population. Beyond this diseconomies may creep in any time in case the city sizes are too big.

Against these benefits, unemployment tends to be higher in urban areas in comparison with rural. In a survey of 14 developing countries, only one Iran had a higher rural unemployment than urban unemployment rate. In six other countries the urban unemployment rate was more than twice the rural rate. Surveys confirm that air pollution, congestion, social disturbances, crime and similar problems are also on an increase disproportionately with city size. But these problems are often aggravated by poor urban management. Typically, governments reduce the absorptive capacity of cities by intervening in labour markets and by pursuing in appropriate pricing policies for public services.

Housing and Urbanization

Urbanization has brought enormous problems of housing in its wake besides adding to the problems of health and scarcities of various kinds. A characteristic feature of urbanization in India has been that population tends to concentrate heavily in the larger cities. This has also created problems of town planning. The Ministry of Works and Housing looks after all the problems relating to urbanization and public health through its various wings and organizations namely Town and Country Planning Organization, Urban Community Development Cell, Urban Development Cell for Metropolitan Cities and other areas of national importance, Central Public Health and Environmental Engineering Organization.

Metropolitan Cities of India

A central scheme has been taken up during the Fifth Five Year Plan period which aims at providing central assistance to State Governments to supplement their resources for integrated urban development of metropolitan cities and other areas of national importance. The main reasons for the deteriorating situation in cities have been lack of integrated approach to the problems of urban development and urban administration, lack of funds, multiplicity of authorities in the fields of local government and urban planning and absence of any unified authority to formulate, coordinate, finance and supervise the implementation of various schemes and programmes.

The Fifth Five Year Plan has made a tentative outlay of Rs. 230 crores (excluding Rs. 20 crores specifically earmarked for the development of National Capital Region). Detailed guidelines have been issued to the State Governments for preparing projects for assistance under the scheme and they have been asked to take steps to draw up a comprehensive urban development plan. Table 20.5 for population of Metropolitan cities in India reveals that Mumbai is the biggest metropolis in India which has a population of 99 lakhs. This is the primate city of India. The Greater Mumbai along with its conurbation has a population of 126 lakhs. The metropolis of Calcutta has a population of 44 lakhs but the conurbation has a population of 108 lakhs. Similarly, Delhi has a population of 72 lakhs and Chennai 54 lakhs. The causes of such a huge concentration of population are commercial, administrative and port facilities found in these metropolis besides industrial and educational development. The other important metropolises are Ludhiana, Kalyan, Bhopal, Indore and Vishakhapatnam. The population of these metropolises are just above 10 lakhs.

TABLE 20.5 : Metropolitan Cities in India, 1991

Sl.No.	*Name*	*Population*	
1.	Hyderabad	30,05,496	5
2.	Vishakhapatnam	10,51,918	14
3.	Ahmedabad	28,72,865	6
4.	Bangalore	26,50,659	7
5.	Cochin	11,39,543	11
6.	Indore	10,86,673	12
7.	Bhopal	10,63,662	13
8.	Greater Mumbai	99,09,547	0
9.	Kalyan	10,14,062	15
10.	Ludhiana	10,12,062	16
11.	Jaipur	14,54,678	10
12.	Chennai	53,61,468	3
13.	Kanpur	19,62,750	8
14.	Lucknow	15,92,010	9
15.	Calcutta	43,88,262	4
16.	Delhi	71,54,755	2

Source : Census of India, 1991.

The most striking feature of urban development in India has been significantly the growth of metropolitan cities as compared to the small and medium size towns (Fig. 20.5).

India has twenty cities with a population of more than 1.5 million or more in 1991. In order of size these are Mumbai, Calcutta, Delhi, Chennai, Hyderabad, Bangalore, Ahmedabad and Kanpur. Calcutta conurbation has a population of about 108 lakhs, Greater Mumbai 128 lakhs, Delhi 93 lakhs and Chennai 54 lakhs.

There are more than 400 city centres in India which have a population of more than 100,000 each and more than 150 million people live in these cities.

The current State of India's urbanization can thus be characterized by five major distinguishing features :

(a) The level of urbanization in India is relatively low when compared to the rest of the world and three-fourth of India's population continues to live in rural areas which are pools of poverty and social backwardness;

(b) The pace of urbanization which had started picking up during 1971-81 has slowed down substantially, the current annual gain in urban population being just 0.24 per cent;

(c) There is a very high disparity in both the levels of urbanization as well as the pace of urbanization among various States;

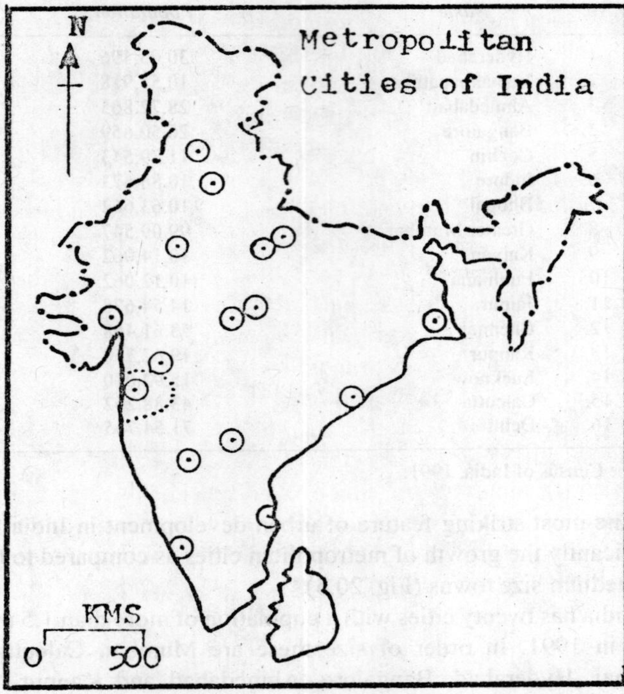

FIG. 20.5 : Metropolitan Cities of India

(d) In some of the more urbanized States, the urban growth has been halted leading to stagnation in urbanization and the implied stagnation in economic development; and

(e) Even in less urbanized States, the rate of growth of urban population has been arrested.

Urban Potentials and Future Estimate of Population in India

Benefitting by the development of untapped natural resources, the urban centres have increased rapidly and have filled up even the marginal and forested lands of India. Transportation, agriculture, location of industries and energy resources play an important role in the growth of towns and in increasing the potentiality of a place.

The urban population of India has shown a marked increase from 156 million in 1981 to 312 million in 1991. The country having an area of 3,787,782 kms. contained 4,689 urban centres with 25.72 per cent of total urban population in 1991 in comparison with 23.54 per cent in 1981. This is due to natural growth and migration as well as by expansion of urban areas themselves by addition of new towns and increase in the areas of some old towns.[6]

In India the highest 89.93 per cent of population is recorded in Delhi centrally ruled state followed by 89.09 per cent in Chandigarh. The least urbanized areas are Dadra & Nagar Haveli, Sikkim and Himachal Pradesh. In the plains of northern India due to rich agricultural tract the potentiality of growth is relatively high. At the state level the variations are even more pronounced. In the country as a whole the state of Bihar ranks 27th with 13.17 per cent urban population. West Bengal with 27.39 per cent and Kerala with 26.44 per cent are the thirteenth seventeenth in rank respectively. So far as the percentage of urban population to total population of state is concerned, this emphasizes the overwhelmingly rural character of the greater part of India. The purpose is to analyze the potentiality concept of Warntz (1960) for urban population in India.

Method of Study

The concept of 'potential' as developed by Warntez (1960)[7] and 'gravity model' developed by Reilley (1931)[8] appear in many forms in social science literature but here we are only concerned with the distance formulation and finding the potentiality of urban centres in India.

The potentiality at any point is defined as —

$$\frac{1}{r} DdA \qquad \text{or} \qquad \frac{\left(\sum_{i=1}^{n} A\ Pi/r\right)}{1} \qquad \text{or} \qquad \frac{Pi}{t}$$

where dA is the small element of the area, and r is the distance from each area to the point concerned. Similarly, the point at which $\frac{1}{r} DdA$ is minimized the integration defines the median centre.

In the second formula A is the total number of small sub-areas involved, Pi is the population in each sub-area and r is the distance from the particular point to a representative point in each sub-area. Court

(1969)[9] has also pointed out the usually simplifying assumptions which are made in computing the contribution of an area's own population that is 'self-potential' to the total potential existing at the central point of an area.

With the help of this potential value isopleths may also be used with effect in the plotting of potentials of population.[10] The influence of any point or population can be expressed as its number of people divided by its distance to the spot at which its potential is being assessed. "It has cultural, social, economic and political significance."[11]

Here an example is given in Table 20.6, to show the method of linear distance calculation for urban population potential of Patna.

In this way, the population potential for all the other 24 state capitals have been calculated and the result is given in Table 20.6.

According to the Table 20.6 and Fig. 20.6 the highest potential is recorded for Patna (836,004) while the lowest is for Palamau (Daltonganj, 97,340). The second lowest figure is for Saharsa, third for Champaran and the fourth for Santhal Parganas. This leads to the conclusion that Palamau, Saharsa and Santhal Parganas have least connections with the

TABLE 20.6 : Urban Population Potential of Patna (1991)

Points of Measurement	*Distance in kms.*	*Population Potentials or Per km. density of Population*
Patna — Gaya	82.68	4,112
Patna — Shahabad	124.80	2,515
Patna — Munger	121.68	3,800
Patna — Muzaffarpur	59.28	4,284
Patna — Darbhanga	85.80	2,704
Patna — Daltonganj	198.12	356
Patna — Dumka	238.68	769
Patna — Saharsa	135.72	784
Patna — Purnea	218.40	1,145
Patna — Saran	82.68	2,171
Patna — Motihari	118.56	1,584
Patna — Bhagalpur	171.60	1,292
Patna — Hazaribagh	168.80	2,307
Patna — Dhanbad	218.40	2,921
Patna — Chaibasa	327.60	1,953
Patna — Ranchi	235.56	1,514
Patna Own		9,16,980
Total Potential for Patna		9,51,291

centre of civilization, i.e., 'Patna' (in terms of social, cultural, economic and political relations). This is also because Palamau and Santhal Pargana have a large forest cover in the State of Bihar while Saharsa is liablle to floods and is covered with marshy land having least transport development.

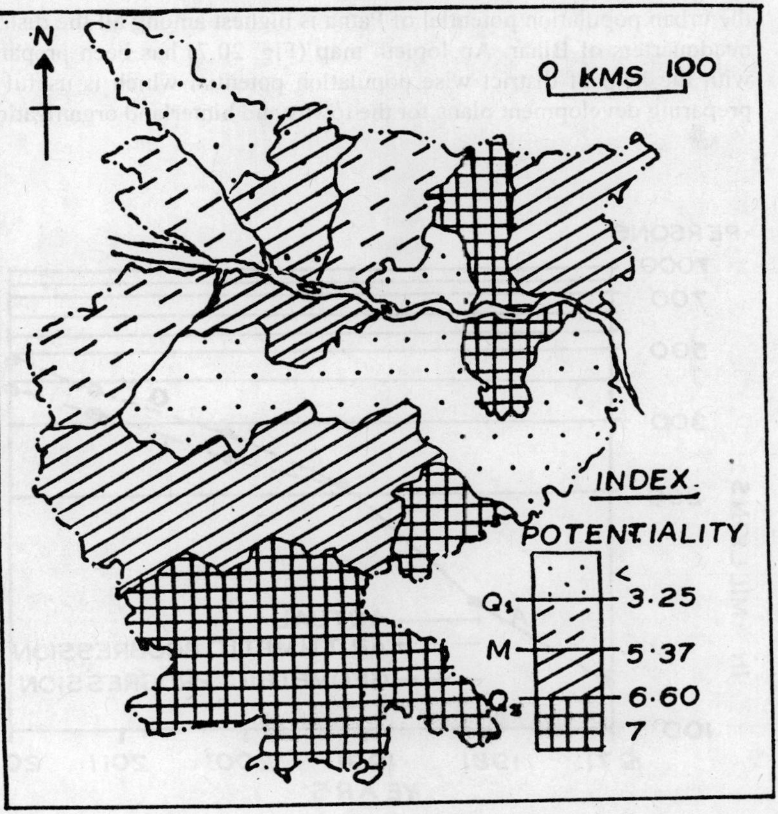

FIG. 20.6 : Bihar : Urban Potentiality (1991)

On the other hand, the districts of Dhanbad, Hazaribagh, Patna and Ranchi have a developed transport system and the growth of mining towns makes them economically significant. The transport, industrial and administrative activities have contributed towards high population potentials in these districts. In the immediately surrounding areas of Patna, the statellite urban centres of Hajipur, Pahleza Dinapur, Phulwari

Sharif and Fatwah are rural service centres and all of them have added to the population potential around Patna. Patna also serves as the metropolitan centre and capital of Bihar.

The hexagonal arrangement of cities of Muzaffarpur, Darbhanga, Munger, Biharsharif, Gaya, Arrah and Chapra around Patna provide their service to the respective zones of its hinterland. On account of this the urban population potential of Patna is highest among all the district headquarters of Bihar. An Iopleth map (Fig. 20.7) has been prepared with the help of district wise population potential which is useful in preparing development plans for the towns and hinterland organization.

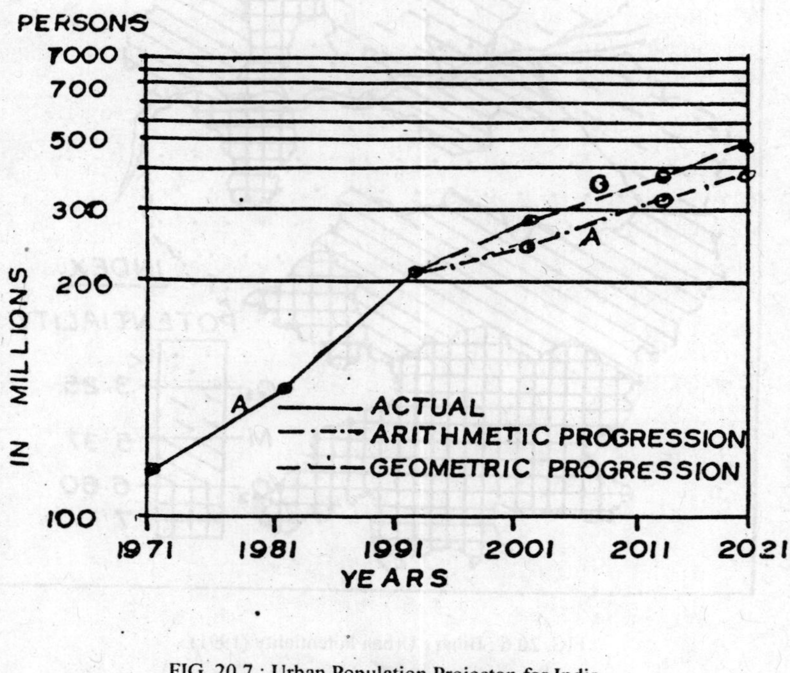

FIG. 20.7 : Urban Population Projecton for India

Future Estimate of Urban Population

Any attempt to make population projection is fraught with uncertainties as there is no fixed law of population increment and various techniques of projections are apt to give a generalized result. Being a number of techniques of population projection, it has been decided here to use simple arithmetic and geometric progression methods which are concerned with the fitting of a straight line, non-liner or logistic curve to the census figures available.

Urban population of India from 2001 to 2021 have been estimated on the basis of arithmetic and geometric progressions.

Arithmetic progression is based on the basis of natural growth of population per year, where as the geometric progression is calculated with the help of the formula :

$$Y = ar^{n-1}$$

where *Y* is the year chosen; *a* is the first year; *r* is the common ratio; and *n* is the interval of years.

In case of applying arithmetic and geometric progression methods, it will be desirable here to consider only 1981-91 census figures. Data for the earlier census periods show a fluctuating trend of growth and are likely to vitiate the result if taken into consideration.

According to Table 20.7, it seems that arithmetic progression is not sufficient to show the increase of urban population of India but geometric progression is likely to give the correct picture. Since India is a developing nation both from the point of view of economic progress and urbanization, the geometric progression will give a correct picture as India is just trying to establish itself on the industrial and regional development map of the world. On these considerations the urban.

TABLE 20.7 : Urban Population Projection in India

Arithmetic Progression

P. 1991	= 212,860,000
P. 1981	= 156,180,000
Difference	= $\dfrac{56,680,000}{10 \text{ years}}$ = 5,668,000
Therefore P. 2001	= P. 1981 + (*r* × 20)
	= 156,180,000 + 5,668,000 × 20
	= 156,180,000 + 113,360,000
	= 269,540,000 Population
Therefore P. 2011	= P. 1981 + (*r* × 30)
	= 156,180,000 + 5,668,000 × 30
	= 156,180,000 + 170,040,000
	= 326,220,000 Population
Therefore P. 2021	= P. 1981 + (*r* × 40)
	= 156,180,000 + 5,668,000 × 40
	= 156,180,000 + 226,720,000
	= 38,290,000 Population

Geometric Progression

$$y^{ar\ n-1}$$

where y is the year chosen for progression; a is the first year; r is the rate of population growth; n is the interval of years; and -1 is the constant.

---

r = P. 1991 = 212,860,000

P. 1981 = 156,180,000

(10th year from 1981)

Thus 212,860,000 = $156{,}180{,}000 \times r^{11-1}$

or $10 = \dfrac{212{,}860{,}000}{156{,}180{,}000}$

or Log r = 1/10 (log 212,860,000 − log 156,180,000)

= 1/10 (8.32808 − 8.19557)

= 1/10 (0.13251)

$= \dfrac{13251}{1000000} = 0.01325$

r = 0.01325 (rate of population growth)

Therefore P. 2001 = Log P. 1981 + r^{21-1}

= Log 156,180,000 + (0.01325 × 20)

= 8.19557 + (0.01325 × 20)

= 8.46057 (See Antilog)

= 288,980,000 (Population)

Therefore P. 2001 = Log P. 1981 + r^{31-1}

= Log 156,180,000 + (0.01325 × 30)

= 8.19557 + (0.01325 × 30)

= 8.59310 (See Antilog)

= 391,830,000 Population

Therefore P. 2021 = P. 1981 + r^{41-1}

= Log 156,180,000 + (0.13251 × 40)

= 8.19557 + 0.530040

= 8.72561 (See Antilog)

= 531,620,000 Population.

Population of India is likely to more than double by 2011 over 1991 figures.

Source : Calculated on the basis of 1981 and 1991 census data.

The urban potential by Warntez method shows the urban population per km. over the space of an urban centre in relation to different outlying nodes over greater part of the country.

Urban population potential is the sign of economic development and cultural advancement of an urban centre in comparison with the primate city of that country. Thus higher potentiality and higher commuter relationship increase the rate of rural population migration within the zone of influence of a particular urban centre. Percentage increase and concentration of population in a particular state depend upon the number and size of urban centres in the area.

FOOTNOTES

1. M.S. Thacker, *India's Urban Problem*, Mysore, 1965, p. 7.
2. J.P. Singh, "Demographic Attributes of Urban India," *Dimensions in Geography* (eds. Sinha, V.N.P. and Mandal, R.B), Associated Book Agency, Patna, 1979, pp. 171-94.
3. K. Davis, *The Population of India and Pakistan*, Russell and Russell, New York, 1951, p. 133.
4. R.B. Mandal, "Population and Settlements of Manipur," *Indian Geographical Studies*, R.B. No. 5, September, 1975, p. 21-22.
5. G.J. Lewis, "Commuting and the Village in Mid-Wales," *Geography*, Vol. 52, 1967.
6. R.B. Mandal, "Growth of Urban Population in Bihar," *The Geographical Knowledge*, 1975.
7. W. Warntez and Neft, I., "Contribution to a Statistical Methodology for Areal Distributions," *Journal of Regional Science*, Vol. 2, 1960, pp. 45-46.
8. W.S. Reilley, "The Law of Retail Gravitation," New York, Quoted by Berry, B.J.L., *Geography of Market Centres and Retail Distribution*, Prentice Hall, Inc., 1945, p. 40.
9. A. Court, "Population Distributions and Self-Potential," Mimeographed Paper, Department of Geography, San Fernand Valley State College, quoted by King, L.C., *Statistical Analysis in Geography*, 1969, pp. 96-97.
10. J.Q. Stwart, "Empirical Mathematical Rules Concerning the Distribution and Equilibrium of Population," *Geographical Review*, Vol. 37, 1947, pp. 461-85.
11. F.J. Monkhouse, and Wikinson, H.R., *Maps and Diagrams*, Methuen and Co. Ltd., London, 1971, p. 359-61.

21

URBAN REGIONS OF INDIA

THE RECENT progress of urbanization in India has been achieved during the last 50 years. In 1901 only 10.9 per cent population resides in urban centres. In 1941 this has gone upto 14.1 per cent. In 1951 it was 17.6 per cent, in 1961—18.3 per cent, in 1971—20.21 per cent, in 1981—23.7 per cent and in 1991 it was about 26 per cent. This overwhelming growth of urban population is due to the increased migration of population from rural to urban areas besides the rise of industrialization and revolution in transport and business sectors (Fig. 21.1). But the pace of urbanization in India is still slow in comparison with U.K., U.S.A. and Japan.[1]

On the basis of population density in cities and their relative distance K.L. Dutta has divided India into five urban zones.[2]

1. Urban zone of North Indian plains.
2. Urban zone of Southern India.
3. Urban zone of Mumbai-Ahmedabad.
4. Urban zone of Krishna-Godavari Delta.
5. Urban zone of Upper Krishna Basin.
6. Island Regions.

The description of above urban zones are as follows :

North Indian Plains

It includes the North Western part, Central part and and North-Eastern part of the country.

North Western Part

It include the urban centres of Delhi, Chandigarh, Simla, Srinagar, Jammu, Jaipur, Ludhiana, Fatehgarh, Sonipat, Alwar, Kota, Dwarika and Jodhpur. This area is known as the Core of urban expansion in India. Here Delhi is the biggest metropolitan centre having a population of 84 lakhs in 1991.

Central Part

It includes the area from Delhi upto Varanasi. The nucleus are Allahabad, Kanpur, Agra, Mathura, Meerut, Gorakhpur, Hathrus, Chhattisgarh and Varanasi. Among these Allahabad and Kanpur are metropolices whereas others are administrative and commercial centres. Most of these cities are megalocephalis which are necessarily multi-functional in nature.

North-Eastern Part

It includes the urban areas of Patna, Bhagalpur, Munger, Dhanbad, Jamshedpur, Ranchi, Biharsharif, Darbhanga, Muzaffarpur, Asansol, Calcutta, Burdwan, Bhubneshwar, Bokaro, Gauhati, Shillong and Imphal. Here the biggest urban concentration is found from Calcutta conurbation upto the Damodar Valley. This area is developing in the form of megalopolis. The industrial complexes of Durgapur, Kulti-Burnpur, Bokaro, Bhilai, Rourkela and the mining centres of Raniganj, Jharia and Ghatshila are the nucleus and nodal points of urban expansion and the whole area is developing as a biggest urban zone in India. On the other hand in Assam, Manipur, Tripura, Arunachal Pradesh and Nagaland the urban centres are few and far between. Calcutta as the primate metropolis of this region has a pollution of 108 lakhs in 1991 census.

Southern India

It includes the urban centres of Chennai, Mysore, Bangalore, Hyderabad, Trivandrum, Nagpur, Cochin and Vishakhapatnam. Some of them are ports, e.g., Cochin, while some others are religious centres as Madurai, still others are industrial as Madurai. Some are industrial centres as Bangalore whereas others are administrative-cum-commercial centres. Chennai the primate metropolis of this region which has a population of 53 lakhs in 1991 census.

Mumbai

Ahmedabad Urban Zone

This area specializes in cotton textile industry and the important centres are Ahmedabad, Baroda, Ankleshwar, Varouch, Jalgaon, Amravati, Surat,

Nagpur, Pune, Akola and Nasik. The conurbation of Mumbai is the nodal point, port town, administrative centre, service centre and collection point of raw materials. Mumbai the primate city of this region has a population of 128 lakhs in 1991 census.

Urban Zone of Krishan and Godavari Delta

This has an area of 8900 km^2 and the number of urban centres are 10. Some of the important centres are Vishakhapatnam, Machhlipattam, Kakinada, Vijaywada, Rajmahendry, Eloru, Guntur, Kotagudam, Vijaynagar, Salem and Tainali. The above three are port cities whereas others are industrial centres.

Urban Zone of Upper Krishna River

Due to industrial development and multi-purpose project many urban centres have cropped up in this region. Here more than five cities are such which have a population of more than one lakh each. These urban centres are Bellary, Garag, Raichur, Arohi, Hospet, Anantpur, Praudator and Ichalkaranjee. Recently, the establishment of iron and steel industry in Hospet has accelerated the pace of urbanization due to migrational accretion.

Island Regions

In Islands of India Goa has 5 lakhs urban population in its 26 urban centres. Some of the important ones are the capital Panji having a population of 85,199 and Marmogoa having 91,285 persons. Andman and Nicobar Island has only one town of Port Blair having a population of 74,810. In Dadra & Nagar Haveli Silvasa has a population of 11,720 and Daman & Diu has a urban population of 47,538 in 1991. In Lakshadweep—Kavrati has a population of 8,664 and Minicoy has a population of 8,313 in 1991.

Himachal Pradesh

It has a total urban population of 444,824 in 1991. There are 55 urban centres of small population size in Himachal Pradesh. The small town is Nainadevi with a population size of only 500 and the largest one is Simla with 109,840 persons. Chamba is highly congested town with 4,000 persons per km^2 while Narkanda has only 300 persons per km^2. Urban population is only 10 per cent of the total population of Himachal

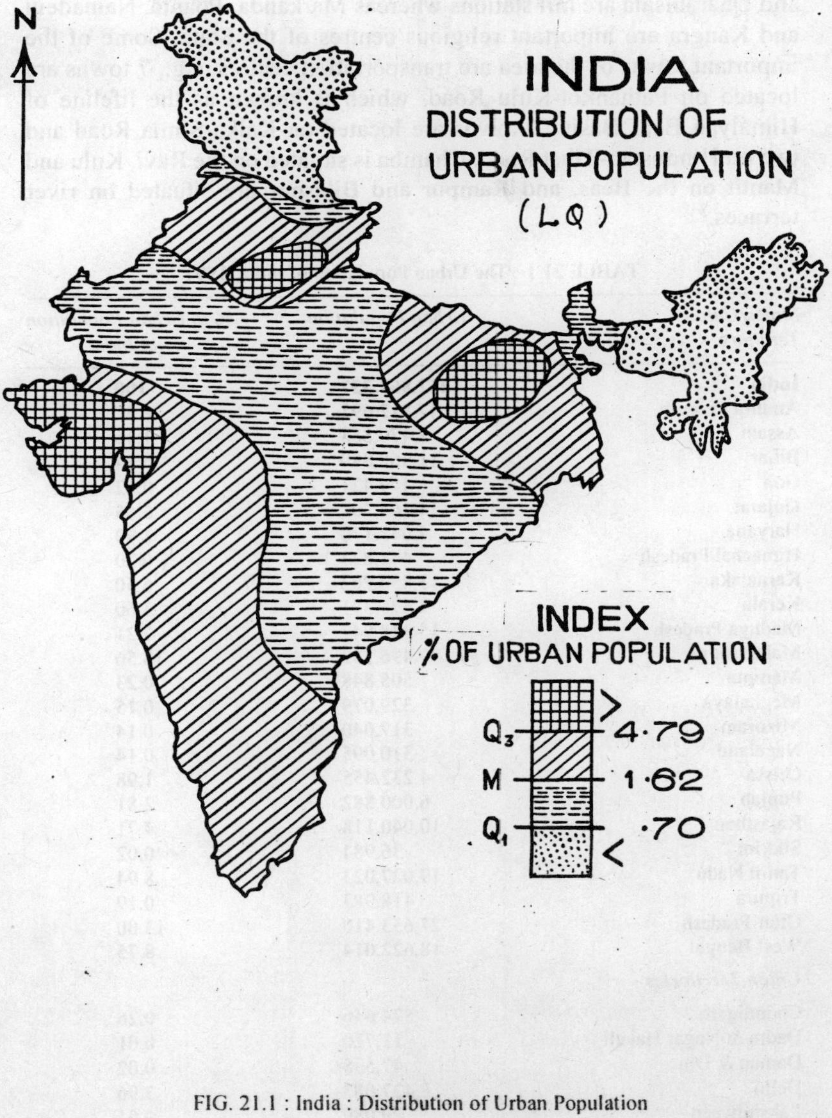

FIG. 21.1 : India : Distribution of Urban Population

Pradesh. Here the concentration of non-agricultural population is found only at the nodal point of transport, industry and various other services. Most of the towns are district headquarters such as Sabathu, Dagshai, Jutosh, Kassuli, Solan, Yol, Dalhousie and Bakloh. Simla, Dalhousie and Dharamsala are hill stations whereas Markanda, Poanta, Nainadevi and Kangra are important religious centres of the State. Some of the important towns of the area are transportation centres, e.g., 7 towns are located on Pathankot-Kulu Road, which is known as the lifeline of Himalyan Beas Basin; 7 towns are located on Kalka-Simla Road and two on Hindustan-Tibet Road. Chamba is situated on the Ravi, Kulu and Mandi on the Beas, and Rampur and Bilaspur are situated on river terraces.[3]

TABLE 21.1 : The Urban Population of India, 1991

States/Union Territory	*Urban Population*	*% to Urban Population of India*
India	**212,867,337**	**100**
Andhra Pradesh	17,812,693	8.37
Assam	2,470,888	1.16
Bihar	11,368,889	5.34
Goa	479,421	0.22
Gujarat	14,164,301	6.65
Haryana	4,045,170	1.90
Himachal Pradesh	444,824	0.20
Karnataka	13,850,702	6.50
Kerala	7,676,371	3.60
Madhya Pradesh	15,348,047	7.21
Maharashtra	39,496,352	18.56
Manipur	505,848	0.23
Meghalaya	329,079	0.15
Mizoram	317,040	0.14
Nagaland	310,095	0.14
Orissa	4,232,455	1.98
Punjab	6,000,882	2.81
Rajasthan	10,040,118	4.71
Sikkim	36,984	0.02
Tamil Nadu	19,027,023	8.94
Tripura	418,983	0.19
Uttar Pradesh	27,653,410	13.00
West Bengal	18,622,014	8.75
Union Territories		
Chandigarh	574,646	0.26
Dadra & Nagar Haveli	11,720	0.01
Daman & Diu	47,538	0.02
Delhi	8,427,083	3.96
Lakshdweep	29,089	0.01
Pondicherry	516,934	0.24

TABLE 21.2 : Size Class of Towns in India, 1991

States	*Above 100,000*	*100,000 to 50,000*	*50,000 to 25,000*	*25,000 to 10,000*	*10,000 to 5,000*	*Below 5,000*
Andhra Pradesh	32	34	91	39	14	3
Assam	4	4	20	32	15	12
Bihar	17	38	79	53	29	5
Goa	0	3	1	8	10	4
Gujarat	21	27	50	74	44	9
Haryana	12	9	17	30	20	2
Himachal Pradesh	1	0	4	7	9	34
Karnataka	21	17	82	70	40	24
Kerala	14	9	46	34	6	0
Madhya Pradesh	23	29	69	177	130	5
Maharashtra	27	28	103	83	40	9
Manipur	1	0	3	5	17	4
Meghalaya	1	0	2	2	2	0
Mizoram	1	0	2	3	5	11
Nagaland	0	2	2	3	2	0
Orissa	7	10	26	51	22	3
Punjab	9	18	25	44	16	7
Rajasthan	14	20	71	87	22	0
Sikkim	0	0	1	0	0	7
Tamil Nadu	25	42	71	80	35	7
Tripura	1	0	4	7	4	2
Uttar Pradesh	42	45	129	236	210	40
West Bengal	23	18	46	33	34	6
Union Territories						
Andman & Nicobar Islands	0	1	0	0	0	0
Chandigarh	1	0	0	0	0	0
Dadra & Nagar Haveli	0	0	0	1	0	0
Daman & Diu	0	0	2	0	0	0
Delhi	1	0	0	3	3	2
Lakshdweep	0	0	0	0	4	0
Pondicherry	1	1	1	2	2	0
India	299	355	947	1,164	735	196

Source : Census of India, 1991.

Simla is the largest capital town of Himachal Pradesh. It has an area of 15 km^2 and a population of 109,840 (1991). It is situated on a high ridge at an elevation of 2,205 metres. It is accessible by road from Kangra, Mandi and Nahan. It has been made the Summer Capital of India in 1861. It is an hill resort town.

Madhya Pradesh

Indore is a planned city of India. It has a population of 11 lakhs in 1991. Roads and bridges are constructed at proper intervals to keep pace with future development and industrial establishments which lie in the fringe area. As the city is centrally located it has specialized commercial markets like the cotton market, M.T.C. cloth market, Nandlalpura vegetable and fruit market, Lakshminagar grain market, Surjnaganj and Jawaharmarg are hardware and market transport goods; Navlakha and Labria Bhern—saw mills and timber stalls. The industrial zone is located in the north and the expansion of the city is going on along Agra-Mumbai National Highway No. 3.

In Bundelkhand, Jhansi is an important city of unplanned narrow and zig-zag lanes and are lined by old fashioned but the built up of houses are strong.

Important Cities	*Population 1991*
Indore (MC)	1,086,673
Bhopal (MC)	1,063,662
Jabalpur (UA)	887,188

In Madhya Pradesh the number of towns are 433 in 1991 and about 15 millions population live in urban areas which comes to about 15 per cent of the total population of the State. Jabalpur is a city level centre followed by Murwara, Mhlda, Balaghat, Narsinghpur, Ambikapur, Umaria and Kaimur town. All these are mining and industrial towns.

In Chhattisgarh region the important urban centres are Raipur, Bhilai, Rajhara and Nandini. Raipur is an unplanned city but some of its colonies are planned, e.g., Panchsheelnagar, New Panchsheelnagar, Shankernagar, Vivekanandnagar and Chaubey colony.

Maharashtra

It has a total urban population of 40 million living in 290 urban centres in 1991. About 25 per cent population of Maharashtra lives in towns of different sizes. The earliest towns are Paithan, Junnar Karad and Deogiri all of which were governed by administrative and defence considerations in their location. Sholapur and Barsi in the Bhima Valley, Satara in Krishna Valley and Nandurbar and Dhulia in Tapti Valley are important market centres. Nasik, Pandharpur and Trimbak are religious centres. The towns of Maratha chieftains are Sangli, Kurundwad, Kagal,

Ichalkaranji, Miraj, Phaltan and Bhor to mention a few. Some other important towns are Aurangabad, Kolhapur, Poona and Mumbai. The Greater Mumbai has a population of 126 lakhs followed by Kalyan 10 lakhs, Thana 8 lakhs, Ulhasnagar 4 lakhs and New Mumbai 3 lakhs.

Orissa

It has a urban population of 42 lakhs living in 119 urban centres. Rourkela is a steel plant in Sundargarh district and Hirakund in Sambalpur district. There are associated with industrial development and hydro-electric project. In north western part, the important urban centres are Bolangir, Sambalpur and Rourkela. In Boudh-Khondmal district the urban population is only 2 per cent. In Dhenkanal and Keonjhar about 5 per cent population is urban and in Mayurbhanj about 3 per cent population is urban. Other important towns are Sonepur, Bellanguntha, Bargarh and Jharsuguda as industrial, commercial and service centres. In Kalahandi and Koraput districts the number of towns are 12. In Koraput Salm, Pargatipuram, Sirkakulam and Yeypore are important towns. Other parts of Orissa is less urbanized and most of the towns are administrative centres. Both Bhubneshwar and Cuttack have a population of 4 lakhs respectively in 1991.

Tamil Nadu

It has a urban population of 190 millions living in 260 urban centres. In Tamil Nadu the concentration of towns is found in the Nilgiris and north-western hilly tract due to the growth of plantation and other allied industries. In the south towns are found in the Valley of Cumbum, the Vaigal and the Tambraparni. In Coimbatore upland the industrial development has favoured the uneven distribution of urban centres.

In Tamil Nadu-Palmi and Bhavani are religious towns, Vellor and Hosur are health and defence towns, Pollachi-bridge head, Salem-industrial town, Ootywater-point, Coutalam-waterfalls are some of the important factors responsible for the growth and development of urban centres. Chennai is the primate city which has attained a population of 54 lakhs in 1991.

Gujarat

This State has a urban population of 14 million living in 225 urban centres. About 50 per cent urban population of Gujarat resides in 15 large towns and cities of the State (50,000-100,000). The majority of towns of the area are Baroda, Rajkot, Jamnagar, Bhavnagar, Ahmedabad and Surat. Most of the towns are industrial, administrative and port location, e.g., Kandla. On the other hand, Dwarka, Somnath and Dakor have developed a religious centres of the State.

Ahmedabad is a primate city, walled city of the State of Gujarat is located on the bank of Sabarmati river. The university campus helds site in the central part of the western segment. Gandhinagar is a planned capital city of Gujarat located along the Sabarmati river 24 kilometres north of Gujarat. It has the facility of National Highway No. 8 and airport. In 1991 Ahmedabad has a population of 29 lakhs.

Karnataka

It has a urban population of 14 million in 1991 living in 254 urban centres. The oldest towns of Karnataka are Vatapi, Pampapur, Gokaran and Singeri. Talakad, Halepid and Belm are temple towns. In Karnataka—Belgaum, Bijapur, Gulbarga, Kalyani, Jalikot, Devanballi, Paravada, Shinaripur, Hosadwrja and Hosanagar are all fort towns. Bangalore has also developed as fort town and these days it is an industrial centre of HAL, ITI and HMT, etc. In Mandya and Godag, industrial towns occupy about 10 per cent and 5 per cent of the total town area whereas in Belgaum 15 per cent land is under administrative zone. In Bangalore and Athani the central business district is located near Chauk Bazar.

Mysore is a garden city of open spaces which developed as metropolis of the region. It has ridge-head location and salubrious climate. In Bangalore the number of well-planned religious neighbourhood have developed, e.g., Jayanagar, Gandhinagar, Shantinagar and Malleswaram, etc. Pattach is still the main commercial hub particularly along cross-roads. The wholesale market is located in the Old and New Tharangupet. The Lalbagh and Cubbon form the heart of the city of Bangalore. The metropolis of Bangalore has attained a population of 41 lakhs in 1991.

Andhra Pradesh

It has a urban population of 18 million living in 213 urban centres. Urbanization gets momentum since independence in Andhra Pradesh. The metropolis of Hyderabad has developed into a large administrative, business, industrial and educational centre. Kurnool is the old capital of Andhra Pradesh. Mothagudem has developed as urban place in 1941, followed by Sirpur, Kagajnagar, Mancheriyal, Markapur, Warangal,

Khammam, Nalgonda, Nizamabad, Anantapur, Cuddapah and Adoni. Most of the towns of Andhra Pradesh are religious, administrative and industrial centres where retail business developed, just to serve the local people.

The metropolis of Hyderabad covers an area of 225 km^2. This is the primate city and serve as the capital of Andhra Pradesh. It comprises the following important units namely—Alwar, Attapur, Bowenpalle, Fatehnagar, Hyderabad and Secunderabad Divisions of Municipal Corporation, Kandikal, Lalguda, Macha Bolaram, Malkazigiri, Osmania University, Secunderabad cantonment and Zamistanpur which evolved and included in the metropolis in different phases. In 1991 the population of the metropolis of Hyderabad was 30 lakhs and Vishakhapatnam 10 lakhs. The other important cities are Vijayanagaram, Guntur, Warangal and Rajamundry.

Uttar Pradesh

Uttar Pradesh has a urban population of 28 million living in 702 urban centres. In Uttar Pradesh the highly urbanized sector is Yamunapur (30.4%) followed by lower doab (24.3%), upper doab (18.11%) and Rohilkhand. The level of urbanization increases from east to west and the few well marked pockets of urban concentration are noteworthy, e.g., Gorakhpur, Varanasi, Kanpur, Allahabad, Rae Bareli, Lucknow, Haridwar, Hathras, Meerut, Mathura and Vrindavan, etc. the western part of Uttar Pradesh is highly urbanized due to historical importance and industrial development (Table 21.3). Besides, the ease of roads and railway network, irrigation channels and electricity grid paved the way for an economic prosperity of an area, and gives impetus to the urban nuclie to grow.

TABLE 21.3 : Population of some of the Important Cities of U.P. (1991)

Cities	*Population*
Kanpur	1,962,750
Lucknow	1,592,010
Varanasi	929,089
Agra	899,195
Allahabad	820,092
Meerut	752,078

Source : Census of India, 1991.

Lucknow

The city of Lucknow is the capital town and the garden city of Uttar Pradesh. Here the area under parks and gardens are a bit more than others. The old settlements found south of the Gomati river as Machchi Bhawan in order to have safety from floods. *Chauk* is the earliest developed business quarter in Lucknow. Here lie the famous gold and silver markets of the city and other retail business. As in the south the land is low, the built-up area extended towards south-west as Aminabad market. It is an important business area of the city. The construction of Bundh along the Gomati river favours the location of Civil Lines and administrative zones. Here lies Hazaratganj the most modern and fashionable market of Lucknow. The construction of bridges across the Gomati favour the location of Lucknow University, New Hyderabad, Nishat Nagar and Badshahnagar, etc. Lucknow is at present capital, cultural and industrial city of the State of Uttar Pradesh. Lucknow has attained a population of 16 lakhs in 1991.

Punjab

It has a population of 60 lakhs living in 119 urban centres in 1991. The Punjab has 29.72 per cent urban population in towns of varying sizes which range from 5,000 to more than 1,00,000. Ambala has an urban population of 34 per cent and Amritsar 30.2 per cent and in between these two ends, the moderately urbanized districts are Ludhiana (29.2%), Jalandhar (28.5%) and Patiala (26.1%). On the other hand, Gurdaspur has only 19.5 per cent urban population followed by 10.8 per cent in Hashiarpur. In Haryana, Karnal has 17 per cent urban population followed by 16.6 per cent in Gurgaon, 15.6 per cent in Hissar, 13.2 per cent in Rohtak, 12.2 per cent in Jind and 9.7 per cent in Mahendragarh. In 1991 the population of some of the important cities of the Punjab are Ludhiana 10 lakhs, Amritsar 7 lakhs, Jalandhar 5 lakhs and Patiala 3 lakhs.

Chandigarh

The capital city of Chandigarh is wholly planned on a grid-iron pattern (1950) and revised in 1951. Initially the plan was made by Albert Mayer but later revised by Corbusier. It has altogether 30 sectors out of which city centre lies in 17 and the University lies in Sector 14. Most of the other sectors are either residential or industrial. The first plan was made for 1,00,000 persons but later it has been extended to accommodate

3,50,000 persons. In 1991 the urban agglomeration of Chandigarh attained a population of 574,646 (Fig. 21.2).

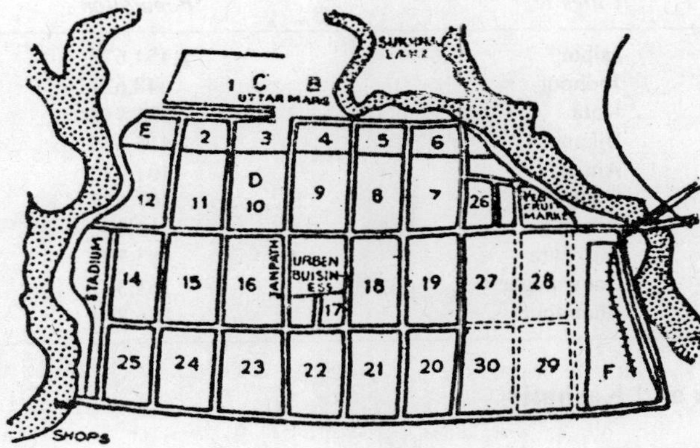

FIG. 21.2 : Patterns of Development of the City of Chandigarh
A. Chandigarh Railway Station, B. High Court, C. Secretariat,
D. Mount View Hotel, E. Engineering College, F. Industrial Area

Rajasthan

It has a total urban population of 10 million living in 214 urban centres in 1991. There are 62 towns in Rajasthan Plain which are very unevenly distributed. In general the towns are concentrated in the northern part of the region and thin out towards the desert. The concentration is more marked towards the availability of water, lines of communication, high density of population and economic advantages. The districts of Ganganagar, Churu and Jhunjhunu have merely 50 per cent of the total number of towns of the region. There are several towns around the city of Bikaner but other districts have only 3 to 4 towns located wider apart. The population size of the towns are also smaller, Jodhpur and Bikaner are the city level centres while others are towns of class II and IV levels.

In Aravalli region most of the towns are connected with the medieval period of Indian history and are invariably Walled towns. Being a State capital their fortification had been a common feature, e.g., in Udaipur, Ajmer, Jaipur, Alwar, etc.

Jaipur city as the capital of Rajasthan is known as the Pink City of India, the Paris of India and a dream in stone. It is built on a checkerboard plan of Hindu architecture. It is a nodal point in the central Aravalli which enjoys the commercial, industrial and educational facilities. The population of some of the important cities of Rajasthan are as follows (1991).

Cities	*Population*
Jaipur	1,454,678
Jodhpur	648,621
Kota	536,444
Bikaner	415,335
Ajmer	401,930
Udaipur	307,682
Alwar	211,162
Bhilwara	483,791
Ganganagar	161,377
Bharatpur	156,844

Jammu and Kashmir

Out of 43 urban centres only Jammu and Srinagar have originated before 1901. The number of towns gives a fluctuating trend since then due to change in the criteria for considering town. Most of the towns are concentrated in the Srinagar Valley, e.g., Anantnag, Baramula, Sopore. The other important towns of the area are Mirpur, Udhampur, Poonch and Leh in the western sector. Class VI towns occupy 70 per cent urban area with only 14 per cent urban population of the State. Leh is the only town in Ladakh district.

Srinagar as the capital city of the State has a leading position in the valley of Sutlej. It is situated amidst Dal and Wulur lakes and the scenic beauty of Moghal gardens. The city of Srinagar has been divided into eight wards and 392 *mohallas*. It has incorporated the following suburbs in its fold, e.g., Thud, Chasmashahi, Dar Bag, Zeethyar, Nowpora, Gupkar, Hasthil and Karapora. Population of the city has increased more than 200 per cent between 1901 and 1991. The sex ratio is only 365. Lal Chauk is the central business district along the Hotel and Residency roads. It has university, hospital, aerodrome, park, pologround and golf course. In 1991 about 23.83 per cent population of the State were living in urban centres.

West Bengal

In West Bengal 27.39 per cent population is urban. The total urban population is 19 million who live in 160 urban centres of different population size. The urban centres (215) are mostly over-grown villages with predominance of non-agricultural occupation (75%). The polarization of towns and cities have taken place along the Hooghly side (105) and around Asansol. In West Bengal the functions of administrative, trade, commerce, transport and industry all have contributed in the growth and development of urbanization. Nowadays due to political and economic changes the sweeping changes have been noticed in the distributional and functional character of urban places.

Contai and Digha are coastal towns whereas in the north, Siliguri is a railway junction, Jalpaiguri is a plantation centre besides the development of Cooch Behar, English Bazar, Shantipur and Nabadwep are the centres of trade and commerce besides Arambagh, Murshidabad, Baruipur, Dinhata and Burdwan.

Most of the towns are unplanned where the primary activity still dominates in the urban environment leading to haphazard growth of the urban centres in the surroundings of Calcutta conurbation. Some of the important urban centres of West Bengal are Calcutta 109 lakhs and Asansol 8 lakhs in 1991.

Manipur

About 27 per cent population of Manipur is urban. They live in 30 urban centres located in the Imphal Valley. Imphal has a population of 2 lakhs in 1991. Imphal is the capital town of Manipur, which is located in the Imphal river valley over a vast plain. As a planned town it could easily be divided into three zones—the municipal, cantonment and other areas. The business centres are known as Maxwell Bazar, Sadar Bazar, Purana Bazar and New Bazar located in the form of a continuous zone about the Naga-Turel and Nambul rivers. The administrative zone has civil hospital, government schools, pologround, library, etc. which lie to the east of the business zone. The cantonment is located in between the municipality and the Imphal river valley.

The Garhkum region of U.P. Himalaya has 25 towns including a city of Dehradun. In this area towns occupy five types of site, e.g., ridges—Musoorie, hill station, Ranikhet, cantonment—Chakrata; cantonment—Lansdowne, district headquarter—Pauri and Almora; Valley—Dehra dun and Naini Tal; river terraces—Kirtinagar and Uttar Kashi; confluence point—Tehri, Devprayag, Rishikesh, Kalsi, Kotdwara, Hardwar, Kathgodam and Tanakpur. Badrinath and Kedarnath are religious towns of the area which lie in Chamoli district. Pithoragarh and Dharchula are progressive towns of the area. The earth-quake of 20th October 1991 has ravaged the towns of Uttar Kashi and the nearby areas.

Assam

It has a urban population of 25 lakhs living in 221 urban centres. In Assam the urban centres are found along the Brahmaputra river which was highly contributed in the cultural and commercial development of Assam. Railway has favoured the growth of towns as Jigighopa, Rangiya, Furkating and Simalgur. Due to the exploration of oil, the following towns have developed: Dhuliajan, Namrup, Morang, Nahorkatia, Margherita, Rangapara, Dibrugarh, Tinsukia besides Gauhati and others. Some of the important commercial centres are Jorhat, Nowgong, Tejpur, Dhubri and Barpeta. Most of the town are of small population size 5,000 to 50,000 except Gauhati which is a city level centre (1,50,000 persons).

In North-Eastern India Imphal is the capital town of Manipur, followed by Shillong for Meghalaya,[4] Aijal for Mizoram, Agartala for Tripura, Kohima for Nagaland and Gangtok for Sikkim. Due to rough terrain the urban development is on its lowest ebb and their spacing is very high.[5]

Some of the important urban centres of Assam in 1991 are as follows :

Cities	*Population*
Gawhati	577,591
Dibrugarh	123,885
Silcher	115,045
Jorhat	111,584

Haryana

The State of Haryana has a urban population of 40 lakhs distributed in 90 urban centres. Here about 24.79 per cent population lives in urban areas. Some of important cities are as follows along with its population in 1991.

Cities	*Population*
Faridabad	613,828
Yogendra Nagar	219,642
Rohtak	215,844
Panipat	191,010
Hissar	180,774

Bihar

The State of Bihar has a total urban population of 114 lakhs who resides in 221 urban centres. In Bihar as a whole only 13 per cent population is urban and the rest is rural. Some of the important urban centres are Patna the primate city of Bihar followed by the industrial centres of Jamshedpur, Dhanbad, Bokaro, Muzaffarpur, Bhagalpur and Ranchi.[6]

Cities	*Population, 1991*
Patna	955,889
Jamshedpur	834,535
Ranchi	598,498
Muzaffarpur	240,450
Bhagalpur	254,993
Dhanbad	151,334
Bokaro	350,540
Darbhanga	219,274
Gaya	291,220
Mango	110,024
Arrah	156,871
Purnea	114,189
Chhapra	136,824
Katihar	154,101

Source : Census of Bihar, 1991.

Conclusion

In India due to industrial development the process of urbanization has stepped up in many areas especially in mineral rich Damodar Valley which is known as the iron and steel belt of India. It has been felt that most of the transportation lines which connect metropolitan cities, e.g., Delhi to Mumbai, Ludhiana, Meerut, Calcutta, Chennai and Ahmedabad are full of urban centres. Most of these centres are either transportation nodes, commercial centres or industrial centres. Hence, the future of urbanization in India is very bright but due to lack of proper facilities in urban areas one can call it a sort of psudo urbanization. The different urban regions of India have different characteristics of development which is based on the local circumstances other than commercial and transport development.

FOOTNOTES

1. A.K. Sen, "City Regions of India," *Geographical Review of India*, 1959, Vol. XIX, pp. 32-42.
2. Kusum L. Datta, "Urban Zones of India," *The National Geographical Journal of India*, 1967, pp. 97-109.
3. A.G. Noble, (ed.), *Indian Urbanization*, Tata McGraw Hill, New Delhi, 1977.
4. N. Hazarika, *Urbanization in Meghalaya*, M. Phil Dissertation, Department of History, NEHU, Shillong, 1983.
5. G. Gogri, *Urbanization in Assam 1901-1971*, Ph.D. Thesis, Dibrugarh University, Dibrugarh, 1986.
6. S.M. Karimi, *Urbanization of Bihar Plain*, Aligarh Muslim University, Aligarh, 1967.

22

URBAN DEVELOPMENT IN EAST EUROPE

THE URBANIZATION of Eastern Europe shows decentralizing tendency of social elites and resources from focal points for the better performance of secondary and tertiary functions and the production of consumer goods through manufacturing industries for social satisfaction. Although the growth and development of urbanization in Eastern Europe is relatively new in comparison with Western Europe, but it has been found that urban culture diffused from Rome as early as the 4th century B.C. like the Western Europe.

On the whole the East European urbanization has been influenced by the communist revolution in 1917, which marked the beginning of urban development with special reference to former Soviet Union in particular and East Europe in general. The revolutionaries had great faith in the power of the government to transform society for the betterment of common man. Since then, urban development in Europe was sought in what Lenin had called 'a new pattern of settlement for mankind' the city of a socialist man. According to Marx-Lenin planning was to create cities without social or economic division. There should be a commitment to the socially integrative value of housing and a wide range of social services. City planning has to be responsive to economic planning, which could help in determining industrial location and settlements to the rate of urbanization in developed regions and major urban complexes. Hence in the communist society city planning should be restricted to a basic physical engineering, architectural profession and providing high density new apartments in approved styles.

East European Urbanization Pattern

In communist countries of Eastern Europe little venture has been done about the socialistic urban society and urban policy. In these countries central planning has been over whelmingly concerned with narrow conceptions of economic growth only through agriculture and industry. For urban areas :

1. "There is no way to house people equally in physically unequal housing. There can be some revolutionary improvements as the bigger houses are allocated to bigger families or subdivided or shared, according to the principle of Marx. But when that is done, many inequalities will remain in housing location, size, quality, condition and maintenance, internal equipment, access to services and so on. This shows that no principle can house people very equally or fairly in the short run, and the revolutionary governments have to decide which principles of equality to apply."[1]
2. At least for the time being efforts should be made to improve and equalize the housing stock even if the inadequacies are increasing by doing so. In future, if economic growth continues and new housing standards go on improving they will have to decide which class of households should get the better stock and which class the worse stock.
3. What should socialist cities be alike? As the country industrialists, what pattern of cities should be build for a few big and many poor, or what mixture, at what locations? And in the cities, should there be dense apartment blocks, or dispersed suburban houses and gardens or what mixture and choices of housing types? Some of the questions are related with the class and sectional interest, town and country, old and young, men and women, married and single, pedestrians and vehicles, public and private resources and diverse arehitectural and social taste. The new Russian and east European planners faced those questions and others, and scarcely any socialist theory to guide them. So some of the questions got socialist answers, and several others got thinly-disguised Western answers, while some of the socialist policies are found to be inefficient and self-defeating in nature.
4. When the planners decide where to locate their industrial and urban developments. In such a situation most governments aimed at 'strong decentralization'. Industry could go to such places only as fast as towns could actually be built or expanded to serve the factories and house their labour. In this way, the centres chosens for industrial development must get major

shares of all new housing, urban roads, water supplies and sewers, power and transport installations, school and health services and so on. They would therefore, have great and growing advantages, over all the 'no growth' towns and villages with their neglected old housing and inadequate services. In this way, even in East European communist countries, the acceleration of industrial growth would increase, many inequalities amidst present consumption pattern and future economic growth.

Brief Historical Development

In 5th century B.C. the Greeks began to found colonies on the Black Sea Coast. Tyras, at the mouth of the Dniester, Olbea, near the mouth of the Bug; Chersonese (Kherson) and Theodosia in the Crimia, were among the more important on the north shore. Because of their favourable position for trade by land and sea in many directions, they grew rapidly into important trading cities. They also maintained slave market and were themselves centres of varied handicraft industries.[2]

The resemblance of pre-historical development in urbanization is portrayed well in Czechoslovakian pre-historical settlements occurred in sparsely distributed primitive groups during the Stone Age. They were strategic in nature and were found to be situated on higher hill tops or ridges.

Rudimentary town like centres were established by the Roman conquest in south western Slavakia. These were more or less military camps connected by simple routes to the major Roman centres located along the Danube river. They mainly functioned as administrative centres. Fortified settlements were built by Slavic inhabitants in the same areas during the Great Moravian Empire. They were walled encampments of several clusters of dwellings surrounded by a castle. Slight improvement in urban functions occurred for primarily the fortified settlements served as tribal centres, market places and cross-roads of long distance overland trading routes. Therefore, there were some form of urban planning.

The other example in Yugoslavia shows that types of settlements and their historical origin were similar to Czechoslovakia. During the Roman period for instance rudimentary urban places occurred along the Adriatic coast which served as administrative centres too.

In 9th century, the Polyanians were settled on the Dnieper and their city of Kiev was on that river. The Severyans had their homes along the Desna, and their chief town of Chernigov lay on the Desna. Farthest north were the Illmensky slave who lived among the swamps and forests surrounding lake Illmen and whose city was Novgorod.

Medieval Period (10th-18th Centuries)

In A.D. 1090 commercial exchanges with central Europe developed: 'the commercial route crossed Hungary towards the Danube, where *Regensburg* became the main clearing house for Kievan goods imported into Germany and France. East European urbanization over centuries is taken as a process resulting from an inter-play of internal socio-economic conditions and the geopolitical external influence. Profound external influence was due to its central location between Western Europe on the west, Russia on the east and Turkish influence in the South.

In Eastern and Central Europe there are many such towns which usually situated at the place where earlier trade routes crossed. Augsburg, on the Lech, commanded the great Danube route from Basle to Viena and these roads led to the mines at Salzburg and into the Inn Valley. By late medieval times Salzburg become a powerful trading town, with merchant connections in the east, and subsequently with the New World.

In 12th century, between the Upper Volga and Oka rivers many towns grew up, of which Vladimir became the chief, others were Suzdal, Rostov, Yuryev, Moscow and Tuer.

In 12th century Russia, of the approximately three hundred towns, many were small in population size, but some were very large and flourishing cities and the urban population can hardly have been less than a million. The average town would have sheltered several kinds of people—Tradesmen and artisans of all sorts—carpenters, boat makers, masons, potters, blacksmiths, goldsmith, tailors, tanners, weavers and labourers, made up the rest. Towns which were capitals of principalities were Kiev, Smolensk, Novgorod, Polotsk, Turov, Chernigov, Pereyaslovl, Galich, Ryazan, Suzdalian and Volynians. Moscow was first of all developed in 1147 and fortified in A.D. 1156.[3]

During Medieval period urbanization had been influenced by external military forces, political pressures, technological examples and economic consequences emerging mainly from Germany and USSR. Although much of the Eastern Europe adopted planning concept from Western Europe. But the southern parts like Yugoslavia and Czechoslovakia between 14th and second half of the 19th century had been under Turkish feudalism. Towards the peak of the Turkish feudal rule in the 17th century slight improvement like hierarchy of towns appeared in the Balkan region of Polaka, Varosica and Varos but after the fall of the

Turkish empire followed by internal wars of liberation in the 18th and 19th centuries caused a decline in urban places.

The southern steppes of Russia and the lower Danube lands saw the appearance of many towns in the 17th and 18th centuries.

The urban population of Russia in 1125 A.D. was only about 1/2 million or 4 per cent of the total.

Post Russian Revolution : 1917-1948

Like the Medieval period in the early A.D. 1900, East European urbanization process was fully externally influenced by the communist orientation under the leadership of Soviet Union after the 1917 October Revolution.

In the post war period of the urban development of the countries of Eastern Europe, the increase in industrial employment reached a total of more than 6.1 million excluding the growth of employment in the first year which was due to the reconstruction and the preliminary reorganization of industry[4] (Table 22.1).

Thus in 1946-65 industrialization was the leading urban process that changed East Europeans urban spatial economic development. In 1960-80 pattern of urban and industrial development shows that they are highly correlated (Table 22.2).

TABLE 22.1 : Changes in Employment in Industry in some East-European Countries

Country	*Area in '000 km²*	*Population*		*Employment in Industry in '000*		
		(1965) million	*(1995) million*	*1949*	*1965*	*Increment*
Bulgaria	111	8.2	16.5	324	954	630
Czechoslovakia	128	14.2	28.6	1540	2476	926
Yugoslavia	256	14.5	29.0	500	1375	875
Poland	312	31.5	63.3	1755	3675	1920
Rumania	236	19.0	39.2	809	1676	867
Hungary	93	10.1	22.0	562	1484	922

Note : Based on author's own estimate.
Source : Kuklinski, *Urban System and Rural Development*, 1972.

Thus after the revolution town planning had a strong ideological cornerstone that the socialist city should be significantly different from that of the West. It should be uniform in its characteristics, and consistent to the goals of social equality. Under this, some of the socialist new towns designed to represent complete departure from Western experience.

Many of them are show places of the new socialist city. These towns include : Nowa Huta in Poland, Dunanivaros in Hungary and Titograd in Yugoslavia. Another factor was post-war reconstruction when housing was to be built as quickly as possible and basic industries were constructed. This attracted many people to be employed in the new industries which is pre-requisite for urbanization.

TABLE 22.2 : Pattern of Urban Development in East Europe (1960-1980)

Countries	*% Share of Industry in G.D.P. 1978*	*% of Urban Population*		*% Average Annual Growth*		*Number of Cities over 500,000 Persons*	
		1960	*1980*	*1960-1970*	*1970-1980*	*1960*	*1980*
Yugoslavia	45	28	42	3.2	2.9	1	3
Greece	31	43	62	2.6	2.4	1	2
Rumania	40	34	48	2.8	2.5	1	1
Bulgaria	64	39	64	3.8	2.5	1	1
Hungary	59	40	54	1.7	2.2	1	1
Poland	64	48	57	1.8	1.7	5	8
U.S.S.R.	62	49	65	2.7	2.2	25	50
Czechoslovakia	72	47	63	2.1	2.0	1	1
German Democratic Republic	69	72	77	0.1	0.2	2	3

Source : World Development Report, 1980, Washington, D.C., p. 149.

Further more in Eastern Europe urban growth has been led by economic policies. Economic plans in different countries have brought growth and its associated urbanization to regions successively more remote from major city centres of Moscow and Warsaw.

In former USSR the State Planning Commission determines the economic norms for the city on the basis of employment figure. The city planners role is simply to implement existing norms as laid out in the 'basic books of standards'. Since 1958, it has been officially recommended that new industries be located in small (population 50,000) and medium sized towns (50,000-100,000), but the population must be more than 250,000-300,000 (Berry, 1976, p. 34).[5] Later, the State Planning Commission recommended more than 5,000 medium sized towns suitable for location of new industrial enterprises. For instance, the 1971-75 plan called for the location of approximately 75 per cent of the large new industrial enterprises in cities with population under 250,000. In major cities, the establishment of new industries or the expansion of existing one is not allowed unless they are directly related to the municipal organizations or consumer services. The possibility of exercising this kind of control has been seen by the communist planner as a means of avoiding many perceived negative consequences of uncontrolled urban expansion which have occurred in other parts of the world.

The application of socialist principles in east European cities have significant impact on their spatial forms and social structures. For example, the effect of Russian style socialism in the cities of East Germany differentiating it from the counter part of the West has been to reduce areal specialization and segregation by substituting for market processes directly planned by the State.

The form of socialist city is simple, a square of administrative—cultural composition at the centre, with radiating streets flanked by massive residential developments, comprising 5 to 9 storey apartment buildings, containing 2 to 3 rooms for families, and dormitories for unmarried workers. Adjacent to the complexes is the economic unit that provides employment in apartments. The significant density differences among parts of the same city have been ironed out and social segregation have been largely abolished. Housing policy is to give priority in new dwellings to young families with children employees in key jobs, and families living in hard conditions. Thus, there is a strong correlation between age structure of inhabitants, producing socially good status as a differentiating factor. Only party leaders and intellectuals reliance on automobile, all residential areas are closely reliance on automobile, all residential areas are closely related to public transportation facilities (Fisher, 1967).[6]

While the functional character in West European cities is of mixed type like multiple-nuclei model or sectorial growth model, but on the other hand East European cities reflects a planned character of national and social needs grounded in the following four principles as summarized by Banyikwa (1978).[7]

Standarization

In place of residential segregation, socialists plan for residential uniformity. Uniform standards to large areas and norms of space per individual, are established by the State rather than relying on market forces. This results in large areas with housing units devoid of architectural designs of individuality.

Proper Size of the Urban Places

This principle attempts to integrate urban planning with regional and national planning, to avoid the megalopolitan explosion. This control avoids problems associated with the megalopolis, and attempts to distribute national resources equitably within towns, between towns and between rural and urban areas.

The Role of the City Centre

While the C.B.D. forms the hub of economic activities in capitalist town planning, the same is accorded more of agricultural role in the socialist countries. This is possible because town planning in socialist countries attempts to converge place of work and place of residence.

The Neighbourhood Unit Concept

This has been created by planners in creating a series of apartment houses of varying sizes sited around a core of shops and a cultural centre. By doing this, socialists achieved mass consumption of goods, without creating unnecessary pressure on the cross-city transport facilities.

These four principles together reinforce each other in creating cities of pre-determined sizes with internal residential uniformity, mutual involvement, and mass consumption of manufactured goods. While the capitalists plan for a differentiated society, the socialists plan for an egalitarian one. Consequently, while capitalist urban space become socially segregative, socialist residential structure emphasize uniformity.

Urban Growth as Part of Overall Economic Planning

The U.S.S.R.

The U.S.S.R. presents some unique aspects of urban development, largely because the implementation of a series of economic plans involve very large number of people and because special relations exist between the State and the individual.[8]

The new towns had to be populated principally from Europeans and Russias for the Asiatic lands, controlled by the Federal Republic. Soviet urban centres have a population of diverse race, tradition, language, and environment, many of whom live in scattered tribal areas, and were widely dispersed. Nowadays, many new towns have been established east of Ural mountain where the growth of population is less rapid in old industrial centres in comparison with the new towns. In 1959 it rose to 99.8 million or 48 per cent. During the same period the population of the biggest cities, those with over 100,000 grew dramatically from 23.1 to

33.5 millions in the western regions and from 5.7 millions to 14.8 millions in the eastern regions of Soviet Union.

Thus in U.S.S.R., the country's industrial development, especially in the last two decades, has been followed by an intensive process of urbanization. Together with the emergence of new urban centres, urbanization includes a considerable development and reconstruction of old ones, as well as their expansion, and a rapid growth of population (Table 22.3).

Since 1959, the annual growth of urban population has been over 3,000,000. According to preliminary estimate for the development of/ urban settlements in the U.S.S.R., which have been worked out by the Gosplan of the U.S.S.R. by 2001. Urban population of the country will be over 157,000,000, that is, about 65 per cent of the country's total population. By A.D. 2000 the urban population of the U.S.S.R. will reach 233-240 million or 75 per cent of the total population.

TABLE 22.3 : U.S.S.R. Growth of Urban Settlements and Urban Population (1926-2001)

Census Year	*No. of Cities*	*Total Urban Settlements*	*Urban Population (Millions)*	*Urban Population as % of National Population*
1926	709	1925	26.3	18
1939	923	2373	56.1	33
1959	1679	4619	100.0	48
1970	1935	5504	137.0	56
2001	3000	10,000	400.0	65

Source : Pravda, April 19, 1920.

Moscow

Moscow is the premier city of the former Union of Soviet Socialist Republic and the East European Communist nations. It has a population of some 6.5 million and is the hub of a urban complex which totals about 9 million people. It is the greatest urban complex of continental Europe, after the Rhine-Ruhr region, and the biggest metropolitan region based on a single city.

In the surrounding area of Moscow more than half the population lives in urban areas, and Moscow is the dominant transport, cultural and industrial centre. Eleven main rail lines and three major airports—Vnukovo, Sheremetrevo and Domodedovo—are located there. It is the centre of culture and entertainment, e.g., the Bolshoi Ballet and the

Moscow Arts Threatre, mass media, tourism, and pilgrimages to the tombs of Lenin and other Soviet heroes. Moscow University, the Academy of Sciences, the Lenin Library and Museum, and several hundred research institutes are located here as centres of Soviet knowledge industry. Moscow's growing engineering and consumer goods industries account for over 20 per cent of the national industrial output.

As a geographical area Moscow may be defined as : at the core is a central area of typical European style buildings surrounded by a ring of boulevard—lined with commercial and residential blocks. Beyond this range are the suburban areas up to the new motorway belt which incircles the city at a radius of about 18 kilometres from the centre. Outside, there is a green belt, some 10 kms. wide and beyond this satellite towns are located such as Gorky and Tula. Some of these are of recent origin, but most of them are pre-revolutionary towns which expanded with the growth of engineering industries after 1917.

Hungary

Following the liberation of the country in 1945, and as a consequence of industrialization, the rate of migration into the towns has increased and the population of several agricultural areas reduced in number. Between 1840 and 1960, Budapest's share of the national population increased several fold, and the national ratio of the urban population doubled during the same period.

TABLE 22.4 : Percentage Distribution of Rural and Urban Population in Hungary

Years	*Budapest*	*% of Urban Population*	*% of Rural Population*
1840	2.7	18.4	81.6
1869	5.4	16.4	83.2
1890	8.1	25.8	74.2
1900	12.6	31.8	68.2
1910	14.6	34.1	65.9
1920	15.4	35.3	64.7
1930	16.6	36.2	63.8
1941	18.4	38.2	61.7
1949	17.3	36.5	63.5
1960	18.2	39.8	60.2
2001	25.4	50.0	50.0

Source : Pecsi, M. and Surfalvi, B., *The Geography of Hungary*, 1964, p. 163. 2001 estimated by R.B. Mandal.

In 2001 about 50 per cent of the national population lives in towns, evenly distributed between Budapest, on the one hand, and the other towns of Hungary on the other.

By the many-sided functions and comparatively large populations, Miskolc, Debrecen, Pecs, Szeged and Gyor stand out among the provincial towns of Hungary. Miskolc and Gyor are the leading provincial centres of industry. Besides Budapest, there are five university towns, namely Szeged, Debrecen, Pecs, Miskolc and Veszprem. Characteristically, the 'farmers town' registered the lowest pace of growth.

Budapest located on the banks of river Danube served as an important frontier and defence for the Celts and Romans. The planned development of Budapest was begun after Second World War. The most ancient historic town nucleus situated on the top of castle Hill and on its slopes. Later on, this nucleus was expanded through the addition of districts settled on the strip of the river banks and in valleys through which passes some major trade routes.

"Greater Budapest"[9] was formed in 1950 by joining to the capital 7 Satellite towns and 15 villages, which increases the number of Budapest districts from 14 to 22. So far as the growth of population in Budapest is concerned the capital city is composed of less of natural growth than of a large-scale influx of migrants. Only within 2 years, 1957 and 1958 alone, for instance, immigration boosted the population by more than 125,000.

Yugoslavia

Along the Adriatic Coast one can trace the continuity of almost all significant urban places back to the Roman and even Greek Periods. This continuity was strengthened during the Middle Ages and the urban tradition has continued until the present day. In the interior of the Balkan Peninsula, the Turkish invasion in the 14th century caused a discontinuity in the development of Serbian, Macedonian and Bosnian towns, which until then had grown under strong Byzantine influences. During Turkish feudal era many flourishing towns came into existence in the Balkan region. With the decay of feudal system and the wards of liberation in the eighteenth and nineteenth centuries caused a decline of urban civilization and even the disappearance of some towns. Indigenous Serbian city planning, after the liberation in the nineteenth century, is of historical interest; it went hand in hand with the reconstruction of towns, and internal colonization of forested areas.

In the north-western part of Yugoslavia, in central and western Croatia and in Slovenia, the urban tradition developed in the early Middle Ages. Because of the economic crisis, many of them were founded deliberately as military bulwarks, e.g., Novo Mesto, Karlovac and Nova Gradisca. The towns in this part of Yugoslavia began a more vigorous growth only in the second half of the nineteenth century, after the peasants bondage was abolished, commerce revived, transportation improved and manufacturing industries were started.

In Yugoslavia 2000 population limit was set as a minimum for a town. The average degree of urbanization 28.3 per cent in 1961 is fairly representative for most parts of the country. The largest increase has occurred in big cities, and urbanization in their vicinity is quickest and most intensive. Medium sized and small cities grow more slowly and this is one of the major problems of Yugoslavian urbanization.

TABLE 22.5 : Yugoslavia : Degree of Urbanization (1921-1991)

Year	*Total Population '000*	*Urban Population '000*	*% of Urban to Total Population*
1921	12,545	1,642	13.1
1931	14,534	2,144	15.1
1948	15,842	2,743	17.3
1953	16,999	3,688	21.7
1961	18,549	5,252	28.3
1967	19,949	6,261	31.7
1991*	25,000	10,000	45.0

* Estimated by author.

A characteristic feature of Yugoslavia is sub-urbanization, in the sense that the non-agricultural population stays in the rural areas and commutes to work in towns. Due to this the degree of urbanization in Slovenia is only 28.9 per cent. The basic characteristics areas follows :

(a) The degree of urbanization on the average is low (28.3%), but varies among regions (republics) between 19.5 and 34.9 per cent.

(b) The slow urbanization before the war and quicker urbanization after the war are revealed by the indices for 1921-48 (167) and 1948-61 (102).

(c) A high percentage of the urban population lives in larger towns of over 50,000 inhabitants (41.4%) and even in those of over 100,000 inhabitants (31.6%).

(d) In 1991 Crotia and Slovenia are engaged in enfighting with the Federal government and hence this has given a jolt to the urban development in Yugoslavia.

Poland

In Poland there are two types of urban units namely 'cities' and 'urban settlements'. There are no defined criteria upon which an area may be called 'city'. Pearson is usually historical and traditional, in the case of newly created cities or urban settlements the criteria are political, social and economic.

TABLE 22.6 : Poland Structure of Urban Settlements by Size, 1972

Size of Settlement '000 Persons	*Number*	*Population '000 Persons*	*% of Urban Population*	*% of Urban Population of the National Total*
Below 5	346	1,053	6.0	3.2
5—10	218	1,549	8.8	4.7
10—20	170	2,364	13.4	7.1
20—50	100	3,095	17.5	9.3
50—100	28	1,989	11.3	6.0
100—200	13	2,044	11.6	6.1
Above 200	11	5,545	31.4	16.7
Total	886	17,639	100.0	53.1

Source : Statistical Year Book, 1973, Warsaw.

There are two characteristic features associated with the urbanization of the Polish lands upto the year 1945; large scale emigration, both seasonal and permanent to foreign countries, and secondly, over-population in rural areas. The first had reached its peak at the end of the nineteenth century and was marked by limited in the inter-war period, particularly as a result of the United States immigration policies introduced after 1917. The second factor got importance during the great economic crisis of the early 'thirties' with mass unemployment in industry and cities.

The re-establishment of the Polish State in the inter-war period brought new additional factors into the process of urbanization among which the most important were :

(a) The creation of an integrated national organization of economic, cultural and political life which served as a basis for the new and unified system of national and regional administrative urban centres.

(b) The development of an industrial structure adjusted to the needs of the national economy; this involved building a new large deep-water seaport (*Gdynia*) and creation of a new industrial region centrally located and easily accessible from all parts of the country.

(c) The integration of transport networds, especially rail and roads by constructing lines connecting Warsaw, the national capital, with the provinces, as well as those between Upper Silesia, with its rich coal-fields and the Baltic parts.

TABLE 22.7 : Poland : Total Population and Urban Population (1810-1991)

Year	*Total Population (millions)*	*Urban Population (millions)*	*Urban Population as % of Total*
1810	9.0	2.0	22.2
1850	14.0	3.1	22.2
1870	17.5	4.1	23.2
1900	23.7	6.3	26.6
1931	29.8	10.6	35.5
1950	25.0	9.6	39.0
1960	29.8	14.0	48.2
1970	32.6	17.0	52.2
1972	33.2	17.6	53.1
1991*	51.0	25.0	65.0

* Estimated figures by author.

Source : *Small Statistical Yearbook*, 1973, Warsaw, Central Statistical Office, 1973, Table II(28), p. 23.

Romania

Urbanization in the Socialist Republic of Romania develops under the decisive influence of socialist industrialization along the following two lines :

(a) an increase in the number of towns with its implied numerical increase in the number of town dwellers; and

(b) the techno-economic development of towns and villages, with its implicit extension of the characteristics of the modern urban way of life.

The returns of the census of March 15, 1966 showed that in the country's 183 towns and the 238 rural communes whose population was included in the urban environment, there were 7,303,503 inhabitants, representing 38.5 per cent of the country's population.

The increase in the number of town dwellers is also due to natural increase and to net immigration. Among various factors, net immigration and the transition of localities from the status of village to that of town are the most significant.

Early in 1970, Romania's 236 towns contained 7,000,000 inhabitants, that is 36 per cent of the country's population at that time. If we add the population of suburban communes, two-fifth (40.2%) of the population are involved in the urbanization process.

Czechoslovakia

The major factors promoting urbanization was industrialization in the Czech Republic especially in second half of the nineteenth and the first half of the twentieth centuries. Other factors of urban development are the development of tertiary activities, levels of mechanization and collectivization of agriculture which have released rural man-power. However, a limiting factor has been the poor amount of house-building in towns.

In Czechoslovakia, typical forms of urbanization include :

(a) the dense network of small towns;
(b) the quality of population is semi-urban; and
(c) the absent of cities with more than 1,000,000 population, except the capital city of Prague.

In Czechoslovakia, National Planning proposals anticipate that by 2000 A.D. the population of Parague exceed 1.5 million and that of Bratislava 0.5 million, and at the present time it is the medium-sized and large cities that are growing most rapidly. It has been hoped that the presence of small towns will show positive advantage under conditions of planned national development, because it could provide relatively dense and differentiated systems of central places.

Conclusions

The basis for transforming urbanization to central socialist planning as exercised in U.S.S.R. stated that, town planning in a socialist outlook, is concerned with the landuse it must deal with socio-economic issues besides physical expansion. As physical changes do not necessarily solve social problems like large future employment, settlements pattern and life style. Social planning as used in Marxist-Leninist literature is designed to impose a particular type of society, a society of equals.[10]

The existence of the central planning system in East European urban places show Soviet influence and consistent application of Soviet procedures, norms and plans as laid down on 'basic book of standards', which guided the socialist new towns, designed to represent complete departure from the western experience.

FOOTNOTES

1. J.C. Fisher, Urban Planning in the Soviet Union and Eastern Europe, in *Taming Megalopolis* by Eldredge, H.W. Fredrick, Praeger Publishers, New York, Vol. 12, 1967.
2. D.C. Money, *Introduction to Human Geography*, University Tutorial Press, London, 1967.
3. W.H. Parker, *An Historical Geography of Russia*, University of London Press Ltd., 1968.
4. Kuklinski, *Urban Systems and Rural Development*, Mysore University Press (Ed. R.P. Misra), Mysore, 1972.
5. B.J.L. Berry, *Urbanization and Counter Urbanization*, Sage Publications, London, Vol. II, 1976, p. 34.
6. J.C. Fisher, 1967, *Op. cit.*
7. W.F. Banyikwa, "Recent Changes in Urban Residential Land Use Policies in Tanzania and Their Spatial repercussions," *The Journal of Geographical Association of Tanzania*, No. 16, 1978.
8. Ronald Jones (Ed.), *Essays on World Urbanization*, George Philip and Son Limited, London, 1975.
9. M. Pecsi and Sarfalvi, B., *The Geography of Hungary*, Corniva Press, 1964, p. 163.
10. Hugh Stretton, *Urban Planning in Rich and Poor Countries*, Oxford University Press, Oxford, 1978.

23

ROLE OF URBANIZATION

MODERNIZATION and urbanization are closely correlated and hence they are in the forefront of research in social sciences. Most of the nations of the world have energized their effort for industrial development and it has a power to attract more and more money, men and material to the urban areas. This has also given impetus for overcrowding in urban centres thereby leading to population explosion and population implosion other than the growth of fabroblast and sacroma settlements.

In urban geography the concept has not developed to such a merit as nationalism in geopolitics and theories of industrial location in industrial geography. But recently, the concept of growth pole developed by Perroux and the central place analysis by Christaller (1933) in Southern Germany and A. Lösch (1954) in U.S.A. are some of the conceptual ideas which have revolutionized research in the field of urban and regional sciences especially removing the transportational bottlenecks and provide equal opportunity through developed means of transport to the urban areas.

The rapid growth of population in urban centres through natural increase and migrational accretion have attracted attention not only to the geographer's but economists, sociologists, demographers, planners and politicians too. Nowadays in most countries of the world the rapid growth of population has nullified the planning prospects of the national government. This has also dampened the environmental situation through carbondioxide build up and increasing the occupancy rate in urban areas.

The excessive migration of labourers in the industrial sectors have created the problems of slums and squatter settlements and they are the fungus of urban environment which has attracted the attention of urban geographers to the farthest limit so that the consciousness may be aroused among the public to ameliorate the situation.

The population growth in urban areas has the problems of housing shortage in most of the developing countries. In India in most of the urban centres at least 10 per cent people are homeless who resides on the roadside, another 30 per cent people lives in slum and squatter settlements whereas the middle class families are struggling for residential space due to high occupancy rate.

The Purpose of Study

This chapter aims to highlight the role of urbanization in modernization and economic upliftment of a nation. The discussion will be done for some of the developed and developing nations of the world in order to make international comparison about the levels of urban and national development.

Analysis of the Problem

All over the world cities really function as the centres of modernization and economic change, because there lie the seats of government, and the main educational, financial and industrial centres. Cities also act as the foci of political and economic activities for the surrounding zone of influence.

Urbanization has a number of meaning attributed to it in terms of economic development. One meaning is "improved possibility of using information and facilities available to the urban community inexpensively and without time lag in any part of the society, polity or economy." Urbanization may also be taken to stand for relatively greater development of non-agricultural production and greater movement of urban used goods and services to rural areas. Urbanization also mean a change from farm process to an instrument of the process of change. Thus we can look upon urbanization both as an instrument of change and a criterion for regional economic development.

The urbanization of sub-urban and urban areas has a definite and strategic role to play as they provide inter alia, nuclei for what in economics is called external economies, because urbanization may be a source of such industrial development as characterized in Japan and China, where some parts manufactured in rural areas of dispersed location and assembled in urban areas.

Role of Urban Places in Economic Growth and Development

Problem Area

Two views haunt scholars who seek to understand the relationship between urbanization and regional economic development.

First View : The first vision held that the process of urbanization lead to regional development. This view is propounded by authors from western industrial nations who propounded that the growth and development of urban centres was a necessary and sufficient condition for regional economic progress.[1]

Urban Situation

E.E. Lampard pointed out that the specialization of a business which generated in the urban centres later diffused to the countryside.[2]

W.L. Kolb[3] stressed the need for a universalist-achievement of value structure for the rationalization of production and hence industrialization, because urban environment provides a necessary condition for the evolution of such value in terms of rural development.

Davis and Golden re-echo the two views by showing that the concentration of activities within a narrow geographical space result in important savings for the urban based firms.[4]

Rural Situation

Urban places which are conveniently located stimulate agricultural production and induce people into the exchange economy. Thus the expansion of urban places serving as market centres are regarded as a favourable indicator of a sound change in the economy because near urban areas commercial agriculture develops.

After all, urban places can only exist after agricultural productivity has progressed enough to sustain people who do not grow their own food.

Rural-Urban Interaction

In the build up of the transactions between the rural and urban areas, urban areas supply manufactured goods, services and innovations of the rural areas, which in turn supply agricultural products, labour and raw-materials for industries to the urban places. These two interactions are reciprocal and mutually supporting.

Evolution of Occupational Structure

The natural course of economy will be one in which the labour progressively shift from primary to secondary and finally to tertiary sector.

Meanwhile urban areas would generate more demands which propel the entire economy forward.

Second View : The second vision advocate that the emergence and growth of urban areas lead to regional economic deterioration.

The generative aspects propounded from the western industrially developed areas are more accurate as a statement of desirable effects of urbanization than a description of the actual situation. Urban places especially in the developing world exert effects which retard rather than propel the economy forward.

Urban places skims of the cream of the able-bodied labour force and orients the bulk of its commercial activity abroad.

The backward linkages which could have spread to the rural areas essentially leak away through the consumption of luxurious imported commodities which are capital intensive.

Meanwhile the energies of the rural population are diverted from productive employment in the rural areas to an unproductive urban service and casual labour.

In this process indigenous industries are replaced by imports and the urban centres became service and administrative centres.

Unfortunately in India there does not exist any evidence on which the kind of generative urbanization propagated by the plans stand.

Approach : An investigation into this are a demand an analysis of the effects of urbanization which can be sub-divided into two sub-groups.

The first category denote the demand stimulus while the second can be called the *supply stimulus*. Let the supply stimulus be called link I, while the *demand stimulus* be called link II.

Link-I : Supply Stimulus

(a) We can begin by examining the effects of urbanization radiated to the economy as a whole, especially on the production of food. The more growth in population number in the urban places demand that the agricultural sector produce enough to feed both the rural and urban population.

(b) The second area require an investigation into the relationship between urbanization and expansion of diversification of urban based occupations.

(c) Thirdly the investigation must address itself to the question relating urbanization and development and diversification of industrial activities.

(d) The fourth area of the investigation would focus on the relationship between urbanization and the pattern of national investments.

These four areas shed light on the demands made on the national resources by the urban centres.

Link-II : Demand Stimulus

Another set of consideration would be an examination of the dynamic efficiencies of urbanization to the economy. Focus may be made on two main areas.

(a) Firstly, the investigation can assess the contribution of urban infrastructure in deducing costs of production for the urban based firms.
(b) Secondly, the investigation would focus on the effects of urbanization to total performance of the rural sector.

In order to operationalize the two links raised five tentative hypotheses associated with the urban growth and urbanization and regional development which are as proposed :

Hypothesis-I : Urban growth is closely associated with the increase in food production in rural areas.

Hypothesis-II : Urbanization is closely associated with the expansion of urban productive employment.

Hypothesis-III : The dynamic efficiencies of urbanization significantly contribute to a reduction in costs of production both to the firm and to the industry in urban areas.

Hypothesis-IV : Some innovations diffused from the urban to the rural areas result in an increased rural productivity.

Hypothesis-V : Urbanization tend to induce government investments in those sectors of the economy that are directly associated with foreign exchange earnings.

Role of Urban Places in Economic growth and Development

Experience of Western Europe and U.S.A.

Path of Development : Once upon a time all societies were primitive and traditional. Then due to technological and cultural innovation power inherent in the society of Western Europe grew fast than the rest of the World.

The stages were primitive communalism, pre-conditions for take-off, which drive to maturity and finally fully developed capitalism.

Role : The role of urban places was to provide a spatial framework on which this pattern of development could develop unabated.

The role of urban places can be broken down into four aspects :

(I) Market Centres;
(II) Industrial Centres;
(III) Administrative Centres; and
(IV) Cultural and Sociological Centres.

Over space these roles were connected by transport and other means of communication.

· The roles which have been received from urban areas are market and manufacturing functions.

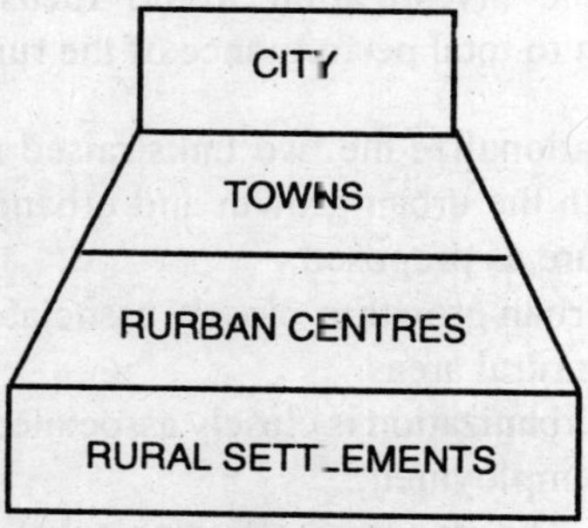

FIG. 23.1 : Market Centre or Central Place Functions

— These acted as the centres for the exchange of rural produced agricultural products.
— These acted as centres responsible for the circulation of rural surplus and other commodities.
— These acted as centres for the distribution of urban produced manufacturing products to the rural population.
— The availability of the market centres helped in the rationalization of agriculture from subsistence to commercial agriculture.
— The access to the market entail the adoption of principles of comparative advantage and translation into agricultural specialization both at crop level and landuse intensity.
— The net result was the relative increase of peasant's income.

The attractiveness of the market centres as areas for spatial equity led to some scholars do research on them. Walter Christaller put forward the spatial structure of the central place concept.

Assumptions

1. There should be an isotropic land-surface both in physical and economic characteristics.
2. The landscape should have easy and uniform means of transport in all directions.
3. The inhabitants of such an ideal landscape should be an economic-men who—
   — wanted to maximize profits,
   — wanted to minimize costs,
   — wanted to have optimal profit from economic enterprises.

What could be the structure of Economic space?

Triangle

Operating from the centre of the square the interprise would not achieve the principle of equity. Some people would have to travel very long distances to the market.

$$a^2 + b^2 = c^2$$

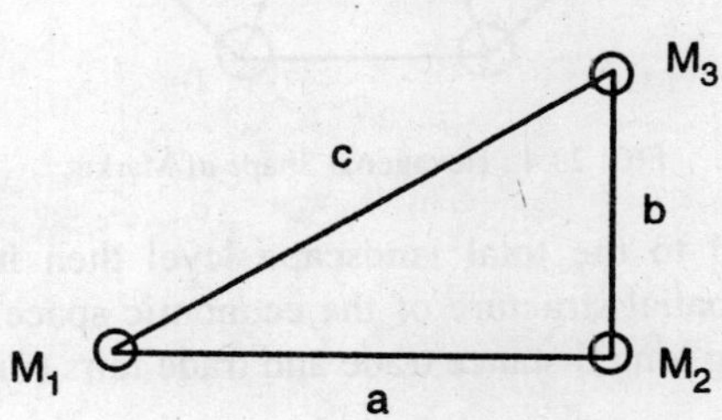

FIG. 23.2 : Triangular Shape of Market

Circle

Operating from the centre of the circle was the most ideal way the economic space will function.

But when the whole landscape is considered not all the customers are covered. So the circle also does not provide the best results.

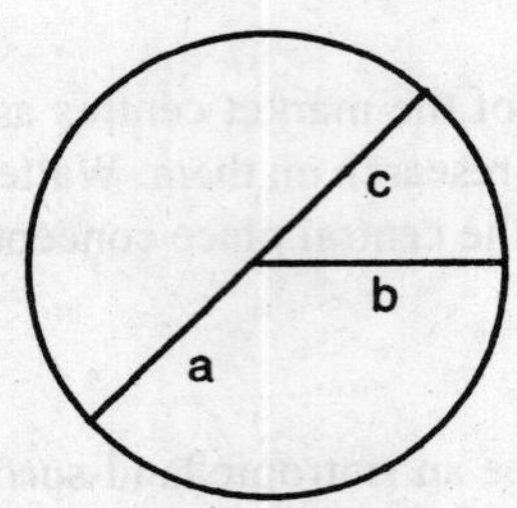

FIG. 23.3 : Circular Shape of Market

Hexagon

Operating from the centre of a hexagon was certainly not very efficient at the unit level.

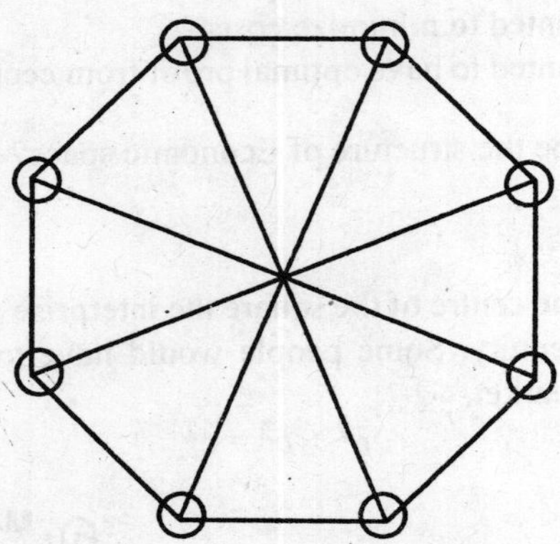

FIG. 23.4 : Hexagonal Shape of Market

When raised to the total landscape level then it becomes very efficient. As a spatial structure of the economic space it encompasses everybody during long distance trade and trade fairs (Fig. 23.5).

Spatial Structure of Central Place System

1. **Each central place has its umland which is composed of the effective demand area.**
2. **The size of the central place determined by the order of service afforded.**
3. **The customers at the farthest end of the demand area are indifferent as to where the services come from.**

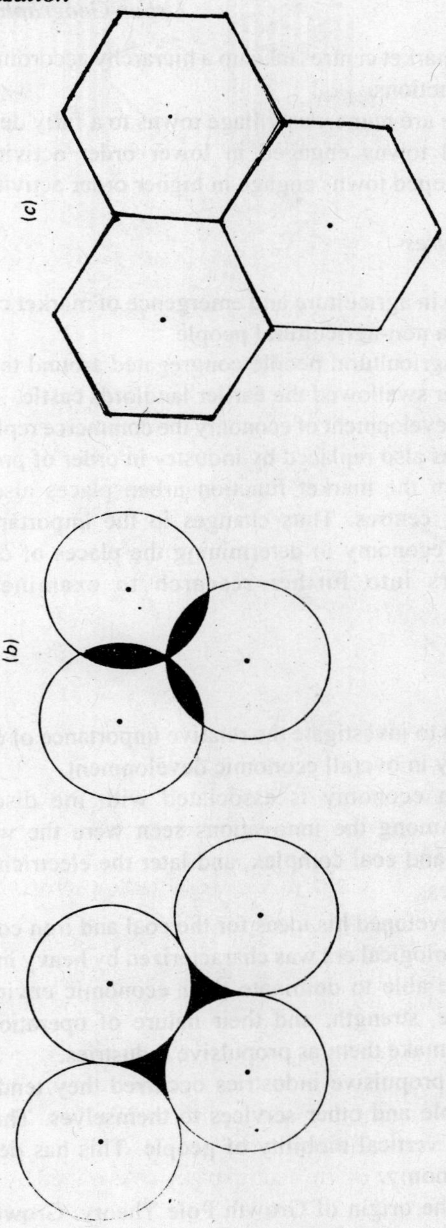

FIG. 23.5 : Alternate Trade Area Shapes

Christaller chose the hexagon as the ideal trade area shape because it closely approximates the qualities of a circle (all areas on boundary are equal distance from the centre) and avoids a problem of underlap or overlap as shown

4. The market centres take up a hierarchy according to complexity of functions.
5. There are numerous village towns to a fully developed town.
6. Small towns engaged in lower order activities while fully developed towns engage in higher order activities.

Industrial Centres

Improvements in agriculture and emergence of market centres helped in the growth of a non-agricultural people.

The non-agricultural people congregated around the towns formed guilds and later swallowed the earlier landlords castle.

With the development of economy the commerce replaced agriculture but lastly it was also replaced by industry in order of preference.

Apart from the market function urban places also functioned as manufacturing centres. Thus changes in the importance of different sectors of the economy in determining the places of development led some scholars into further research to examine the essential characteristics.

Perroux

Perroux set out to investigate the relative importance of different sectors of the economy in overall economic development.

Growth in economy is associated with the discovery of new innovations. Among the innovations seen were the water and wood complex, iron and coal complex, and later the electricity, dynamo and alloy complexes.

Perroux developed his ideas for the coal and iron complex.

This technological era was characterized by heavy industries. These industries were able to dominate their economic environment due to their large size, strength, and their nature of operation. These three characteristics make them as propulsive industries.

Wherever propulsive industries occurred they tend to attract raw materials, people and other services to themselves. This indicates the horizontal and vertical mobility of people. This has developed as the pole of the economy.

This was the origin of Growth Pole Theory. Growth Pole did not differ to any geographic area but referred to economic space. It shows that growth is inherently imbalanced, but Perroux believed that this was a temporary aberration which would be restored to equilibrium effect by the process of reverse flows in products, ideas and finance (Fig. 23.6).

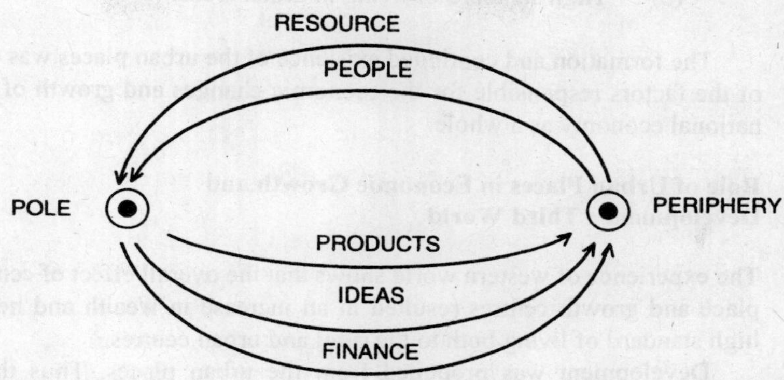

FIG. 23.6 : Perroux Shape of Market

Growth Centre

Growth centre is a geographical interpretation of the growth pole concept over space.

Effects of Growth Centre and Central Places

Central Place : Brought

— commercialization of agriculture;
— release people from agriculture;
— introduction of improved agricultural technology;
— raise in rural incomes; and
— improve standard of living to rural people.

Growth Centre : Created

— Technological focus in industries.
— Technology for diffusion of rural areas.
— Manufacture products for consumption both in the urban and rural areas.
— Urban employment for the people who migrated to urban places.
— Higher and regular income to urban workers.

Result

(a) High standard of living in rural areas.
(b) High standard of living in urban areas.

The formation and continued existence of the urban places was one of the factors responsible for the economic changes and growth of the national economy as a whole.

Role of Urban Places in Economic Growth and Development : Third World

The experience of western world shows that the overall effect of central place and growth centres resulted in an increase in wealth and hence high standard of living both to the rural and urban centres.[5]

Development was propelled from the urban places. Thus those variables from the core influence and finally replace those in the periphery.

Glance on Space and Domination

Before systematic European contact with Third World, it was primitive and traditional in relation to the developed economy and civilized culture. There exists a dichotomy, the modern and traditional sectors of the space economy.

The role of modern space economy is to transform the activities of the traditional space economy. The modern economy has first contact with the urban centres. The role of the urban places in the third world would be to modernize the rest of the countryside.

Friedmann

1. Regional Development was first realized when Venezuela got contact with Europe.
2. Coastal towns grew up in importance and even one became capital city. These were areas from where administrative, market and industrial structures grew up fast and then diffused to the hinterland.
3. With the intensity in metropolis—coast hinterland interaction ideas of specialized agriculture and the service offered by the market centres created the impetus for higher productivity.

4. When this pattern developed Venezuela became independent and cut-off the link with the metropolis. But the modernizing elements were already established in Venezuela.

Result

(a) Industry grew up fast on the coast which then became the core of development.
(b) Agriculture and commercialized farming support the core.
(c) The same mechanism of core-periphery development was set on.

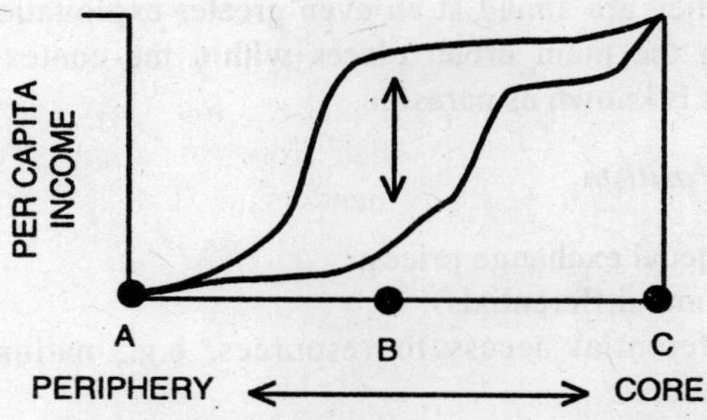

FIG. 23.7 : Core-Periphery Development
A. Primitive and Traditional, B. Transitional, C. Fully Developed

The gap which is great at stage *B* is just a temporary feature and has been ignored out at point *C*.

General Roles

1. Contact points for modernizing element.
2. Areas of control and specification.
3. Market centres for the exchange of rural and urban based imported products.
4. Large areas for industrial development.
5. Centres of cultural change.

Role of Urban Places in Socialist Construction

The emergence of urban centres rest on the concentration of a social surplus products over limited geographic place. These areas of concentrated surplus product becomes core of development.

Within the context of the west urban places had an internal logic to the course of national development.

Socialist Opinion

It is true that the emergence of urban places requires an extraction of social surplus products.

But the existence of rural-urban distinction is a spatial reflection of the exploitation of society.

The temporary nature envisaged in the imbalanced development with time is a misinterpretation of the real world situation. Wherever, flows exist they are aimed at an even greater exploitation of the rural economy. In the main urban places within the context of capitalist development is known as parasitic.

Forms of Parasitism

(a) Unqeual exchange prices;
(b) Income differentials;
(c) Differential access to resources, e.g., national investible resources;
(d) Differential access to services and fruits of science; and
(e) Differences in the distribution and utilization of mental *vs.* manual labour.

Bourgeois Opinion

These five forms of inequalities put forward are true but they are essentially irrelevant after mature urban development. This has been demonstrated by Friedmann's graph.

When society becomes 80 to 90 per cent urbanized, and continue to be urbanized, slowly and slowly the differences withered away leaving behind socio-economic and spatial equilibrium.

Socialist Opinion

The Bourgeois defence is defective in that it overlook the class based society. Capitalist development in individualistic and profit motivated.

The profit accrued to the people owning urban activities are not shared out equitably by all the citizens in urban places. History of urban places show that initially urban places exploited the rural areas but later urban areas exploited even those who worked for the urban centres.

Urbanization is not equal to overall development. Even today the inconsistency is shown in the continued existence of shelters in the developed world urban centres and to Bourgeoing squatter settlements in the Third World. The process of urbanization within the context of capitalist framework of development is a mere transfer of rural to urban poverty.

The Bourgeois arguments are abstracted from reality and are essentially a mystification of the development process.

Favourable development must achieve balance between the urban and rural areas. These must be translated from the overall national development plan.

Role of Urbanization in India's Economic Development

In developing countries of the world the focus of attention remains in the rural sector but the haphazard growth of cities has brought in its train several problems like deterioration in housing and public health facilities, not to speak of the growing crime rate.

Urbanization is a process of population increase in urban area following some non-agricultural activities. The percentage of urban population to the total population of a country reveals the level of urban population growth. Higher is this percentage, greater is the level of urbanization and *vice versa*. Similarly, the per capita income is a clear indicator of the level of economic development achieved by a nation. A country which is highly developed therefore, enjoys a very high per capita income and this bring a high standard of living for its people.[6]

The economic development of a region always leads to greater urbanization. A high standard of living always leads to greater demand for goods and services. The need for greater production results in mechanization, division of labour, specialization of jobs and large scale production. In short, economic development induces further urbanization through industrial development. It can even be stated that the extent of industrialization acts as a common indicator in understanding the level of economic development and urban growth achieved by the economy. In this way, urbanization becomes a part and parcel of economic development.

This should be understood from the fact that in recent times, with the development of our economy, there has also been an increase in the number of people living in urban areas from 2.58 crores in 1901 to 10.90 crores in 1971 and 22 crores in 1991 for India as a whole.

The effect of economic development on urbanization is always positive in the sense that it always results in greater urbanization, the reverse is not always true in the case of India. This Chapter tries to analyze the positive and negative impact of urbanization on the economic development of our country.

The growth of small and medium sized urban centres always lead to expansion of existing facilities. This is reflected in the improvement in transport, communication, housing, education, employment facilities, trade and commerce, civic amenities, etc. These improvements induces the 'pull factors' in attracting immigrants towards these regions. The population shows an upward trend. Correspondingly the demand for goods and services increases. The need for a greater supply of commodities and services to meet the increasing demand encourages investment activity in the economy. The multiplier effect becomes significant. This result in more opportunities for greater investment and the pull factors become dominant and the region continues to attract more and more immigrants. In such a situation, urbanization acts as a stimulant in furthering economic growth.

However, such a situation cannot continue for ever. After a particular stage, which one may call the optimum, the negative effect of urbanization on economic development starts operating. The optimum level can be stated to be that level where the burden on civic finance as a result of huge inflow of immigrants just equalizes the beneficial results of urbanization. The optimum, in a way, shows that the civic finance can do only that much and no more with regard to the provision of all the necessary amenities to the public up to the level of optimum. Though the burden on the civic finances for providing the basic amenities to the public increases. Undoubtedly, this happens due to immigration. The greater burden under such circumstances is confidently met by the administrative machinery because of high level of industrialization and greater taxable capacity of the people. At this stage, urbanization does not contribute to further economic development.

This situation only serves as a warning signal for the negative effect of urbanization. Sooner or later, the downward trend starts exhibiting all the evils of over-urbanization. With the gaining momentum of full factors, the administrative machinery finds it burdensome to maintain the quality and quantity of public utilities and other amenities to the people. As the inflow of people towards these regions increases, the administrative machinery becomes less capable of meeting the additional requirements of the people and to maintain the desirable standard in the case of all civic amenities. This is the situation one comes across in big metropolitan cities like Calcutta in India. Over-urbanization in such a metropolitan cities reflects all the evils like housing problem, water scarcity, lack of medical aid, employment problem, spread of slums, increase in the number of beggars, lot of pavement dwellers, higher crime rate, atmospheric pollution and so on. This affects the economic development in two ways.

Firstly, the rural migrant who comes to the metropolitan city with the hope of getting a decent job is disillusioned once he realizes the situation there. Having left his village in search of a job, he does not like to go back with a feeling of defeat. He struggles hard to find a job. Sometimes, he fails in his attempt and becomes one among the jobless and houseless poor. Even when he gets a job, it may not be to his satisfaction and may not be equivalent to his qualification. Poverty, lack of job satisfaction, under-employment, lack of proper accommodation and other minimum comforts of life become the root cause of frustration.

With this anticipation about the outcome of over urbanization, the administrative machinery is compelled to change the pattern of expenditure. The primary concern of the administration becomes the provision of basic civic amenities and other requirements to the people and the multiplier starts working in the reverse direction. Urbanization, beyond optimum level, acts as an impediment for further economic development.

The optimum level of urbanization depends upon the capacity of the region in providing all the amenities to the public in a fair manner. Grater capacity always raises the level of optimum and a lower of optimum is an indication of lesser monetary fiscal, administrative and other capacities of the administrative machinery. The following factors are responsible for this trend. The first factor relates to the attitude of the migrant. The migrant in most cases desires to move towards a big metropolitan city, ignoring all small and medium-sized towns and cities, because of the desire to live in a well known metropolitan city. The natural outcome is the over crowding in these big cities. This is evident from the data available. It can even be pointed out that urban growth in India has been mostly due to the urbanization of very few big cities. For example, Calcutta, has a population of 91.65 lakhs as on March 1, 1981 and has reached 108 lakh in 1991. The corresponding figure for the year 1971 was 70.31 lakhs. This is followed by Greater Mumbai, Delhi, Chennai, Bangalore, Hyderabad, Ahmedabad, Kanpur, Pune, Nagpur, Lucknow and Jaipur. In 1991 Mumbai has reached a population of 128 lakhs which is the primate city of India.

The growth of small and medium sized towns and cities is not quite significant. In fact, there has been a degeneration of smaller towns. The period from 1901 to 1991, there has been a decrease in the number of Class V and VI towns. Partly, this may be due to the tendency of the smaller towns to go up to the higher classes. Still, the fact remains that the smaller and medium sized towns and cities are not given their due importance in our development programmes.

The other factors responsible for such a situation relates to the lack of proper planning of the towns and cities in India. It is even said that many cities and towns in India are nothing but overgrown villages. Further, the investment decisions are biased by other considerations rather than favourable economic factors. The third factor which is of no less importance relates to the tremendous increase in population. The 1991 census shows that the population of India has doubled to 85 crores since 1947. It was 54.81 crores in 1971. The net addition of 12.56 crores of people shows a 24.75 per cent growth in the last decade compared to 24.80 per cent in 1961-71 period. Though the percentage of growth has been more or less steady, there has been an absolute increase in the total population. This is mainly due to improved medical aid, lower death rate, increased life expectancy, etc. The birth rate has actually declined from 41 per thousand in 1971 to 36 in 1981 and 29 in 1991. The death rate however decreased from 15.7 per thousand in 1971 to 14.8 per thousand in 1981 and 13 in 1991. The life expectancy has increased from 40 years in 1971 to 62 years in 1991. The density of population as a result of absolute increase in population has risen to 221 people per square kilometre as against 177 in 1971. All these have led to over crowding everywhere.

On the other hand, the supply of land cannot be enhanced, the additional labour force in rural areas as a result of the population increase cannot be absorbed in agriculture. The surplus labour force will have to engaged in occupations other than agriculture and in the absence of sufficient non-agricultural occupations in the villages the villager has to turn towards the city for a job. If urbanization has to induce further economic development, then the present trend should be reversed and measures which will improve the situation have to be undertaken. They include :

(i) Rural development programmes which reduce the rural-urban dichotomy and minimize the push and pull factors leading to over-urbanization of a few cities. This will also lead to balanced regional development which is necessary for achieving a high level of economic development for any nation;

(ii) Proper planning of towns and cities, particularly relating to the land utilization pattern, industrial development, educational development, cultural and civic amenities development;

(iii) Development of smaller and medium sized towns and cities for diverting the attention of the migrant towards these places; and

(iv) Further efforts should be taken to reduce the birth rate in a significant manner through various methods of family planning have been enforced.

FOOTNOTES

1. B.F. Hoselitz, "The Role of Urbanization in Economic Development : Some International Comparisons,"in *The City in the Third World*, (ed. D.J. Dwyer), Macmillan, London, 1971, pp. 169-90.
2. E.E. Lampard, "The History of Cities in Economically Advanced Areas," *Economic Development and Cultural Change*, Vol. 3, 1955.
3. W.L. Kolb, "The Social Structure and Functions of Cities," *Economic Development and Cultural Change*, Vol. 3, 1955.
4. K. Davis and Golden, H.H., "Urbanization and the Pre-industrial Areas," *Economic Development and Cultural Change*, Vol. 3, 1955.
5. D. Soja, *Geography of Modernization in Kenya*, Nairobi, 1972.
6. S.N. Nair, "Impact of Urbanization on India's Economic Development," *Southern Economist*, Vol. 20, No. 3, June 19, 1981, pp. 15-16.

PART-II

URBAN LIFE

24

URBAN CULTURE

Importance

URBAN CULTURE is believed to be as old as agricultural one. The phenomenon of urbanization is both ancient and modern and hence the urban culture is also ancient and modern. The ancient is identified with Asian civilization and the modern is identified with western metropolitan civilization. Industrialization is key to the present urban culture[1] and city civilization. Today problems in the city are enormous and "the ultimate solution of city problem is to abolish the city and solve the city problems by leaving the city." Only urban planners are thinking in terms of discontent with urban life, but people are so much indulged even in dirty urban affairs that they want to live in the same just like the insects of a rotten *gular* fruit. The urban culture in wetland areas are affected by flood.[2]

The joy of being an urban resident is ephemeral. Even if the concentration of all powers of administration, business and education in urban areas have failed to usher a healthy life and the urban centres continue to be dirtier and polluted with each passing day. Heaps of garbage lie around carelessly and discarded manner. Mosquitoes abound and infest most localities in the capital. The residents remain unbiased and unconcerned at the present level of hygiene.

Should cleanliness be restricted to the confines of ones home? One should consider a holistic approach towards cleanliness. Clean society reflect its culture. The human consciousness is moulded by the culture, and one is born in and one in turn can reflect back on it.

Process of Cultural Development

The cultural development in every walk of life, in language, literature, fine arts and even in religion and society passes through three stages namely :

FIG. 24.1 : Dress Material as Cultural Variations in Urban Areas

1. Confrontation and conflict;
2. Mutual Appreciation; and
3. Assimilation.

In urban areas there is a vast difference between people of various caste, income groups and religion. All these have increases the sharpness of a culture. So far as the development is concerned it could not be achieved without appreciating each other and understanding their ideas. With these conflict, and appreciation comes the assimilation or inter mingling of various ideas which ultimately given rise to a new culture—either in terms of dress, food habit, education, painting, music, architecture including other ideas of urban living.[3]

The Indian City : A Conglomeration of Culture

According to Anjana P. Desai (1992) is a conglomeration of culture. The individual characteristics of cities and their style of working can be understood if one can place them within the framework of a certain cultural milieu which they belong to. The way the space is organized or the way human interactions take place depends on the shared values, specific needs, behaviour, and adaptive strategies of groups of residents. The aim of this chapter is to throw light on an Indian city which represent "Indianness," having a "conglomeration" of culture for maintaining diversity which is cemented in the base of traditional unity. The main objectives are : to understand the nature and rationale for socio-cultural interactions of interstate migrants in an Indian Cities, and to find out to what extent migrants have adapted themselves or have faced barriers to the local system.

In Indian situation, a city is not just an agglomeration of communities, but a "conglomeration" where the identity of all communities is visible and yet a diffusion of culture and integration taken space. These processes depend on migrants' adaptation, and on cultural barriers of local residents. Demographic and socio-economic variables, level of exposure and period of stay, cultural consciousness and ability to communicate, food habits and general lifestyle have affected the behavioural pattern of all communities in different ways in terms of location and interactions on space.

Ahmedabad is an industrial-cum-commercial city in the State of Gujarat in western India has been selected as the study area. In the census of 1991, it had a provisional population of 3.2 million with 30 per cent as interstate migrants. Here interstate migrants have been classified my mother tongue, and seven major language groups, which are not similar to the local one, have been selected for the study. Sindhi and Punjabi groups have been interviewed from the northern side; Bengali from the east, Tamil, Telegu, Kanada and Malayalam from the south. Sampling has been done on a random and stratified basis, with variable proportion by mother tongue and also by different geographical areas of the city. A total of 1,500 migrants and 300 local residents have been interviewed with a field questionnaire.

Results show that spatial conglomeration and the socio-cultural network system have developed among migrants depending on differential levels of attachment towards cultural roots on one hand and barriers of acceptance posed by local residents on the other.

Definitions

Culture is a way of life of people integrated into small bands, tribes, communities, cities and civilizations. Similarly, culture is a concept that touches the very core of human existence and survival and the cleanliness is the turn imbibed from our culture.

There's something definite to do with culture, when it comes to maintain a certain standard of cleanliness around us. We are not adverse to the idea of living by filth and stench. An American would be penalized for not cleaning ice after snow-fall during the night from his surroundings. This reflects an *administrative culture*.

Heaps of Garbage

In urban centres most people do not look out for a garbage bin when they want to throw household refugee. It is rather thrown out of sight may be on the road or on the footpath or a park. In Delhi, the urban authorities have to be a mute witness to the over flowing garbage bin the everemanating stretch from public conveniences and the clogged open drains. The garbage collectors are very trouble-some lot as they do not come regularly for performing their duties, besides which they do not throw the garbage on the fixed place but discard it in one open corner of the colony. This is very unhygienic. The situation aroused as there are no garbage bins in and around some of the colonies.

Multifarious Cultural Events

In urban areas the musical concerts, exhibitions, drama, films, poet's symposia, art galleries, zoo and museums are the representatives of cultural enigma. Besides this, the fashion show, disco hall, night club, video centres and zymnasiums are some of the basic places of cultural events which should be adopted and marked as culture of a society.

In this way, we can say that urban culture is the sum of different aspects of urban life which are made by the ingenuity of man besides technological levels of an urban centre.

Urban Areas are Pioneers in Ready to Wear

Urban people are the pioneers in ready to wear because they perceive the innovation earlier than others. They always perceive the developed fashions of the society. This makes strong base of urban culture. Very accurate market assessment and thereafter tailoring to this need has been the forte of urban businessmen. The young urbanites are becoming more fashion conscious and they are better exposed to quick change. So, urban people in Delhi, Calcutta, Mumbai and most of the capital cities are sure that whatever is prevalent in Europe comes to India almost immediately.[4]

The urban area provides a complete family fashion store with something for everyone, and now we want to emphasize on the word complete. So the stores are found full of bags, belts, shoes, costume jewellery, perfume and even shoe laces. Hence, urban culture prefers the specialization of shops, i.e., caps, pen, hospital, lady's wear, gent's wear, toys, sports shop, ornaments shop and others.

The Sequence of Cultural Development

The development of urban culture is a stage of the sequence of social development. It is a corollery of food gathering, hunting, herding, agriculture and urban development. According to Jacob's model this has been linked by three processes domestication of animals, the permanent cultivation of crops and the trading of goods with respect to urban centres. The position of these stages and processes with respect to the time is given along with the archaeological evidences.

Jacob's Model of Urban Cultural Development

Jane Jacob's emphasized the increasing evidence of highly specialized and long range trade among human populations in the forest gathering and hunting stage among cities of the early dates. Jacob's linear view of urbanization in terms of early trade and exchange the permanent agriculture is a by-product of food requirements of the hybridized environment of the city.[5]

Huxley's Model of Culture

The British Biologist Julian Huxley had considered three components of culture while comparing cultural and biological evolution of man. The three components are mentifacts, socio-facts and artifacts.

Mentifacts

These include religion, language, magic, folklore, artistic tradition and others. These are crucial and durable elements of culture, because they relate to human ability to think, forge ideas and images against which other aspects of culture could be measured.

Socio-facts

These aspects of culture create links between individual and groups. At individual level they include family structure, sexual behaviour and child rearing. At group level they include political and educational systems.

Artifacts

These are called cultural freight of material manifestation of culture. These include groups of material technology, e.g., food, shelter, transport besides agricultural production, landuse and clothing of a particular design are known as artifacts.

Cultural Hearth's of the World

According to archaeological evidences and historical events the world has been divided into many cultural areas which could be seen in Table 24.1.

The Table 24.1, clearly shows that urban culture spread in different parts of the world sometimes before 3000 B.C. in Mesopotamia and Nile Valley.

Urban Cultural Regions of the World

Even the urban culture of different parts of the world has dissimilarities which favours the demarcation of cultural regions. More or less the

TABLE 24.1 : Main Urban Cultural Areas of the World

Zone	*Location*	*Early Urban Culture*	*Representative City*
Middle East	Mesopotamia	Sumerian (2700 B.C.)	Uruk
	Nile Valley	Egyptian (3000 B.C.)	Memphis
	Indus Valley	Indus (2500 B.C.)	Mohanjodaro
East Asia	Hwangho Valley	Shang (1300 B.C.)	Auyang
America	Yucatan Peniusula	Mayan (500 A.D.)	Tikal
	Central Mexico	Aztec (1400 A.D.)	Tenochtitlan
	Peru	Inca (1500 A.D.)	Cuzeo
West Africa	Niger Valley	Yoyuba (1300 A.D.)	Ife
Southern Europe	Aegean Island	Aegean (2000 B.C.)	Knossos
	Italian Peninsula	Etruscan (1400 B.C.)	Rome
South-East Asia	Mekong Valley	Khmer (1100 A.D.)	Angkor
	Japan	Yanato (600 A.D.)	Naniwa
	Central Myanmar	Pya (700 A.D.)	Sri Ksetra
	Ceylon	Sinhabese (1000 A.D.)	Polonnaruva
	Ganges Valley (India & Nepal)	Chechar (1000 B.C.)	Vaishali
		Nalanda (400 A.D.) (Buddhist Culture)	Nalanda
		Hindu (5000 B.C.)	Kedarnath
			Badrinath
			Janakpur
			Rameshwaram
			Deoghar

world could be classified into the following regions :

(a) Western Culture;
(b) Eastern Culture;
(c) African Culture; and
(d) Mid-Eastern Culture.

The description of these cultural regions are as follows :

Western or American Culture

This culture is most modern and covers urban areas of London, Berlin, Paris, Stockholm, Viena, New York, Ottawa, Waterloo, Los-Angeles, Aukland and others. In all these areas the food habits, dress material, religion, art, using condition, trade, commerce, transport, language and technological innovations are more or less the same. The western urban culture is considered as modern metropolitan culture.

Eastern or Indian-Chinese Culture

This is known as ancient urban culture of the orient. The massing of millions of people on a limited resource of housing, water supply, congested roads, over-crowded buses, trains and the abundance of spontaneous settlements[6] but a definite change towards western culture is quite discernible fact in India, China, Myanmar, Malaysia, Japan, Korea, Pakistan and Indonesia.

African Culture

In different countries of Africa and South America the urban centres are over pressed with western culture but due to lack of resources they are lagging behind in many respects. The water supply is poor, spontaneous settlements abound, demarcation of Class-A zone, Class-B zone and Class-C zone are distinct in urban areas, e.g., in Dar es Salaam, Class A zone is fixed for ministers and high officials, Class B is reserved for Asians and Class C for Africans. More or less the same culture is found in South America also, because the lack of resources has forced the urban centres in these areas to function according to the donors choice of planning and resource supply.

Mid-Eastern Culture

The urban culture of Beghdad, Riadh, Tehran, Dubai, Muscat, Abudhabi, Tripoli, Cairo, Ankara and Kabul are more or less the same. The dominance of Muslim culture, desert environment, production of oil, and the *purdah* system of Muslim women favours an altogether different culture in comparison with other parts of the world. These are the areas of ancient city culture like Mohanjodaro and Harappa in Indus Valleys, Memphis in Nile Valley and Uruk in the Valleys of Tegress and Uphrates.

It could be concluded that the culture areas are homogeneous regions in terms of religion, language, dress material, food habit, living conditions and environmental make up.

Changing Lifestyle in Asian Cities

A new affluence has creptup into Chinese, Indian and Malaysian cities where people are spending a big amount of their budget for the purchase of jewellery, musical instruments and furrning shoes that cost a month's salary.

Most of the people in Asian cities live in tiny cramped homes that rarely have proper ventilation and 24 hours supply of running water. But in cities like Beijing, Mumbai, Bangkok and Calcutta[7] most people own colour television that can be operated by remote control. In these cities most of the families are now acquiring Video-cassete Recorders (VCR) which have become a status symbol.[8]

A recent survey of 500 urban households by the Shanghai socio-economic survey organization revealed that by 1992, out of every 100 households 60 will have VCR's, 23 telephones, 13 air conditioners and 4 motor cars.

Most of the families considered for survey own a colour T.V., Refrigerator and Washing Machine. In these cities a good doctor earns the equivalent of a months salary within five to six hours. Some of the workers who are engaged in two jobs or part-time jobs have contributed substantially in changing the lifestyle in Asian cities.

Bare cement floors and white washed walls are now being transferred with wallpaper and paint. Young couples scrimp and save to buy European and Japanese style furnitures for their homes. China's youth now described the changing lifestyle as 'interior decoration fever'.

A survey revealed that urban families with per capita daily income less than Rs. 130 spent on an average Rs. 2,000 to decorate their homes.

Recently, the extensive use of fashionable dresses especially by college girls and computer revolution in day today's life have made the cultural advancement truly dynamic and ultra-modern.[9] Innovative street theatres are now commonly seen on festive occasions in many urban centres of the Asian realm.

FOOTNOTES

1. L. Mumford, *The Culture of Cities*, Secker and Warburg, 1938.
2. P. Self, *Cities in Flood*, Faber, 1961.
3. P. Barley Migerd Nutgens, *Living in Towns*, B.B.C., 1966.
4. R.E. Dickinson, *City Region and Regionalism*, Routeledge and Kegan Paul, 1947.
5. C.S. Fisher, "Towards Sub-Cultural Theory of Urbanism," *American Journal of Sociology*, Vol. 80, 1975.
6. U. Singh, "Cultural Zones of Allahabad," *National Geographical Journal of India*, 1966, Vol. 7, pp. 95-104.
7. A.B. Chatterjee, "Cultural History of Howrah City," *Geographical Review of India*, Vol. XXVI, 1964, pp. 23-24.
8. S. Sinha, (Eds.), "Cultural Profiles of Calcutta," *The Indian Anthropological Society*, Calcutta, 1972, p. 283.
9. A.K. Lal, *The Urban Family*, Concept Publishing Company, New Delhi, 1990.

25

URBANIZATION AND INDUSTRIALIZATION

URBANIZATION is the cause of industrialization and hence they are highly correlated. Nowadays urbanization is a great concern for the local government because the rapid growth of urban population has created environmental problems of various types.

Location of Industry

For urban development the location of industry is quite favourable because it brings road, people, public services, houses, raw materials besides money, men and material to the place which give rise to urban centres. .

The agriculture may not be able to give so much of services to the people, but by the establishment of industry the population supporting capacity of the region increased manifold and hence industrialization brings progress and prosperity to the city.[1] This may have some social and economic repercussions and their delineations are as follows.[2]

(a) *Development of Capitalism* : Concentration of money in a few hands.

(b) *Increase in Classism* : The labourers organizes and make demand through strike and lock-out.

(c) *Labour Problems* : As the number of labourers increases their health decreases.

(d) *Industrial Disputes* : The accidents give rise to industrial disputes.

(e) *Economic Crises Increases* : Industrialization brings luxurious items into the market but the cost of edible items increases.

(f) *Set Back to Rural Industries* : The industries located in rural areas cannot prepare such a finished goods as industries located in urban areas.

(g) *Division of Labour* : Industrialization favours division of labour and specialization.

(h) *Progress in the Business* : Due to industrialization there is a substantial progress in industrial production and the trade of carriages, telephones, engines, aeroplanes, etc.

(i) *Large Scale Production* : The industrialization favours population increase as well as large scale production of iron, cement, fertilizers, steel and other manufactured goods in lesser costs.

Urbanization and Industrial Complexes

Urbanization include economies of agglomeration of transport and industrial raw materials, scale of production and several other considerations.[3] Urban centres provide a better economic point not only to the industrial firm or related industries but to the entire hinterland of urban centres. Such urban centres become self-sustaining with economic and social aspects. Hence, industrial complexes have strong measure especially in the developing societies, which help in eliminating the economic disparities. The growth and development of industrial complexes on the urban periphery are more disadvantageous, for the local people in terms of increasing the traffic congestion, haphazard growth of colonies, environmental pollution, high cost of living, working inefficiency due to noise, unionism besides accident and prostitution. All these problems are subject to improvement through different planning measures.[4]

Patterns of Urban Industrial Growth

In urban centres of different population size, different types of industries predominate the scene which in turn affects the industrial environment and linkages as well.[5]

Industries of Metropolitan Centres

The metropolis of Kanpur (1,962,750 of 1991) is an industrial capital of Uttar Pradesh.[6] It is located on the right bank of river Ganges. Besides cotton mill, Kanpur has several industries, i.e., rolling mills, engineering, chemical, plastic, medicines, paints, oil furniture, hosiery, bakery and dairy products. The growth of large scale industries also encourages small scale industries due to linkage effect. Currently the manufacturing activities developed haphazardly all over the city which presents diversity in the industrial landscape of Kanpur.

The industrial concentration in the city has led to the economic prosperity and rise in the standard of living of the city dwellers by providing employment opportunities in industries and business in trade and commerce.

Industries of City Level Centres

Bokaro Steel City (350,450 of 1991), ninth new industrial city of Bihar is located in Dhanbad district at a distance of 260 kms. from Patna and 300 kms. from Calcutta.[7] Bokaro is a monotype steel complex with six blast furnances in the middle Damodar Valley. Bokaro gets washed coal from Dugda, Kathara, Giddi and Kargali Washeries. Iron ore is obtained from Kiriburu and Meghahatuburu mines of Bonai-Keonjhar range. Limestone is available from Bhavnathpur of Palamau district. Water is available from Tenughat Dam and power from Chandrapura Thermal Power Station.

Industries of Small Towns

Revelganj in Saran district of Bihar is a small town whose hinterland was noted for salt-petre besides indigo production and collection centres of the district. With the development of its entrepot character and transhipment led to the growth of some associated industries, i.e., boat making, blacksmithy, carpentry, pottery and the making of agricultural implements are important. Wood is an important raw material and it is easily available in the nearby forest areas of Gorakhpur and Basti.

Effect of Urbanization on Industrialization

Although urbanization in India is connected with industrialization, many of its peculiarities can be traced to the special features of towns in the preceding centuries. The most important of them is the fact that concentration of urban population was typical of the period of Indian feudalism. Feudalism in India showed signs of crisis already in the 15th century and by the 18th century decline had already set in. When Europeans began investing India they found prosperous towns and cities here. As they began to flood South Asia with factory manufactured goods, the local artisans found it difficult to compete with cheaper foreign products and crafts in India began to die out.

With new modes of production increasingly taking roots in India, towns began to spring up everywhere, causing a spurt in rural migration. Consequently, during the 19th century the population of Indian towns and cities swelled as people flooded them, seeking a means of living in the railways, in trading offices, banks and so forth. The number of inhabitants of India's towns grew ten fold over a hundred years.

The towns were founded not only as bourgeois proletarian centres but as a concentration of intelligensia who played such an important role in the history of the national liberation struggle. The role played by the urban population in the revolutionary events of 1905-1908 graphically illustrates the social and political results of urbanization at the turn of the century.[8]

One aspect of urbanization that needs special attention is the lack of assimilation of immigrant rural elements into urban life. Significant number of immigrants, not necessarily of the same ethnic origin, tend to form a sub-culture of their own, while living in slum areas or 'shanty towns' on the fringes of urban life. They are neither urban nor rural and tend to exhibit the worst features of both rural and urban life.

The bulk of industrial labour and low paid workers in the service sector tends to be sucked into these fringe zones of the Indian city. These amorphous twilight zones tend to develop into an under-world, culturally as well as politically. Social stratification, sharp enough in the villages, is also a sharp reality in these twilight zones.

The twilight zones in the cities can therefore become an explosive element, unless the basic contradictions are easily resolved. In our country the social distance between the 'upper' and the 'lower' classes is much more pronounced than in the West and class consciousness is also sharper. An awareness of enormous social differences, a feeling of belonging to a certain class and having little or nothing in common with a 'higher' or 'lower' class is now very much a part of urban life.

The urban population in our country will rise rapidly in the next three decades, with at least one third of them living in slums. Therefore, new needs generated by urbanization and industrialization call for the development of infrastructural facilities, social assistance schemes to prevent the breakdown of even tolerable standards of urban existence.[9]

Urban Share of Industrial Workers

In developed countries of U.S.A., France, England and Germany about 75 per cent workers are engaged in industries. In U.S.A. most of the

Industrial workers are concentrated in the urban centres of New York, Philadelphia, Boston, New Jersey, Marryland, Cleveland, Duluth, Garry, Chicago and Pittesburg. Similarly, in England the concentration of industrial workers are found in London, Lancashire, Birmingham, Sheffield, Glasgow and Barbickshire.

In India out of total workers hardly 4 per cent engaged in industrial activities but most of these industrial workers are concentrated around the industrial centres of Calcutta, Kanpur, Delhi, Mumbai, Chennai, Ludhiana, Jamalpur and Ahmedabad. The workers reside in rural areas and daily come to their respective factories from far flung places as commuting workers. In most cases the industries located at urban centres are polluting the local environment more on the cost of facilities of roads, electricity, higher education and generation of employment.

Industrialization and Social Disorganization

Urbanization creates several allied problems in the area such as :

1. Decline of mortality;
2. Unemployment;
3. Increase in crimes;
4. Change in families social values;
5. Disintegration of joint family system; and
6. Juvenile delinquency.

The description of these points are as follows :

1. The decline of mortality favours increase of population and disorganization of an individual. The labourers family live away from them which give rise to alcoholism, prostitutes and other cheap means of recreation.
2. Although industrialization relegates unemployment but the massing of people favours unemployment. This also happens due to mechanization of the industrial sector.
3. Industrialization favours the increase in quantity and quality of crimes. The incident of stabing, murder and suicide are common feature in urban areas.
4. In urban areas elderly people won't get much respect. Wife and husband both have to do work in the factories in order to get their livelihood.
5. This is also an invitation to the breakdown of joint family system. The children's develop all sorts of habits and most of

them get good education due to close proximity of educational institutions.

6. Juvenile delinquency is a common feature in urban areas. Industrialization brings modernity on the one hand and free-mixing of young turks on the other. All these favours sexual indulgence of young people in urban areas.

Industrial Sectors of Munger

The industrial zones of the city of Munger have been found in dispersed manner, e.g., the cigarette factory of Basudeopur, Gun Factory of Fort Area, Threasher making and grill factory near Neelam Road, Chauk Bazar, Bakapur, Nimtala, etc. Other industries of the city are mostly household, *bidi* making, hand spinning, basket weaving, oil crushing, medicine and saline water making near Khankah of Munger and other places.

The cigarette factory (I.T.C. Ltd.) is equipped with modern technical machinery which helped in the development of Munger from 1921 upto the present. There are 2,500 workers in A.D. 1900. This factory has its own printing press as a separate department. There is also a departmental Saw Mill which fabricates all types of cigarette box.

I.T.C. Ltd., Munger is an oldest and largest cigarette factory of Asia, which runs three shifts daily. There are about 35 gun manufacturers which perform their job under a shed inside the industrial Estate, Karn Chaura, Fort, Munger and the space has been allocated by the Industrial Department, Government of Bihar at the Fort. Although Munger-made gun is sold all over India but currently this factory is facing competition with several other gun manufacturing centres of the country.

Besides these there are several leth factories in the city of Munger which are located in different sectors such as Bekapur, Neelam Road, Maksuspur, etc.

FOOTNOTES

1. Rana Pratap, *Growth and Regional Pattern of Industrial Complexes*, Concept Publishing Company, New Delhi, 1983.
2. S.S. Azam, *Industrial Landscape of the Lower Sone Valley in Bihar*, Unpublished Ph.D. Thesis, Ranchi University, Ranchi, 1968.
3. A.J. Wright, Manufacturing Districts of the United States, *Economic Geography*, Vol. 14, 1938, pp. 195-200.
4. E.W. Gilbert, "The Industrialization of Oxford," *Geographical Journal*, Vol. 109, 1947.

5. G.S. Kulkarni, "Industrial Landscape of Greater Poona," *Bombay Geographical Magazine*, 1961, pp. 8-9.
6. S.N. Singh, "Impact of Industrialization of Urban Society : A Case study of Kanpur," *Urbanization and Regional Development*, Concept Publishing Company, New Delhi, 1982, p. 13.
7. Ram Prabesh Singh, "Bokaro : The Future Steel Town," *National Geographical Journal of India*, 1961, Vol. 7, pp. 245-56.
8. U. Singh, "Industrial and Commercial Landscape of Allahabad," *National Geographical Journal of India*, 1960, Vol. 6, pp. 227-41.
9. S.M.A. Ziaurab, "Report on the Industrial Survey of the Agra District, Allahabad, 1924.

26

URBAN WATER SUPPLY

WATER is one of the three basic needs of man. In urban areas the concentration of population is so high that the timely and regulated supply of water is considered necessary. In India the supply of clean drinking water and sanitation are State subject, but the Government of India provides technical assistance to the State Government and the Union Territories for formulation, implementation and maintenance of urban water supply and sanitation programmes through the Ministry of Urban Development. This Ministry organizes conference, seminars, provides training, research and management of the Information System through technical personnels.[1]

Need for Water Supply and Sanitation

1. The water in wells, ponds and rivers are contaminated and for this fact the supply of safe drinking water is essential.
2. The garbage and night soils are dangerous, hence their proper management is essential in order to usher an era of cleanliness in our cities and towns.
3. In urban areas due to congregation of population, huge rush of traffic especially during day time ejects such a huge amount of carbon monoxide, obnoxious gases and solid wastes that they are polluting the sources of ground water and giving rise to slums in urban areas where the maintenance of sanitation is quite essential.
4. The congregation of large number of people require quick and timely supply of water because they cannot move to far away places to fetch water for safety reasons and convenience of work.

Indian Situation

The provision of safe and adequate water supply for the ever increasing population and hygienic disposal of wastes is a basic necessity for the healthy living of the urban community. The urgency for water supply and sanitation measures assume special importance in a tropical country like India which is a subject of periodic wars with water and filth borne diseases, viz., cholera, dysentry, typhoid and gastro-enteric fevers. An organized water supply system also provides an opportunity for speedy removal of human excrete and other offensive wastes through water flushed sewers. Unfortunately, in the sub-continent of India—no one cares to discuss the methods of fecal disposal. This task has been entrusted to sweepers whose job throughout his life is to remove nightsoil—has earned him the ultimate epithet of social degradation. The sweepers are considered here as untouchables. This is a very sad state of affairs in the way of social development.[2]

The science of environmental hygiene is of recent origin. It gained importance with the shifting of emphasis from the curative to the preventive aspects of public health. This again was a result of the growing industrialization and urbanization in the more advanced countries in the West which pin pointed problems relating to environmental hygiene—which deals with town and country planning, housing, safe water supply, collection and disposal of community wastes, prevention of water and air pollution, control of industrial wastes, food, sanitation, etc. Hence, water supply and sanitation measures got priority in their nation building programme as an essential prerequisites. It has been emphasized by the Bhore Committee and the Environmental Hygiene Committee (1948-49) which aims to supply safe water upto 90 per cent population of urban areas within 40 years. The programme also carried out 100 per cent grant for rural investigations and Public Health Engineering Training Programme, based on urban centres.

During the Seventh Five-Year Plan an outlay of Rs. 2,935.64 crores under the state sector and Rs. 30.11 crores under the Central Sector have been provided for urban water supply and sanitation as against the projected requirement of Rs. 5,797.7 crores. Keeping in view the outlay provided, the decade targets in respect of urban water supply and urban sanitation were scaled down to 90 per cent and 50 per cent respectively and these revised targets are expected to be achieved by March, 1991. It has been envisaged that the progress during the Seventh Plan period in the Urban Water Supply and Sanitation Sector would be as follows :

TABLE 26.1 : Population Coverage

(in million)

Water Supply and Sanitation	*As on 31.3.1985*		*As on 31.3.1990 (Envisaged)*	
	Population	*%*	*Population*	*%*
1. Urban Water supply	127.23	72.88	169.96	86.40
2. Urban Sanitation	49.56	28.40	87.91	44.70

The working group formulating the Eighth Five-Year Plan (1992-1997) has envisaged a requirement of Rs. 14,094 crore for Water Supply, Sanitation and Solid Waste Management. This includes a provision of Rs. 44 crore as Central Sector outlay for the support programmes. It has been decided by the working group of Eighth Plan to provide water supply facilities to 100 per cent and sewerage and sanitation facilities to 75 per cent of the urban population by the end of the Eighth Plan. The Government has introduced use of Micro-Computer for planning and designing of water supply distribution and sewage systems besides imparting training to public health engineers.

The Urban Development Ministry has developed a Management Information System (MIS) which has been disseminated to State Public Health Engineering Departments and Water Supply and Sewage Boards for application. It gives technical advises and clearance to urban water supply and sanitation schemes for external assistance. It has helped in completing six World Bank Aided Development Projects in Maharashtra, Uttar Pradesh, Punjab and Rajasthan. It has recently taken up the programme of liberation of Scavengers' in collaboration with the Ministry of Welfare in 500 towns in the first phase within a period of two to three years. It is also responsible for monitoring progress besides improvement of sanitation, roads, drainage, street lights, slaughter houses and community facilities under the Sri Nagar Core Area Development Project.

In the State of Bihar upto 1947 only 8 urban centres, e.g., Patna, Gaya, Arrah, Munger, Bhagalpur, Muzaffarpur, Sultanganj and Jamshedpur had piped clean water supplies. Upto 1966 about 60 per cent urban population of the State of Bihar enjoyed this facility in more than 30 urban centres. A sum of Rs. 2,500 crore has been estimated to fulfil the task but during Fourth Five-Year Plan of the State of Bihar only 4.86 core given for the purpose. This clearly points out that the problems of safe water supply will stay for a longer time not only in the State of Bihar but throughout the Third World countries.

Urban Water Supply and Sanitation in Karachi

The shortcoming of sanitation and water supply have been highly felt by the urban poor who are living specially in slums, shanty towns and squatter settlements which are mushrooming all over the developing world.

Out of Karachi's six million population two million live in the slums. Their houses are known as Karachi Abadi (temporary dwellings). These settlements have no drainage, no water supply and they are over-crowded insanitary and unhygienic, and sometimes they lack even the legal right to exist.

One out of five children born in the slums dies before the age of five—due to poor arrangement of human waste disposal and a lack of health consciousness. The problem is common in most of the Third World big cities. It is very difficult to deal with. Flush, toilets and water taps are out of question. Even if there is no shortage of water in Karachi, the costs of laying sewers connection to every house would be prohibitive.

In the city of Karachi, the Baldia project aims to install soak-pit latrines. The starting point was the design by a Dutch Advisory Mission of low-cost household latrine. In consists of a soak-pit 14 feet deep and 5 feet in diameter connected to a squating plate by a four-inch pipe with a water seal. The cost per soak-pit is Rs. 200 ($ 120) and it has a life of 20 years for nine people.

Initially, the Karachi Municipal Corporation was not interested with water supply and sewerage on a mass scale. The project of Baldia will take at least ten years to supply water. The slums of Baldia are not at all uniform in layout, buildings and inhabitants—and hence, for the time being it is futile to think about the supply of safe drinking water and avail the facility of soak-pit latrines in the near future. The migrants people of Sind and Punjab adopt rural life style even if they reside in urban areas.

In urban areas rivers and pools used for drinking water can be polluted by industry, or by washing and bathing. Urban water pipes are also contaminated by sewage, especially when the pressure drops, like dirty water, the inadequate sanitation causes disease. In too many developing countries the municipalities still have to wash inner drains contaminated with sewage. Therefore, it has been felt that safe latrines are an essential for basic health.

Situation of Water Supply in Bangkok

The provision of safe water to the inhabitants of a metropolitan city of Bangkok—which has a population of 8.5 million spread over an area of 6,400 km^2 is really a gigantic task. The authorities of Thailand's capital have been working on it for the last 75 years. The huge immigrant population to achieve economic book and industrial development itself also have to provide the clean water.

Today about 68 per cent population of Bangkok's have access to piped water, and some 700,000 households are connected to a metred and charged for the supply of safe water.

The expansion of such services has been recognized to be of utmost priority for health and to facilitate continuous development. The Asian Development Bank has lent the Metropolitan Water Works Authority $ 130 million for the Third Phase of Water Treatment Plant Construction and this project has completed at the end of the International Drinking Water Supply and Sanitation Decade, 1990.

Water Management in Delhi

The State Government of Delhi is concerned over the declining per capita availability of water in the National Capital Territory of Delhi. The Planning Commission has called for better water management and conservation to tackle water scarcity on a sustained basis.

At an internal meeting of the Planing Commission presided over by the Deputy Chairman, the then Mr. Pranab Mukherjee, recently, the commission felt that while tackling the present water shortage problems of Delhi on sustainable basis, the demand management including conservation measures were more necessary than turning merely to supply side of the problem.

"It is estimated that with the technological innovations available today, water consumption for cities could be curtailed by at least one third with no sacrifice of economic output or quality of life," the meeting felt.

The installed capacity of Municipal Water Supply of Delhi is 600 million gallons per day (MGD); out of which 15 per cent of transmission losses are excluded, leaving the net quantity of water available to be about 450-475 mgd., the commission observed.

With this level of water availability, per capita water in Delhi may be about 56 gallons per day (GPD) as against about 55 gpd in Mumbai, 30.5 gpd in Calcutta and 15.5 gpd in Chennai.

The Commission noted that there was substantial variation in the availability of Municipal Water Supply in different areas in Delhi which at times created law and order issue.

The long-term strategy to fulfil the ever-growing water needs of Delhi was through few multi-purpose projects like Tehri Dam (U.P.), Kishau Dam (U.P.) and Renuka Dam (H.P.), whose estimated costs are Rs. 5,058 crores, Rs. 1,054 crore and Rs. 372 crore respectively.

"Going by the present experience, these projects may take quite long time and benefit from these may not be expected at least in this century," the Commission felt.

The Commission also suggested few "action points" as part of demand management of the water supply problems in the national capital.

A vigorous and extensive mass information campaign through airing radio and television spots should be taken up to enhance public awareness about city's water plight and to let people know how they can conserve water, it said.

The water pricing structure should have adequate built-in financial penalty to act as an effective restraint on consumer to remain within the allotted quotas, it said adding that the restructuring of water pricing as recommended in the study report was very much needed in as much as only about one-tenth of one per cent of the annual expenditure of an average family was spent on water.

Commercial establishments particularly big hotels and automobile service stations should be compelled to meet most of their water demands for various uses other than drinking.

International Drinking Water Supply and Sanitation Decade

The United Nations declared the period 1981-90 as the International Drinking Water Supply and Sanitation Decade with a view to provide safe drinking water supply and adequate sanitation facilities in all the member countries by 1990. The Decade Programme was launched in India on 1st April 1981 with the objective of increasing the urban population coverage with water supply facilities upto 72.25 per cent.

Remedies of the Problem

1. The remedies of the Water Supply and Sanitation lie in the hands of the State Government and the residents of the town. The regular cleaning of roads and timely supply of water is quite essential.

2. On community basis it could be decided to through the garbage at a definite point so that the Municipal Van could collect it easily.
3. The other way is that the residents of the area could clean their surroundings as this system is practiced in Tanzania.
4. The construction of water tanks, laying of pipes, regular cleaning and maintenance of tanks and pipes are essential to supply clean drinking water.
5. The town administration should restrict themselves in throwing used dirty water through regulated channels to the nearly rivers and wetland areas as this can also pollute our water and environment as well.

FOOTNOTES

1. "Water Supply and Sanitation," Ministry of Urban Development, New Delhi, Government of India Report 1988-89, pp. 12-14.
2. T. Lee, and Burton, J., "Future Water Supply for Indian Cities," *Deccan Geographer*, 1965, pp. 3-4.

27

URBAN TRANSPORT

THE LINKAGE of urban centres is possible due to the development of road, rail, waterways and airways. Transportation lines act as the life line for urban development without which the concentration of population and economic progress of the urban areas are not possible. Transport is necessary to satisfy the function and utility of a place. Without transport the territorial specialization is impossible, and the communities must exist solely on local resources and the density of population is dependent upon these resources. Therefore, the efficiency and ease of transport and its capacity to generate high level of urbanization should be emphasized. The transport lines act as conveyers of trade, commerce and mutual understanding and innovation between places have their roles properly played in localization of facilities, concentration of specialized population and this in turn led to the location and growth of urban centres in the region.[1]

Nowadays the existence of cities depend on the costs of transport because we need the close link with transportation lines in terms of living space, location of shops, industries, educational institutions and accommodate large-scale activities of modern civilization. In fact transport generate various activities and this in turn generate transport.

It is difficult and expensive to provide transport facilities in cities, and the larger the city, the arrangement of transport is more difficult. A bad transport system can make the city life intolerable. A rapidly growing population of the world living in cities with steadily deteriorating transport systems.

Network Analysis

Figure 27.1 analyses the linkage of urban settlement system which represents an interaction between each place in the system with every other place. This may be useful to apply graph theoretic constructs in order to explain the degree of connectivity of the system, and the structure of linkage patterns. Matrix algebra could also help in the identification of urban hierarchy and in isolating areas which are poor in urban growth. The dynamics of the settlement system can also be examined in case the same matrix algebra is applied for successive census periods.[2] For Fig. 27.1 the graph theoretic *Td* matrix is as follows :

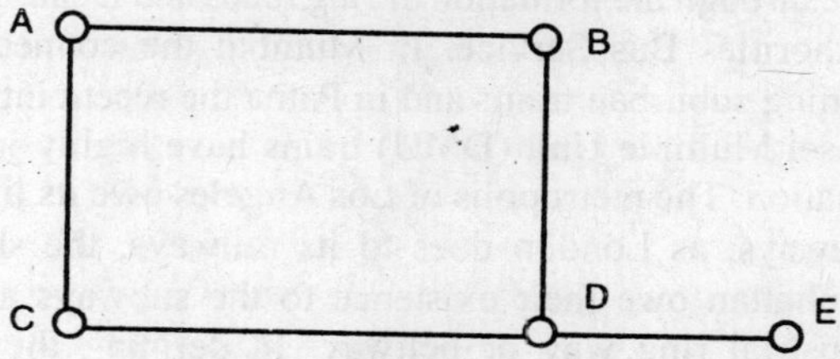

D Matrix

	A	*B*	*C*	*D*	*E*	Total
A	0	1	1	0	0	2
B	1	0	0	1	0	2
C	1	0	0	1	0	2
D	0	1	1	0	1	3
E	0	0	0	1	0	1
Total	2	2	2	3	1	10

FIG. 27.1 : D Matrix of Transport Network

For row and column the highest value comes for centre *D* and hence *D* point has high accessibility among the urban centres in the system.

The Function of Urban Transport

(a) Transport is not only serving the city but it is a part of the city. It occupies nearly 30 to 40 per cent land near the city centre and about 20 per cent land near the suburb. Cities are made up of buildings and transportation lines.

(b) The houses and the urban activities could not be properly planned without transport system. The level of the development of transport system varies from one part of the urban centre to the other, and hence we could say that the urban way of life is wholly dependent upon the nature and quality of its transport system.

(c) The interaction between transport and the nature of the city defines the function of transport. Calcutta owes its life style through metro-railway and tram services. Delhi owes its life style through the formation of ring roads and Delhi Development Authorities Bus Service. In Mumbai the connection of fast running suburban trains and in Patna the recent introduction of Diesel Multiple Unit (DMU) trains have highly improved the situation. The metropolis of Los Angeles owe its life style to its freeways, as London does to its railways, the skycrapers of Manhattan owe their existence to the subways and Boston's industrial ring way or beltway. In defining the function of urban transport one has to drive back to the question of what sort of city one wants, what size, and for what class of people? Thus the transport in urban areas should be organized according to the demand set by the people in urban areas.

Problems of Urban Transport System

The problems of urban transport system could be divided into seven groups (Fig. 27.2). There may be the problem of public transport, problems of the pedestrian, environmental problems, parking difficulties and high charges. These nine problems are very different from one another but as they are closely related so they cannot be solved in isolation.

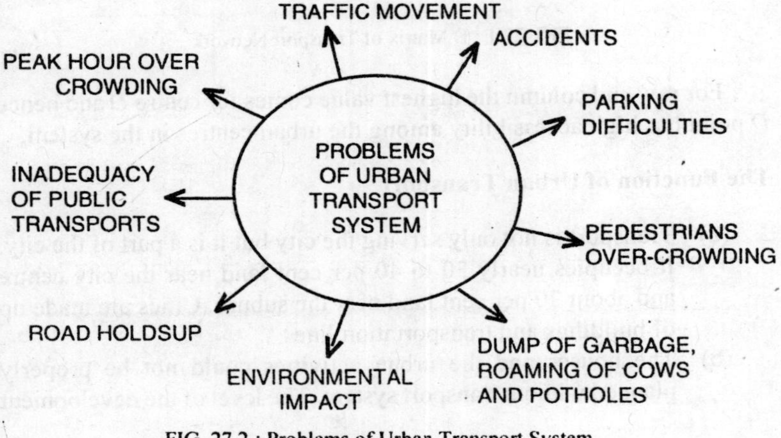

FIG. 27.2 : Problems of Urban Transport System

Patterns of Urban Transport Development

Some urban centres have strong transport base, some have weak transport base, in others the centre is hardly distinguishable from the suburb, whereas in some other cities the linear fashion emerged as in Patna and in others the transport lines are distributed haphazardly.

The patterns of urban transport is wholly dependent upon the structure of cities. The structure of cities depend upon four factors which ultimately govern the development of transport—Geographical features of the area, relative accessibility, developmental control and dynamic process. A solution to the urban transport problem must be based on a stable practicable equilibrium between these different forces.

Five types of solution of urban transport problems are as follows :

Archetype A—For Full Motorization

This type of road has been designed to provide free-flow of motors in all parts of the city. Non-car users may favour buses which travel mainly on the secondary roads in high speed. For cyclist there is no problem as

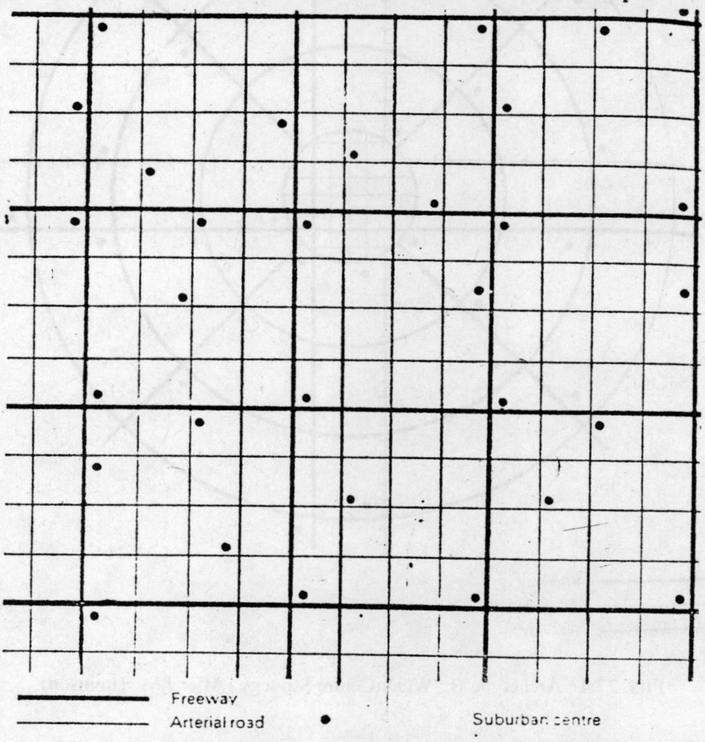

FIG. 27.3 : Archetype A : Full Motorization (After J.M. Thomson)

they can select routes along subways with a signal controlled interaction. For pedestrians the wider road is hazardous, unless special facilities are provided and the road crossing are frequent from pedestrians point of view. It's examples are Los Angeles, Detroit, Denver and Chandigarh (Fig. 27.3).

Archetype B—Weak-Centre Strategy

A weak-centre strategy portrayed schematically in a city with a radial road network serving a small city centre to which a relatively high proportion of city-centre workers travel by car. The majority of jobs are located in sub-urban and peripheral places and are served by car with the benefit of high-capacity ring roads. It does not provide a self-

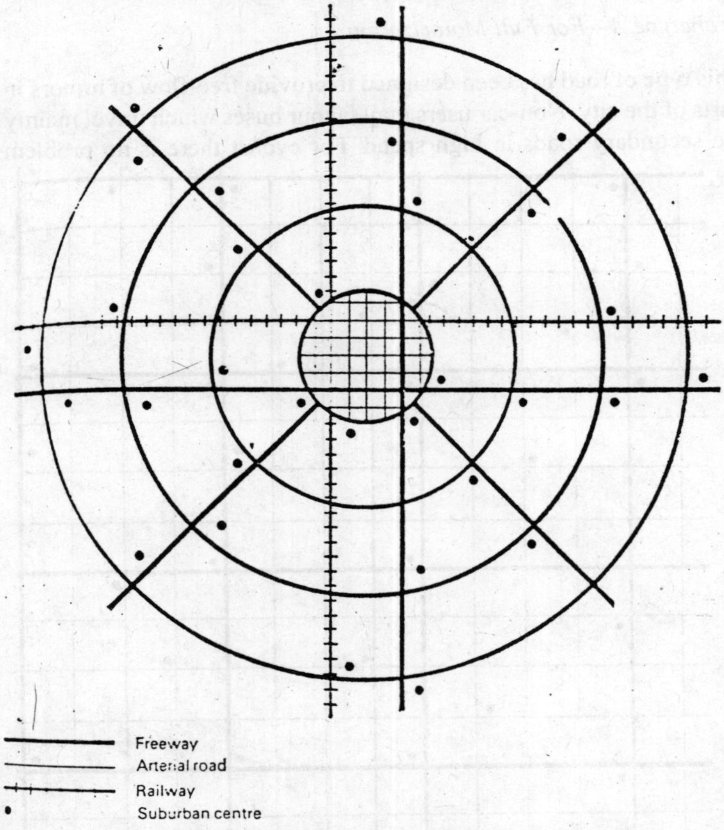

FIG. 27.4 : Archetype B : Weak-Centre Strategy (After J.M. Thomson)

regulating strategy in which the market forces tend to move outside the centre. Hence, planning controls are necessary to maintain the optimal balance between the centre and the sub-centre in terms of volume of employment, shopping and other activities should be kept at the right level. Melbourne, Copenhagen, Chicago and Boston are examples of this type (Fig. 27.4).

Archetype C—Strong Centre Strategy

The concentration of urban activities are found in most great cities of the world, e.g., London, Paris, New York, Tokyo, Moscow, Rome,

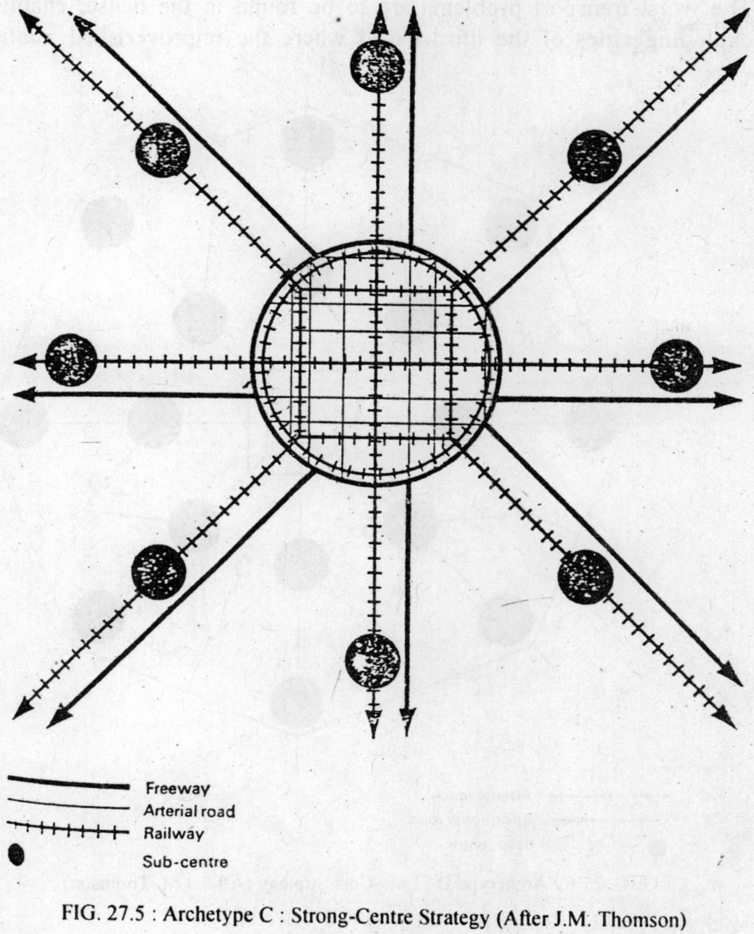

FIG. 27.5 : Archetype C : Strong-Centre Strategy (After J.M. Thomson)

Shanghai and Berlin. In these cities the continuance of strong centre, the transport system designed with superior accessibility in all parts of the city from the core. It must be based on focalization of public transport network. In case new roads are built which make other parts of the metropolis more accessible than the city centre has to depend upon private motorists, a strong force with the introduction of such a transport has pulled activities away from the city centre. The strong radial lines of transport system determine the location and creation of strategic sub-centre outside the lines of combination (Fig. 27.5).

Archetype D—Low Cost Strategy

The worst transport problems are to be found in the dense, chaotic exploding cities of the third world where the improverished public

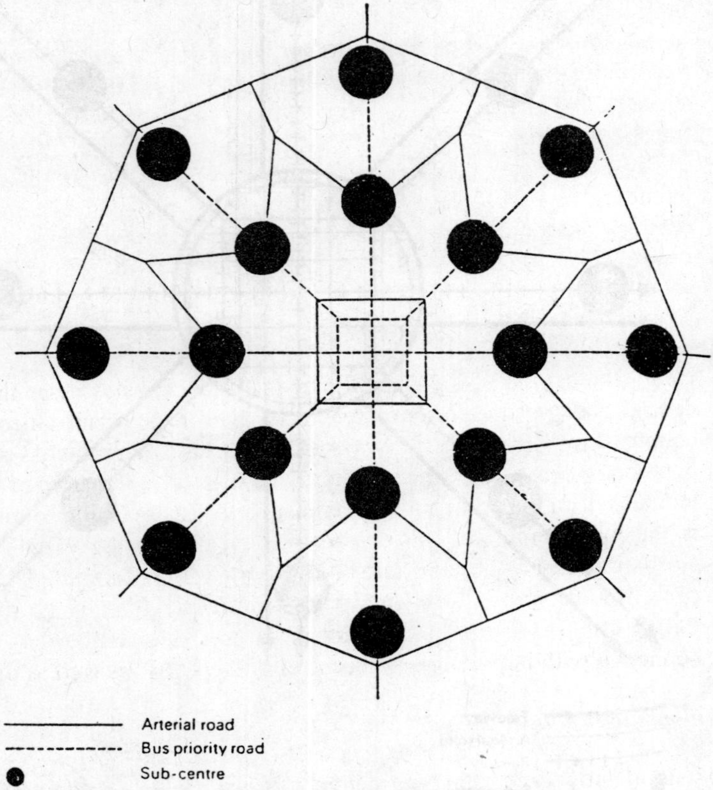

FIG. 27.6 : Archerype D : Low-Cost Strategy (After J.M. Thomson)

sector is quite incapable of providing the infrastructure requirements of even a low vehicle-owning private sector. This situation is found in the urban centres of Mumbai, Calcutta, Karachi and Lagos.

In low-cost strategy the transport system must be confined to ordinary roads, since free way investment is ruled out. Buses and trams must carry the greater part of the traffic and the roads must be managed to make this possible. In this plan, the city centre must contain of high density of population served by buses. Trams can provide a better accessibility, because a fixed guideway can take long vehicles. This could not be used on an ordinary road (Fig. 27.6).

Archetype E : Traffic Limitation Strategies

The essence of a traffic limitation strategy is that it aims deliberately to limit the volumes of traffic, not in desperate attempt to avert chaos but in positive recognition that appropriate restraints upon demand are needed for transport just as for other goods or services. It is the only strategy that attempts to deal directly with the root cause of the urban transport problem (Fig. 27.7).

It must posses a strong city centre well served by public transport including rail. The city centre will be at the top of the hierarchy of centres, consisting of sector centres, suburban centres, and neighbourhood centres.

Transport Development in Delhi

Delhi, the capital of the Republic of Indian Union possesses the most central position due to its closer link from all important cities of the country through rail, road, airways and waterways. Hence, it has strategic position in between the desert country of the west, plain land of the eat, Himalayas on the north and the Peninsular landmass on the south. Delhi is the most important railway centre, being the headquarters of the Northern Zone of Indian Railway and is extremely served by all other parts of the country. Delhi enjoys the benefit of two modern airports of Palam (Indira Gandhi International Airport) and Safdarjang which connect it with important urban centres of the World as well as of India.

Connaught Place

It is an important shopping centre of New Delhi. It acts as Central Business District or Down Town of Delhi along with circular roads and

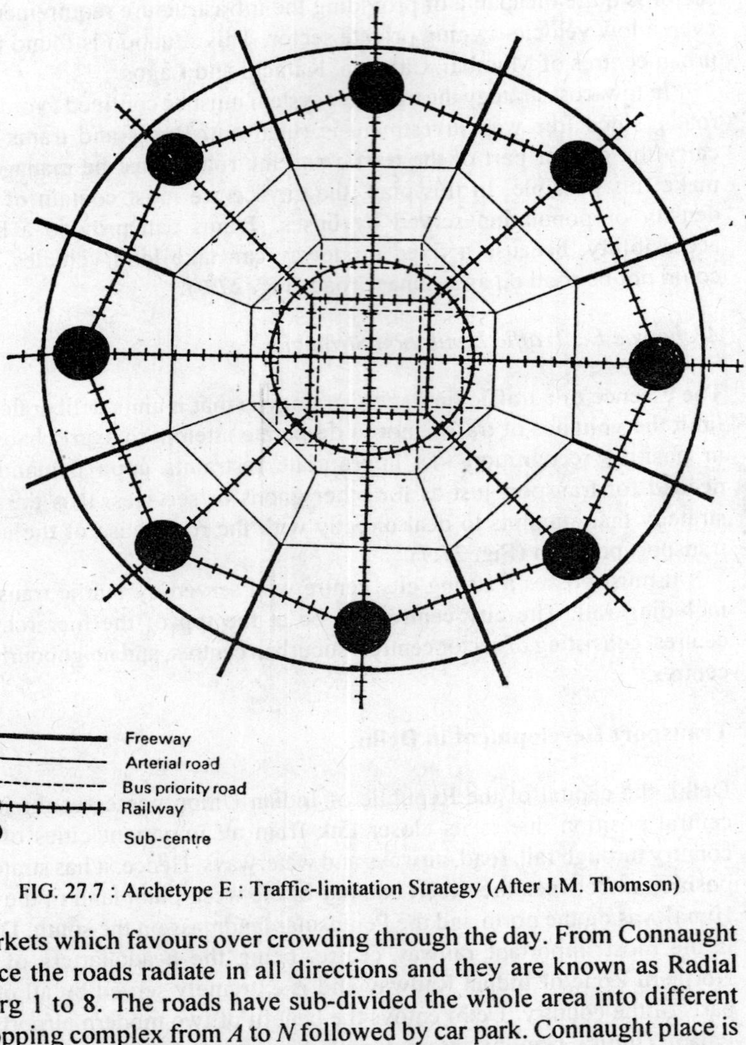

FIG. 27.7 : Archetype E : Traffic-limitation Strategy (After J.M. Thomson)

markets which favours over crowding through the day. From Connaught Place the roads radiate in all directions and they are known as Radial Marg 1 to 8. The roads have sub-divided the whole area into different shopping complex from *A* to *N* followed by car park. Connaught place is surrounded by Din Dayal Upadhyay Marg on the north-east, Puchkuin Marg in the north-west, Jai Singh Marg in the south-west and Hailey Marg on the south-east (Fig. 27.8).

Mudrika Ring Service

Rashtrapati Bhawan is known as Kendriya Terminus and it is well linked by circular ring services of Mudrika. The buses bearing numbers

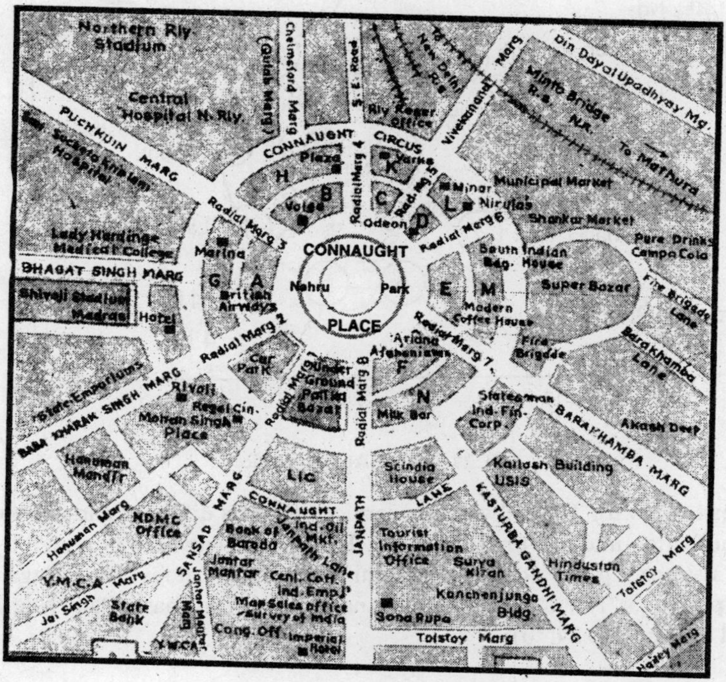

FIG. 27.8 : Connaught Place

of 100 to 900 run from Kendriya Terminus to different directions linking it with—Azadpur, Morigate Terminal, Indraprastha Depot, Lajpath Nagar Terminal, R.K. Puram, Dhaula Kuan, Raja Garden, Karolbagh Terminal and Punjabi Bagh. This ring service has provided a very high level of connectivity to different parts of the city (Fig. 27.9).

Delhi Ring Railway

The railway line (Broad Gauge) runs in different directions from Delhi Junction, e.g., in the north-west it goes to Panipat and Rohtak, in the north-east, it goes to Saharanpur and Tundla and in the south-east, it

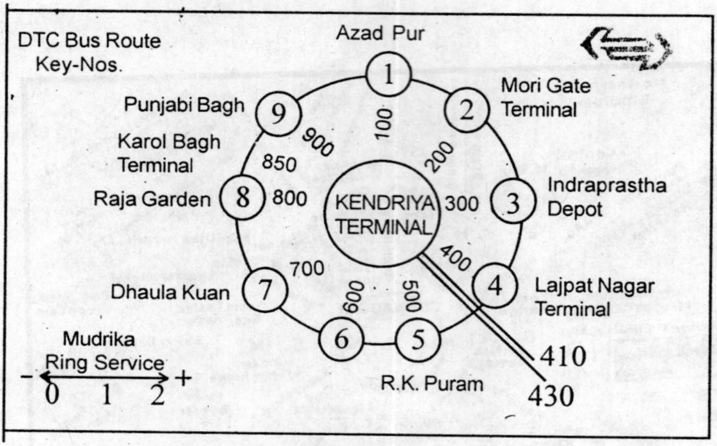

FIG. 27.9 : Mudrika Bus Service

goes to Rewari. The railway ring service connects different parts of the conurbation into one unifying bond. The linkage has been provided by the stations, e.g.—Safdarjang, Sarojini Nagar, Lodi Colony, Sewa Nagar, Lajpat Nagar, Hazrat Nizamuddin, Anand Vihar, Dayabasti, Patel Nagar and others (Fig. 27.10).

Calcutta Traffic System

Calcutta's bedevilled traffic system which has earned from the users many more disquieting appellates than any other public utility is poised for a turn around in the next five or six years.[3]

The optimism stems from the fact that an investment of Rs. 120 crore is expected on a transport infrastructure development project for Calcutta from the Japanese Overseas Economic Cooperation Fund (OECF).[4]

The project includes construction of 10 vehicular fly-overs mainly for smoother north-south traffic, 14 pedestrian walk overs at busy intersections in and around the Central Business District (CBD) and construction of two multi-storeyed under-ground parking plazas in the CBD.[5]

FIG. 27.10 : Delhi Ring Railway and other Services

The Japanese Government has entrusted the feasibility study of the project with the Japanese International Cooperation Agency (JICA). After a JICA team had made a preliminary inspection in December 1989, the decks were cleared for further negotiations in December 1990, a memorandum of understanding was signed among the Government of India, the State Government and JICA over the areas of the study including construction schedule, project cost estimate, economic analysis, project evaluation and preliminary surveys including socio-economic activity in the project areas. JICA started its study in September and is expected to finalize the report by the end of 1992. Only then can the OECF be approached for the release of funds. This means actual work on the field will have to wait till 1994-95.

But those who know are ready to bear with this gestation period. The reasons are plenty. First, the Government has no money in its kit's to finance such big schemes which aim at real structural changes rather than casual tinkerings attempted from time to time. Secondly, the OECF fund looks more or less certain with the involvement of JICA which will be conducting the study here and valuate the findings back in Japan. There is an awareness among the transport planners here that basic structural changes in the traffic profile are a must in a city which is going to have a population of at least 10 million in the metro core area by another five years. In the city almost all vehicles pass through the central area and the number of vehicles per km road space averages 740 against 364 in Mumbai. Calcutta, again, has the largest volume of pedestrian traffic will have to add to this encroachments of the road space by vendors and presence of slow moving vehicles including bullock carts and rickshaws. Calcutta, a survey shows, has an extremely limited road space which works out to only 4 per cent of the total land area. In a modern city, the average should be 16 to 18 per cent. Compared to other metropolitan cities, Calcutta also suffers from another disadvantage. It has little scope to expand in all directions. The north-south movement of traffic is far greater than the east-west traffic.[6]

The previous efforts to improve the traffic system in the city included funding of Rs. 108 crore under the World Bank—aided Calcutta Urban Transport Project (CUTP) and under Calcutta Urban Development Project (CUDP) which was also partly financed by the World Bank but most of the CUTP fund went on the acquisition of buses and tram cars. Whatever was left for the so-called traffic engineering was used in widening some of the major roads, putting up dividers on some roads in the CBD and supplementing street-lighting efforts. The CUDP funds were utilized in setting up garages, passenger shed and workshops.

Transport Development in the City of Munger

The city of Munger has an awkward location with special reference to the transport route network because it lies far off from the railway tract of Kiul-Bhagalpur loop line. Therefore, Munger-Jamalpur rail line only acts as a sub-urban railway. Inside the city Purabsarai and Munger Junctions are located besides Bus depot in Shastri Nagar and Bus stand near Town Hall. Fort Gate, Purabsarai and Munger Junction in Laldarwaja are some of the important transport nodes of the city which act as the nerve centres because they provide life line to the city people, in terms of moving to the offices or other sectors of occupations.[7]

The city of Munger has the facility of labourers train (Colli Gari) which serve the commuting population of the city as daily commuters to the railway workshop at Jamalpur and cigarette factory in Munger.[8]

On northern fringe of the city the facility of ferry point serves people in crossing the river for onward movement towards North Bihar. In urban area the maintenance of roads are very poor and the sanitary and lighting of roads are in a very poor state due to lack of fund and stealing activities of people not only for electricity but also for bulb point etc.[9] (Fig. 27.11).

Traffic Flow in Urban Areas

Table 17.1 show the traffic flow on the Chauk Bazar of Munger counted on the 15th March, 1991 with the help of 12 students just as helpers. The counting has been done for 24 hours starting from 12'O Clock in the night back to the same time next night. The counting has been for the crossing of traffic, e.g., pedestrian, cyclist, rickshaw pullers, scooters, bus and cars which are the main traffic of the Chauk Bazar area.[10]

TABLE 27.1 : Name of the Traffic

Hours	*Pedestrians*	*Cyclists*	*Rickshaw*	*Scooter*	*Bus*	*Car/ Trecker*
0-1	200	158	216	20	8	30
1-2	142	90	185	16	3	25
2-3	40	82	83	10	5	16
3-4	69	120	65	11	3	12
4-5	156	192	98	29	10	41
5-6	225	236	205	34	19	70
6-7	506	252	206	41	15	68
7-8	550	285	229	54	11	71
8-9	693	354	299	54	12	34
9-10	844	340	480	116	12	51
10-11	927	400	535	152	9	60
11-12	1055	540	320	81	7	38
12-13	1306	609	379	98	9	49
13-14	955	501	260	62	5	40
14-15	1320	575	395	100	8	50
15-16	1512	839	597	182	7	75
16-17	1850	977	610	226	8	69
17-18	1983	822	465	220	6	58
18-19	1975	588	520	202	7	56
19-20	1467	595	586	204	8	32
20-21	946	640	474	237	5	31
21-22	658	657	42	125	5	22
22-23	295	432	257	94	3	11
Total	19,673	10,284	7,655	2,396	186	1,005

Source : Calculated by R.B. Mandal, 1991.

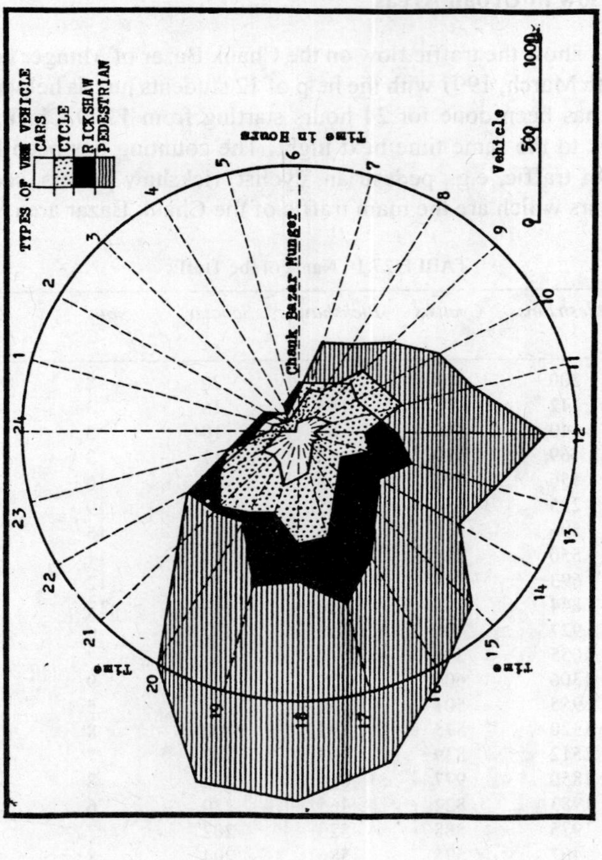

FIG. 27.11 : Types of Vehicle in Munger

With the help of Table 27.1, it is quite apparent that Chauk Bazar, Munger has about 20,000 pedestrians per day, 10,000 cyclists, 8,000 rickshaw pullers, 2,000 scooters, 186 buses, about 1,000 cars and 168 bullock carts passes through the nucleus of Chauk Bazar cross road. From the figure above it is quite clear that from 10 p.m. in the night upto 9 a.m. in the morning the number of traffics are generally very low during night hours. Two peak hours are 10 to 11 a.m. in day and again 4 to 8 p.m. in the evening because during these hours the number of pedestrians and cyclist surpass all other traffic on Chauk Bazar.

FOOTNOTES

1. J.M. Thomson, *Methods of Traffic Limitation in Urban Areas*, OECF, Paris, 1972.
2. A.K. Sharma, "System Analysis is Technique : A Tool for Transportation Location Decision," *Urban and Rural Planning Thought*, Vol. 17, 1974, pp. 214-225.
3. M. Guha, "Transport in and Around Calcutta," *Geographical Review of India*, 1955, Vol. 17, pp. 4-8.
4. J.M. Thomson, *Great Cities and Their Traffic*, Penguin Books, 1977.
5. U. Singh, "Growth of Transport and Communication in Allahabad," *National Geographical Journal of India*, 1959, Vol. 5, pp. 186-204. Traffic and Transportation Plan of 1966-1986 of Calcutta Metropolitan Planning Organization," Saraswati Press Ltd., Calcutta, 1967.
6. D.R. Miller, (Ed.), *Urban Transportation Policy : New Perspectives*, D.C. Health, Lexington Mass, 1972. Automobile Safety Foundation, Urban Freeway Development in 20 Major Cities, 1964.
7. R.J. Pryor, "Definition of the Rural-Urban Fringe," *Social Forces*, Vol. 47, 1968, pp. 202-15.
8. J.W.R. Whiteland, "Fringe Belt : A Neglected Aspect of Urban Geography," *Transactions of the Institute of British Geographers*, Vol. 41, 1967, p. 223.
9. M. Young, "Some Geographic Features of the Urban Fringe," *The South Eastern Geographer*, Vol. 2, 1962, pp. 1-6.
10. R.L. Morril, "Expansion of the Urban Fringe: A Simulation Experiment," *Papers of Regional Science Association*, Vol. 15, 1965, pp. 185-99.
11. R.R. Singh, Jaunpur and Its Rural-Urban Fringe : A Study in Settlement Geography, Unpublished Ph.D. Thesis, B.H.U., 1977, pp. 150-60.

28

URBAN HOUSING

HOUSING is one of the three basic needs of mankind. In India, the house is generally constructed with either brick or mud as material of wall thatched with tile or R.C.C. roof. The modern concept of housing is the home in which all the basic needs of a family are fulfilled. In India most of the population in urban areas live in slums. At present there is acute shortage of houses in urban areas and much of the available accommodation is qualitatively of sub-standard type. Due to unplanned industrialization, the problem of housing has become grave in urban areas. This is due to the excessive rural-urban migration, and natural growth of urban population. Although, the population is growing fast but the houses are not built in the same proportion. As a result there is a grave housing problem in urban areas. According to 1991 census the shortage of houses in urban areas is about 40 million in India.

Role of HUDCO in the Field of Urban Housing

HUDCO in its 25 years service of the nation provided loans for urban housing schemes to housing boards, development authorities, city improvement trusts, municipal corporations, public sector undertakings, State Governments and other such organizations for the construction of houses/flats and development of serviced plots including site and services, skelatal housing and care housing schemes. This has also incourages the development of serviced plots and part of loan also given to general public for easy hire purchase instalments or on rental basis by the agencies depending upon actual need and demand of the desirous persons.

Conditions of Houses in Urban Areas

The housing condition of poor class people in urban areas is really pitiable. The characteristics are closed houses in the neighbourers, the high price of land and the labourers remaining near industries have highly increased the occupancy rate of houses. The houses are made very close to each other so that the full utilization of land could be made. There is lack of fresh air in many houses. There is utter lack of windows, the presence of only one door and privacy is the first priority of such houses.

The houses of industrial workers are made of mud wall in which there is a dark kitchen room, followed by unhygienic living and sleeping room. Actually the housing problem has been of recent origin because the construction of houses does not cope with the increase of population.

In Mumbai, for low income group people, one room flat of three storeyed have been constructed and there condition is so pitiable that there dismantling has been suggested instead of improving their lot. These flats are called *chawls*. The same poor condition is of the houses in Kanpur, which are labourers houses known as *Ahata*. Almost the same condition prevails in almost all urban centres in India. Although in some sectors the houses are constructed but no care is paid to the health and efficiency of the people and their residents.

TABLE 28.1 : Demand-supply and shortage of Houses in Urban India

Year	*Demand*	*Supply*	*Shortage of Houses*
1951	12,500	10,305	2,195
1961	15,780	14,063	1,717
1971	21,640	18,413	3,227
1981	28,732	25,542	3,290
1991	34,569	29,647	4,922

Table 28.1 shows that shortage of houses is an acute problem in urban areas. The progress in house construction is so slow that housing shortage seems to be an unending problem with the current rate of population growth. In 1991 the Government of India has announced to construct 2 crores houses in the 8th plan but this will not solve the problem.

Causes of Bad Housing in Cities

Industrialization and urbanization go side by side. The causes of bad housing are as follows :

(a) Increase in population;
(b) Unplanned Industrialization;

(c) Income differential gap between population increase and the construction of houses; and
(d) Profit Motive.

(a) In a developing country like India several towns have developed in different parts of the country even without proper amenities. The population of cities have increased from 24 per cent in 1981 to 27 per cent during 1991. This has been followed by the influx of people enbulk from rural to urban areas.

(b) Unplanned industrialization is continuing in Indian cities, as there is no control over the private sector. In some areas industries had been established in places, where there had been no pressure of population and it has been effected into housing shortage in urban areas.

(c) During the last several decades the construction of houses is not going on in the same proportion as population increase in urban areas. This has resulted into extreme shortage of houses.

(d) In urban areas, the houses are generally constructed from profit point of view and no attention paid to the health and other amenities. Most of the labourer's houses are of small size and people are forced to live in those unhealthy houses because there is a shortage of houses in almost all urban centres of the country.

The effects of bad housing are lowering of efficiency, less production and the increase of social evils in the urban society.

"In the slums of industrial sectors, manhood is unquestionably brutalized, womenhood dishonoured and childnood poisoned at its very source."

Steps Taken by the Government to Solve the Problem

The Government of India and local bodies have been given the task to solve the housing problem. Some of the steps taken by the government are as follows :

The Subsidized Industrial Housing Scheme

This scheme has been started in 1952. The aim was to provide houses to the labourers. Under this scheme the government has given loan to the extent of 65 per cent to the industries along with the State Government, Legal Housing Construction Societies and Cooperative Societies for the construction of houses to the labourers. The labourers may purchase these houses according to rules framed by the government. These houses could not be sold or changed without prior permission of the government. In the Third Five Year Plan, it was made obligatory for mill owners to provide housing facilities to their labourers. The Central and State Governments have also formed various Housing Boards and Societies to provide houses to the needy families.

Low Income Group Housing Scheme

This scheme has been started in 1954. The persons constructing houses could get a loan upto 80 per cent who have no house of their own and have income less than Rs. 6,000 annually. Even on this basis the local bodies and Cooperative Societies are given loans.

The Slum Clearance and Improvement Scheme

This scheme has been started in the year 1956. The aim of this scheme was to give financial assistance to the State Government and local bodies for improvement of slum areas. It has been estimated that about 12 lakh houses are not fit for human living. Hence the long-term and short-term housing schemes have been started. But as it could not be possible to provide houses to all the people living in slum areas, this scheme could not succeed well.

The Middle Income Group Housing Scheme

There is a provision for middle income group people to have loans for the construction of houses in urban areas. The Life Insurance Corporation of India and the State Government also give loan under this Scheme.

The Rental Housing Scheme

This scheme was started in 1959. The aim was to provide house on rent to the State Government employees. The Life Insurance Corporation of India also gives loan on this scheme which could be paid in 20 years.

Land Acquisition and Development Scheme

It has been felt by the government that low income and middle income group people could construct the house if land on reasonable price was made available to them. With this view a plan was set up that the State Government could acquire land and plots at suitable site, develop them and give the same to the people who are in urgent need of it.

TABLE 28.2 : Housing and Population in Million Cities (1991)

Name	*Area*	*Houses*	*Houses per km²*	*Population*	*Persons per km²*
Calcutta	104.00	556.478	5,351	4,388,262	42,195
Ahmedabad	92.98	289,382	3,112	2,872,865	30,891
Chennai	128.00	338,414	2,644	5,361,468	41,886
Mumbai	603.00	1,203,126	1,995	9,909,547	16,434
Bangalore	174.71	262,926	1,505	2,650,659	15,147
Delhi	446.26	599,358	1,343	7,174,755	16,087
Hyderabad	169.31	219,661	1,297	3,005,496	10,052
Kanpur	298.98	243,378	814	1,962,750	6,564
Pune	324.52	160,190	494	2,055,034	6,323

Among the Indian metropolitan cities having more than a million population of Calcutta has the highest density of houses and population per unit area. In 1991 for example, Calcutta had a staggering number of nearly 5,351 houses and 42,195 persons per km^2.[1] Ahmedabad and Chennai have also very high housing density 3,112 and 2,644 respectively in 1991.

Housing for the Urban Poor

The problem of housing of urban poor is really finding a land in any locality at reasonable price and helping them to arrange houses with their own resources. Many State Governments have launched programmes in this direction. The slum clearance scheme and the scheme for construction of housing for low income groups and economically weaker sections of society have helped to add to the housing stock for the urban poor. All the efforts have been inadequate considering the number of people. Therefore, new approaches are called for the provision of "sites and services" scheme which is one of the methods by which the poor can be given land for putting up housing by using their own resources. In Delhi, nearly 70,000 families have been accommodated in 27 newly created resettlement colonies which have been provided with adequate services. The environment that one creates in these colonies is something which is very important. Housing is only one of the factors which affect the environment, provided the families get a feeling that the land on which construction of houses will be allotted to them on long-term basis. They should mobilize all their efforts in saving and putting efforts for housing. Infact, it should be possible to guide and channelise their resources into construction of reasonable houses. With greater involvement and guidance, even architectural control is feasible. In fact in some colonies, even in Mumbai the individual who have been allotted land for occupation in resettlement areas have joined together to construct 'row' housing which is subject to certain architectural control and which gives a uniqueness and beautify the whole colony as it was before.[2]

Integration of Housing

There is one point which Indira Gandhi stressed is that we are paying so much attention to housing. I don't approve of separate housing colonies for people of separate social status. This is what is keeping us down because the weaker people, when they are separated from others, can never raise their standard of living. One of the great things that the authorities in Singapore have done while providing housing is that they have taken care to see that in every building there are people of all classes, doing different types of work, and belonging to all races. You know, they have a majority of Chinese in Singapore. Next comes a mixture of Indians from north, south and so on. Then they have the Malays, and Eurasians and various other communities. They went out of their way to see that in every building there should be a mixture of the population because they reasoned that it is only this way that integration could be achieved. We must also see that we don't segregate people. It is no use saying "We are providing these houses to *Harijans*, so let us put them two miles out of the village." The result is, they will not live there and all the effort and money will go waste by leaving the problem unsolved.

Mumbai

The majority of the workers in Cotton Textile Mills in Mumbai have been living in *chawls*. The *chawls* consists of back to back tenements with illventilated, dark, insanitary and overcrowded rooms. Facilities of water supply, sanitation, etc. are usually inadequate and unsatisfactory. The *chawls* are built by the Municipal Corporation and the Improvement Trust. The housing conditions are a bit different. But even there overcrowding persist due to taking of lodgers by labourers for adding to their income. The *chawls* were also provided with roads, water, lighting and shops. Schools and dispensaries were also established in the *chawl* areas. Still the houses constructed under the scheme were unpopular due to their distance from the mills and lack of cheap transport to the mill areas. The inadequacy of markets and shops, defective medical facilities, etc. are the persistent problem since long. In 1937, the Government appointed an Industrial Housing Committee and subsequently on its recommendations certain structural alternations and improvements were made in the existing tenements. Thereafter, an enquiry conducted by the Mumbai Labour Office revealed that one-room tenements constituted 81 per cent of the total tenements in the city and 74 per cent of industrial workers live in these tenements. Later, the Mumbai City Improvement Trust constructed over two thousand tenements on behalf of four Textile mills in the city, besides building tenements and semi-permanent sheds for rehousing persons dehoused by the slum clearance operations. In addition of this the Mumbai Municipal Corporation built some *chawls* in the estates acquired by the old Improvement Trust, which house a number of textile workers. Some member mills of the Mumbai Millowner's Association, Mumbai have also provided quarters for their workers in tenements consisting of single as well as double rooms. There are common arrangements for water supply and sanitation. A group of mills which have built *chawls* for its workers provide electric light in the tenements. Usually, three to four persons occupy one room. Thus, the housing accommodation for workers in Mumbai, on the whole, has been inadequate and unsatisfactory. Recently, the Mumbai Housing Board, constituted under the Housing Act of 1948, has undertaken the construction of a large number of tenements for the workers. The State Government has also launched some special schemes to improve the conditions of workers.

Ahmedabad

In Ahmedabad, houses for industrial workers have been constructed by the Municipal Board, mill-owners, the Textile Labour Association and Co-operative Societies besides the private landlords. However, all such houses are exceedingly small in number. Consequently, a considerable majority of workers live in *chawls* owned by private landlords. It is not uncommon to find more than one family living in a room. The floor in the majority of cases is of mud and hence they are damp and dirty. Despite the fact that a majority of the tenements have windows, the rooms are dark and illventilated. Generally no light is provided in the compounds of these tenements. Water supply is inadequate. In most cases no bathrooms and toilets are provided. There is no proper system of drainage for all the chawls. In recent years there has been some improvement in the position through the implementation of the housing schemes of the Government.

Kanpur

Proverbially, Kanpur, the industrial metropolis of Uttar Pradesh, is known as "The Manchester of India." It is the most congested area in U.P. Besides having a large number of cotton spinning and weaving mills, it is an important centre for leather tanning and manufacturing of leather goods. There are also a number of engineering works, foundries, glassware factories, oil and flour mills, jute and cotton mills, ice cream factories and chemical works. It is, therefore, densely over-crowded and insanitary. Until recently, very little attention was paid to the question of housing for the large and rapidly increasing population in the city. There has been haphazard industrial development and expansion of factories in the very heart of the city with the result that there is serious over-crowding for the working class. A very large number of workers in the city live in *ahatas* owned by private landlords. Most of these *ahatas* are extremely insanitary and over-crowded and lack of adequate sunshine and ventilation, to say nothing of water, light and conservancy that's better. "A night visit to these areas for a stranger is a risky venture—a sprained anckle is almost a certainty, while a broken neck by stumbling into a blind well or a goodly-sized hole, would not be an impossibility." Prominent among the employer's settlement are MacRobertganj Settlement, Kokomiganj Settlement, Allenganj Settlement, Elgin Mills Settlement and Maxwellganj Settlement. The J.K. Groups of Mills have also constructed a large colony for housing most of their workers. However, not more than 10 per cent of the total industrial workers of the city have been housed in these places. Naturally, the vast majority of the workers have been living in the slummy dwellings owned by private landlords and for that too they often have to pay high rents. In 1943-44 the Government of U.P. granted an interest-free loan of Rs. 30.3 lakhs to the Kanpur Improvement Trust for the construction of 2,400 family quarters. Towards the end of the year, the Trust constructed these quarters in the Juhi area which were available to workers on moderate rent but the sanitary conditions could not be very satisfactory and materials used in the construction of these quarters do not seem to be of good quality. In the year 1945, "the Kanpur Development Board" was constituted by the State Government which constructed a number of houses for the workers. The Board was also asked by the State Government to take up the work of slum clearance and improvement of *ahatas*. Some recent surveys conducted in this respect at Kanpur reveal that the housing conditions have deteriorated considerably in the post-war years and over-crowding, insanitation, caste-segregation and social degradation remain the common features of the *ahatas*." No doubt, due to the implementation of various schemes like the subsidized industrial housing scheme, some improvements in this regard have been effected during recent years.

Calcutta

Calcutta is another industrial centre showing the heaviest concentration of industrial workers. While there was a rapid development of industries in the city, no attempt was made to provide housing facilities for the workers. A considerable majority of the workers live in *bustees* constructed by private landlords. The degree of over-crowding and congestion found in these *bustees* around Calcutta is probably unequalled in any other industrial centre in the country. During recent years some good housing colonies have sprung up for jute workers in Calcutta. The jute mills in and around Calcutta have provided houses to nearly 50 per cent of their employees. As regards textile workers, it has been estimated that 40 per cent of the workers have been provided with housing by employers. There are also some quarters provided by Port Commissioners to nearly 50 per cent of the workers. A few tanneries in Calcutta also provide free housing to some of their workers. Despite this the housing conditions at Calcutta have been quite unsatisfactory. There is acute congestion and over-crowding and the conditions of sanitation, water supply, etc. are also bad. In recent years, there has been some improvement in this respect with the implementation of Government Housing Schemes.

Other Industrial Centres

The housing conditions at other industrial centres of the country are similarly most unsatisfactory. Over-crowding is common and the conditions of ventilation and sanitation are bad. In Chennai, most of the workers live in single room with or without verandah, with no window of even a ventilator. The buildings are *pucca* with brick walls, and each house is split up into a number of small portions, each rented by one family. Toilet arrangements are highly unsatisfactory. Usually, there are no bathrooms and water-taps. The rooms are mostly ill-ventilated and dark. Chennai Municipal Corporation has also provided houses to some of its employees in the conservancy department. Conditions regarding lighting, ventilation and water-supply however, remain unsatisfactory.

Most of the workers in Chennai are housed in colonies popularly known as *cheries*. There are small colonies of thatched huts and have been built with no provision for any sanitary and other facilities. They are dirty, damp and unhealthy. The mud huts often leak during the rainy season. These huts are erected by the workers themselves with borrowed money on a site for which they pay ground rent. The Buckingham and Carnatak Mills, Chennai provide housing accommodation for nearly 10 per cent of their workers. The usual type consists of a living room, a kitchen, a washing place, a front verandah and a courtyard. Pipe water is supplied. A nominal rent is charged, though, for better accommodation the rent is a bit higher. At Jamshedpur, the housing accommodation is short of the demand and over-crowding is very common. However, some good houses have been provided by the Tatas. In Delhi also the slum conditions prevail. There are *katras* where over three lakhs of people live in sub human conditions. About 40 per cent of the workers in Delhi have been provided houses by the employers. At Sholapur the housing conditions seem to be satisfactory. The town is not congested and most of the workers live in two-room tenements. At Madurai also, the housing conditions are quite satisfactory and the housing colony consists of two *pucca* houses with sufficient accommodation per family. The colonies are almost self-sufficing from the point of view of amenities such as schools, markets, dispensaries, etc. The Empress Mills, Nagpur and the Bangalore Silk, Cotton and Woollen Mills have provided good housing colonies for the workers in which the conditions are satisfactory. In sugar industry, the employers provided housing for about 30 to 50 per cent of the labour force. In industries like engineering, paper, chemicals, tanneries, etc. Only the larger units provide housing but to a small portion of workers only.

Modern Trends in Urban Housing

Shelter is the prime need of human being. The housing congestion and polluted urban environment have favoured the construction of housing apartments in most of the cities of the world. The examples may be given from Aakansha Homes in Lucknow, Devlok Apartment in Meerut, Heritage City and South City Apartments south of Delhi.

The shortage of residential space in urban areas, high occupancy rate, skyrocketing house rents and ever increasing population are some of the factors which forced people to live in beehive like pattern houses or urban housing apartments. Thus recently in urban areas the continuous growth of population have created the problems of space and housing

shortage. With this view the vertical expansion has started coming up in the form of housing apartments. For example, some of the housing apartments of Patna could be seen in Fig. 28.2 and Table 28.3.

FIG. 28.1 : An Exclusive Estate, South of Delhi

- Luxury apartments on free-hold land.
- 2, 3-bedroom and 3-bedroom duplex units.
- 8 kms. from Indira Gandhi International Airport.
- Approachable from NH-8 and Mehrauli-Gurgaon Road.
- Self-sufficient residential complex with shopping centre, club, school, swimming pool, parks and more.

TABLE 28.3 : Housing Apartments of Patna, 1992

Sl.No.	*Housing Apartment*	*No. of Flats*	*Location*
1.	Grand Apartment	80	Fraser Road
2.	Shanti Apartment	85	Off Fraser Road
3.	Baivabh Apartment	36	Buddha Marg
4.	Gagan Apartment	80	Exhibition Road
5.	Udaigiri Apartment	225	Museum Road
6.	Yamuna Apartment	80	Boring Road
7.	Himgiri Apartment	20	Boring Canal Road
8.	Nilgiri Apartment	20	Boring Canal Road
9.	Pioneer Apartment	20	Boring Canal Road
10.	Narmada Apartment	10 storeyed	—
11.	Kaveri Apartment	10 storeyed	—
12.	Ashiana Chambers	8 storeyed	—
13.	Vimla Apartment	4 storeyed	Boring Road
14.	Sumita Apartment	4 storeyed	Rajendra Nagar
15.	Lotus Apartment	3 storeyed	New Patliputra Colony
16.	Govind Enclave	7 storeyed	Kankarbagh Colony
17.	Aangan Apartment	4 storeyed	J.N. Road, Kadamkuan
18.	Mandan Apartment	4 storeyed	West Boring Canal Road
19.	Nawal Kala Enclave	4 storeyed	Kadam Kuan

Source : Collected by R.B. Mandal, 1992.

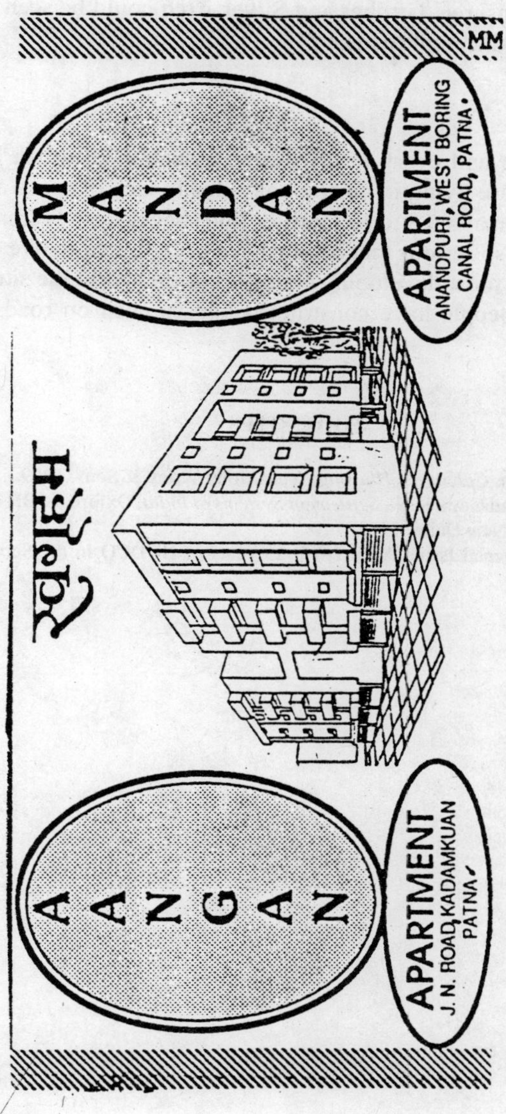

FIG. 28.2 : Housing Through Housing People

In Ferojshah Road of New Delhi the housing apartments like Adishwar, Diwonshree, Girdhar and Silver Arch could be seen besides several others in different roads.[3]

Conclusion

A cover on each and every head is quite essential. In urban areas of Calcutta people are living in pigeon holes due to lack of space and high rents. In one room about 8 to 10 families find shelter from bottom to top. The condition is so serious that people have been forced to live in slums and squatters. In most developing countries of the world the situation is so serious that people have constructed houses even on roadside and tramways.

FOOTNOTES

1. J. Houseton, *Calcutta's Housing Needs*, John Wiley & Sons, 1969.
2. M. Alam and Gopi, K.N., *Settlement Systems of India*, Oxford & IBH Publishing Company, New Delhi, 1982.
3. *Shelter*, Special Issue, April 1995. 25 years of HUDCO in the Service of the Nation.

29

URBAN GROWTH AND INTERNAL MIGRATION

MIGRATION is the periodic movement of population by breaking social and cultural ties from the original place of living. It is the Third component of population growth after fertility and mortality. It may be emigration or outmigration and immigration or in migration. All migrants are movers but all moves are not migrants. The factors governing urban migrations could be :

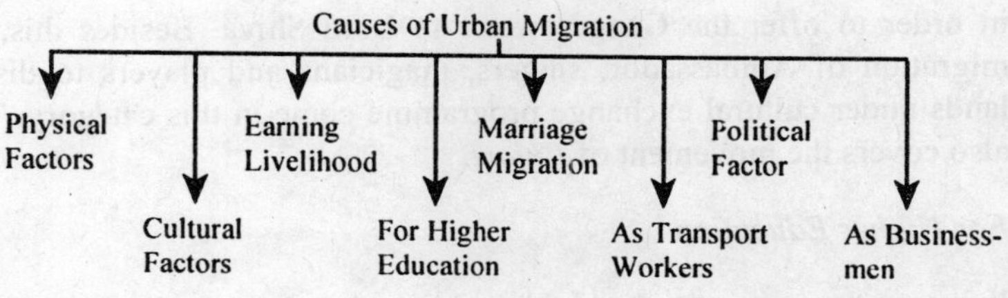

Physical Factors

In urban areas the gold mining in Bogota or Columbia, in Tanzania resettlement schemes in Kilimanjaro, Mbeya and Mwanza and in India the flood problems in North Bihar and the establishment of colonies have help in migration of people from one area to the other.

Employment

Earning livelihood is a prime force in urban development. People get attracted towards urban centres for business, as rickshaw puller, to join different kinds of services and for higher education. Employment in industrial establishments, printing press and transport sectors are common issues to have the growth of urban centres.

Marriage Migration

It may be denoted as forced migration of women folk from one urban centre to another or town to the village. This type of migration is compulsory even if one has no desire to move. It is a sort of forced migration. Each and every married woman has to follow this suit in the urban areas.

Political Factor

When the leaders of the Government decide to shift people from one town to the other in order to reduce the pressure and building of new towns or colonies. Moghul emperors were taking such a decision quite often in order to change the capital. In 1971 about one lakh military of Pakistan surrendered before Indian military in Dacca and brought them to the Bhagalpur, Patna and Tihar Jail of Delhi to have safe confinement.

Cultural Factors

Every year lakhs of pilgrims moves from Sultanganj in Bihar to Deoghar in order to offer the Ganges water to Lord Shiva. Besides this, the migration of Ambassador, singers, magicians and players to distant lands under cultural exchange programme come in this category. This also covers the movement of *Pujari.*

For Higher Education

It covers the university level education and technical education at city level centres where people go to achieve some sort of specialized training, e.g., fruit technology, printing technology, textile technology, medical and engineering education as well.

As Transport Workers

Most of the migrants in urban areas are meant for pulling rickshaw, taxi and motor drivers besides garden collies, maid servants and night guards. These migrants are poor and hence they want to reside in slums and squatter areas where the houses are cheaply available.[1]

As Businessmen

These days there is a migrants boom in the business sector in urban areas. This has happened due to increase in unemployed youth in the society who wants to start a shop in order to have some sort of employment. In 1969 when the author counted the total number of shops in Patna it was 9,000 but it has increased to double 18,000 in the year 1985 when it was recounted. In this way, business sector has attracted large number of migrants in urban areas.

Intervening Obstacles Concept of E. Lee

Lee considered[2] that each and every migrant faces with negative factors at the place of origin and positive factors at the place of destination even if he has to face intervening obstacles of physical, monetary, ignorance, political and cultural relations, etc. (Fig. 29.1).

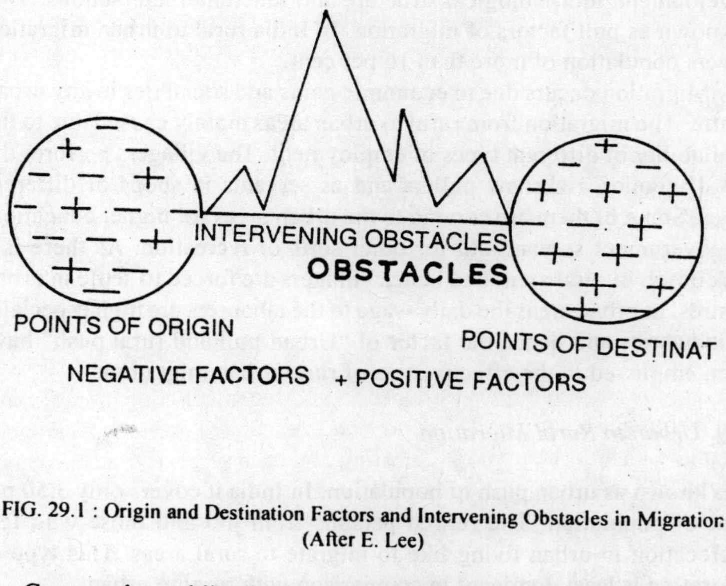

FIG. 29.1 : Origin and Destination Factors and Intervening Obstacles in Migration
(After E. Lee)

Commonly maldistribution of population is described in terms of 'over urbanization' caused by excessive migration. Urban life requires a complicated set of services, housing, traffic, sewerage, water and so on that cannot quickly be scaled up as population grows. Municipality and Corporations are usually short of money and may lack the managerial skill to cope with a city that doubles its size in a decade. Wherever, this happens the results are unemployment, substandard housing, deteriorating public services, congestion, pollution, crime and so forth.

Types of Migrations

Migration is a factor of population and economic growth of an urban centre. Migration may be of three types, e.g.

(a) Rural to Urban Migration

Rural-urban migration is a complex process and its effects are multi-dimensional. In case of heavy and continuous flow of selective migration of young males from rural to urban areas, stagnation of economic growth may take place in rural and economically depressed areas due to the loss of active labour force. But rapid growth occurs in urban destinations in terms of population and labour force, economic development, morphological structure and functional interactions. This is known as pull factors of migration. In India rural to urban migration covers population of more than 16 per cent.

Migration occurs due to economic gains and social ties in any urban centre. The migration from rural to urban areas mainly caused due to the availability of different types of employment. The villagers perform the job of coolies, rickshaw pullers and as servants in shops of different types. Some of them are coming to the urban areas for higher education, as government servant and for other sorts of recreation. As there is a lack of job in rural areas and hence villagers are forced to settle in urban centres. In urban areas the daily wage to the labourers are high especially in industries and hence the factor of "Urban pull and rural push" have been employed in the effectiveness of rural-urban migration.

(b) Urban to Rural Migration

It is known as urban push of population. In India it covers only 3.50 per cent of population. The retired persons from job and those who feel suffocation in urban living like to migrate to rural areas. This type of migration is least dominant in comparison with rural to urban.

(c) Urban to Urban Migration

Most of marriage migrations are of this type. It also happens due to technology transfer and joining of service in centres of higher learning besides, industry and business establishments. Nowadays, this type of migration is more prominent as people living in urban areas cannot adjust easily in rural areas. In urban centres the establishment of big libraries, theatre or cinema houses, Yoga training and the nodal centres of administration and medical facilities attract people of diverse fields to urban centres.

Patterns of Urban Migration in India

The pattern of rural-urban migration is just 16 per cent in India followed by 3.50 per cent urban to rural migration. In a recent survey by the author it come to light that about 85 per cent population came to urban areas from the countryside. This has increased the frequency of rural-urban migration throughout the world. In India the migration of labour force took place in a big way from the rural areas of Bihar, West Bengal, and Madhya Pradesh to Calcutta, Durgapur, Bokaro, Jamshedpur, Raniganj and Bhilai. These areas also supplies labour force to Ludhiana, Jalandhar, Delhi, Jaipur, Amballa and Mumbai (Fig. 29.2) through

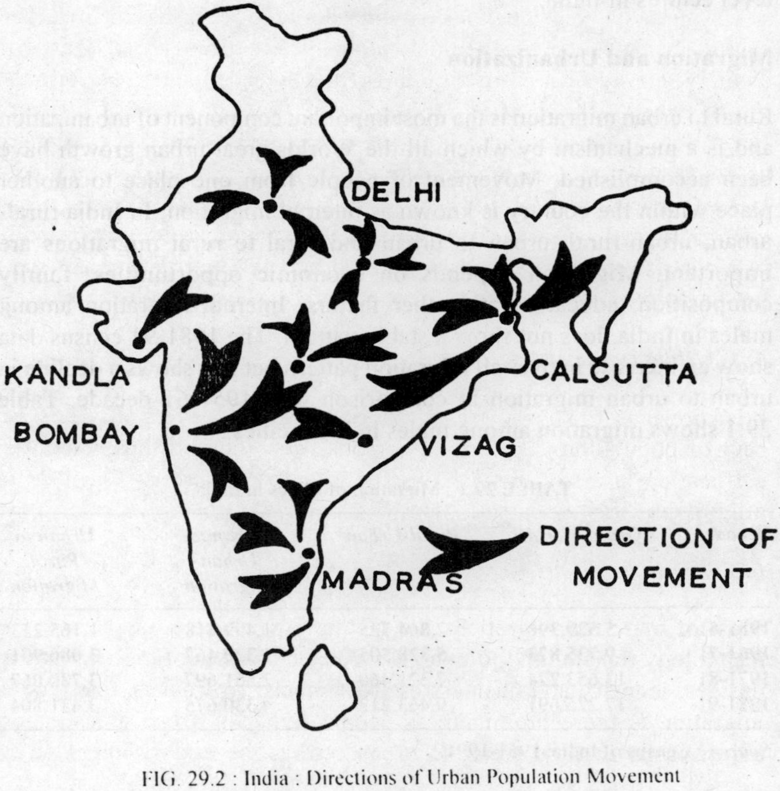

FIG. 29.2 : India : Directions of Urban Population Movement

fieldwork it has come to light that about 85 per cent population of Delhi, Calcutta, Patna, Dar es Salaam and Munger come from rural areas of neighbouring countryside. In the city of Munger migrants come from Khagaria, Begusarai, Kharagpur, Barahiya, Lakhisarai, Jamui, Bariyarpur and the neighbouring villages. The people living in slums and squatter settlements in urban areas have migrated to the urban places either due to economic stringency or social boycott. In the urban centre of Hajipur north of Patna 76 villages have been included in the town due to re-demarcation of urban boundary. Similarly, near the city of Munger the villages of Bank, Heru-diara, Sandalpur, Raisar, Gulzarpokhar, Basudeopur and Chaintoli have been included in the urban fold. The students taking higher education have contributed substantially in rural-urban migration. For example in 1962 the city of Muzaffarpur has only two colleges but in 1998 after 36 years the number of colleges has rose to 22. Similar, is the case of rural-urban migration near most of the city level centres in India.

Migration and Urbanization

Rural to urban migration is the most important component of urbanization and is a mechanism by which all the Worlds great urban growth have been accomplished. Movement of people from one place to another place within the country is known as internal migration. In India rural-urban, urban-rural, urban to urban and rural to rural migrations are important. Migration depends on economic opportunities, family composition, education and other factors. Internal migration among males in India does not show a stable pattern. The 1981-91 census data show an increase in over all migration pattern but this shows a decline in urban to urban migration in comparison with 1951-61 decade. Table 29.1 shows migration among males in India cities :

TABLE 29.1 : Migration of Males in India

Decade	*Gross over All*	*Rural-Urban*	*Urban to Urban Migration*	*Urban to Rural Migration*
1951-61	5,529,396	2,864,725	1,499,418	1,165,253
1961-71	9,735,873	5,328,505	1,320,467	3,086,901
1971-81	13,653,224	7,321,460	2,551,697	3,780,067
1981-91	17,225,691	9,463,212	4,330,675	3,431,804

Source : Census of India, 1961-1991.

Further, the author feels that by 2001 the net migration to urban Class I areas will be 22 million. This means that there will be 31 per cent growth in Class I urban areas in 1991-2001. This happens due to migration and the remaining 69 per cent due to natural growth. In this way, it is quite clear that even in future the over-crowding in large urbar centres would be due to growth in the indigenous population and not migration.

Every year, thousands of young men come from rural areas to cities and towns in search of job or in the hope of getting work. Many of them are married and they leave behind their wives and children at home. This 'lone' migration of adult males underlines the tragic reality of the process of urbanization in India. The longing of youth for the cities is deep, pervasive and real. It cannot be dismissed as a mere fancy of youth for the glamour of urban living.

The high growth of urban population is mainly due to the rural-urban migration. In urban areas the lower rate of mortality and higher rate of fertility are the causes of higher population growth. Table 29.2 clearly presents this view. On this point Davis felt that "In India-Pakistan' as in other areas, the city population has a lower birth than the rural population. We cannot, therefore, account for the growth in the urban portion of the population on the basis of a greater natural increase. The only other demographic explanation lies in migration."[3]

TABLE 29.2 : India : Estimated Birth and Death Rate by SRS, 1969-91

Year	*Birth Rate Urban/Per '000*	*Death Rate Urban/Per '000*
1969	32.6	17.6
1970	29.7	15.7
1971	30.1	14.9
1981	29.2	11.0
1991	24.0	7.7

Source : B.L. Agarwala, "Sample Registration in India," *Population Studies*. Vol. XXIII, No. 3, Nov. 1969, p. 379-94 and the figure for 1991 is estimated.

The causes of migration may also be discussed in terms of 'push' and 'pull' factors. Rural poverty and unemployment 'push' people to urban areas while high wages and better living conditions 'pull' people to urban area. The underlying assumption in such analysis is that the push is from rural areas and the pull is from urban areas. But it is not always true. Under conditions of rapid population growth and all round-scarcity of resources the push is from every where, whether it is urban or rural.

There is also another type of urban push. In the absence of comprehensive social security measure, there is a constant push factor in urban area which operates whenever a migrant in a city falls sick, or unemployed or retires from service. This is specially true of the migrant who comes to the city in the expectation of a job but fails to get it. He is inevitably push back to rural area or to some other city.

In big urban centres of India the majority of workers are migrant. The condition in some of the leading metropolitan centres of India is as follows (Table 29.3).

TABLE 29.3 : Percentage of Migrants to Total Workers (1991) Age Group (15-34)

Urban Centre	*Percentage of Workers*
Greater Mumbai	83.7
Calcutta	71.8
Delhi (M.C.)	84.3
Chennai	51.3
Ahmedabad	69.0
Hyderabad	39.5
Bangalore	58.3

Hence, it is interesting to note that in India, much of the female migration to the cities is 'marriage migration' or 'associated migration'.

National Policies and Directions of Migration

National economic policies—which provide fiscal incentives and low interest loans to promote capital—intensive industry, for example may also exacerbate urban problems by encouraging rural-urban migration without creating enough new urban jobs.

Whatever the cause, the drift from countryside to city is a concern to governments. A 1983 U.N. Survey of 126 governments of developing countries found that only 3 considered the distribution of their population 'appropriate'. Moreover, all three were government of small island : Barbados, Malta and Nauru. Concern was greatest in Africa, the Middle East, and low income countries of Asia.

Between 1925 and 1950 at least 100 million people in the developing countries about 10 per cent of their rural population in 1925—migrated from the countryside to their nearest urban areas.

During the following 25 years, the number rose to 330 million, equivalent to almost a quarter of the rural population of the developing countries in 1950.

The Role of Internal Migration

Current high rates of urban growth in developing countries are only partly due to rural-urban migration. Natural population growth is estimated to about 60 per cent of the rise in urban populations, according to a UN sample of 29 developing countries. Perhaps another 8 to 15 per cent is attributable to the reclassification of rural areas to urban status. Additional evidence from India, Kenya and several West African countries confirms this pattern.

On an average the fertility rate in urban areas is lower than rural and hence the effect of urbanization on aggregate fertility is limited in the short run, especially because migrants tend to be of childbearing age, raising the number of births in cities even if the fertility is lower. Migrants, therefore, put greater strain on the city life to cope with rapidly growing numbers. In broad perspective the shift of people from rural to urban areas mainly reflects the process of industrialization and the changes it brings in the demand for labour. Some conditions in rural areas—unequal land distribution, landlessness, agricultural mechanization, natural calamities and in the past, forced labour migration—have strongly influenced population movements in many countries. But, by and large, people move to towns and cities for higher income and better job opportunities.[4]

For individual family, those attractions could be considerable. A move from the rural northeast of Brazil ιo Rio de Janeiro, for example, can roughly triple the income of an unskilled worker, the family income of a manual labourer in Sao-Paulo is almost five times that of a farm labourer in the northeast. The higher cost of urban living may borrow rural-urban wage differentials in real terms, but urban dwellers also generally have much better access to basic public services, e.g., in rural areas of sub-Saharan Africa only 10 per cent of population have access to a safe water supply, compared with 66 per cent of the urban population.

Generally migrants are considered as assets to the urban economy.[5] Evidence from Brazil, Colombia, Kenya, Korea, India and Malaysia suggests that migrants increases the urban income. A World Bank study of Bogota, Columbia found that migrants earned more and hence there are more migrants at all educational levels.

Role of Migration in Urban Development

1. It assimilates the culture of rural and urban areas.
2. Migration makes people more active in terms of economic development.

3. Most of the marriage migrations from rural to urban areas are forced one but on the other it helps in the increase of urban population.
4. Migration brings money, men and material into the town/city.
5. Rural to urban migration increases population density in urban centres and decreases population density in rural areas.
6. Migration helps in change of food habit, living environment and adjustment to a new place about which he or she is unknown since times immemorial.
7. Migration breaks the cultural and social ties at the place of origin and established a new living environment at the place of destination.

Conclusion

Migrants usually carry their culture to its new place and zealously guard it by creating an imbalance to maintain their identity. The emergence of the forces of social and regional development through rise in population movement acts as a catalytic agent for the growth of science and technology as well as for the process of urbanization and social transformation. People who have been inhabiting on foreign lands for hundreds of years have got a history of their own culture. The geographical and ecological conditions of their living, race, social and cultural life and their participation in the development of other countries are the problem of this study. Ultimately, it could be said that urban migrants are selective and better placed in comparison with the economy of rural migrants.

FOOTNOTES

1. J.W. Simmons, "Changing Residence in the City : A Review of Intra-Urban Mobility," *Geographical Review*, LVIII, No. 4, pp. 622-51.
2. G.J. Lewis,"Commuting and the Village in Mid-wales," *Geography*, Vol. 52, 1967.
3. R. Lawton, "Problems of Population Mobility in Contemporary Britain," *Geography*, Vol. 49, 1964.
4. D.J. Bogue and Zachariah, K.C., "Urbanization and Migration in India," *Indias Urban Future*, (ed. Roy and Turner), University of California Press, Los Angeles, 1962.
5. B.P. Bimal, "Demographic Structure of Urban Centres in the District of Munger," Unpublished Ph.D. Thesis, T.M. Bhagalpur University, 1993.

30

POLLUTION OF URBAN ENVIRONMENT

ENVIRONMENT is the total effect of air, water, land and biosphere on the human being whether living in towns and cities. It is called physical environment in case related to land, air and water. The cultural environment is controlled by the human being such as houses, roads, industries, etc. Nowadays, with the increase of population in urban areas the pollution of environment has increased substanially and hence its control is necessary.

The Guardian of Pollution Control

In India the Central Pollution Control Board was constituted in 1974 which has been instrumental in spreading the message of pollution control across the country. Its function is to promote cleanliness and restore whole-someness of water in wells and streams on the one hand, and to improve the quality of air in the urban centres.[1]

The Central Pollution Control Board has a plan to press into action specially equipped mobile laboratories to monitor ambient air, water quality in highly pollution affected areas. It will also check on emissions of affluents coming from industries of urban centres (Fig. 30.1).

Model of Urban Environment

There are two types of feedback cycles in the system : positive feedback and negative feedback. Positive feedback is often known as the vicious cycle, whereas negative feedback is self-regulating and inducing the system to approach an equilibrium or steady state. Fig. 30.2, illustrates the concepts of positive and negative feedback for the urban environment. Many processes in nature also exhibit feedback cycles. For example, off-road vehicle use may be a positive cycle because as vehicle use

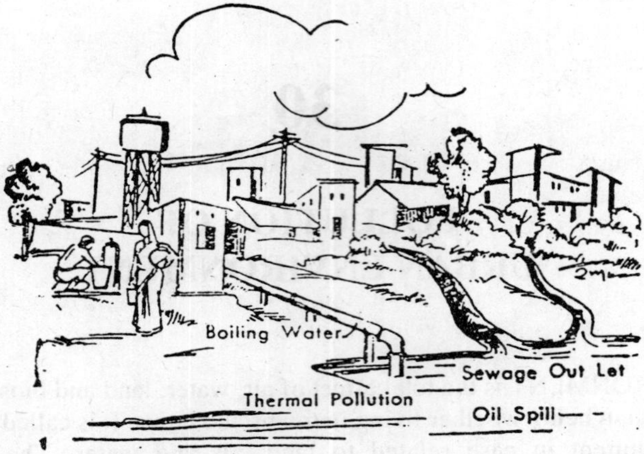

FIG. 30.1 : Sources of Urban Environment Pollution

increases the number of plants that are uprooted, increases the rate of soil erosion. On the other hand, systems such as rivers often display a negative feedback because a steady state is formed. That is, as rivers change in response to an increase in regional rainfall the channel and flood plain system will make changes to accommodate the new, increased amount of water and sediment within a short time so that a new steady state may be established (Fig. 30.2).

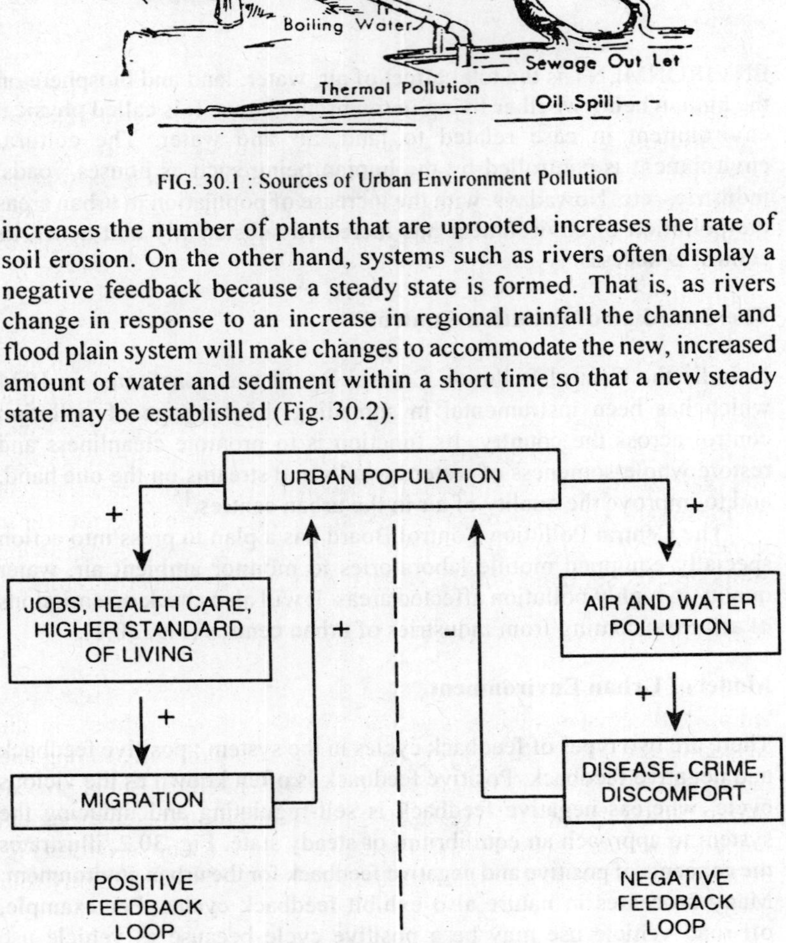

FIG. 30.2 : Model of Urban Environment

Urgency of Solution of Problem

In the last quarter of 1984 the leakage of cyanide poisonous gas from a pesticide factory in the city of Bhopal, India, killed 2,500 people and made blind to 50,000. Near about the same time in Mexico City, Mexico, 350 people died with break out of fire in a gas depot. All these are the 'frightening tragedies' whose instance effect is not limited to our urban environment only.

Nowadays, with the overwhelming increase of population, the burden on the available residential houses, cultivable land, roads, railways, water taps and industries located in urban areas have surpassed all previous levels in terms of harnessing the resources and creating environmental hazards.

Some of the important hazardous places on the Ganges river are Barauni, Mokameh, Patna, Bhagalpur, Buxar and Sahibganj, whereas on the Damodar river Bermo, Chandrapura and Bokaro, and Dehri on Sone river besides Jamshedpur on the Kharkai and others. In most of the urban areas the chocked drains, the predominance of service latrines, dump of garbage on roads, open drains and unsanitary lanes are crux of the issue. Throughout India, urban areas pose a deserted look on the environment which is serious for a sound socio-economic development of the country.

The dumping grounds of pollutants and waste materials coming from urban areas either in the form of solid or gaseous in the Abiotic Environment (Non-living World), e.g., earth surface (lithosphere), water (hydrosphere) and air (atmosphere) is disastrous for the human health. But their combined harmful effect is revealed on the Biotic environment or biosphere (living world) also (Fig. 30.3). Currently, the rapid urbanization is beyond our control, which resulted into overgrowing shortage of food, energy crisis, social conflicts and economic drudgery.

Stages of the Development of Urban Environment

In India, urban environment has been passed into different stages from simple cities of Paleolithic stage upto the modern mega-cities. In the beginning the urban life was very simple full of natural surrounding and hence, the environment was clean in those ancient cities. But at present due to rapid population increase and industrialization the environment has been polluted many times in relation to the ancient scenario (Fig. 30.4). We can divide the historical evolution of cities into roughly four stages.[2]

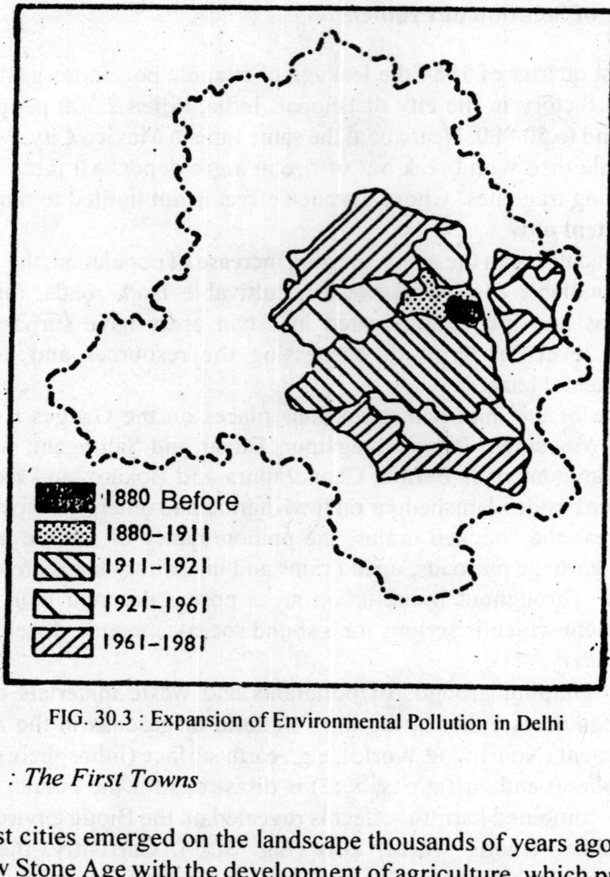

FIG. 30.3 : Expansion of Environmental Pollution in Delhi

Stage 1 : The First Towns

The first cities emerged on the landscape thousands of years ago during the New Stone Age with the development of agriculture, which provided the excess of food resources that is necessary to the maintenance of a city, Lewis Mumford writes that these neolithic villages, "lacking the size and complexity of the city," nevertheless exhibited the essential features of a city," the encircling mound or palisade, setting it off from the fields, permanent shelters, storage pits and bins, with refuse dumps and burial grounds.[3]

In the first stage per square kilometre the density of population is much higher than in the surrounding countryside, but the density is still very low to cause rapid, serious disturbance to the land. In fact, the city dwellers and their animals provide dung, which is an important fertilizer for the surrounding farmlands.

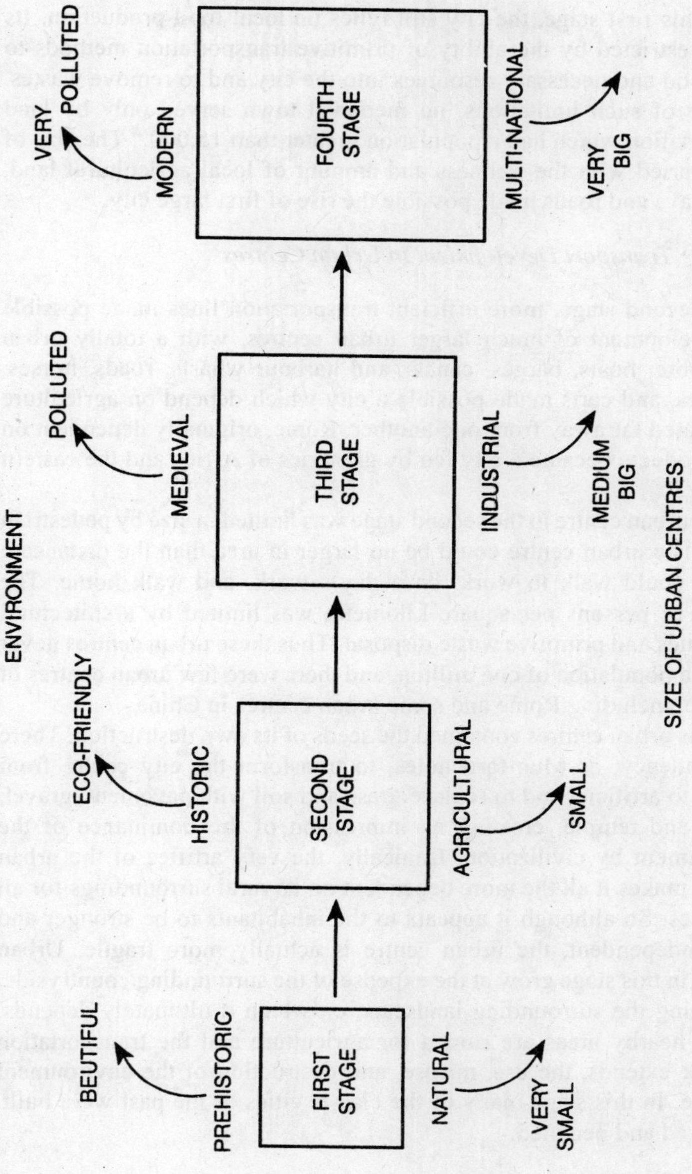

FIG. 30.4 : Stages of the Evolution of Urban Centres

In this first stage, the city still relies on local food production. Its size is restricted by the ability of primitive transportation methods to bring food and necessary resources into the city and to remove wastes. Because of such limitations, no medieval town served only by land transportation which had a population greater than 15,000.[4] The size of a city varied with the richness and amount of local agricultural land. Waterways and roads made possible the rise of first large city.

Stage 2 : Transport Development in Urban Centres

In the second stage, more efficient transportation lines made possible the development of much larger urban centres, with a totally urban social core, boats, barges, canals, and harbour wharfs, roads, horses, carriages, and carts made possible a city which depend on agriculture and located far away from one-another. Rome, originally dependent on local produce, became a city fed by granaries of Africa and the eastern world.

An urban centre in the second stage was limited in size by pedestrian travel. The urban centre could be no larger in area than the distance a worker could walk to work, do a day's work, and walk home. The density of persons per square kilometre was limited by architectural techniques and primitive waste disposal. Thus these urban centres never exceed a population of one million, and there were few urban centres of this size, including Rome and some urban centres in China.

This urban centres contained the seeds of its own destruction. There is a tendency, as Mumford notes, to transform the city centre from natural to artificial and to replace grass and soil with pavement, gravel, house, and temple, creating an impression of the dominance of the environment by civilization. Ironically, the very artifice of the urban centres makes it all the more dependent on its rural surroundings for all resources. So although it appears to the inhabitants to be stronger and more independent, the urban centre is actually more fragile. Urban centres in this stage grow at the expense of the surrounding countryside, destroying the surrounding landscape on which it ultimately depends. As the nearby areas are ruined for agriculture and the transportation network extends, the use, misuse, and destruction of the environment increase. In this stage many of the classic cities of the past were built, prospered and declined.

Stage 3 : The Urban Centres of the Industrial Revolution

The Industrial revolution allowed greater modification of the environment and greatly increased the ease and capacity of transportation. Two technological advances that had significant effect on the environment were improved sanitation methods, which led to the control of many diseases, and modern transportation methods, including the improved method of sailing ship and the steam engine.

The size and density of urban centres increased rapidly. While the urban centre of Rome in A.D. 274 occupied about 13 square kilometre, modern greater London occupies about 1685 square kilometres and metropolitan New York City more than 6475 square kilometres. Prior to Second World War only 25 cities had populations over 1 million, by 1954 there were 64, 83 in 1960, more than 100 in 1975 and 250 in 1995.

In the third stage rapid transportation allowed the development of suburbs, and industries intensified pollution of their air and water. These changes increased the urban dwellers sense of separateness from their "natural" environment. Some attribute the recent interest in hiking, wilderness, and environmental conservation to the ease of transportation, which allows the urban dweller brief experiences with a landscape which is not urban and the increased artificiality of the city itself. In the end of the third stage, the city became more concentrated and artificial with skyscrapers reaching nearly a kilometre upward and surface and underground transit system moving people long distances to and from work. In this stage the urban park achieved a new importance.

Stage 4 : Telecommunications and Air Travel

We are perhaps at the beginning of a fourth stage in the development of cities. With telecommunications (telephones, television and computers linked by satellite and internet), much communication and commerce (which is much of the life of the city) can be done by people isolated from one another, at home, or long distances apart. The aeroplane has also freed us even more from the traditional limitation of situation. We now have urban areas where previously transportation was poor—in the Arctic (Fairbanks, Alaska) or on islands (Honolulu, Hawaii). The construction of multistorey apartments have changed the urban housing scenario at this stage. The negative effects of urban sprawl have led many back to the urban centres. The drawbacks of suburban commuting and the destruction of the landscape in suburbs have brought the city centre a new appeal. Perhaps, as telecommunications free us from the necessities of certain kinds of commercial travel and related activities, the city can return to a cleaner, more pleasing centre of civilization.

This optimistic future for cities requires a continued abundance of energy and material resources, which are certainly not guaranteed, and a wise use of these resources. However, if energy resources are rapidly depleted, modern mass transit may fail; fewer people will be able to live in suburbs and the cities will become crowded, reliance on coal and wood will increase air pollution, and the continued destruction of the land within and near cities could compound transportation problems, making local production of food impossible. The future of our cities depends on our ability to plan and to conserve and use our resources judiciously.

Environmental Problems at Large

Under such a circumstances it has been pointed out that by the year A.D. 2000 most urban centres of the world will face an acute ecological crisis in case some urgent improvement measures are not taken.

Hazardous Waste

It has been indicated that 5 million chemical substances have been identified for the benefit of human society but they brought a new danger through the generation of hazardous waste in the process of manufacturing in urban centres. These waste materials have a bad consequences of causing cancer, delayed nervous damage, malformations in unborn children and mutagenic changes. Uptil now many hazardous wastes were disposed of without proper evaluation of the environmental consequences such as fire, explosions, air and land pollution, contamination of food and drinking water, damage to people's health, harm to plants, animals, birds and fishes.[5]

The most worst incident happened was that of the "Minamata Disease" in Japan and Indonesia (1950's) where the discharge of methyl-mercury to the sea caused the contamination of fish in Jakarta Bay, which caused high fever, mental disorders to 2,000 people out of which 400 died. Similarly, in Northern India (1981-82) about 3,000 people died with the outbreak of Minamata (encephalites). In U.S.A. an area of about 65 square kilometres was contaminated with wastes from manufacturing of defoliants, pesticides and chemical warfare agents, causing crops to get dry and livestock drinking from the well to perish.

Now the same condition is prevailing in other developing countries where water bodies are so pulled by urban based factory waste and chemicals as in Sahadra, Delhi that they become sources of mass epidemic diseases. In this way, developing countries are the first victim through oceans, into which industrialized western countries continues to dump waste of household and industries. This crisis of environmental pollution between man and the biosphere is getting increased day by day because the big urban corporations of the western world, intensively transfer ecologically, harmful industrial goods which highly pollute the urban environment in developing countries.

Acid Rain

Acid rain is polluting the urban environment due to burning of fossil fuel in power station, factories and motor vehicles which brings down diluted sulphuric and nitric acid (Fig. 30.5). This has the great power to kill fish and other creatures living in water, besides corroding buildings and ancient monuments of the world. In American and European countries due to tall chimneys of the factories they got success in warding of the problems of pollutants in the nearby areas but "the sulphur and nitrogen compounds emitted by burning fossil fuels can be blown thousands of kilometres by the winds, to cause acid rain in countries far from their points of origin."[6] According to an estimate about 284 million tonnes of sulphur oxide and 90 million tonnes of nitrogen oxide the world is getting yearly which have already caused environmental pollution on regional scale in industrialized countries. The dry deposits of sulphur

FIG. 30.5 : Killer Chemicals : As Acid Rains

and nitrogen oxides pollute the environment just near the source but the longer the gases stay in the air, they are going into complex change and the repercussions are disastrous for human being, fish, animal and vegetation of urban places. Each and every country is generating such pollutants and sending some parts of it through winds to the neighbouring countries. In this way, they are both 'net exporter' and 'net importer' of air pollutants.

Lakes and rivers are the worst victims of acid rain. According to an estimate in Southern Norway out of 5000 lakes—1750 have lost their fish and other 900 are seriously affected. Similarly near Shahadra in Delhi, India the nearby ditches are so much polluted that vegetation dried off, fishes died and it is quite unfit either for animals or men to touch it and how much dangerous it may be, through acid fall and breathing of such polluted air which is beyond estimate.

It has been estimated that the opening of a large metal smelter in northern Mexico in the last quarters of 1985 may export acid rain enbulk to U.S.A. parts of Rocky Mountains as nowadays the U.S.A. is doing this for Canada.

But now the problem may be more vigorous as the new Mexican smelter—at Nacozari, 100 kilometres from the U.S. border has opened up at the end of 1985. This is the second largest emitter of sulphur in North America after the Canadian Sudbury complex in Ontario. Gaseous sulphur and nitrogen oxides are the destructive constituents of acid rain[7] (Fig. 30.6).

Energy Farms

The world has enormous amount of biomass, but its regional distribution is quite uneven. It has been seen by experience that all energy productions present environmental problems even if it is attractive. some plants have been discovered which may be the alternative sources of our energy requirements.

The production of ethnol in sugar factory has also great power of polluting the environment in case it is discharged without treatment into inland or coastal waters. Under energy farm many countries are planting trees for industry, field wood, charcoal besides soil protection and reducing the problems of flood. Such plantations cover 2.7 million hectares in Indcnesia, 2.6 million in India, 40,000 in Bangladesh, 300,000 in the Philippines and over 200,000 in Thailand, but we should keep a watch as deforestation may not aggravate further because it may pollute more, to our already polluted urban environment.

Alarming Antibiotic Resistance

Nowadays lot of reports are coming about the increased ineffectiveness of a range of antibiotics which have prompted World Health Organization (WHO) to identify the main factors contributing to anti-microbial

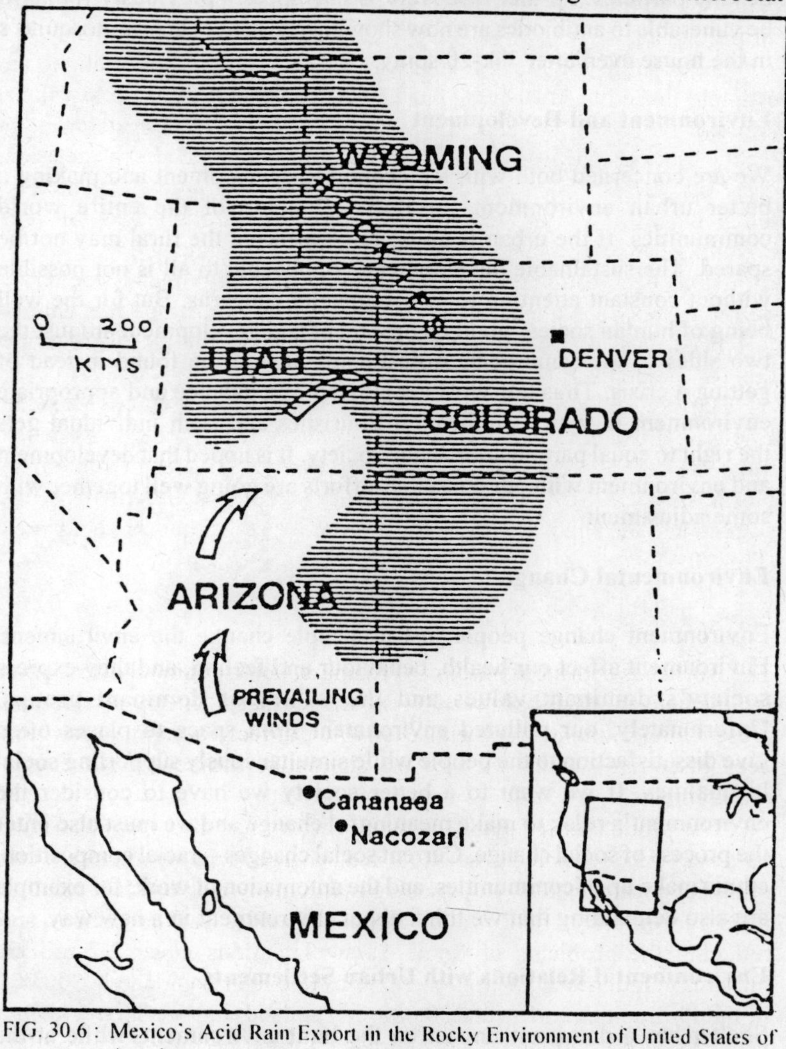

FIG. 30.6 : Mexico's Acid Rain Export in the Rocky Environment of United States of America

Source : John McCormick. 1985.

resistance. It has been found that the misuse of antibiotics in hospitals and private practice, and unwise self-treatment have contributed much to the problem of resistance. It has been investigated that the antibiotic resistance is now passing from one bacterial species to another by genetic particles. Further discovery is that species previously found to be vunerable to antibiotics are now showing resistance such as mosquito's in the house even after Tik-20 spray.

Environment and Development

We are concerned both with the economic development and making a better urban environment for the well being of the entire world communities. If the urban is going to be affected the rural may not be spared. The sustainable development—applicable to all is not possible without constant attention to environmental concerns. But for the well being of human society good environment and development are just the two sides of the same coin. Adjustments have to be found instead of getting a crisis. Thus the World is in need of a balance and appropriate environment in which civilization flourishes and each individual gets the right to equal participation in the society. It is hoped that development and environment with our conscious efforts are going well together with some adjustment.

Environmental Change

Environment change people just as people change the environment. Environment affect our health, behaviour and feeling, and they express society's dominant values and the values of dominant groups. Unfortunately, our polluted environment from space to places often give dissatisfaction to the people while simultaneously supporting social inequalities. If we want to a better society we have to consider the environment's role : to make meaningful change and we must also enter the process of social change. Current social changes—racial composition, ethnic make up of communities, and the automation of work; for example are also demanding that we think about environment in a new way.

Environmental Relations with Urban Settlements

Shelter has a dominant impact on the built environment, while at the same time its characteristics are influenced by the natural environment. Urban settlements can influence the environment by consumption of natural resources; by adding physical objects to the environment and by acting as an intermediary between human being and the nature.[8]

The form of urban settlements is generally determined by social requirements and the physical environment. Settlement is an intermediary linkage between human and social organization and nature. It may be a place of refuge, privacy, safeguards against climatic hazards, prestige and status symbol of a person in the society besides reflecting the technological, economic and social conditions. On the other hand, the origin of urban settlements starts depleting natural resources, on the other, its metabolic processes pollute the land, air and water at the same time, because all these are related with the process of feedbad mechanism. Such a grouping of houses on a particular location may guide the functions in a positive or negative way but that certainly pollute the environment by noise, congestion of houses, movements of people, atmospheric and water pollution besides the degradation of social norms. The level of environmental pollution may be varying in different physical and social conditions. It has been realized that the modification in environmental pollution is possible by way of judicious preservation and protection of resources, increase in renewable resources, i.e., afforestation and recycling. With the growth of population the current situation may deteriorate further but in order to save the urban environment from further deterioration the local government should make rules of human behaviour in relation to settlement, society, production programmes and the environment.

Air Pollution in Delhi

In Delhi, there are some 800,000 vehicles, which daily discharge more than 325 tonnes of poisonous gases into the atmosphere. Above 250 tonnes of carbon monoxide, 40 tonnes of hydro-carbon, 30 tonnes of oxide of nitrogen, five tonnes of sulpher dioxide and large quantities of particulate matter including lead are released. A joint study by the Indian Institute of Technology and Directorate of Transport, Delhi Administration pointed out that the traffic emissions account for around 50 per cent of the total pollution. The deadly combination of carbon monoxide and sulphur dioxide has brought in its wake respiratory problems, which affect nearly 30 per cent of the city's population (Table 30.1).

Past experience has shown that legislation will not help much in reducing pollution. Legally, almost any person whose vehicle emits excess smoke should be prosecuted, but in 1983 the Delhi Police were

TABLE 30.1 : Ambient Air Quality in Major Cities

(Ambient air quality—1989)

	Sulphur dioxide				*Nitrogen dioxide*				*Particulate matter*			
	Min.	*Mean*	*Max.*	*N*	*Min.*	*Mean*	*Max.*	*N*	*Min.*	*Mean*	*Max.*	*N*
Delhi (Nizamuddin)	BDL	13.0	41.5	109	BDL	17.5	30.5	109	56	331	1062	111
Calcutta (Municipal Corpn.)	4.7	87.5	229.8	100	33.1	108.9	362.7	100	90	554	1093	100
Chennai (Parrys Corner)	BDL	6.5	53.5	112	1.4	22.1	69.3	119	86	169	302	123
Bangalore (Anand Rao Circle)	7.2	26.9	104.4	39	7.7	22.7	69.7	40	21	141	478	43
Ahmedabad (GIDE, Navoda)	0.8	24.7	123.2	162	6.7	50.1	187.5	165	72	232	514	168

Note : Recently, in November 1998 in Delhi according to the courts order the 15 years old vehicles have been removed from the road, and the vehicles on the road has been advised to use unlaided petrol to save the city of Delhi from pollution.

able to catch only 487 motorists. The study reveals that carbon monoxide in concentrations as low as 30 ppm (parts per million) causes headache, diziness and lassitude, and prolonged exposure beyond 100 ppm can be fatal. Unborn hydro-carbons and oxides of nitrogen cause photo-chemical smog while lead pollution leads to brain damage. Suspended particulate matter cause respiratory ailments bronchitis, asthma and lung diseases Nowadays, the situation has reached to such a pass that not only people but vegetation and animal life are threatened. Traffic hazards due to reduced visibility—losses worth millions of rupees every year are also causing concern.

It has been felt that near Delhi the Yamuna river is so polluted that a single bath can make one sick. But since 1977-98 the reduction in pollution to the extent of 35 of 40 per cent had been achieved on account of various measures taken by the Municipal Corporation. In fact Najafgarh rivulet is the source of some 49 per cent of the total pollution load of Jamuna in Delhi. In 1985 May, the Delhi Corporation decided to carry out the programme of pollution reduction in the Yamuna with immediate effect by utilization of 90 cusecs of effluent from the Keshopur Sewage Treatment Plant for irrigation. This effluent was being allowed to return to the Yamuna via Najagfarh rivulets. About 30 cusecs more sewage will be picked up from Najafgarh rivulet subsidiary and diverted to coronation Sewage Treatment Plant. Thinking is going on the redevelopment plan of both banks of Yamuna river so that the river became an integral part of the life of people of Delhi.

On the basis of Table 30.1, it could be said that in Calcutta Metropolitan Corporation the mean sulphur dioxide present is 87.5 in comparison with 6.5 of Chennai, 26.9 of Bangalore and 24.7 of Ahmedabad in 1989. The mean level of nitrogen dioxide present were 17.5 in Delhi, 108.9 in Calcutta, 22.1 in Chennai, 22.7 in Bangalore and 50.1 in Ahmedabad. This clearly shows that in Calcutta the level of pollution is more in comparison with other cities.

Siri Siri Bags the Choke Cities

Polythene shopping bags and wrappers are a potential threat to urban environment. Once you have discarded them after use, they again return to you in a variety of ways, though you do not realize it. For example, they choke your drains and provide breeding facilities to deadly germs.

A recent study has shown that about 250 tonnes of plastic wastes come out of the various colonies of the major cities alone, every day.

This disrupts the sewer system, those essential arteries of the city-life, choke the land-mass and clog the pores of the wetlands. The consequences will soon be there for all to see.

Unfortunately, even the villages and small towns are not free from this danger. Millions of people returning to their home towns every day, carry their shopping in those colourful bags. This pleases their family and children, who after preserving them for a time dispose them off in wells, rivers, tanks and drains. Many throw them off into the fields. They do it with a sense of ride, to show off. When their neighbours see that their men from the cities regularly send them those "good things of life," they are impressed.

In Delhi, the worst offenders are the upper income groups of the so called posh colonies. Those educated, the residents of these affluent areas are unaware of the damage done by plastic bags. Nearly a million children in Delhi's schools carry their lunch boxes in plastic bags. They callously throw them away and cause an unhealthy environment.

The use of plastic bags is a habit forming phenomenon. It is not only convenient, but also fashionable. Cane, bamboo and cloth bags are used only a handful of ethnic minded shoppers. This limitation of the Indian society is exploited by the manufacturers to their great advantage. Sadly, no attempt is being made by the Government to create public opinion against it. On the contrary, more and more makers of these "siri siri" bags are appearing on the scene, as the demand increases.

The Indian makers of the "siri siri" bags, as they are natefully called in Sri Lanka, are callously indifferent to the problem, and have virtually succeeded in making it into a social necessity. They give them different shapes and colours to suit the consumer's fancy. When they came first decades ago, they were accepted as an innovation. No thought was then given to the question of their being ecofriendly, or otherwise.

The Government of India, unfortunately, has failed to consider the problem seriously. The massive use of polypacks by the Mother Dairy and the Delhi Milk Scheme is a glaring example. The packing culture of the public sector Super Bazar and the private department stores of Delhi, Mumbai and Bangalore is also polythene-oriented. The manufacturers of bread have luckily as yet kept themselves away from this temptation.

In Mumbai and Bangalore, the highly sophisticated provision stores sell their wares, tea, coffee, jaggery and cookies and a hundred other varieties of things in pre-packed plastic packets. When in Bangalore sometimes ago, this practice was resented, Nandini, Bangalore's version of Mother Dairy, promptly decided to sell its highly popular *peda* and milk-pudding in cardboard boxes. The environment-conscious Bangaloreans welcomed it with a sigh of relief.

The steps that Sri Lanka has taken to stop the use of "siri siri" are worthy of emulation. In that small country, the people call them the "plagues of shopping." No customer accepts them. Instead, they are pleased to carry their shoppings in grey cotton bags made of cheap materials. If in the schools even a single polythene cover is spotted in the fields, the authorities immediately swing into action. The parents of the child who brought it are called and admonished for that "public nuisance."

Not that the Indian environmentalists are not aware of this menace. But, we cannot take any effective legislative action against the offender. Not so in Sri Lanka, where polythene is legally banned.

In addition to this, in Sri Lanka a number of voluntary social organizations ever engaged in the business of educating the people. The members of the "Women's Education Centre" move in the streets in groups and tell the people about the damage the "siri siri" bags may cause to their health.

The famous "Ruk Rakagan no," a society calling itself the "Survivors of the Trees" has waged a virtual war against them. They picket the departmental stores and book the "culprits" for prosecution. They are often accompanied by teachers, school children and shop-keepers. Likewise the "Fernando Group" of social workers is promoting denim bags. "Save The Earth from Polythene" is their clarion call.

The media too are not lagging behind. The popular papers of Colombo are publishing regular theme pages on the subject. The lead in this was given by *The Island* whose *Earth Watch* is widely read and appreciated.

Back home, there is no evidence of any concerted move to check the menace. It is, therefore, not unrealistic to visualize a time a decade or two hence, when our earth will contain layers of plastics.

FOOTNOTES

1. Central Pollution Control Board, East Arjun Nagar, New Delhi-110032.
2. D.B. Botkin, and Keller, E.A., *Environmental Studies*, Charles, E., Merill Publishing Company, Columbus, 1982.
3. L. Mumford, "The Natural History of Urbanization," in *The Ecology of Man*, New York, 1972, pp. 140-52.
4. Leibor, *et al.*, *Transportation and Town Planning*, M.I.T. Press, Cambridge, 1970.

5. "The State of the Environment Report." prepared by the Environment Liasion Centre, Nairobi on World Environment Day, 5th June, 1983.
6. John McCormick, "Mexico's Acid Rain Exports," *Daily News*, Dar es Salaam, 7th April, 1985.
7. J.C. Wheeler, "Dawning Challenges Ahead," *Ecoforum*, Vol. 8, No. 6, 1983, p. 13-14.
8. A.L. Mabogunje, *et al.*, *Shelter Provision in Developing Countries*, John Wiley and Sons, New York, 1978, p. 4.

31

URBAN EMPLOYMENT

EMPLOYMENT is defined as the gainful engagement of people in some occupations in urban areas. The work in urban areas may be classified into cooking by house wives, administrative job, transportation workers, industrial workers, household industrial workers, educational workers, recreational works and people engaged in trade and commerce.

The urban employment comes under work culture. Although in most of the urban areas the number of childrens or non-workers dominate over workers. It means in urban areas the dependency ratio of non-workers over workers is high. This also gives us an idea that the future growth of population in urban areas will be higher.

The income generated by urban people through urban employment base is known as non-basic income. The income generation by primary functions is known as basic where as the secondary and tertiary sectors provide employment of non-basic nature.

Index of Surplus Urban Workers

The national proportion employed in any occupation is obtained and the number expected to be so employed in a town of any size is calculated. In case the actual employment when compared with this expected total shows a surplus then this represents basic employment. Based on this J.M. Mattila and W.R. Thompson proposed an index of surplus workers :

$$S = ei - \frac{et}{Et} Ei$$

where ei is the regional employment in the industry concerned; et is the total regional employment in all industries and economic activity; Ei is the national employment in the industry concerned; and Et is the total national employment in all industries and economic activities.

Structure of Urban Employment

In urban areas the functions performed by its resident is known as urban functions. In case the employment is in industrial sector it is known as 'secondary employment' and in case the employment is in trade and commerce it is known as 'tertiary employment' and most of the training and teaching works come under quarternary employment and the works of directors come under quinary function.

TABLE 31.1 : The Occupational Structure of Urban People in India (1991)

Sl.No.	*Occupations*	*Percentage*
1.	Cultivators	17
2.	Agricultural Labourers	10
3.	Household Industry	15
4.	Manufacturers	6
5.	Trade and Commerce	25
6.	Transportation Workers	5
7.	Storage and Office Works	22

Source : Estimated by R.B. Mandal, from previous censuses.

Cultivators

In urban centres people living on rural-urban fringe are mostly agriculturist who own land and grow vegetables to sell in the city. They virtually grow crops to sell in the market and hence they are known as 'commercial crops'. This comes under primary function and it adds rurality in an urban centre. Cultivators control the vegetable market of the city who push their products of potato, brinjal, ladies finger and tomato in the market throughout the year.

Agricultural Labourers

The landless people who does the work in the fields as cultivators is known as 'agricultural labourers'. These 5 days, the labourers are getting Rs. 50 for 8 hours of work. Their job is to cultivate the land, do the job of weedicide, pesticide besides harvesting and selling of products in the market. These days with the increase of literacy, cultural and economic development it is not easy to find agricultural labourers. They are rarely available and their charges have gone up.

Household Industries

It includes flour mill, rice mill, pulse mill and cottage industries, stone cheaps industry, brick making industry and saw milling. The work of house wives, child rearing, sauce industry, candle industry, jam industry, milching of cow and dairy and poultry industries also come in this group. In most parts of the world it does not exceed 15 per cent of the total population.

Manufacturers and Industrial Labourers

In urban areas only 6 per cent people engaged in industrial activities. In case the town specializes in a particular industry than it may exceed 25 to 50 per cent such as in Jamshedpur and Bokaro and so on. In Jamalpur railway workshop of Bihar most of the 4000 workers are commuters who daily come to the city on labourers train from different directions and again they go back to their respective home after 8 hours of service. The situation is more or less the same in Delhi, Calcutta, Mumbai and Kanpur where sub-urban trains help the labourers in commuting.

Trade and Commerce

Trade and commerce is considered as most gifted sectors of employment in urban areas. It covers about 25 per cent labour force of the urban areas where people are engaged in various types of commercial activities. For example, the employment structure of Raja Bazar vegetable market in the city of Munger is given in Table 31.2.

This clearly shows the micro level employment structure of urban people. Besides this, nowadays there is a flood of photostat shops, computer centres, book stalls, publishers, printing press, readymade dress, shoe stalls, tailors, saloon and cards, toys, musical stores and banks, cinemas and theatre complex, etc. In most parts of the world this sector has thrived so much that it always extract money from the consumers of the surrounding hinterland.

The urban business elites have connections with the neighbouring bigger urban centres who have business links on wholesale level. For example, in North Bihar-Muzaffarpur is a very big centre of cloth sale on wholesale basis. Lakhisarai in Munger is such a wholesale grain *Mandi* that its one day sale can equalize the whole month business of the city of Munger.

Transportation Works

Transport is considered as life line. Most of the rickshaw puller, bus and taxi drivers, railway employes and tempo-auto drivers who get livelihood from the means of transport come in this category. Only 5 per cent workers of urban centres come in this category of employment. These days unemployment has forced people to adopt transport as means of their livelihood. Because even graduates do not get a job of Rs. 1500 per month but most of the rickshaw pullers are earning this amount by pulling rickshaw.

TABLE 31.2 : Employment Structure of Raja Bazar Vegetable Market, Munger 1992

Sl.No.	*Types of Shops*	*No. of Shops*	*Persons Employed*
1.	Vegetable Retail	57	57
2.	Vegetable wholesale	8	28
3.	Fish	10	10
4.	Meat	8	16
5.	Eggs	6	6
6.	Motel	3	12
7.	Betel	3	3
8.	Rice + Pulse	3	9
9.	Grocery	2	3
10.	Duck and Hen	2	2
11.	Cloth	1	1
12.	Tea stall	1	1
13.	Watch Repairing	1	1
14.	Stationary	1	2
15.	Sports	1	4
16.	Electronics	1	2
17.	Indian Gas	1	10
18.	Medicene shop	1	2
	Total	110	169

Source : Field work, February, 1992.

Storage and Office Works

In urban area the location of Food Corporation of India godowns, warehouses and the storage of water, medicene and other items favour storage sectors employment. The office works include the job of executive, judiciary and police administration. Besides these, educational and personal services come in this category.

Plan to Develop Samastipur Through Employment

The district administration has launched an ambitious plan for the beautification of towns and the development of surrounding rural areas at an estimated cost of Rs. 34 crore in 1995.

In the financial year of 1995-96, a target for constructing 13,000 Indira Awas had been fixed at an estimated cost of Rs. 20 crore. Elaborating the scheme, he said that about 4,500 Indira Awas houses would be constructed under Jawahar Rojgar Yojna while 9,000 such houses would be constructed with the help of HUDCO and other Banks' cooperation. About 1,100 such houses had been completed. The rest will be constructed in two phases.

Under Jawahar Rojgar Scheme, bridges, dilapidated Panchayat Bhawan, roads and primary school buildings would be constructed at an estimated cost of Rs. 8 crore. It could be hoped that after such construction, there would be no primary school left without buildings. There are 158 such schools. The rest 6 crores will be spent on beautification scheme.

Under TRYSEM scheme about 1,411 persons had been selected to make them self-dependent. Under trial women prisoners and juvenile delinquents of the district have been included in this scheme. Vocational training in embroidery, spices preparation, and typewriting is being imported to them to earn their livelihood in a decent manner and be a good citizen of India.

Under DWACRA, (Development of Women and Children in Rural Areas) about 30 groups of women will be formed, each group consisting of 15 to 20 such women getting training in sewing, embroidery, cutting, masala preparation, carpet and candle making and so on. About 450 such groups under DWACRA which have become defunct during 1994-95, is now being rejuvenated.

A proposal had been sent to the Central and the State Governments to ensure them employment. Under this proposal every family of unemployed persons will be provided with 'family cards' which will give them an opportunity to work for 100 days in a year.

Conclusion

As against general impression a sizeable number of migrants remain unemployed or under employed in the cities. A recent study, sponsored by the International Labour Organization (ILO), reveals that about 20 per cent of the slum dwellers in the city of Mumbai are unemployed and a third of them have been unemployed since more than five years. Most of the street hawkers, domestic servants and rickshaw pullers are under-employed.

REFERENCES

Alexander, J.W., "The Basic—Non-basic Concept of Urban Economic Functions," *Economic Geography*, Vol. 30, 1954, pp. 246-61.

Berry, B.J.L., *The Human Consequences of Urbanization*, The Macmilan Press Ltd., 1973.

Coale, A.J. and Hoover, E.M., *Population Growth and Economic Development in Low Income Countries*, Princeton University Press, 1958.

Rao, V.K.R.V., "Population Growth and its Relation to Employment in India," *India's Population* (ed.) S.N. Agarwala, London, Asia Publishing House, 1959.

32

SLUMS AND SQUATTERS UPGRADING IN THIRD WORLD COUNTRIES

SLUMS and squatters are considered as problem areas for urban development in Third World countries. They are illegally occupied houses and creating a nuisance of environmental pollution and degradation of urban living conditions. Environmentally, these areas are not fit to stay in our urban society, because squatters provide a millieu of negligence of social order, economic backwardness, drudgery of leading a life besides politically unconsciousness. Slums are the dirty urban areas where the drains are full of mud and refuge water of the houses, full of filth and flies, mosquitoes and insects abound. The roads are either unmetalled or full of potholes where even passers-by have to wade through muddy water and creating a sense of danger in walking even during day time what to talk of night. The slums have mostly no road lights, prevalence of service latrines and the road corners are full of heaps of garbage. Sometimes, the roaming of pigs and stray cattles are creating trouble for the passers-by. Most of the houses are occupied illegally without land ownership and houses have been constructed without obtaining legal permit by the municipal authority.

With the growth of urban population nowadays slums have created a panic for cities development. It has been a serious issue for the town planning not only in India but in Europe and North America as well- Slums are the marginal areas of cities where the service facilities are poor, people are illiterate, drinking water supply is miserably on the lowest ebb and the vice of underworld's predominate the scene. The study of slums and squatters is meaningful for the environmental modification and ameliorating the problems of urban areas for providing more amenities to the elling masses.

The Concept of Slum

Slum dwelling signifies as a common phenomenon of degraded life of urban dwelling itself. A slum represents a micro-habitat unit within a large framework of urban built up space. The presence of a slum as a part of urban habitat refers to a condition of defective physical, social and economic environment. It means physically poor quality sub-standard housing characterize slum areas which pose danger of health and life of slum dwellers. The socio-economic substandard denote an area of degenerated living "an abode of half starved, a place of poverty, wretchedness and vice, the house of racketeers and criminal" and so on.

Most urban centres of the worlds irrespective of location and size have invariably developed as slum. Slum emerges within a relatively big urban centres and always tends to grow and multiply.

The concept of slum as given by R.E. Dickinson connotes "as extreme condition of blight in which the housing is unfit as to constitute a menace to the health and the morale of the community." It exists in all towns and owing to insufficient housing facilities it is mostly occupied by the poorest class of society and new coming migrants. These are the areas of high incidence of crime, mortality, illiteracy and disease. In slum, the vice of underworlds and the incidence of juvenile delinquencies are found to be in developed form.

Due to extreme conditions of overcrowding slums and squatter settlements have proliferated in non-western cities, occupying about 10 to 80 per cent of any available space illegally. Such de facto settlements are built within can in Algeria, reed in Baghdad and mud and straw in India, Kenya and Tanzania. The slums and squatters lack basic services and pose enormous risks of disease, fire and range of social problems. The causes for the growth of squatters are poverty, frustrated demand for housing, bulk migration of people from rural to urban areas for employment which serve the interest of the political-business elites and bad elements in the towns. The urban housing schemes are often very expensive for the poor.

The comprehensive improvement of housing, living and environmental conditions is the main aim of the various pilot projects undertaken in the urban squatter communities in the Third World. An important finding is the strong integrating role played by site planning, reblocking and housing in effective improvement and development of slum and squatter areas. The component is applicable to a composite strategy involving squatter upgrading on the one hand, and site and service or resettlement on the other. Other components of strong integrating effect are the employment generation by job training, job placement as housing schemes are often very expensive for the poor.

Problems of Slums and Squatters

In most of the Third World countries, the rapid urban growth occurs through squatter settlements. These uncontrolled settlements often lie around the periphery of the built up area, and are made up of temporary buildings with few public services. The names of squatter vary from country to country, as in Latin America they are called ranchos or favelas; in Asia *bustees* or kampongs and in Africa bidonvilles or shanty towns.

According to an estimate of United Nations Agencies squatter settlements cover from 10 to 80 per cent of the city population. For example, about 25 per cent of Rio de Janiero's population live in squatters. In 1970's squatters were described as a 'spreading fungus' and 'excessively squalid and deprived'.[1] But nowadays a more tolerant attitude prevails as the shanty towns are recognized as performing an important transitional role in urban evolution.

The studies of squatter settlements show that they range enormously in the standards of housing, amenities and access to safe drinking water supply. Mostly, the risks of public health problems are correspondingly high. Clearance of such squatter settlements or slums and rehousing, the population in more distant areas are problematic. It has recently happened in Manila. It has been expensive for the city corporation and has pushed the displaced population even farther from job opportunities which lies near the city centre.

This chapter aims to investigate the problems and prospects of slums and squatter upgrading in the Third World countries with special reference to comparative analysis of these aspects in terms of the prevailing socio-economic conditions and measures adopted by the local government, international organizations and development on the self-help basis. This study will highlight the problems in analyzing the slum conditions partly available in India, Tanzania and Kenya. This will also analyze the spatial designs for squatter upgrading with special reference to Indian city's.

Factors Contributing to Slums and Squatters

Slum exists because they serve the interests of the political and business elites. According to McAuslan, "It is the land owners and entrepreneurs who sub-divide the land illegally including businessmen who use shanty town population for cheap labour, officials who use this illegality to extract bribes and politicians who exchange promises of improvements for votes and who benefit by the system."[2] The people from rural areas often move to cities looking for jobs and then squat illegally on unoccupied but polluted land near railway station, bus stand, court compound and factories because there is no transport to move outlying areas.

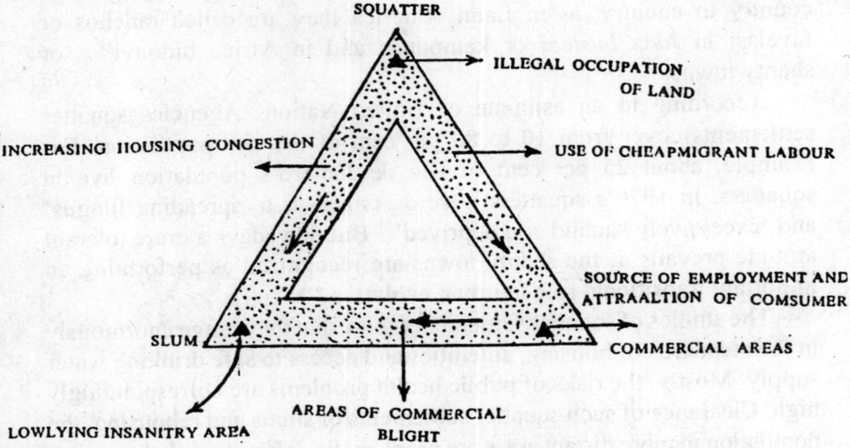

FIG. 32.1 : Dimensions of Slum and Squatter Settlements

Governments are finding themselves in a difficult position financially to eradicate slums because Turner has written in scathing terms of—'tin can cities' infect metropolitan centres of every developing country from Cairo to Manila, and their inhabitants live as "...displaced persons" who have been displaced from agricultural sector without being incorporated into modern industrial life.[3] Turner consistently emphasizes the positive aspects of the role of spontaneous settlements in contribution towards the solution of immense problem of urban shelter.[4]

As an architect, Turner make what may prove to be a highly significant distinction between 'improving' or 'consolidating' squatters those with blue-collar jobs who have come to the spontaneous settlements in a inner city tenement areas and show a high degree of dedicating to the physical improvement of their dwellings along with very poor strata of the city besides recent migrants from the countryside (Fig. 32.2).

Squatter settlements which have quickly assumed regular layouts lies in eastern Asia with high densities of population. The building patterns are too chaotic to permit of their easy rationalization. In such a

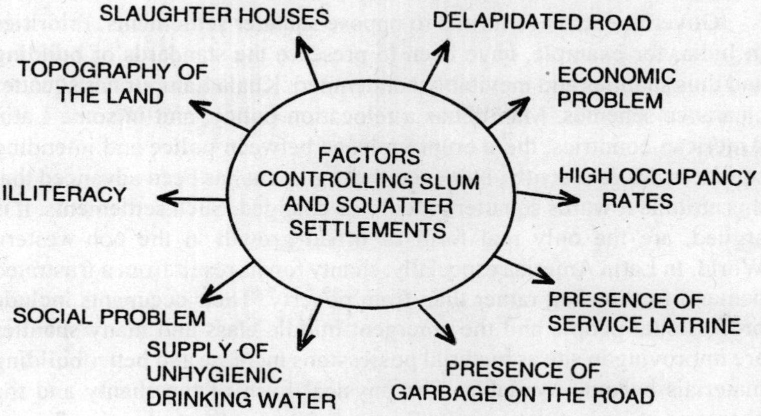

FIG. 32.2 :Old Remnant of Broken Houses

situation, the official programmes based upon the creation of new areas for spontaneous settlement might face formidable control problems.

In Calcutta, the migration of working-age males to squatters create an absentee house owner. This form of tenure is also important in case of the squatter yards of Kingston, Jamaica, where the government control is assured as in Hongkong, where intentions to squatter structures may be demolished by the authorities and the materials confiscated. It is usually not worth trying to improve a structure whatever the family income.[5]

In Cairo one million people squat illegally in a country. In Mumbai, the only home for upto half a million is the pavement and in Lagos more than a million live in illegally developed shanty towns. The city of Cario has been considered as the 'city of the dead' by McAuslan,[6] where the city's garbage collectors live in country amid discarded food scraps and rubbish.

D.T. Herber[7] has distinguished between the incipient squatters of recent origin, the semi-legal squatters, where local custom enable their initial establishment and plots are subsequently bought and sold as real estate. In Hongkong squatter population grew from 30 to 300,000 in 1947-1949, and it is around 10,000,000 in 1990. One-third of the population of Mexico city live in shanties, as do over two-million people in Calcutta conurbation. In Latin America, space around any new housing project is rapidly covered with shanties which seek to share the services made for others. The contrast between the rich and the poor become sharp and immediate.

Governments have turned to oppose squatter settlements. Priorities in India, for example, have been to preserve the standards of building and thus shanties and inevitably condemned. Kuala Lampur has squatter clearance schemes. Manila has a relocation policy, and in some Latin American countries, there erupt violence between police and intending squatters. More recently, however, the argument has been advanced that this attitude towards squatters should be changed. Such settlements, it is argued, are the only real form of urban growth in the non-western World. In Latin America especially, shanty towns result from a frustrated demand for housing rather than from poverty. Their occupants include professional people and the emergent middle-class and many shanties are improving in site as material possessions increase and better building materials become available. The physical form of the shanty and the character of its population vary from one country to another, but Turner has argued that shanties are essential where governments cannot supply the housing need. Again in Latin American cities, distinction could be drawn between the poor central shanties, occupied by recent migrants who needed to be close to possible sources of work, and the peripheral shanties with their more established population. In western world slum has features of sub-urban development. Colin Roser, from his close involvement in the planning of Calcutta, supported Turner's case.[8]

Calcutta's *bustees* existed because the need for low-cost housing could not be met elsewhere and housing programmes with idealistic objectives were futile. Besides providing cheap housing, *bustees* served as reception areas for new migrants, provided cottage industries and catered for mobility. The planned programme for Calcutta is a massive bustee improvement scheme, new experimental building techniques, and the construction of 'model bustees' to house on million people in ten years. Turner would advocate this role of government aimed at control and improvement, rather than prevention of shanties. There are other examples of the acceptance of those principles, such as self-help housing of Bogota's Ciwded Kennedy, Peru's housing associations and the Chilean government attempts to provide 'sites and services', whilst people build their own houses. In this way, the advantage of shanties can be maintained, whilst their worst expressions can be avoided.

In some non-western countries, with advantages of organization the investment of capital have attempted more ambitious solutions to their urban problems. Hongkong new town of Tsuen-Wan had reached a population size of 223,000 in 1968; Singapore built 700,000 units public housing in the 1960's in resettlement schemes through which they forced to accept high density of occupance; Venezuela has a programme to build three new houses at Cinded Cuyana, Tuz Medio and El Talbazo. Western prototypes are not directly transferable for the shanties of developing countries.

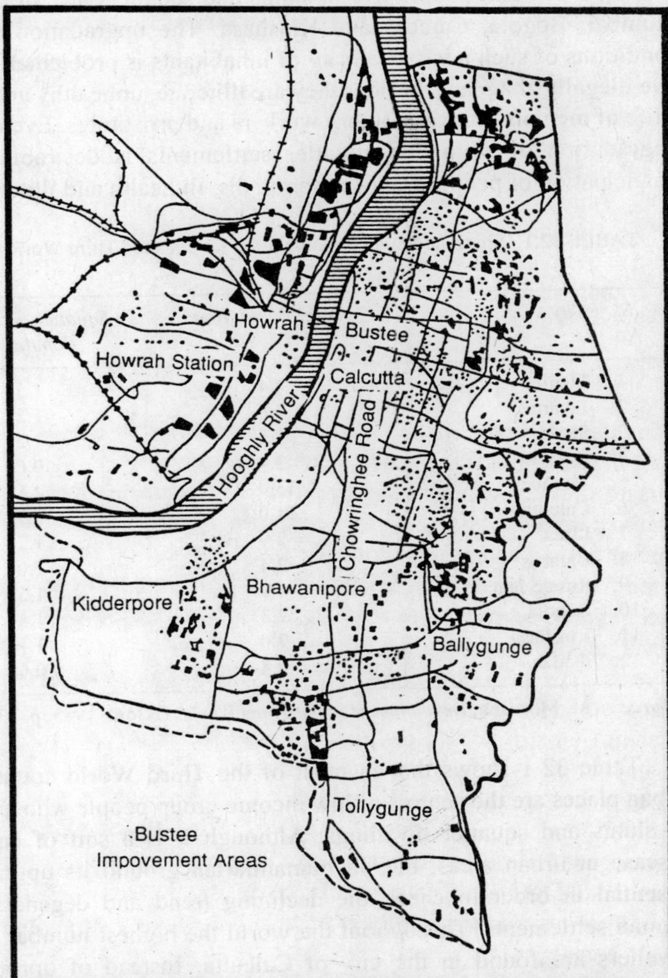

FIG. 32.3 : Bustee Improvement Areas of Calcutta

Table 32.1 and Fig. 32.4 show that in some cities of the world more than 80 per cent people are living in squatter houses, i.e., in Addis Ababa, Yeounde and Duuala, in the continent of Africa whereas in Calcutta, Rabat, Ibadan, Accra, Lusaka and Santo-Domingo (West Indies) about 60 to 80 per cent people are living in squatter houses. The cities where 50 per cent people live in slums and squatters are Mexico City, Mumbai, Bogota, Caracas and Kinshasa. The upgradation of living conditions of such a vast majority of inhabitants is problematic as they live illegally at a place, besides they are illiterate, unhealthy and leading a life of menials, thieves, factory workers and prostitutes. Even then the upgradation of slum and squatter settlements is desirable for the imancipation of people from economic ills, ill health and illiteracy.

TABLE 32.1 : Slums and Squatter Population for Selected Third World Cities

(*millions*)

Sl. No.	*City*	*Total Population*	*Squatter and Slum Population*
1.	Manila	4.4	1.5
2.	Jakarta	4.6	1.2
3.	Seoul	5.5	1.2
4.	Karachi	3.4	0.6
5.	Mumbai	13.0	2.5
6.	Calcutta	11.0	5.3
7.	Lima	2.9	1.2
8.	Caracas	2.4	1.0
9.	Rio de Jenerio	4.9	1.5
10.	Bogota	2.3	1.4
11.	Kinshasa	2.0	1.1
12.	Ibadan	0.8	0.6

Source : M. Morrisch, *Development in the Third World*, Oxford, 1983, p. 135.

Table 32.1 shows that in most of the Third World countries the urban places are the centres of low income group people who are living in slums and squatter dwellings. Although it is a sort of cancerous disease in urban areas, but on humanitarian ground its upgrading is essential in order to check the declining trend and degeneration of human settlements. Throughout the world the highest number of slum dwellers are found in the city of Calcutta. Instead of uprooting or demolishing squatters, people should give an alternative thought about their permanent settlement.

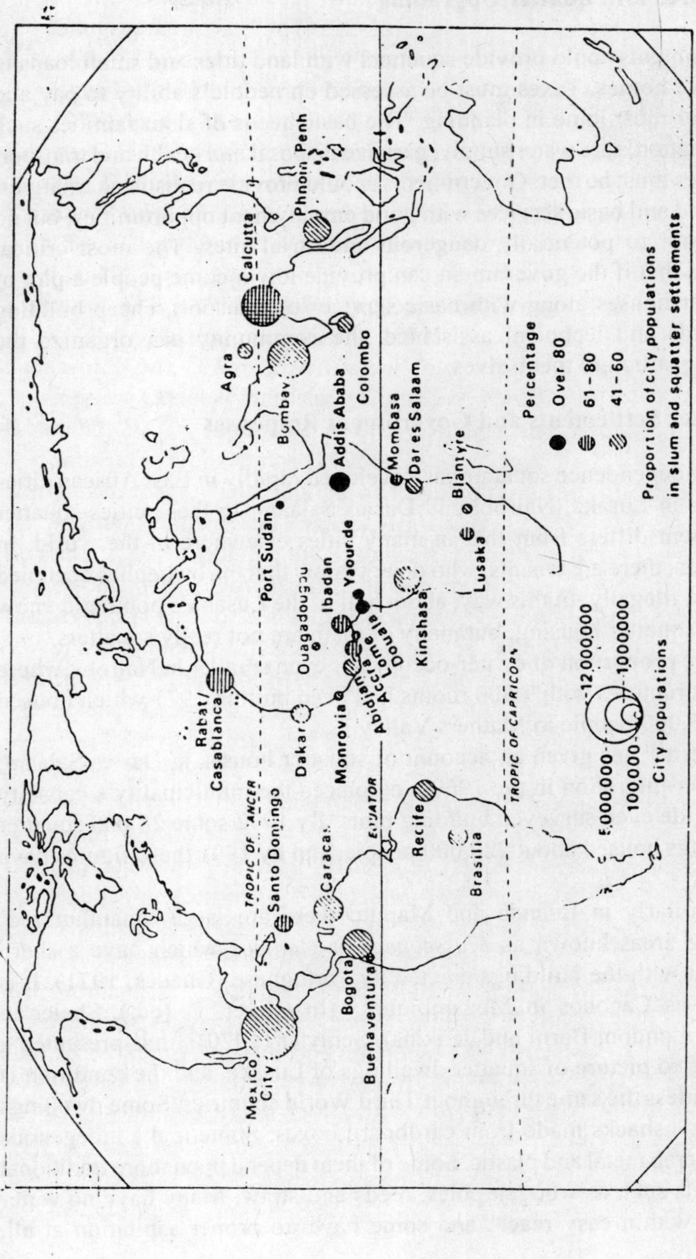

FIG. 32.4 : Distribution of Slum Areas in the World

Measures for Squatter Upgrading

Governments could provide squatters with land titles and small loans to improve homes. Taxes must be assessed on people's ability to pay and the poor must joine in planning. The basic needs of slum families such as sanitation, safe water supply, garbage disposal and health and transport facilities must be met. Governments should provide realistic alternatives, i.e., land and basic services with good employment opportunities, but no next door to potentially dangerous industrial sites. The most critical thing is that if the government can provide low income people a plot of land for houses along with basic services of transport, cheap building materials and technical assistance, the community can organize the housing structure themselves.

Squatter Settlements and Government Responses

After independence squatter has developed rapidly in East African cities such as in Lusaka, Nairobi and Dar es Salaam. In these cities squatter settlement differs from that in many cities elsewhere in the world. In squatters, there are tenants, who do not know that the owner has occupied the land illegally. In this way, about half of the Lusaka's population now live in squatter housing, but many of them are not really squatters.

The proportion of owner-occupies is even smaller in Nairobi, where 1,600 structures with 7,600 rooms had been built in 1971 which housed over 35,000 people in Mathers Valley.[9]

Stern[10] has given an account of squatter houses in Dar es Salaam, where in-migration in the 1960's outplaced the municipality's capacity to provide even surveyed building plots. By 1972 some 28,000 squatter dwellings housed about 220,000 people, and by 1991 these figures have trebled.

Similarly in Luanda and Maputo there are large 'unauthorized' housing areas known as *Mussechees* or *Canicos* which have a sharp contrast with the buildings erected by Portuguese (Guedes, 1971). It is known as Canocos in Mozambique[11] (In Oliver, P. (ed.), Sheler in Africa, London, Barni and Jenkins). Schlyter (1920)[12] has presented a diversified picture of squatter dwellings of Lusaka, and the condition is more or less the same throughout Third World countries. Some dwellings are crude shacks made from cardboard boxes. Some make indegenous use of scrap metal and plastic. Some of them depend upon more traditional materials such as wooden poles, reeds and straw. Many have no water supply within easy reach, and some have no proper sanitation at all.

Some squatter houses are as substantial as those of most urban dwellers having pit latrines and illegal hook-up to an electricity supply. The material conditions are more or less the same whether it is a Latin American barrics or North African bidonvilles.

The Situations in India

In Delhi the capital city of India and Chennai of Tamil Nadu the situations are worst to be judge in the Table 32.2.

TABLE 32.2 : Growth of *Jhuggis* in Delhi

Year	*No. of Clusters*	*No. of Squatter Jhuggis or households*	*Growth Index 1951 = 100*
1951	199	12,749	100.00
1961	544	42,814	235.80
1971	1,124	115,961	820.00
1973	1,373	141,755	1,001.18
1991	2,005	275,900	2,013.32

Jhuggi-Jhopri squatter settlements in Delhi carried out by the Sociology Division of the Town and Country Planning Organization of the Government of India in 1993 gave the following figures of growth of such *Jhuggis* between 1951-1991.

The figures for 1971 and 1991 do not include 50,000 families hut had already been re-settled in *Jhuggi-Jhopri* colonies during the decade 1961-71. Even the above figure alone represents 18 to20 per cent of the population of Delhi.

Authentic data about such colonies in other cities is meagre and no always comparable because of varying norms and definitions. According to a list of blighted area prepared by the Municipal Corporation of Mumbai before the Second World War there were 85 slum localities covering an area of 93 hectares. In 1956-57 this had increased to 144 localities over an area 355 hectares. An enumeration in 1969 listed 206 *zopadpatty* colonies with 108, 273 huts accommodating 631,888 persons. The report on the Regional Plan drawn up by the Mumbai Metropolitan Regional Planning Board records—"It is a shocking knowledge that this population of over 6 lakhs has amongst than only 1353 wcs and 482 water taps. The Regional Plan for Poona gives the figure of similar *zopadpatty* hutments having increased from 800 in 1937 to 6,300 in 1951 and 17,982 in 1968."

The Basic Development Plan for Calcutta Published in 1966 stated that about one-quarter of the total population of Calcutta and one third of the population of Hawrah and Bally approximately 912,000 lived in *bustees* under the most degrading conditions. A study of slums in Ahmedabad conducted by two economists of the Gujarat University school of social sciences in 1972, found roughly 80,000 households residing in slums as against 18,652 huts in 1960-61, an increase of 329 per cent in a decade.

In Chennai a complete census of the slum population was carried out by the Slum Clearance Board set up under the Tamil Nadu Slums (Clearance and Improvement) Act 1971. The results are summed up in the Table 32.3.

TABLE 32.3 : Slums in Tamil Nadu

Year	*No. of Slums*	*Families*	*Slum Dwellers*	*Increase per cent*
1961	548	27,851	412,168	—
1971	1,202	163,804	737,531	78.9
1991	1,957	225,554	1,232,337	82.0

The general increase in the population of Chennai city during 1961-71 was 42 per cent, so that, the rate of growth of the slum population was almost twice as fast. The proportion of persons living in slums increased from 23.8 per cent in 1961 to 29.9 per cent in 1971.

Rehousing the Slum Dwellers

It was in 1952 that the Government of India launched the subsidized scheme for Housing of Industrial Workers with 50 per cent capital subsidy and the balance as loan to State Governments and local bodies. The scheme was extended in 1957-58 to cover the rehousing of slum dwellers in general. The progress of slum clearance and rehousing in various towns has been very uneven and far from adequate. The Maharashtra Housing Board for instance was able to construct under the scheme only 24,190 units upto 1974 according to a paper placed before the seminar on Slum Clearance held at Chennai in June 1975. The Tamil Nadu Slum Clearance Board under took a most spectacular crash programme of re-housing all the slum dwellers in the city of Chennai in seven years. After five years of dedicated effort it was possible to provide only 16,000 units of minimum house, when on an average 15,000 families were being added to the city every year.

The difficulties of land acquisition the paucity of finances and the rising costs have made it practically impossible to provide houses to the slum communities. New approaches and innovations to cut costs have been tried with the support of Housing and Urban Development Corporations (HUDCO) by combining commercial projects with housing for the economically weaker sections, planning and designing cheaper houses such as the experiment re-housing of the Sabarmati squatters by the Ahmedabad Study and Action Group at Vasna and a number of other schemes in various states. Inspite of these admirable efforts we are no nearer bridging the yawning housing gap. What are then the prospects for the year 2000? The whole urban India appears to be moving inexorably towards the slum city.

Environmental Improvement in City Slums

A Central scheme called the Scheme for Environmental Improvement of Slum Areas was taken up in April 1972, with a view to giving hundred per cent grant to State Governments for undertaking schemes to provide amenities to slum-dwellers, in those area which were not to be taken up for clearance and rehousing for the next ten years. The scheme was initially undertaken in respect of 10 cities with a population of 8 lakhs and above and was later extended to 9 more cities. This scheme has been continued during the Fifth Five Year Plan period and is to be extended to all cities with a population of 3 lakhs and above.

Negative and Positive Responses

The government response to squatter settlements during colonial period has brought immense criticism due to inhuman behaviour and ineffective way of tackling it. For squatters rarely was an alternative accommodation offered, so those evicted hastened to set-up even flimsier shelters, in even more unhygienic conditions, on another patch of the land.

In Dar es Salaam legal rights have been granted to many who built illegally in the 1960's (Stern, 1979) and basic services have been provided for 70,000 people in the Manzese area. Lusaka has also embarked upon a massive programme of squatter upgrading (Pasteur, 1979) which involves granting of legal occupancy rights and provision of services such as roads, water supplies, schools and clinics. But such schemes have been criticized either from political spectrum or as forestalling the revolutionary force of the poor.

Site and Service Scheme

In African countries the government cannot build enough houses, but they cannot permit totally uncontrolled settlement. Site and service schemes offer a compromise something should be provided to many people, rather than everything just for a few. It has been introduced on self-help basis of the individuals who cannot do for themselves.

Generally, the government determines the location of the scheme, layout, individual plots, and provides access roads, piped water and sewerage while the new residents undertake the actual house construction. In some schemes the government has also constructed a coie of each house in the form of toilet and shower, and provide a concrete foundation for two or three rooms.

Pumwani Redevelopment Scheme, Kenya

In 1967, a Slum Clearance Housing Project was started at Pumwani.[13] The idea was that the old and ugly slums and squatters would be demolished and the residents accommodated in new modern multi-story and single story dwelling units. The new units were constructed to the entire satisfaction of the administrators and policy-makers. But the scheme only succeeded in increasing the housing stock and failed to accommodate tenants of the slums simply because the rents or mortgage payments were just too high for them. This example highlights the need to be extremely careful in planning housing schemes. If a proposed scheme is to succeed, realistic objectives must be defined and adhered to by strong administrative measures. It is not enough to give lip service to ideas.

Sites and Services for Slums in Mumbai

The metropolis of Mumbai is the commercial land industrial capital of India. It offers vast employment opportunities making it heaven to work in, but it is hell to live in. Due to sky high prices of land rent and house rent, the newcomers to Mumbai have to live in shabby human colonies.

The Mumbai Metropolitan Region extends over an area of 4370 square kilometres, holds a population of 1.19 crore. Of this 84 per cent of the population lives in 437 square kilometres called greater Mumbai. About 45 per cent of Maharashtra's urban population is concentrated in Mumbai Metropolitan Region and on an average 70,000 households are added to it annually. Only 35 per cent housing requirements are met. Per capita monthly income is high Rs. 800. Even then the Mumbai Metropolitan Region has 8 lakh hutments in over 1700 squatters and slum areas. About 45 lakh population is living without land and proper services in slums and on pavements. About 10 to 15 per cent population of Mumbai hudded in decaying houses where per capita space is just 3.50 square metres. For these facts, the metropolis of Mumbai has earned the nickname "slum polis" and Mumbai is called "slum bay."

Since 1946 only one lakh people have given environmentally sound houses and for this fact the hutments have made Mumbai a big slum. In order to plan and improve the housing situation World Bank has given a Rs. 282.3 crore to Mumbai Urban Development Project. Initially a beneficiary will contribute only 26 per cent of the cost of 'his shelter' and the balance given as loan. The basic facts recognized by Mumbai plan are given infrastructure and tenurial rights on land.

(a) People are motivated to build environmentally acceptable and socially satisfying colonies; and
(b) Though illegal and environmentally unacceptable, the slums do provide some kind of shelter to about 50 per cent of the population. As the complete eradication of slum is not possible, the quality of life should be improved by providing reasonable amount of civic amenities. The slogan is "do not end the slum, mend it to live."

Mumbai plan envisages upgrading the environment in the slums, granting tenure to the occupants and regularizing the slum households. One part of the project is the "Sites and services programme." In the International Year of Shelter for the homeless, the government agencies handled a programme for 40,000 dwelling units. The private and cooperative sectors have made 40,000 units. Hence, 80,000 dwelling units have been handed over to the new comers. The second part of this project involves redevelopment of slums at a cost of Rs. 100 crore. This fund has been given by the Prime Minister to the Government of Maharashtra. It will redevelop 65,000 units. Over 10 large plots have been identified in Mumbai for creating sites and services clusters. Work has already started in Charkop-Kandivili which has an area of 93 hectares.

In Mumbai the illegal growth of slums cannot stop unless there is an effective official organization to see that new huts are not constructed. It has been hoped that with the improvement of 40,000 hutments a year, we will provide a shelter to about two lakh people against an annual influx of 1.25 lakh persons. Even then Mumbai will mitigate the housing distress of its huge homeless population in the near future.

Slums and Squatters Housing Schemes in Madhaya Pradesh

The Government of Madhya Pradesh has decided to provide 50 per cent subsidy to the Housing Board in cities to develop land for building houses for low-income groups. The State Housing Board has got 45 hectares of land for building low cost houses in three industrial growth centres. Pothanpur in Indore district and Mandideep and Govindpura in Bhopal district.

The State Housing Board is also undertaking the construction of 1,000 single room tenements for the resettlement of slum dwellers. They have to be shifted from their present locations near the Bhopal railway station. The Board has built two model houses with *Khaprail* (tile) roof in Bagh Sewania, at the rural-urban fringe of Bhopal to test how far these units would be acceptable to the *Jhuggi* dwellers. The object is to keep the cost of each unit at about Rs. 6,000.

The Board is progressing on another scheme of building 'skeleton houses' in six cities including Indore, Ujjain, Bhilai and Durg, under a World Bank Scheme. The Board is constructing only the core section with pillars without roof and allot these houses to the prospective owners for completing the construction. They would be free to put up a second story, and such a skeleton house cost about Rs. 14,000.

The State Town and Country Planning Organization (TCPO) has also formulated similar low-cost housing schemes for 16 smaller cities including Amarkantak, Sanchi, Panna and Beladila. The cost of construction is Rs. 15,000 per unit and the funds available from Housing and Urban Development Corporation (HUDCO), New Delhi.

Indore Habitat Project

Britain has contributed Rs. 344 million to the Indore Development Authority's Habitat Improvement Project which provided funds for the physical, health and socio-economic improvement of 183 slums which directly benefit to some 40,000 people.

The funds provided under the British bilateral aid programme to India has been utilized for :

(a) to improve basic infrastructure within the slums such as roads and pavements, water supply, street lighting, storm water drains, sewerage, toilets and community halls; and

(b) to provide improved health programmes—including primary health care and family planning and socio-economic services such as adult and primary education, job oriented skills, training and education against drug and alcohol abuse.

The project will be implemented over the next five years by the Indore Development Authority supported by the Indore School of Social Work, the MP State Government Health Department and some non-government organizations. After completion of the project, responsibility for maintaining the benefits achieved will rest jointly with the Municipal Corporation of Indore and the local community.

Where the government disregards the cost affordability situation, squatting flourishes. Squatting has several advantages and disadvantages. The advantages are as follows :

Squatter settlements provide seven functions of major importance. They are the reception centres for new migrants. They provide housing within the means of the very poor besides a variety of small scale employment. Their social and communal structure provide a cushion for residents during times of unemployment and other periods of difficulty. They encourage self-help in improving the standards of houses. At the last, they provide a location within easy reach of possible work places of the city (Fig. 32.5).

(a) Shelters are provided to people who would otherwise have to live under bridges in sewers or culverts, etc.;
(b) If shelters are built by private entrepreneurs, the process introduces investment into the socio-economic programmes. Such funds might otherwise have been utilized in other fields not necessarily meant for the betterment of slums and squatter areas;
(c) Squatting expands the effective boundaries of the city without making much demand for services; and
(d) Any one willing to invest in shelter though limited in resources, implies a willingness to work within the framework of an orderly society.

The disadvantages are as follows :

(a) Without planning, the haphazard erection of shelter makes the area inaccessible to servicing traffic and vulnerable to fire attack and create grave health hazards, etc.;

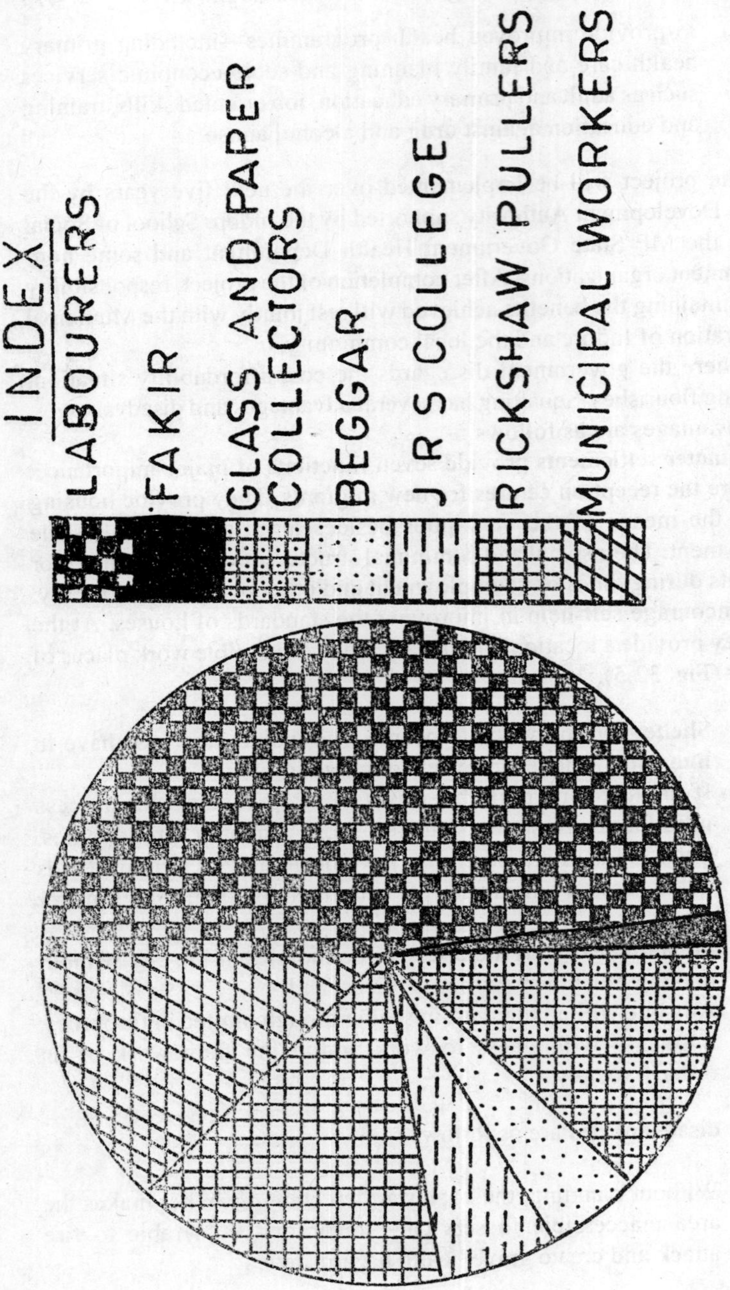

FIG. 32.5 : Occupational Structure of Migrant Living in Squatter Settlement of Purabsari, Munger

(b) The structure, being built hurriedly (12 to 48 hours) by inexperienced hands, is of expanding horizontally and vertically safely and conveniently (especially when subjected to repeated additions and alterations); and

(c) With more funds available to the household, the shelter cannot be systematically improved to express the new socio-economic status of the occupants. Hence, prosperous households leave the area with the result that an area will always remain a sore in the city.[14]

A Dream Comes True for the Shelterless

In Patna, Bihar, the *rain baseras*, financed by HUDCO became a reality here today, when six baseras were inaugurated. The Government is committed to uplift the weaker sections in the urban society the poor to avail themselves of the benefits.

Built at a cost of Rs. 70,000 each against the sanctioned amount of Rs. 90,000 the rain baseras would provide shelter to 18 persons each. Another *rain basera* was also nearing completion in Patna district. The cost of production was around Rs. 110 per square feet. The HUDCO has sanctioned 58 *baseras* for Dhanbad and Bokaro, which have to be constructed by Mineral Area Development Authority.

Conclusions

Problems of slums and squatter settlements are well known for the society in general because they are illegal homesteads with dirty surroundings. It is a sign of economically bankrupt and socially discarded community due to their misdeeds, un-employment, poverty, disease, dishonesty and habit of the vice of underworld's. Therefore, one can find all the characteristics in the urban centre of today where slums and squatter homesteads have proliferated.

Whatever may be, if we want to develop our urban community, we have to give land titles to such illegal homestead permanently by providing a proper housing facility in the cities where they are residing.

FOOTNOTES

1. P. Haggett, *Geography : A Modern Synthesis*, Harper and Row Publishers, New York, 1983, p. 406.

2. P. McAuslan, *Urban Land and Shelter for the Poor*, London, 1985.
3. D. Turner, "Comparative Analysis of Processes of Modernization," *The City in Modern Africa*, 1967, pp. 21-38.
4. J.F.C. Turner, "The Squatter Settlement," *Ekistics*, 23, 1967, pp. 123-82.
5. D.J. Dwyer, *The City as a Centre of Change in Asia*, Hong Kong University Press, 1972, pp. 166-78.
6. P. Auslan, *Op. cit.*
7. D.T. Herbet, *Urban Geography*, David and Charles Ltd., 1972, pp. 47-50.
8. C. Roser, "Housing for the Lowest Income Group : The Calcutta Experience," *Ekistics*, 183, 1971, pp. 126-31.
9. R. Martin, *Self-Help in Action : A Study of Site and Service Schemes in Zambia*, National Housing Authority, Lusaka, 1975.
10. R.E. Stern, *Urban Policy in Politics and Public Policy in Kenya and Tanzania*, Praeger, New York, 1979.
11. D. Pasteaur, *The Management of Squatter Upgrading*, Farnhorough, 1979.
12. A. Schlyter and Schlyter, T., *The Development of a Squatter Settlement in Lusaka*, Swedish Council of Building Research, Stockholm, 1980.
13. W. Adlerstein, *Management of Large Cities in Africa*, DSE, Nairobi, 1973, pp. 162-66.
14. Indubala Poddar, "Slums and Squatter Settlements in the City of Munger," *East-West Geographer*, Vol. 1, September, 1990, pp. 61-68.

33

PROBLEMS OF URBAN DEVELOPMENT

IN INDIAN CITIES the problems are manifold and the solution lies only in the present framework of our urban society, administration and technological knowhow and the availability of resources. The local planning body aims at limited improvement of physical landscape, construction of better drains, roads, water towers and safety latrines. It also aims at better utilization of the available human resources along with the housing improvement and creation of jobs and may be attempted on self-help basis to solve the urban problems.

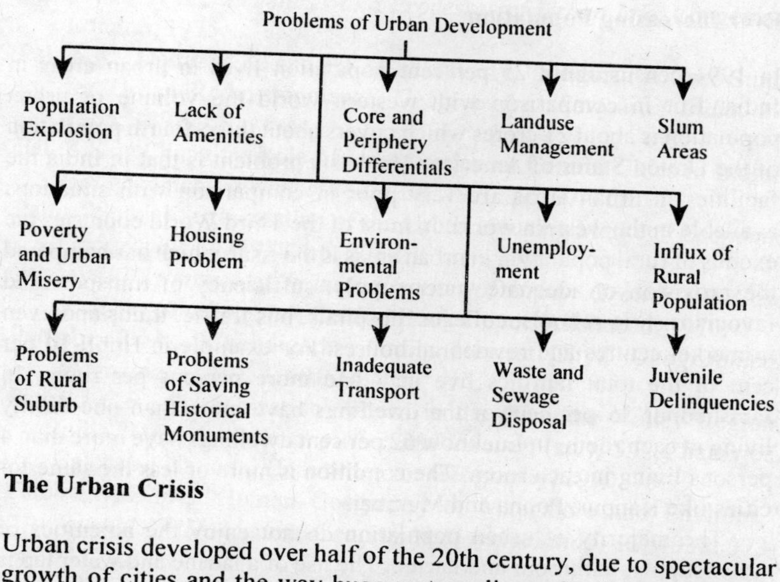

The Urban Crisis

Urban crisis developed over half of the 20th century, due to spectacular growth of cities and the way huge metropolises of millions of people have mushroomed in almost all sectors of the Globe. Nothing could stop the expansion of cities, particularly in the Third World, where they are growing at a rate of 3.6 per cent per annum. In only 35 years (1960 to 1995), the urban population of the world went from 300 million to 3 billion and it is expected to exceed the rural population for the first time at the end of the 20th century.

Some cities have exploded and imploded both. For example, Nairobi, the capital of Kenya, grew over 625 per cent between 1960 and 1995. To date there are 92 cities of more than 10 million, and there will be 30 by the year 2000, including the Nigerian megacity of Lagos in the developing world.

The miserable life of these megapolises are well known to all of us. The growth of slums and squatters have heightened the levels of pollution and the crime. The traffic jam is a common issue that mark their daily lives. In shanty towns people from the rural areas still migrate and find them irresistibly attractive. The cities are hives of industry which create far more job opportunities and wealth in comparison with rural areas ever do (urban areas supply 60 to 80 per cent of GNP in the developing world) and have the lion's share of all the health, training and cultural facilities.

Ever Increasing Population

In 1991 census about 25 per cent population lives in urban areas in India. But in comparison with western world the volume of urban population is about 21 crores which covers about three-fourth population of the United States of America. The basic problem is that in India the facilities in urban areas are very poor in comparison with situations available in the western world. In most of the Third World countries the exodus of rural population in urban areas is the issue which has hampered the provision of adequate water supply, sufficiency of transport and favours rush in schools, colleges, hospitals, bus traffic, trains and even in market centres and residential houses. For example in Hubli 14 per cent of the total families live in 7 and more persons per room. In Jamshedpur 35 per cent of the dwellings have more than one family living in each room. In Lucknow 62 per cent dwellings have more than 4 persons living in each room. The condition is more or less the same for cities like Kanpur, Poona and Mumbai.

The majority of urban population do not enjoy the amenities of water supply, bathroom and latrine. The use of a latrine and water tap is shared by ten, twenty and upto hundreds of persons. Hence, from 54 to 87 per cent of dwellings in the cities are without water tap. In Hyderabad and Secunderabad cities many people take bath once in a week. In

Jamshedpur 49 per cent people fetch water from masonary well. In Kanpur 35 per cent households have no water tap at all, and in Poona 30 per cent of the households share one tap by 25 and more families.

"The existing large and small sized towns have already grown beyond their capacity and even the present rate of growth is likely to be unmanageable. Acceleration of growth means the infusion of more employment" and worsening the living conditions of people of urban area.

In urban areas due to high density of population (30,000 per km^2) in Calcutta the horizontal expansion should be prohibitive in favour of vertical expansion. Hence, the resulting congestion has been judged to be intolerable by any standard.

Core and the Periphency Differentials

The recognition of the city core and its related areas differentiation is of course a key objective in any urban plan. In core areas, the worsening of traffic problem increased the quantity and complexity of functions to serve a much larger and far-flung population and workers besides increasing the congestion along with mix of small scale industries even in residential areas. On the other hand, in the fringe area of the city the dump of heaps of garbage, residential quarter of low income group people, low siting conditions and less development of transport linkage are some of the basic problems of urban development.[1]

Problems of Land Reserves and Landuse Management

In urban areas the presence of land reserve is essential for effective planning. Planning will take a long period, not less than 10 to 25 years. Unless free land is available in fairly well determined quantities and at favourable locations for specific future uses, obviously there would be no plan except casual development efforts. With the 4 per cent of annual increase in urban population the demand of land for housing colonies has shoots up drastically.[2]

Slum Areas

Slum is highly conjested low income part of the city where a considerable diversity in occupations, caste and religion is found. This is the area of shortage of houses, insanitary condition, retched and muddy lanes besides the utter lack of drinking water supply. Slum clearance is a vast and complex issue. The slum is an extreme illustration of housing shortage, illiteracy, heterogeneity of caste and functions of people. So far as the older slums are concerned extensive redevelopment schemes are necessary for abolishing them. Equally important is the problem of preventing the emergence of new slums.

At present, our cities are mixture of magnificence and squaltor of places and slums. They provide sometimes the best opportunity and sometimes deep dispondency. Recent urban growth has created problems of an unprecedented character. Most of the migrants coming from rural areas to the cities resides into slum areas as they are poor. Houses are deplorably short and found here and there and thereby creates overcrowding. The municipal services are found to be absent. Recreational facilities are non-existent and the agglomeration of houses is communality.

In the cities of Calcutta and Kanpur about 20 per cent of its population ekes out its existence in the worst slums of the world. About 65 per cent house of the labourers are unworthy of human habitation.

Poverty and Urban Misery

Although poverty and misery are basic problems of urban areas but the lack of recreational activities and lack of unity among the people are found everywhere. The plight of the poors, scheduled castes and *Harijans* are still worse especially during rainy season when hutments leak water. The level of literacy for the 90 per cent population is very low. They work mainly as potters and construction workers.

Health Problems

In most of the Indian cities the hawkers stalls have grown up here and there in insanitary areas. The inadequate supply of water favours outbreak of diseases like cholera, small pox, plague and misels. These have visited Calcutta, Mumbai, Patna and almost all urban centres of India regularly. To make matters worse, the over salutation by the Hughli river in recent years and upward push of saline water have made the situation difficult for horticulture, business, waterlogging on roads is a common feature especially during the rainy season as the site of Calcutta is worse among all the urban centres in India.

Housing and Living Problems

The housing problem is due to increase in population, unplanned industrialization, profit motive and increase in gap between population and construction of houses. The metropolis of Calcutta is in desperate need of emergency measures to improve its deplorable living standard. This situation has reached due to arrival of a flood of *refugees* from Bangladesh since 1947 and 1971 wars. About a quarter of population of the city lead a life of single person without their family female counter part and they form more than half of the households. Most of them are male and living alone as married migrant. Many of them are illiterate, unskilled workers and their per capita income is very low. The majority of this large group live in about the worst living condition in the city.

The shortage of houses and insanitary living conditions of people in Calcutta gives somewhat a sence of enormity of the problems. The responsibility of these problems goes to the politicians and the administration of West Bengal Government, Calcutta Metropolitan Corporation and other local agencies. According to an estimate there is a shortage of 6,500,000 units of houses in Mumbai and the shortage of houses in even higher 7,500,000 in Calcutta in 1991.

Economic and Social Crisis

Economic and social issues are the common problems of urban centres in India and especially in the metropolis of Calcutta. The pressure of urban growth in and around the city continue to be most disturbing and critical in India. A very little effort has been made by the concerned government and agencies except a few cases of slum clearance and *refugees* rehabilitation in order to alleviate the burden of incredibly low standards of living experienced by the majority of residents of the metropolis of Calcutta. In order to solve these problems a close coordination in between iron and steel industry in the region, Damodar Valley Corporation, the Departments of West Bengal Government, the Calcutta Metropolitan Corporation, the Calcutta Improvement Trust and other municipalities and trusts in West Bengal remains an urgent necessity to reduce the problems at hand. Recently, the physical and social conditions of Indian cities have deteriorated with an alarming rate due to alarming population increase and environmental pollution.

Problems of Unemployment

Many Indian urban dwellers are without employment. Many more are houseless year by year. Migrating population to the cities and natural increase of population are the causes of housing shortage. In urban areas the mal-distribution of population would jeopardize the whole planning endeavour.

The problems of immediate future is to find an efficient metropolitan design which could be able to find the solution of social and physical problems of the cities. Meier's solution to the design problem is the cellular megalopolis, each cell taking the form of an "urban village." The city surveys almost uniformly report a high degree of unemployment and underemployment.

Influx of Rural Population

The problems of towns are on an increase not only in India but even in metropolitan cities like New York. In India there is special urban problems because of the large influx of rural population to the cities, and this tendency is found in almost all countries of the world. This has happened due to high income potential in urban areas.

The municipalities were set up in urban areas to share the administrative responsibility of the local population. But due to lack of finance their functioning is not upto mark in the current situation of unabated population growth in our urban places. Hence, the growth of urban population and the influx of rural people to urban areas and the inability of the cities to provide the basic amenities which have given some of our cities a bad name. For example, Calcutta, with all the experts coming from different parts of the world, nobody can really figure out how Calcutta can be saved. Now our problem is to see that similar situations do not arise in the other cities. Mumbai is fast moving in Calcutta's direction. The problems of cities which are not complex in their own way. There has been a decline even in the number of smaller towns.[3]

Problems of Migrants

Another problems specially in the metropolitan city of Calcutta is that of over migration to the city and a large number of migrants. Displaced migrants dominate the scene who came to the city due to communal disturbances in Bangladesh. These people created problems of settlement, residential houses and illiteracy. Those who came from urban centres demands greater facilities whereas migrants coming from rural areas have low demand for housing and social amenities. The people living in urban areas are mostly migrants and in some cities they constitute 70 to 80 per cent of the total urban population.

Need of Uniform Urban Policy

There should be an urban policy in our State to act upon. It seems that in Delhi, the growth of the city was quite haphazard until very recently. There are certain rules. But if one breaks the rule, after paying a small penalty the work on construction should not stop but it should continue.[4]

In one State about construction of the road it was told: "Well, we want to do something, but there is a stay order by the court for the past four or five years." These are all matters which must come under some kind of a uniform urban policy in order to ameliorate the scene in true to the sense.

Provision of Infrastructural Facilities

The first necessity is to provide the infrastructure, i.e., adequate water supply, drainage, electricity, roads and means of transport. There is also problems of the slum and the pavement dwellers. We have to remove slums, but we can not remove the people who live in slums; they have to be provided proper livelihood. In Delhi, people have been removed and a very sincere and genuine effort is being made to provide them employment and transport to bring them to and from their places of employment.

Problem of Rural Suburb

There are problems of the rural suburb which constitute the area around the city. They should not be allowed to become a dumping ground for the rubbish of the city. A Delhi street corner abounding with very imaginable shape, colour, pattern and sound, spilling into the pavement below. A slum enclave in the heart of the Capital facing a thermal power station which blackens the sky and deposits spoot inside buildings for miles around. The middle orderly rows of apartments with adequate open space now being built on a large scale by the Delhi Development Authcrity in an effort to combat disorderly growth. In the bottom a low cost model rural habitat designed by the Central Building Research Institute, Roorkee find their place. There have been cases where funds have not been used for purposes for which they were meant. At the same time we must strike a balance between vigilance and interference.

Reform the Functioning of Our Local Bodies

It is high time to rejuvenate our cities to energize and if necessary, to reform or even change the pattern of our local bodies. Perhaps greater use could be made of mohalla committees, special community groups devoted to specific problems such as cleanliness, sanitation, amenities for the poorer localities, slum clearance, beautification, planting of trees and so on. If small groups would take interest in whatever they can be entrusted in, perhaps with a little help lot of things could be improved. Every new area must be carefully planned. We have the system of Master Plans, zoning regulations and these are absolutely essential even for ordinary development. It does involve sensitivity towards the use of space, the necessity for green areas, parks, playgrounds, etc. We must also inculcate better building habits. I have spoken in many places about the need to use local materials. Perhaps it is not so easy now to do that in growing cities because of the need to go upward. It could be adopted in low-cost housing which has been considered essential. We are increasing the cost of later because urban land is very expensive. If you build lot of small houses, you are spending more because you do not count the land cost, than if you were to build flats in cities. Wherever it is possible, people are more closer with nature.

Problems of Saving Historical Monuments

Many cities are destroying historical buildings or dismantling beauty spots. The West is trying desperately now to recover what they had lost and tried to re-build things in the old fashion as far as possible. In fact, during the war the Vienna Opera House was razed to ground and everybody thought to built a new modern building. The money for the Opera House was not given by the Government; it was entirely raised from the public. The public wanted to make it in every detail as it was before. We have, in almost all our cities, some historical buildings are found. The historical sites are not only those which existed in ancient times or in Moghul times but even some of the buildings built by the British are part of our history. We should hesitate before we decide to remove them.

Problems of Inadequate Transport

Naturally, any city planning must take cognizance of the traffic that is likely to exist 20 or 30 years from now. One big problem for instance, which Calcutta and Delhi share is the lack of adequate transport.

Dilapidated Roads

Besides shortage of residential houses there is an utter lack of sanitation, water supply and repair of roads in Indian cities. Only the patchwork repair of large pothole on roads will solve the problem either by the municipalities or by the Government.

Waste and Sewage Disposal

Another important question is of waste and sewage disposal. These should be studied by our local bodies. Some ideas are expensive, but there are others which are low-cost methods and these should be implemented whenever feasible. We must also make an effort on the use of waste, on recycling it. Some of it may be used for manure. Other things such as paper, rags and so on can be recycled. There are other things which are used as sources of energy and similarly there is metal scrap and so on. All these things are treated as rubbish today, are valuable substances.

Environmental Problems

The problems of environmental pollution exists in almost all urban centres of India. Now the government is trying to do something in Delhi with the power houses and cremation ground. But we find that the public transport, the private buses, the power stations and other factories generate a lot of pollutants. Some of our Indian cities having far less industries than the industrialized countries, have a far greater rate of pollution because they have not taken steps to check and clean up their cities. So we must insist from now that any new services, any new industry that comes up, must have those safeguards.

Nowadays, with the overwhelming increase of population, the burden of the available residential houses, cultivable land, roads, railways, water taps and industries located in urban areas have surpassed all previous levels in terms of harnessing the resources and creating environmental hazards.

In Bihar some of the important hazardous places on the Ganges river are Barauni, Mokameh, Patna, Bhagalpur, Buxar and Shahebganj; near Bermo, Chandrapura and Bokaro along the Damodar river and on the Sone river near Dehri, on Kharkai river near Jamshedpur and others are highly polluted. In most of the urban areas the chocked drains, the predominance of service latrines, dump of garbage on roads, open drains and insanitary streets are crux of the issue. Throughout Bihar, urban areas pose a deserted look on the environment which is serious for a sound socio-economic development of India, the country as a whole.

Juvenile Delinquents Threaten Urban Peace

Juvenile delinquency has been a problem of all societies at all times and is fast becoming a social problem in urban areas. Crimes committed by juvenile delinquents are not limited to only the economically backward strata, but is rampant among the elite section of the society as well.

The National Crime Research Bureau (NCRB) statistics show that though proportion wise, there has been an increase in crimes like murder, attempt to murder, sexual assaults, kidnapping and abductions by juvenile delinquents and there is a steady decline in other crimes like burglary and thefts.

As compared to 4.4 per cent crime related to murder and attempt to murder committed by children in 1992, latest data shows that these crimes have risen to 5.2 per cent. Similar is the case with sexual assault, kidnapping and abduction, whereas it was 2.1 per cent in 1992 but now it has gone up to 3.7 per cent. Under the Indian Penal Code, the maximum number of crimes committed by children are related to thefts which is 25.4 per cent followed by burglaries 12 per cent and riots 10.8 per cent. The percentage in these areas have declined as compared to the recent years, whereas it has risen in the major areas.

"There are various sections of the society where juvenile delinquency is prominent," said Dr. S.S. Sanyal, a Professor in the National Institute of Criminology and Forensic Sciences. "One section belongs to the *jhuggies* where poverty and the social environmental leads them on, secondly there are college and school dropouts where frustration plays a major role followed by children of working parents who have no time to answer their queries and children of the elite section who feel they get away with anything," she said. Dr. Sanyal felt that the media also plays a role, "especially the violence in movies, the advertisements which are explicit and magazines too."

A boy comes under juvenile delinquents till the age of 16, whereas it is 18 in case of girls. Thereafter, they are sent to regular jails meant for all criminals. "Juvenile delinquency is the result of an encounter between an individual, his emotional disturbance and the social environment," said Dr. Sanyal. She said slum children saw violence often and drug pedding and bootlegging was rampant which resulted in their involvement.

"As far as children of working parents are concerned, most of them commit sexually-related crimes as they have many queries with no one to answer," she said. Moreover, there is a lot of exposure these days." A susceptible age is during adolescence where the individual is at war with himself and the world," Dr. Sanyal stated adding that "this is when children need a lot of attention." As far as the elite section, "too much money and the thought that they can get away with anything indulges them to take to crime," She opined. Children from this section are involved in smuggling, drug trafficking and in sexual assaults on others, the professor said.

The time of schooling and environment are also responsible for the degeneration of these youngsters. In school teachers should pay proper attention to all students as dropouts often end up as deliquents," she said.

Dr. Sanyal said that children were involved in all kinds of crimes from murders to stealing, drug trafficking, boot-legging and sexual offences." Rape is a common crime as children are left with no answers to most questions and start experimenting," she said. Under the local acts are special laws, the maximum number of children have been involved in the Prohibition Act which is 16.1 per cent followed by gambling which is 11.4 per cent.

"There are juvenile delinquents even today and though there has been a rise in certain crimes like murder and rape, there is a steady decline in juvenile delinquency," stated Mr. L.C. Amarnathan, Director, NCRB." One major reason is the Juvenile justice Act which has changed now," he said. "Initially, these minors were sent to jail with the hardcore criminals where they became more exposed to crimes and once out indulged in these activities again," Mr. Amarnathan said, "though now there are special homes for them where they are not treated like convicts, but are given proper love, affection and a family atmosphere in a bid to reform them."

FOOTNOTES

1. R.P. Beckinsale and Howston, J.M., *Urbanization and Its Problems*, Blackwell, 1968.
2. M.J. Wise, "Problems of Planning for a Changing Britain," *Geography*, Vol. 49, 1964.
3. Bert F. Hoselitz, "Cities of India and Their Problems : A Review Article," *Annuals of the Association of American Geographers*, 1959, Vol. 49, pp. 223-31.
4. Indira Gandhi, "Haphazard Growth of Our Cities," *Yojana*, Vol. XIX, 15 November, 1975, pp. 5-8.

PART-III

URBAN PLANNING

34

RURAL-URBAN FRINGE

IT CANNOTES marginal areas of rural as well as urban. In other words rural-urban fringe is an area which lies at the rear end of the urban centre. It is a transitional zone where the characteristics of both urban and rural settlements are found in a mixed form. Ruraji and Pilatcher (1958) have established difference between rural-urban fringe and suburb. Wissink (1961) pointed out difference between fringe, suburb, Pseudo-suburb, Satellite and Pseudo-Satellite. Some other words used for rural-urban fringe are outer linked areas, sub-urban zone and extended fringe, etc.

Recently, the study of rural-urban fringe has gained importance because the research scholars of urban geography, sociology, land economics, authorities of town administration and planners have given due attention about the study of rural-urban fringe. Rural-urban fringe is a quasi-urban area where the experiments of both rural and urban development have been gaining ground especially for the construction of roads, houses, rise in literacy including the dumping ground of wastes, increased amount of juvenile delinquency and suffer from pseudo form of urban development.

Ideas and Concepts

The concept of rural-urban fringe has been first of all developed by von-Thunen in A.D. 1826 who devised the concentric development of land use zones around an urban centre. In the year 1925 Jonasson, while studying the European cities, revealed that landuse in the suburban area fulfills the need of the town. In 1928, MacKangie first of all expressed in "The New Exploration" that town recklessly uses the surrounding area with the establishment of Bungalow, Godown, factories, Bill-Board, Gas Filling Station and residential houses (Fig. 34.1).

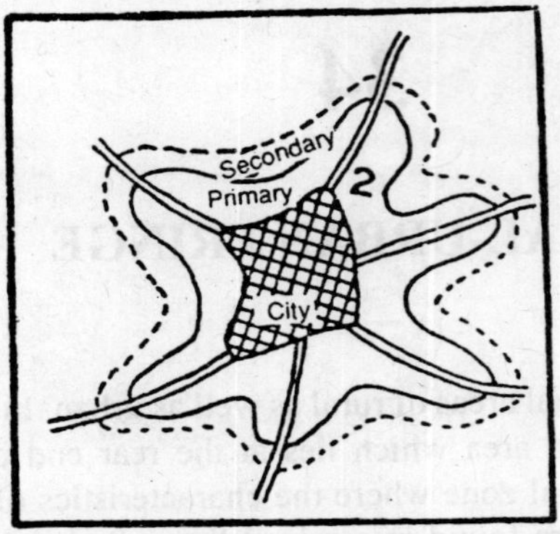

FIG. 34.1 : Rural-Urban Fringe

Some Important Works Done in India

In 1955 R.L. Singh studied about the rural-urban fringe of Varanasi. He tried to show that fringe area is dependent upon the urban centre for some of the urban facilities. In this regard he has done the sample of Sundarpur village.

U. Singh studied about the rural-urban fringe of KAVAL towns in Uttar Pradesh. He observed the construction of residential houses even in industrial area due to expanding city limit. The area under urban functions and fringe zone could be seen in Table 34.1.

TABLE 34.1 : Fringe Area of City Level Centres

City	*Village Considered into the Urban Boundary*	*Area in km² of the Village Included in the Towns*
Kanpur	157	204.80
Agra	34	23.04
Varanasi	93	20.48
Allahabad	55	15.36
Lucknow	47	48.64
Total	386	312.32

The inclusion of villages into the urban fold have covered almost 53.65 per cent of land under rural-urban fringe.

In 1972 R.C. Gupta studied about the rural-urban fringe after considering Sahadara a sub-town of Delhi in this group (Fig. 34.2). He also categorized it as rurban fringe. In the same year Hiralal studied about the concentric development of rural-urban fringe around the city of Bareilly, where he demarcated the Primary and Secondary zone (Fig. 34.3), after considering spatial, occupational and demographic attributes of the fringe area.

FIG. 34.2 : Rural-Urban Fringe of Delhi

In 1980, M.M.P. Sinha studied about the rural-urban fringe of the city of Patna where he demarcated the primary and secondary fringe (Fig. 34.4). The variations in the fringe is due to the physical, cultural and economic differentiation of the area. As the urban facilities are changing their intensity and areal extent so the characteristics of the fringe also gets change.

In 1976 S. Nangia in her books *Delhi Metropolitan Region : A Study in Settlement Geography* studied about the rural-urban fringe of Delhi. Here the fringe zone extended over an area of 212 km^2 and included 177 villages in its fold. This zone is not concentric in nature rather it is polygonal in shape. In western side the fringe extends over 10 kilometres while it extends over 18 kilometres in the east. In north western the higher development of fringe is found towards Sonipat whereas Faridabad in the south-east. The characteristic of the fringe of

Delhi is that it commands a lot of industrial nucleus in its fold such as Nazafgarh, Azadpur, Okhla, Sahadara and Ghaziabad. It commands sewerage treatment plant and recreation centres as well.

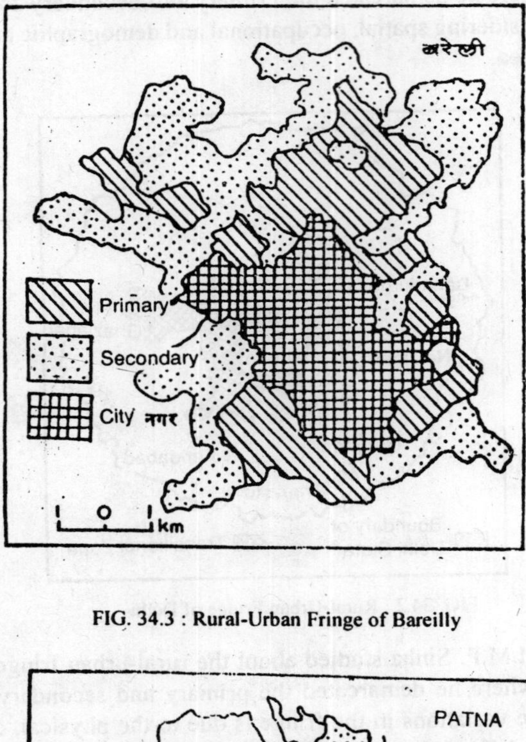

FIG. 34.3 : Rural-Urban Fringe of Bareilly

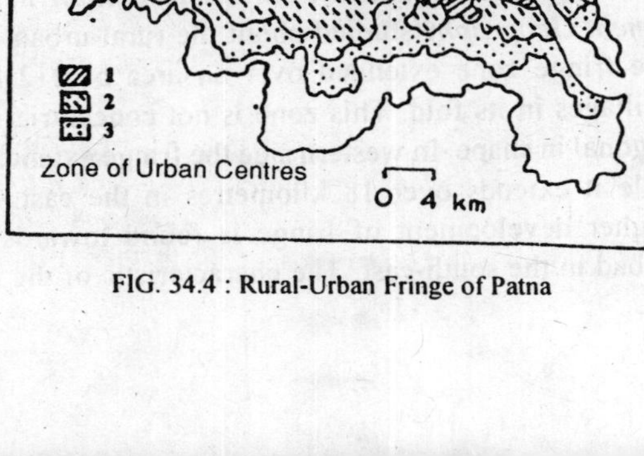

FIG. 34.4 : Rural-Urban Fringe of Patna

Definition

Rural-urban fringe has mixed characteristics of both rural and urban. Some of the definitions are as follows.

1. "Rural-urban fringe is an area with distinctive characteristics which is only partly assimilated into the urban complex, which is still partly rural."—*H. Carter.*
2. "Where the urban influences are essentially mingled with rural forces often without striking a proper adjustment as in the case of unplanned growth designates the nature of the rural-urban fringe."—*R.L. Singh*
3. "The urban fringe form the active expanding sector of the compact economic city and it lies at the periphery of urban areas."—*Andrews*
4. "Fringe Settlement as a two directional movement reiterates that new residents converge upon the fringe both from urban places and rural areas."—*Rodehaver*
5. "In the fringe area there is a mingling of people of some of whom work in and are oriented towards agriculture while at the same time the remainder pursue urban occupations and an urban way of life."—*Dewey*
6. Rural-urban fringe is really an extension of the city itself, present and potential and since the city or cities of a metropolitan area and the suburban or fringe areas are a unit economically and sociologically the entire area should be thought of an planted as a unit."—*G.S. Wehrwein*
7. "The fringe is the zone between the country and the city." —*R.R. Mayer* and *J.A. Beegle*
8. "Rural-urban fringe is that area adjoining the inner fringe outward from the economic city in which there is an intermingling of characteristics of agriculture and urban land uses."—*R.B. Andrew*
9. "The urban fringe is defined as the land surrounding the town which is not considered as a part of it but whose use is influenced directly by the town."—*Lewis Keeble*
10. "The urban fringe is a mixture of land uses—rural and urban and classified it into a series of belts surrounding the city by the analysis of land-use characteristics. Such area is full of serious and complex problems and needs proper planning for solution."—*Solter*

11. "Rural-urban fringe is the rural land with urban phenomenon. The rural land is forced into urban uses prematurely and is almost frozen rarely being restored to agricultural uses." —*R.L. Singh*
12. "Rural-urban fringe is really an extension of the city itself, actual and potential, and since it is an area where most of the land uses are in flux."—*R.L. Singh*
13. "Urban fringe is a zone of cultural development that has taken place outside the political boundaries of central cities and extends to the areas of agricultural activities."—*F. Arpke*
14. "The fringe is the extension of housing estates of buildings along the main arterial roads, and by the location of new factories, golf courses, water-works, ceneteries . . . and the like."—*R.E. Dickinson*
15. "The rural-urban fringe is an area of transition between well-recognized urban land-use and the area devoted to agriculture."—*G.S. Wehrwein*
16. "The rural-urban fringe is a 'suburb which begin where the continuous build-up town ends. First, there is the built-up area of houses with small gardens, forming dormitory communities from which more than half the active populations work in the town."—*Garnier* and *Chabot*
17. "The space into which the town extends as the process of dispersion creates the concepts of rural-urban fringe. The centrifugal forces impel functions to migrate from the central zone of a city towards its periphery."—*Hiralal Yadav*

Characteristics and Problems

It is quite apparent that rural-urban fringe is a transitional zone between urban and rural areas. The urban fringe suffers from the problems of urban expansion where the extension of buildings on road side, the establishment of new industries and several urban functions and characteristics features develop. Sometimes even the cremation ground, burial places, dump of urban refuse, park, garden, water works and golf-course find their place amidst agricultural fields on the fringe area.

A study of American city gives us an idea that a motorway has helped much in fringe development in comparison with railways. This area also favours the construction of human settlements and the establishment of factories at slow pace. In this zone a fast change of land use pattern could be seen which is a sort of premature urban development.

Residents of the fringe area have to manage drinking water, latrines, electricity and gas on self-help basis. Generally, the municipal authorities consider this area outside the municipal limits. Here the plan of industrial establishment could be managed well on the open land. The fringe area is considered as workshop for the construction of new buildings, play fields, *butcherkhana*, petrol depot, cremation ground, aerodrome, sewage plant, dump places of urban refuge besides brewery industries which spells out bad smell. For this fact, G.S. Wehrwein has considered it as an area of present and future development of urban settlement.

Walter Firey has studied the characteristics and problems of Flint city of Michigan State of U.S.A. and he has drawn the following conclusions :

(a) Fringe withdraw the land from agricultural production;
(b) A lot of differences are found in the distribution of plots, and industries cropped up here and there;
(c) One has to pay a heavy taxes in order to manage urban amenities in the fringe area;
(d) Due to construction of houses the price of the land shoots up very high which favours withdrawal of the land from agriculture; and
(e) One could observe a shift change in the characteristics of population.

In 1960 Golledge has shown the characteristics of Sydney city's rural-urban fringe which are as follows :

(a) It is a zone of fast changing ownership of the land;
(b) The plot of the land is of small size;
(c) A zone of intensive agricultural operation;
(d) Population is dynamic but the density is low;
(e) The expansion of residential houses took place fast;
(f) Municipal facilities are almost non-existent; and
(g) The house rent is commonly very high.

In 1965, R.E. Pahl has shown the characteristics of rural-urban fringe under four heads :

(a) Segregation of plots and buildings;
(b) Selective immigration by dynamic people;
(c) It is a commuting zone of workers of industries and commercial centres of the city; and

(d) It is a meeti~g point of geographical and social forces for human occupation.

Pryor (1968) has considered the land closer to the city as social and economic unit of the town. The tax for this rural sector comes from rural areas and facilities of electricity, gas, drinking water and transport and communication have been availed by the urban areas. From the point of land utilization and demographic characteristics the rural-urban fringe availed facilities of both the world. The other characteristics of rural-urban fringe are as follows :

(a) The urban facilities are always inadequate by any means.
(b) The zonal arrangement is not interlinked.
(c) The areal extent covers the land even beyond the city boundary.
(d) There is always chance of the increase of population density.

The residents of rural-urban fringe are wholly dependent upon urban area for employment, retail business along with the dearth of transport and mass communication.

Sudesh Nangia (1976) while studying the rural-urban fringe of Delhi Metropolitan region highlighted the following characteristics and problems :

(a) Rural-urban fringe is full of huts, slums and squatter settlements and the construction of buildings going on unawaited without any proper plan;
(b) A mixed form of land-use is found here;
(c) It is difficult to continue agricultural land-use on a permanent basis;
(d) The area suffers from urban facilities;
(e) It is difficult to provide facilities at this place due to dispersed location of settlements and huge costs involved; and
(f) In rural-urban fringe the living conditions of both rural and urban areas are found.

M.M.P. Sinha (1980) has highlighted the rural-urban fringe of Patna in detail. He concluded that rural-urban fringe has varying width found as a narrow zone at the rear end of the urban centre. It has been found that the rural scenario dominates the scene towards villages and urban scenario dominates the scene towards urban area.

Causes of the Development of Rural-Urban Fringe

In India the expansion of urban core adds dynamic character in the fringe development, whereas in the western world the expansion of fringe area takes place along the transport routes. In 1978 K.N. Gopi has analyzed the *Process of Urban Fringe Development : A Model Centre*, and he observed the similarities of urban fringe of the west and the east in terms of the availability of race-course, industries, sewerage plant along with the open space and land reserves.

Types and Demarcation

The boundary of rural-urban fringe changes continuously along with the expansion of urban limits. It may be of the two types :

(a) *Primary Urban Fringe:* This area is found towards the urban centre. It has been called by different names by different authors.

Author	*Term*
Andrews	Urban Fringe
Reinemann	Outlying Adjacent Zone
Myres and Beegle	True Fringe
Whitehard	Inner Fringe Belt
M.M.P. Sinha	Sub-urban Fringe.
R.B. Mandal	Sub-urban Zone/Rurban Zone

(b) *Secondary Urban Fringe:* This area is found around the primary urban fringe. It is known as rural fringe, suburban zone, partial zone, rurban fringe and outer fringe.

In 1942 Richard Andrews has classified fringe as :

(a) urban fringe—closer to the town; and
(b) rural-urban fringe—closer to the village.

In 1967 Whiteland has classified fringe as : Inner fringe zone, Middle Fringe zone and Outer Fringe zone.

In 1968, R.J. Pryor has classified fringe as : Urban Fringe and Rural Fringe.

For the delimitation of rural-urban fringe the following measures may be considered :

(a) changes in the land use;
(b) changes in the built-up area;
(c) occupational structure;
(d) house types;
(e) distribution of industrial and non-agricultural activities;
(f) limit of essential services; and
(g) distribution of educational institutions.

M.M.P. Sinha (1978) has considered the following facts for the delimitation of rural-urban fringe :

* Time taken in journey to work;
* Urban habit;
* Land value;
* Public utility services;
* Immigrant population;
* Non-agricultural activities;
* Population density;
* Primary activities;
* Built-up area;
* Age-sex ratio;
* Literacy rate; and
* Agricultural activities.

REFERENCES

Andrews, R.B., "Elements in Urban Fringe Pattern," *The Journal of Land and Public Utility Economics*, Vol. 18, 1942, pp. 167-83.

Martin, W.T., "The Rural-Urban Fringe: A Study of Adjustment to Residence Location," University of Oregon, *Studies in Sociology*, Vol. 1, 1953, p. 3.

Mishra, B.P., "The Concept of Rural-Urban Fringe," *Recent Trends and Concepts in Geography*, (Edt. R.B. Mandal and V.N.P. Sinha), Vol. III, Concept Publishing Company, New Delhi, 1980, pp. 143-56.

Morril, R.L., "Expansion of the Urban Fringe : A Simulation Experiment," *Paper of Regional Science Association*, Vol. 15, 1965, pp. 185-99.

Pryor, R.J., "Definition of the Rural-Urban Fringe," *Social Forces*, Vol. 47, 1968, pp. 202-15.

Singh, R.L., "The Rural-Urban Fringe Studies," *The Geographer*, Vol. 14, Aligarh, 1967, pp. 9-21.

Singh, R.R., *Jaunpur and Its Rural-Urban Fringe : A Study in Settlement Geography*, B.H.U., 1973, pp. 150-60.

Srivastava, B. and Ramchandran, R., "The Rural-Urban Fringe," *The Indian Geographical Journal*, Vol. XLIX, 1974, pp. 1-9.

Whiteland, J.W.R., "Fringe Belt : A Neglected Aspect of Urban Geography," *Transactions of the Institute of British Geographers*, Vol. 41. 1967, p. 223.

Wehrwein, G.S., "The Rural-Urban Fringe, Readings in Urban Geography" (ed. Mayer and Kohn). Central Book Depot, Allahabad, 1967, p. 533.

Young, M., "Some Geographic Features of the Urban Fringe," *The South Eastern Geographer*, Vol. 2, 1962, pp. 1-6.

35

RURAL-URBAN LINKAGES

RURAL AND URBAN are not dichotomy terms rather they are closely linked in terms of town as service centre and village as centre of agricultural production. Nowadays such a dichotomy no longer exists. The social process of urbanization has invaded even the rural areas. Summer villages, week-end cottages and the acceptance of urban values and styles of living by those residing in rural areas have completely relegated the old concept of rural-urban dichotomy.[1] But the built environment changes very slowly in rural areas and the urban areas. We inherited this linkages of our cities and towns from villages in the past, and continuing for the future through rural-urban migration, consumer behaviour patterns, umland structure, hinterland organization and core-periphery models.

Conceptual Framework

Rural and urban living are continuum and with the universal availability of radio, newspaper and automobiles, the two rural and urban settlements have become closely linked rather than a dichotomy.[2] Further, the distinction between the two is not often clear-cut, specially in conditions of housing, facilities of shopping, manufacturing, transport, education and the manner of living. For example, in Bihar Plain, Noorsarai, Silao, Baheri, Kusheswarsthan and Baghopur were all designated as rural settlements in 1991 census but they are little different from the smaller towns in the area. Thus the line drawn between rural and urban for statistical or census purposes is necessarily arbitrary.

The functions of the urban settlements constitute the main unifying bond between the people of the town and the zone of influence.[3] By virtue of such a distinct location and layout, the urban centres form a socio-economic hub for the umland, disseminating cultural influence, and acting as collecting and distributing centres of regional produce.

The commuter relations between the urban places as centres of employment and the rural areas as a residence of the employees is another link between urban centres and their hinterland. A third form of inter-relation is provided by outdoor recreation. This has really established a community of interest among the people of urban centres and its zone of influence.[4]

The functions of rural areas are accommodated in agricultural lands and villages and connected with primary and secondary occupations, whereas urban activities are accommodated in shops, workshops, offices, warehouses and public buildings. The natural link between the two is provided by transport routes. The characteristics of rural settlements, their size, functions, spacing, layout and build vary regionally with the physical environment, density of population and with the character of peoples income and culture.[5]

The close relationship between rural and urban areas has been evinced in the daily quick delivery of perishable goods like milk and vegetable to the town dwellers. Such close tie-up depends upon the facilities of transport. Cities like Patna, Muzaffarpur, Bhagalpur, Gaya, Munger, Biharsharif and Darbhanga in India are easily accessible from surrounding villages which form large city regions. On the other hand, the smaller towns with very poor accessibility have very small hinterlands. For example, Barbigha and Barahiya in Munger district, Murliganj and Birpur in Saharsa district, Chakia in Champaran and Revelganj in Saran district are small towns with very small hinterlands.

Urban centres as regional administrative headquarters constitute another reason for the link between the town and the country.[6] The urban centres have a strong impact upon the rural settlements of the surrounding city region due to the expansion of urban built up land for residence, commerce, industry, transport, recreation and market gardening. Thus they affect both the character and structure of rural settlements, especially placing of rural dwellings and changing dress materials of rural folk. Urban centres have grossly eroded the moral values of rural people.

It is, therefore, apparent that both towns and villages are interdependent and cannot be studied in isolation. It is true that each of them have their own characteristics and to this extent both of them have an independent location and functional structure. But the characteristics such as utility services, trade and commerce, administrative and other nodal functions, etc. have to be examined. These two should be studied, together as part of the same continuum. The best example of this link is provided by the urban centres which include both towns and villages.

It can be readily seen that in all the studies of rural service centres, the towns occupy a higher position than that of the village as centres of cultural activity. In this, the general hierarchy of cultural activity emerges. In this, the general hierarchy of cities, towns, rurban centres and rural settlements has a common application over the whole area. In areas where urban centres are lacking or where smaller towns do not surround the parent city, some villages have emerged as major service centres, such as Koilwar in Bhojpur district, Dalan in Katihar district, Sasuamusa in Saran district and Dhaka in Champaran district. These villages are, therefore, most significant in the study region and have the highest potentiality of population and economic growth due to independent hinterlands of their own.

Urban centres are conspicuous by virtue of their services which they perform for the rural inhabitants. In Bihar plain the centres of the highest ranks are Patna, Muzaffarpur, Gaya, Darbhanga, Arrah, Saharsa, Purnea, Biharsharif, Motihari, Barauni and others to name a few. It is apparent that the hinterlands depend upon these urban centres and, in turn, these centres depend on the surrounding hinterland. In this connection G.K. Mishra[7] while delineating the 'rural-urban continuum' highlighted the spatial interaction at various levels of functional hierarchy of central places in Raichur district, India. He applied the principle of 'minimum aggregate travel distance' for making inter-regional adjustments in identifying the existing central places within the hinterland. He considered all the central places as growth foci which will develop into small towns of Raichur rural-urban continuum, bridging the existing gap between towns and rural settlements in Raichur district.

Review of Past Literature

Cities act as catalysts in introducing new innovations in rural areas, perhaps more slowly but nonetheless inevitably. Many rural settlements at the fringe of urban centres are rural or urban just because the municipal limits does not or does include them.

Besides these in India, the social structure is such that separate grouping of urban centres as contrasted with rural areas is less efficient for the purpose of social analysis. This is because even those economic activities which are normally considered to be urban activities---textile manufacture, metal work, construction, the manufacture of chemicals and fertilizer, electric machinery and supplies, cement wholesale and retail trade, weaving apparel and food processing—have segments of their labour force in rural areas.[8]

The social elements such as the prevalence of joint families have in some cities showed little differentiation from the age old rural tradition, and the prevalence of the same in an unanticipated manner in cities.[9] Moreover, some of the studies of neighbourhoods of Indian cities have found such in common with rural social patterns.[10] As for example, in the entire area around the Asansol industrial complex, villages as well as towns might be considered as one unified urban area as rapid urbanization turned the hinterland of each large factory into dormitory settlements for their workers.[11]

R.E. Pahl[12] (1966) in his book *Urbs in Rure* in London's rural-urban fringe of Herfordshire found that 'a housewife who lives in a rural areas, but who does all her shopping in a neighbouring town, may be participating in urban life as much as another housewife living in a peripheral urban estate'. This sharing of urban provisions—work, services, entertainment and so on—by people living in and believing themselves as part of a village community. Ruth Crichton (1964) in her *Commuters' village* found that 'recent changes in the size and shape of her village have discouraged integrations with the Reading town'.

Rural communities can be isolated as separate systems for the purpose of academic study but this is an increasingly unreal exercise. In fact, many areas commonly referred to as rural is more urban than rural in all the more important aspects which go to make up the way of life of the individuals who comprise its resident population.[13]

Opler[14] in his study of *The Extensions of an Indian Village* pointed out the common origin of villages and towns and descent shared with a cluster of surrounding villages. The village exogamy creating kinship ties with other villages and caste assemblies cutting across village lines; customary work, obligations of practical and ritual services in another rural community; pilgrimage to distant shrines and religious centres in towns. These connections also spread over towns, market towns with courts, land registration and record offices, officials of the government, agricultural—implement stores and temporary out-migration for higher education in urban areas.

Besides, there are some examples that many artisans and small manufacturers of the former villages, are becoming urbanized, and hence the villagers must either go into the towns to get carpentry done or hire urban specialists to come to the village to perform the required job.

D.N. Majumdar reports "that one of the most important continuing contacts with the city is in the marketing of village surpluses like fruits, milk, grain and animal products. For this purpose, each individual own or hire his own bullock cart, or he may deal with one or more of the middlemen who collect the materials in the village and sell it in the city."[15]

The villagers like to sell their product in markets nearest to the city or proceed direct into the city as higher price is available there. The "initial effect of the extension of transportation to the rural areas permit urbanites to set-up rural markets rather than to encourage the villagers to come into the cities."[16]

Identification of Rural-Urban Relations

Rural urban relations may be classified into the following types :

(a) Trade and financial relations;
(b) Agricultural relations;
(c) Industrial relations;
(d) Social and cultural relations;
(e) Educational and health relations;
(f) Commuter's relations;
(g) Transport and communication relations; and
(h) Administrative and political relations.

Trade and Financial Relations

Retail trade, wholesale trade and banking services located in urban centres provide a wide range of services to the surrounding tributary area by selling jwelleries, clothes, medicine, sweets, sport goods and educational items. These services have a great impact in regional economic development.

Agricultural Relations

Near the urban centres the intensity of agricultural activities is generally high, and this tendency is decreasing towards the increasing distance in peripheral areas. Besides this, the trade of agricultural commodities and urban centres as the consumption points of regional produce, contribute a lot in regional economic development.

Industrial Relations

Industries are located in urban centres but they are receiving raw materials and labourers from rural areas. This also leads to develop other infra-structure facilities in the regional economic preferment which has taken place, in the region concerned.

Social and Cultural Relations

Social relations are family relations, marriage relations and cultural includes religions, food habit, copy of some of the way of urban living in rural areas, besides urban centres as the centres of sports, cinema, theatre and museum provide a variety of entertainment to the inhabitants living around them.

Educational and Health Relations

The urban centres provide university level education in different post-graduate departments, training institutes and colleges. The students from rural areas also come to these colleges and institutes located in urban areas. In areas of dense population the rush of students in colleges located in urban areas is too much whereas in sparsely populated rural areas the rush is very poor.

Commuter's Relations

In Jamalpur railway workshop commuters train, carries labourers three times daily from Sultanganj in the east, Kajra in the west and Munger in the north. In India other active areas of commuter's movement can be observed near Delhi, Calcutta, Chennai, Kanpur and Jamshedpur.

Transport and Communication Relations

Urban centre acts as convergence point for different modes of transport routes. Without transport and communication relations with the neighbouring towns and villages one cannot think of urban development because these are the nerve centres from where transportation lines radiate like spoke in different directions and they are lifeline for the movement of people and sustenance of urban scenario.

Administrative and Political Relations

Urban centres act as place of state capital, district headquarter, subdivisional headquarter along with anchal headquarter. On these centres the provision for dakbungalow, the establishment of courts, police station and government offices favour the movement of people each and every day from the demarcated administrative indicator of rural-urban relations within a well defined zone of influence.

The Role of Urbanization in Regional Economic Development

Urbanization is the increase of population in urban areas. In most cases it has been observed that urbanization is a component of regional economic development because urban centres provide, inter-alia, a variety of centralized services for the surrounding zone of influence, e.g., marketing of agricultural surplus, products of cottage industries, the supply of fertilizer, engineering goods, pumping set, medicines and specialized skills in a wide variety of situations which are necessary for regional development.[17]

Friedman's Core-Periphery Model

The core periphery model was initially put forward in a study of Venezuelan regional development policy by Friedman.[18] Development takes place from small number of "centres of change" located at the points of highest potential interaction within a communication field. Innovations diffuse from these centres to areas of lower potential interaction. Such 'core regions' are major centres of innovative change, while all other territory consists of 'peripheral regions', which are dependent on core regions for development. It is happening due to the weakening of the periphery by resource transfer to the core, and besides production increased agglomeration of economies at the core[19] (Fig. 35.1).

Fig. 35.1 shows that the peripheral areas sends labourers and consumers, food and raw materials to the core regions of urban areas whereas core regions supply industrial products, finance and innovative information for economic development to the peripheral regions. Lampard[20] also favours this point that the specialization of a business which generated in the urban centres later diffused to the countryside. Further, Kolb[21] stressed the need for a universalist—achievement of value structure for the rationalization of production and hence industrialization. This is because urban environment provides a necessary conditions for the evolution of such values in terms of rural development. These interactions are found because the economy of rural areas are primitive whereas in core regions it is fully developed due to leadership in business, industrial production, nucleation of transportation lines and political favour.

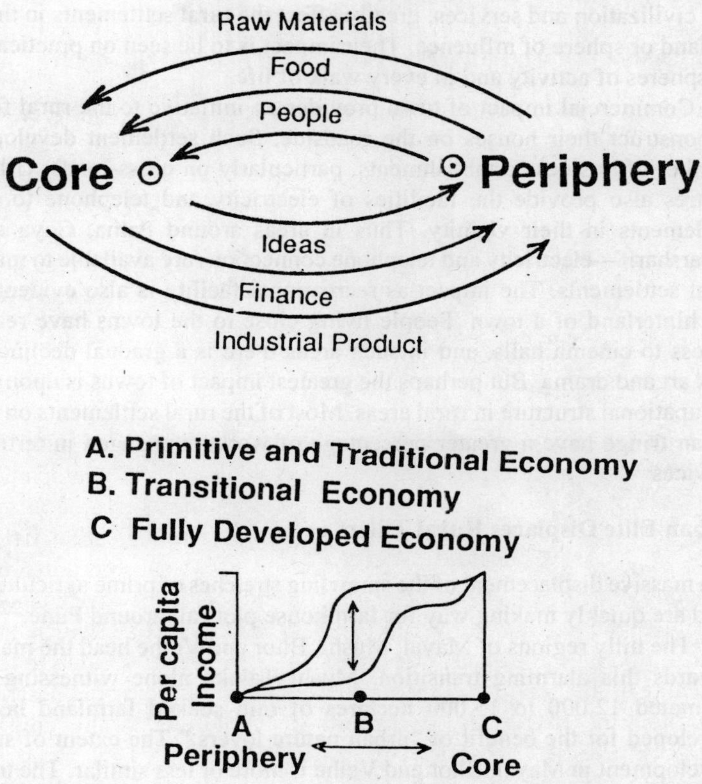

FIG. 35.1 : Core Periphery Model and the Process of Feedback Mecanism

Impact of Urban Centres on Countryside

In the process of urbanization, slow but noticeable changes are taking place in the Bihar Plain. These changes are more in Munger and Patna districts and comparatively less in Darbhanga and Muzaffarpur districts. Existing cities are heading towards a higher level of urbanization. The variety and amount of nodal services are increasing and becoming more complex. The processing and manufacturing activities are being added to rural areas where they had already been in existence. All these changes are not confined to cities only. The impulses which these changes generate are not only restricted to towns only, but carried into the countryside as well. Thus towns and cities as the centres of present day civilization and services, greatly affect the rural settlements in their umland or sphere of influence. Their impact is to be seen on practically all spheres of activity and in every walk of life.

Commercial impact of town provides an initiative to the rural folk to construct their houses on the roadside. Such settlement develop a number of business establishments, particularly on cross-roads. Urban centres also provide the facilities of electricity and telephone to the settlements in their vicinity. Thus in areas around Patna, Gaya and Biharsharif—electricity and telephone connections are available to many rural settlements. The impact as recreational facility is also evident in the hinterland of a town. People living close to the towns have ready access to cinema halls, and in such areas there is a gradual decline of folk art and drama. But perhaps the greatest impact of towns is upon the occupational structure in rural areas. Most of the rural settlements on the urban fringe have a greater percentage of workers engaged in tertiary services.

Urban Elite Displaces Rural Tillers

In a massive displacement of the sprawling stretches of prime agricultural land are quickly making way for farmhouse plots all around Pune.[22]

The hilly regions of Mayal, Mushi, Bhor and Velhe head the march towards this alarming transition. Mushi Taluka alone witnessing an estimated 12,000 to 15,000 hectares of rain soaked farmland being developed for the benefit of "urban nature lovers." The extent of such development in Mayal, Bhor and Velhe is more or less similar. The total number of farmers being affected in the process could easily exceed 1000 at the moment (1998) with the acceleration of the process of urbanization in the periphery of 50 kilometres beyond the civic limits of Pune and Pimpri Chinchwad is the large scale conversion of agricultural land for non-agricultural purposes has been witnessed.[23]

FOOTNOTES

1. J. Eyles, "Inhabitants Image of Highgate Village," Department of Geography, London School of Economics, Discussion paper 15, 1967.
2. M. Aurousseau, "The Distribution of Population," *Geographical Review*, Vol. XI, 1921, p. 567.
3. United Nations Population Studies, No. 8, 1950, p. 2. Quoted by A.E. Smailes, *The Geography of Towns*, London, 1966, pp. 32-33.

4. Hans Carol, "The City Centred Region," *Terra*, Vol. 81, 1969, pp. 117-22.
5. Harold, M. Mayer and Clyde F. Kohn, "Rural-Urban Link," in *Readings in Urban Geography*, Chapter IX, Allahabad, 1967, p. 27.
6. R.E. Dickinson, "The Regional Relations of the City," in *City Region and Regionalism*, p. 167.
7. G.K. Mishra, "Rural-Urban Continuum," *Indian Journal of Regional Science*, Vol. IX, No. 1, 1977, pp. 36-44.
8. R.D. Lambert, "The Impact of Urban Society Upon Village Life," in *India's Urban Future* (ed. Roy Rurner), University of California Press, Berkeley, 1962, p. 118.
9. K.M. Kapadia, "Rural Family, Pattern," *Sociological Bulletin*, Mumbai, Vol. V., 1965, pp. 111-26.
10. A. Bopegamage, "Neighbourhood Relations in Indian Cities Delhi," *Sociological Buttelin*, Mumbai, Vol. VI, No. 1, 1977, pp. 34-42.
11. Vithal Babu, Report on a Preliminary Inquiry on the Growth of Steel Towns in India, UNESCO Research Centre, Calcutta, 1959.
12. R.E. Pahl, "The Rural-Urban Continuum," *Sociologia Ruralis*, Vol. 6, Nos. 3-4, 1966, p. 316.
13. H.E. Bracy, *People and the Countryside*, Routledge & Kegan Paul, London, 1972, p. 139.
14. Morris Opler, "The Extensions of an Indian Village," *Journal of the Asian Studies*, Vol. XVI, November, 1956, pp. 5-10.
15. D.N. Majumdar, *Caste and Communication in an Indian Village*, Asia Publishing House, Mumbai, 1958, p. 329.
16. *Ibid.*
17. B.F. Hoselits, "The Role of Urbanization in Economic Development : Some International Comparisons," in *The City in the Third World*, (Ed. B.J. Dwyer), MacMillan, London, 1974, pp. 169-90.
18. John Friedmann, *Regional Development Policy : A Case Study of Venezuela*, Cambridge, M.I.T. Press, 1966.
19. R.B. Mandal and G.L. Peters (Ed.), *Urbanization and Regional Development*, Concept Publishing Company, New Delhi, 1982, pp. 1-10.
20. E.E. Lampard, "The History of Cities in Economically Advanced Areas," *Economic Development and Cultural Change*, Vol. 3, 1955.
21. W.L. Kolb, "The Social Structure and Functions of Cities," *Economic Development and Cultural Change*, Vol. 3, 1955.
22. Hort, J.F. "Urban Encroachment on Rural Areas," *Geographical Review*, 1976, Vol. 66, pp. 1-17.
23. Agashe, Anand, "Urban Ellite Displaced Rural Tiller," *The Times of India News Service*, March 4, 1993, p. 12.

36

RURAL-URBAN COMPOSITION OF POPULATION IN INDIA

THE MAJORITY of population (74%) in India resides in villages and 26 per cent resides in urban centres. It is so as there is a traditional society in India where there is lesser tendency of movement. Hence, they advanced the slogan of 'son of the soil'. The people resides in urban areas has the main occupation of trade and commerce besides some other non-agricultural occupations, while in rural areas people are mainly farmers and agricultural workers. In urban areas per-capita income is high and each and every person gets some kinds of job and hence, most parts of India observes seasonal migration towards urban areas. In India, the rural population dominates over urban population in terms of its number, areal coverage and political make up of the society.

India is truly a country of villages. According to 1961 census about 82 per cent people resides in villages and only a smaller amount 18 per cent resides in urban areas. According to 1921 census the rural population was 88.8 per cent and the urban population was 11.2 per cent. But after these census decades due to industrialization the urban population is increasing continuously. In 1931 urban population was just 12 per cent, in 1951 it was 17.3 per cent, 1961 it was 18 per cent, in 1981 it was 23.7 per cent and in 1991 it was 26 per cent. This gives us the increasing share of urban population in India in comparison with declining share of rural population (Table 36.1). This happens due to acceleration of rural-urban migration in India.

Factors Affecting Rural-Urban Composition of Population

Urban and rural population are quite different due to their functional specialization. The majority of population who resides in rural areas are engaged in Primary occupation but urban populations are engaged in tertiary occupations. Hence, in villages the main contributing factors are agriculture, rural industry, the works of transportation and running a small shop at few places. But in urban areas industrial development, development of trade and commerce, commencement of roads and new means of transport and the institutional development have highly helped in the enhancement of urban culture[1] (Fig. 36.1).

TABLE 36.1 : Composition of Rural-Urban Population in India

Year	*Total Population in crore*	*Urban Population in crore*	*Rural Population in crore*
1901	23.33	2.6	21.27
1911	23.20	2.6	23.14
1921	25.14	2.8	23.60
1931	27.90	3.4	24.86
1941	31.77	4.6	27.77
1951	36.11	4.8	32.60
1961	43.92	7.9	36.83
1971	54.82	10.9	44.73
1981	68.38	15.6	53.32
1991	84.60	33.0	51.60
2001*	96.00	56.0	40.00
2011*	110.00	82.0	28.00
2021*	125.00	96.5	28.50

Note : * Estimated figures by the author.

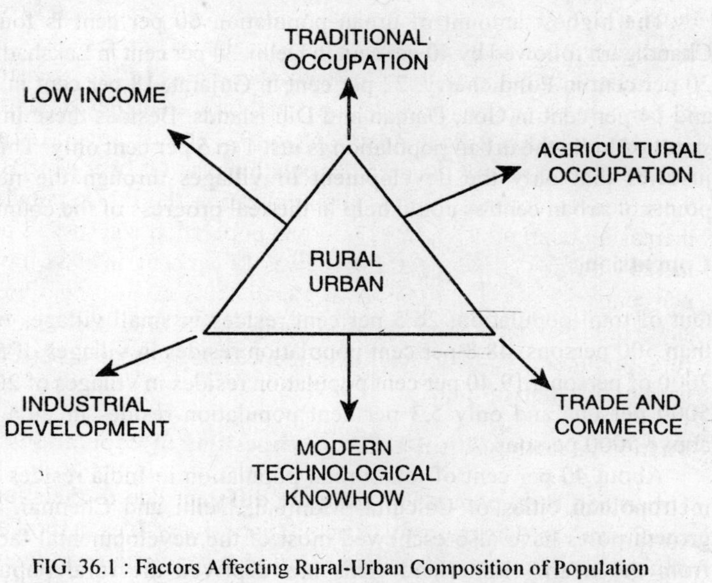

FIG. 36.1 : Factors Affecting Rural-Urban Composition of Population

Distribution of Rural and Urban Population

In India the rural population is much better organized in comparison with any part of the globe. Hence, villages are known as the base of Indian culture. The village is the institution in India which is self-reliant and villagers cooperate with each-other at the time of need. Hence, we can said that "India is Par-excellence a country of villages." Due to the development of big industrial houses in the country, the base of rural industries has been lost to a greater extent. According to 1991 census the rural and urban populations are disproportionately distributed throughout India. In some states of India, e.g., Andman Nicobar Island, Tripura, Nagaland, Mizoram, Meghalaya, Manipur, Madhya Pradesh, Jammu and Kashmir, Himachal Pradesh and Assam about 99 per cent population is rural. On the other hand, the lowest amount of rural population 41 per cent found in Chandigarh a centrally ruled area and this is 60 per cent in New Delhi, 66 per cent in Lakshdweep and 80 per cent in Pondichery. In other states the rural population is 77 per cent in Gujarat and 81 per cent in Bihar (Table 36.2). On the basis of above it could be said that *India lives in her huts and cottages of olden days* and the majority of population resides in rural areas. Only a fragment of total population resides in urban areas as the process of industrialization is slow and the urban development is not well planned.[2]

The highest amount of urban population 60 per cent is found in Chandigarh followed by 40 per cent in Delhi, 34 per cent in Lakshadweep, 20 per cent in Pondicherry, 22 per cent in Gujarat, 18 per cent in Bihar and 14 per cent in Goa, Daman and Diu islands. Besides these in other states of India the urban population is just 1 to 5 per cent only. This also justifies that only the development of villages through the nucleus points of urban centres could help in the real progress of the country.

Conclusion

Out of total population, 26.5 per cent resides in small villages of less than 500 persons, 48.8 per cent population resides in villages of 500 to 2000 of persons, 19.40 per cent population resides in villages of 2000 to 5000 persons and only 5.3 per cent population resides in villages of above 5000 persons.

About 40 per cent of total urban population in India resides in the metropolitan cities of Calcutta, Mumbai, Delhi and Chennai. These growth poles have also eschewed most of the developmental facilities from the nearby hinterland. This has deprived the rural population enbulk to lead a traditional life as they have inherited from their forefathers since millenium.

TABLE 36.2 : Rural-Urban Percentage of Population in India, 1991

	States	*Rural Population in %*	*Urban Population in %*
1.	Andaman & Nicobar Islands	99.83	0.17
2.	Jammu & Kashmir	99.73	0.27
3.	Himachal Pradesh	99.63	0.37
4.	Meghalaya	99.60	0.40
5.	Mizoram	99.50	0.50
6.	Tripura	99.50	0.50
7.	Assam	99.44	0.56
8.	Manipur	99.30	0.70—Q1
9.	Nagaland	99.30	0.70
10.	Madhya Pradesh	99.00	1.00
11.	Rajasthan	98.69	1.32
12.	Andhra Pradesh	98.53	1.47
13.	Dadra & Nagar Haveli	98.57	1.43
14.	Orissa	98.50	1.50
15.	Uttar Pradesh	98.50	1.50
16.	Haryana	98.27	1.73—Q2
17.	Maharashtra	98.00	2.00
18.	Karnataka	98.00	2.00
19.	Punjab	97.60	2.40
20.	West Bengal	97.00	3.00
21.	Kerala	95.40	4.60
22.	Tamilnadu	95.40	4.60
23.	Goa	Q1—95.20	4.79—Q3
24.	Daman & Diu	85.70	14.30
25.	Bihar	81.61	18.39
26.	Pondicherry	79.40	20.30
27.	Gujarat	77.57	22.43
28.	Lakshadweep	65.60	34.40
29.	Delhi	60.00	40.00
30.	Chandigarh	41.00	59.70

FOOTNOTES

1. G.S. Gosal, "The Occupational Structure of India's Rural Population : A Regional Analysis," *The National Geographical Journal of India*, 1959, Vol. 5, p. 137.
2. J.P. Singh, "Some Observations on India's Urbanization : A Third World Perspective," *Urbanization and Regional Development* (Ed. R.B. Mandal and G.L. Peters), Concept Publishing Company, New Delhi, 1982, p. 41.

37

MASTER PLAN OF TOWNS

"MASTER PLAN is a blue print of various proposals that are contemplated to improve the existing conditions of a city and to control its future growth in a coordinated manner." In other words, any scheme of land development in urban area is known as Master Plan. It is a programme of future development of the city. Master plan considers the entire city as an unit. The development of city takes time, and the process of development has been controlled by the Master Plan. Its ideal have definite aims which never fulfilled in true to the sense. Master Plans are generally prepared to reconstruct the old city with planned housing colonies and modify the old areas with advanced service facilities. In India almost all cities which have at the purview of master plan for its modification and future development. Master Plans have prepared for almost all such urban centres. In case one has to develop a new city, the Master Plan also get prepared for such an urban centre, for New Delhi, Gandhi Nagar, Chandigarh, Bhubneshwar, Canbera, Washington are all planned towns prepared on the basis of Master Plans. The Plans of these cities have been prepared by experienced town planners who are well versed in this field.[1]

The Need of Master Plans

1. Before Second World War towns and cities were expanding in a haphazard manner and hence after 1945 it has been realized that there development should be controlled through Master Plans.
2. The development of industries in residential areas of high population density should be shifted to the fringe area of less population density so that they may not harm the health of the people too much by environmental poilution.
3. As the arrangement of utility services was not proper, hence there was a huge rush in the houses.

4. The migration of rural people towards urban areas was going on unawaited and this has favoured the uncontrolled development of cities.
5. In cities there is an utter lack of houses on proper rent.
6. On the side of main thoroughfares of the city there is a huge crowd and the contributing factors were trade and commerce behind the scene.
7. There was acute shortage of school, hospital and social amenities.
8. There is a dirth of open space and hence the air pollution has deteriorated the scene too much.
9. The transport development has virtually crippled the situation.
10. Living standard is very costly in urban areas.

A Case Study of Patna

Physical Characteristics

Patna is a very ancient city. It has grown contiguous to the historic town of Pataliputra, which was built by Ajatsatru, the king of Magadha, in the sixth century B.C.

The city has developed mostly on either side of a main road, the Ashoka Raj Path which runs east to west on a ridge almost parallel to the southern bank of the river Ganges. Along its length, the City now measures about 24 kms. and along its breadth only 3 kms. While the river Ganges always formed a natural barrier for expansion towards the north, any possible southward expansion was also greatly restricted because of the southern areas being liable to floods from the river Poonpoon. This linear character has been responsible for many of the planning problems that confront the city today.

The existing city can be divided into three distinct zones according to their functions and physical characteristics, viz., the Eastern Zone, the Central Zone and the Western Zone. The Eastern Zone is the old city and the ancient capital and is full of derelict buildings huddled along narrow lanes and bye-lanes. The major wholesale markets of the city are located here. The Central Zone is the business and commercial core of the city but also houses the bulk of the institutional, cultural and district administrative buildings. This zone is the most overcrowded and has large pockets of blighted areas. The Western Zone, which comprises mainly of the New Capital built in 1911-12, offers a striking contrast to the other two Zones of the City, in its monumental buildings and a well laid out ring and radial road pattern.

A study of the growth of the city, from historical times till today, indicates that the city has always tended to grow westwards. As a result of this, the Eastern Zone has generally suffered from some amount of neglect and degeneration. There is, therefore, an urgent need to shift the balance of the city towards the east, by locating some of the important functional elements in that area.

Haphazard Development

During the last few decades, considerable building activities have been going on in the city especially in the Western and Central Zones. But in the absence of a Master Plan, most of these developments have taken place in a piece-meal and haphazard manner. For proper growth, it is extremely necessary that all future development in the city should take place according to a proper overall plan for the city.

Population

The population of Patna has been increasing steadily during the last four decades. In 1921 it was 119,975 and in 1961, 364,594 according to the provisional census figures. In other words, the population has risen by about 204 per cent in the last 40 years. In the next 20 years, that is by 1981, the population was 8 lakhs.

TABLE 37.1 : Growth of Population of Patna

Census Year	*Population*	*% Decadal Growth*
1881	170,654	—
1891	165,192	– 3.30
1901	134,785	– 22.50
1911	136,153	+ 1.01
1921	119,976	– 11.90
1931	159,690	+ 33.10
1941	196,415	+ 23.00
1951	283,497	+ 44.32
1961	364,594	+ 28.61
1971	475,300	+ 30.36
1981	811,249	+ 70.68
1991	955,889	+ 17.82
2001	1,200,000	+ 25.53

Seasonal Influx of Population

Patna has a large periodic influx of population, such as during marriage 'seasons', when people from all over the State come here for marketing, or during religious festivals, when people congregate here to have holy dip in the Ganges or pay obeisance to the many shrines and temples of the city. Patna also has an excellent possibility of becoming tourist centre. The existing hotels, *dharamsalas* and rest-houses are too few in numbers to cope with this large volume of seasonal population influx. The city, therefore, needs many more rest-houses, hotels and public conveniences in all parts of the urban area.

Livelihood Pattern

A study of the livelihood pattern of the people indicates that, as compared to other urban centres of India, Patna has a very high percentage of people engaged in agriculture. According to the 1991 census, 16.9 per cent of the total population of Patna were engaged in agricultural operation. A large agricultural community in an urban area creates a perpetual drag on the urban economy. This indeed has been so in the case of Patna. Furthermore, the percentage of people engaged in productions other than agriculture is rather very low. In 1991, this was only 11.1 per cent of the total population of Patna. To promote conditions for a sound economic base for the city, this distribution of population between agricultural production and productions other than agriculture, needs be reversed and re-balanced. Patna has never harboured any big industries, nor there appears to be any immediate potentialities for this in the near future. However, the city has all the necessary prerequisites for the growth and development of the processing industries, which required to be promoted on an extensive scale to increase the employment potentials for both the skilled and unskilled working force.

According to 1991 census, 46.4 per cent of the total population of Patna was engaged in services, 21.3 per cent in commerce and 4.3 per cent in transportation services. These figures are comparable to those of other cities of India. Patna is the seat of the State Government and also the headquarters of the divisional and the district administrations. Government servants, therefore, form the largest proportion of services personnel. The city has many good and old institutions of higher learning, which attract a large number of students from throughout the State. The General Hospital at Patna, with its many specialized agencies, offers medical services that is availed of not only by the people of this city but also by people from the neighbouring areas. These institutions taken together, are the second largest employer of service personnel. The people engaged in business and commerce, most are small traders or employed in medium and small business establishments. Unlike other metropolitan cities, like Calcutta, Mumbai, Chennai, etc. Patna does not have very big commercial houses and employment in such houses are very small. The predominant functional characteristics of the city, therefore, are two-fold, viz., governmental and institutional. The future pattern of the city, as far as possible, has to retain these functional characteristics.

Housing

A survey of the housing conditions, conducted during the 1961 census, indicates that the city has about 52,250 dwelling units. Out of these, about 75 per cent have a very short life left or have no life at all. In the Eastern Zone, nearly 90 per cent of the houses belong to this category. The present housing needs of the City is of about 50,000 dwelling units will be required. To meet the present deficiency and the future requirements, about 5,000 houses would be required to be built every year for the next 20 years. In terms of land area this would necessitate development of 160 hectares of new residential area and renewal of another 40 hectares of existing blighted areas every year.

Slums and Blighted Area

Patna has large pockets of slums and blighted areas in almost all parts of the city. Slums are generally associated with unplanned industrial developments. Patna does not have any large scale industries. The slums of this city have, therefore, generally grown as a result of degeneration of certain areas, partly due to old age and partly due to their incompatibility to keep pace with modern urban developments. To some extent, this process has been accelerated due to people's preference to live in new areas. Slums are bad, both economically and sociologically. Slums that have grown through years and consolidated over generations, cannot perhaps be completely eradicated and brought at par with the new planned developments. But a phased programme of redevelopment for these areas is required to be drawn up and implemented on the principles of what is known as "conservative surgery." This principle implies that all buildings, roads, etc. that can be repaired must be repaired and general improvements in the area should be brought about by a minimum of changes, acquisition and displacement. The principle further implies that all the people who are displaced in the process of redevelopment, should be provided with alternative accommodation in the same area in 'transit camps'. which may later be converted into regular dwelling units. In fine, the 'battle', with slums shall have to be an operation for renewal and not clearance. On the latter premise the 'battle' can never start, for no country can ever afford that.

Density

A study of the ward-wise gross residential densities shows that these vary from 2 to 100 persons per hectare. The lower figure is due to the existence of large pockets of fallow or agricultural land in some of the municipal wards. The higher density is too high indeed, especially when judged in the context of the fact that most buildings in the city are only one or two storeys high. The high density also gives an indication of the intensity of overcrowding. The northern parts of the Central zone, comprising the areas of Sabzibagh, Bakargunj, etc., are most overcrowded. Patna lies in a seismic belt, and it will, therefore, not be very desirable to build tall multistoreyed structures on a large scale. Consequently, it will also not be possible to achieve very high densities for residential developments. Studies on density patterns, with different types and heights of structures, indicate that the city shall have to be developed mostly on the medium density-range with gross density varying between 50 to 75 persons per hectare. This implies re-densification of some of the existing thinly populated areas and decentralization of the overcrowded areas.

Road and Traffic

Patna's worst and perhaps the most apparent problem is traffic and transportation. As mentioned earlier, Patna is a linear city which, over a period of 26 centuries, has grown mostly on either side of a single road (Ashoka Raj Path). This linear character of the City has greatly hampered the development of a proper road network and consequently also of an efficient mass transportation system. This, in turn, has promoted road network and consequently also of an efficient mass transportation system. The preponderance of slow traffic completely paralyses the speed and efficiency of fast traffic and creates tremendous traffic hazards. The problem is further aggravated by the narrow width, improper junctions and defective alignment of most of the existing roads. The Eastern

Railway Main Line, which passes through the City in an east-west direction, has, from the point of commutation, been a great severing agent between the areas lying to the north and south of it. The pressure of development now is mostly in the areas south of the railway line, but there is only one overbridge at Chairayatar serving a stretch of 20 kms. of developed area on its either side. Traffic during peak hours is mostly generated by 'journey to work'. Vast concentrations of residential areas, improperly located in relation to each other and also in relation to centres of trade and employment, create unnecessary congestion on the road which can be avoided by their proper location. The most pressing need for improvement of traffic in Patna is to provide a suitable road network of proper width and alignment which will be consistent with the volume of traffic and would ensure safe and efficient traffic circulation within the city.

Public Open Space

Patna has a very little organized public open space. It has only three major public open spaces, viz., the Harding Park, the Gandhi Maidan and the Mangles Tank. These, along with other smaller parks, comprise about 50 hectares only. On a very modest standard of one hectare per 1,000 persons, the city's present needs of public open spaces work out to 350 hectares. Another 240 hectares will be required for the growing population during the next twenty years. Public open spaces have often been called the 'lungs' of cities. Most cities have some highlights like botanical or zoological garden where people can congregate on weekend or holidays to have a feel of being close to 'nature'. Patna has nothing of this type. Parks and play-fields are required for all age groups and within easy reach of all residential areas. A healthy mind can grow only in a healthy body, and the city should be able to provide the basic groundwork of this, by providing adequate areas for outdoor recreations.

River-Front

Patna is situated on the southern bank of the river Ganges and has 24 kms. of river-front stretching along its entire length. But this unique advantage of the river-front has been wasted and perhaps lost for ever. The best part of the river-front, roughly between the Bankipore Girls' School in the west and Bettiah Girls' School in the east, is so thickly built up, that the presence of the river is hardly felt even a few metres from its bank. Along the rest of the frontage, both towards the west of the Bankipur Girls' School and east of Bettiah Girls' School, the river has a shifting course and until and unless the river is trained, this stretch is difficult to be put to any useful purpose. Nevertheless, the possibility of the river-front being converted into a recreational place the modality of its formation should be explored even if this has to be done on a limited scale.

Bathing Ghat

The river Ganges imparts a holy character to this city. Bathing in the Ganges is a daily ritual with many people. During festivities, a large number of people also come from outside to take dips in the holy river. Unfortunately, the bathing ghats of the city are too few in number and also too inadequate in the conveniences which they provide. More bathing ghats are required to be provided on improved lines and the existing ones are required to be remodelled.

School

There are at present 150 primary, 15 middle and 27 higher secondary schools in the city. The distribution, in respect of numbers inadequate and in respect of locations inequitable. Of the existing schools, about 75 per cent of the primary schools have rented accommodations with no provision of play grounds and other outdoor activities. Most of the secondary schools also do not have proper playfields. More new schools are therefore required with adequate playfields and other amenities and the existing ones are required to be improved upon.

University and Colleges

The University and the Colleges are mostly cramped in one area of the Central Zone. The other institutions of higher learning are required to be provided on a zonal basis. The Eastern Zone and the Western Zone require two more colleges each, at least up to the graduate level. Quite a large percentage of students in the University and the colleges come from outside Patna. Hostel accommodations for these students, which are very scanty, are scattered in different parts of the City, mostly in private buildings. This is not congenial to sound intellectual development. Residential campuses for students should be provided near the university and the colleges.

Hospital

The Patna Medical College Hospital, commonly known as the Patna General Hospital, with over 1,200 beds, offers excellent facilities of specialized treatments, which is availed of not only by the local population but also by people from the neighbouring areas. Besides this, there are also a few other hospitals, notable among which are the Holy Family Hospital at Kurjee in the Western Zone, the City Hospital in the Eastern Zone and Rajendra Nagar hospital in the southern zone. But the City is deficient in the provision of health facilities especially in the lower units of health dispensation, such as dispensaries and health centres. This causes a pressure on the bigger hospitals, which normally should be attending to only specialized treatments. Dispensaries and health centres are required to be provided in every neighbourhood and residential districts.

Government Office

Apart from being the State Capital, Patna is also the administrative headquarters for the division and the district. The Collectorate and the District Offices are located near the river Ganges to the north of Gandhi Maidan in a group of buildings which have now become completely old and inadequate to accommodate the growing needs of these offices. The massive building of the District Court, which is located on the bank of the river Ganges, immediately to the west of the General Hospital, has also similarly become inadequate for its present space left within its own compound. Considering these factors, it appears desirable that the District Court be shifted from its present site and its building be converted into an extension of the Hospital. The present site of the Collectorate and other District Offices is also not very ideal. In view of very old age of the buildings, it would be better if the site is cleared, as far as necessary, and a nucleus for the much needed city-centre established there. By this, a very important part of the river-front can be opened up for use of the public. The Collectorate, the District Courts and the District Offices can all be re-located as a group at one place in a more central location of the city, somewhere near the Eastern Zone to create a focus of employment there.

City's Cattle

The City has a large number of unhygienic and insanitary milch cattle stables which are mostly located in thickly populated areas. The growing cattle congestion, combined with indiscriminate dumping of cattle dung and lack of proper sanitary facilities in the cattle stables, constitute a serious health problem for the city. The cattle population has to be taken out of the city, on its fringes, in properly planned cattle colonies and the keeping of cattle for commercial purposes within the city should be restricted and regulated by the Patna Municipal Corporation.

Utility Services

One of the most important imperatives in the preparation of a City's Master Plan is to provide to its citizens proper conditions for healthful living. Proper drainage and sewerage are the basic pre-requisites for this. Now most parts of Patna has underground drainage system and only a very small part of the Central Zone of the City had night-soil sewers. Topographically, Patna can be called a flat city though the ground generally slopes gently towards the south. But the proper drainage of the City has been made extremely difficult due to two main reasons. Firstly, the city is always threatened by floods, from the north, because the highest flood level of the river Ganga is higher than the average level of the city. Secondly, vast areas south of the by pass road are subjected to annual inundations from the river Poonpoon. But not withstanding these and other difficulties, the city has to be provided with a proper drainage and sewerage system.

Water-supply appears to offer no serious problem for this city. Nature has been specially bountiful to Patna in respect of water resources. Plenty of pure drinking water is available at a sub-soil depth of about 200 metres. The entire water-supply of the city is drawn by tube-wells and works on a decentralized system. This is good, in a way it localizes pollution or breakdown in water-supply due to mechanical failure. However, the present water-supply needs augmentation and the Patna Water Board, which is vested with the statutory responsibility for water-supply in the city, has already taken steps in that direction.

Patna's present demand of electricity (including Danapur and Khagaul) is 13,500 KVA and the supply is about 11,500 KVA. By 2001 the demand is expected to be of the order of 112,750 KVA. The Bihar State Electricity Bard has to make adequate arrangements to meet the present deficiency and the future needs.

Major Proposals

Ultimate Limits of the City

The ultimate extension of the city is proposed up to the ultimate alignment of the Patna Bye-Pass Road in the south and the Patna Canal on the west. The embankment of the Patna Bye-pass Road will serve as a protective bund in the south against the flood waters from the river Poonpoon. The proposed south-ward expansion will help in destroying the present linear character of the city and will facilitate a more balanced distribution of the social service elements and provision of a proper road network.

Within this proposed urban limits, the needs of expansion of the city for the next 20 years is expected to be met. Any further growth shall have to be in the form of satellite towns whose locations may be fixed as and when the needs for these arise.

Green Belt

To define the proposed urban limits and also to restrict unbridled urban sprawl, a green belt is proposed all around the city, excepting along its north, where the river Ganges automatically arrests development in that direction. No building activities is proposed to be permitted in this green belt, but only restricted types of agricultural operations may be allowed.

Industry

About 240 hectares of land has been earmarked in the Western Zone of the City near Digha for small and medium scale industries. Another about 160 hectares have been provided for similar industries in the Eastern Zone, immediately to the south-east of the Patna City Railway Station. These two together will give employment to an estimated 55,000 persons. Sites for processing and service industries are proposed in the Central Zone along the stretch of land between the Eastern Railway Main Line and the Kankarbagh Road. Besides these, it is also proposed to reserve 5 to 10 per cent of land for service and cottage industries in each residential neighbourhood.

Slum Renewal

It is proposed to renew the existing slums and blighted areas of the city in a phased manner involving the minimum possible displacement of people from their places of residence and work. The principle proposed to be adopted for all redevelopment works is one of 'conservative surgery'. As far as practicable, all persons displaced initially in the redevelopment of an area, are proposed to be accommodated in 'transit camps' to be built either in the same area or in its vicinity. Rehabilitation is proposed on a total scale, that is rehabilitation of both the place of residence and place of work.

Residential Density

The average gross density of the new residential area is proposed to be kept at 150 persons per hectare. In some of the existing congested areas, where the density is as high as 500 persons per hectare, it is proposed to reduce the gross density to a maximum of 250 persons per hectare. In thinly populated areas, the density is proposed to be augmented suitably.

Seventeen new residential neighbourhoods covering an area of about 2,000 hectares are proposed to be developed in the city on the general principles of neighbourhood planning. This will meet the city's present housing deficiency as well as its future housing needs. Eight of these new neighbourhoods are proposed in the Western Zone, seven in the Central Zone south of the railway line and two in the Eastern Zone. Some of these neighbourhoods near the proposed industrial estates in the Eastern and the Western Zones, are proposed to be reserved exclusively for housing of Industrial workers. The existing built-up areas of the city have been divided into 27 neighbourhood units, each with population varying from 10,000 to 20,000. Each of these neighbourhood units are proposed to be provided with the minimum, day to day basic needs of the community such as schools, shops, health clinics, etc.

Rest-House and Hotel

To cater to the needs of the large mobile and seasonal population of the city, rest-houses and public conveniences are proposed to be provided in almost every neighbourhood and at all-important places of the city where people congregate in large numbers.

Sites for hotels are proposed to be provided in all the three Zones of the City. Their locations are to be worked out in the detailed plans of the different Zones. One proposal of a hotel that has rentatively been finalized is on the site of the present Dak Bungalow. Another proposal that is now under consideration of the Government, is for a multi-storyed riverside hotel east of the Anugrah Narain Sinha Institute very close to the Sri Krishna Memorial hall.

Roads

It is proposed to provide a road network for Patna somewhat on the 'grid-iron' pattern. This is proposed to comprise of 10 major north-south roads and 7 major east-west roads. Almost all existing roads and road junctions are proposed to be widened and improved upon and some of the roads are proposed to be re-aligned. On all important existing roads, slow and fast traffics are proposed to be segregated by providing separate lanes for these. Separate walk ways for pedestrians are also proposed to be provided on all these roads. For easy commutation between the two parts of the City on the north and the south of the Eastern Railway Main Line, it is proposed to provide five new overbridges and improve the two existing underpasses with the opening of Gandhi. Setu the cities of Patna and Hajipur has been connected with wider roads.

Vehicle Parking

Proper and adequate off-street parking facilities for all types of vehicles are proposed to be provided at suitable places on all important roads. Three sites for parking of public-carriers (trucks) are proposed. First near the proposed Industrial Estate at Digha, the second to the south of the Eastern Railway Main Line near Kankarbagh and the third near Patna City Railway Station east of the Dhawalpura Road.

The central bus station and garage, now located near the Collectorate north of the Gandhi Maidan, is proposed to be shifted to another site south of the Patna Junction Railway Station. The site of the present bus station is proposed to be cleared up and only a bus stand catering to the needs of local routes with queing space and amenities for passengers' waiting, etc., are proposed to be provided in that area. Two similar bus stations are proposed to be provided, on each at Malsalami in the Eastern Zone and another in the Western Zone. Besides, these, properly designed bus shelters are proposed to be provided along all bus routes. At present the place south of Harding Park is the biggest bus stand from where buses originates to different directions.

Public Open Space

Open spaces, parks and playfields for all age-groups are proposed to be provided on the basis of a tiered set-up. Every residential neighbourhood is proposed to have its own toilets, play areas and parks within easy reach of every home. In the next stage of the tiered set-up each of the three Zones of the City is proposed to have zonal parks. The Gandhi Maidan, which is proposed to be linked up with the river-front by clearing up the site now occupied by the Collectorate, District and Police Offices, is also proposed to form as an extension of the city centre proposed in that area. A sports stadium has already been constructed in Rajendra Nagar. Special attractions for children, such as toy-train, children's park, etc., are proposed to be provided at selected places of the City. As a part of this programme a children's toy-train is already being constructed by the Patna Improvement Trust in the Hardinge Park in the Western Zone. A 5 hectare Botanical Garden is proposed in the Western Zone, west of the Canal Road. This will form part of the Patna Improvement Trust's Shrikrishnapuri Scheme.

A 80 hectare Zoological Garden is proposed to be provided near Bulandibagh, in the central Zone, Preliminary works on this have already been taken up in hand by the Patna Urban Development Authority.

River-Front

The Master Plan recognizes the necessity of developing the river-front into parks and promenades, but in view of the practical difficulties and the cost involved to do so, no proposals of any large scale reconstruction of the river-front is being made at this stage. River training shall have to precede any major redevelopment of the river-front. It is, therefore, suggested that as a long-term measure, detailed studies for economic reclamation of at least a part of the river-front may be made. But meanwhile all possible efforts are made from time to time to acquire suitable lands on the river-front for development into parks and riverside promenades. As a beginning, it is proposed to clear up the site of the present Collectorate and District and Police Offices and open up the area to join up the river-front with the Gandhi Maidan.[2]

School

Schools for children for each neighbourhood are proposed to be provided within the neighbourhoods.

University and College

In view of the fact that the University and most of its constituent colleges are situated in a concentrated manner in the Central Zone of the

City, in a strip of land along the north of the Ashoka Raj Path, it is proposed to meet the need of their further expansion also in the same area, so that ultimately the entire area may develop into a consolidated educational campus. There is, however, some obvious difficulty to this as some of the contiguous areas are very thickly built-up and, therefore, it is no immediately possible to meet all the needs of expansion of the University in the northern area alone. Subsidiary campuses located at other places would be necessary, with these considerations the following proposals are made :

(i) All substantial vacant area north of the Ashoka Raj Path, between the Bankipore Girls' School and the Bettiah Girls' School be reserved for educational use only;
(ii) The persmissible use for all further building construction in the areas of Ranighat, Baksariatola, Khan Mirza Mohalla, Sultangunj, Choudharytola, etc. is to be restricted for educational purposes only;
(iii) About 40 hectares of vacant land east of Rajendra Nagar is earmarked for a residential campus for the students of the University and Colleges where the Saidpur Hostel is located;
(iv) About 25 hectares of vacant land contiguous to the present Commerce College in Kankarbagh, is earmarked for a subsidiary educational campus, south of the railway line;
(v) Another subsidiary educational campus is proposed to be located on about 60 hectares of vacant land to the immediate west of the existing Veterinary College; and
(vi) Two new sites for colleges are proposed in the Eastern Zone, one near Maroofgunj, and the other near Mahabir Ghat.

Hospital and Health Facilities

The health facilities in the City, like parks and play-fields, are proposed to be provided on the basis of a tiered set-up. The proposed tiers, reckoning from the bottom are :

(a) Neighbourhood health clinic;
(b) District health centre;
(c) Zonal hospital; and
(d) City hospital.

The locations of neighbourhood health clinics and district health centres are not shown in the Master Plan but are proposed to be worked out in the detailed plans of the neighbourhoods and the residential districts. Proposals for zonal hospitals, are as follows :

(i) One hospital in the Eastern Zone near Gulzarbagh Railway Station on an area of about 25 hectares, a part of this area is also proposed to be used for infectious diseases hospital;
(ii) A 10 hectare at south of the Eastern Railway Main Line near Bulandibagh to serve especially the needs of the proposed new residential neighbourhoods in that area; and
(iii) A 25 hectare site south-west of the Digha Branch Railway Line and north-west of the Secretariat to serve the needs of the entire Western Zone and especially of the proposed new residential neighbourhoods in that area.

It is proposed that the District Courts adjoining the General Hospital be shifted to another site as suggested in a subsequent paragraph, and when this is done the premises of the District Court be utilized for the extension of the General Hospital.

Shopping: Retail and Wholesale

Shopping precincts for the day-to-day needs of the people are proposed to be provided in every neighbourhood. Besides these, three major shopping centres are proposed for the City, one in each Zone. These are :

(i) 20 hectares in the Eastern Zone, in the areas of Maroofgunj and Mansoorgunj, informatory the present wholesale markets located in that area;
(ii) 40 hectares in the Western Zone, south of the Digha Branch Railway Line near the present colony of the Pataliputra Co-operative Housing Society; and
(iii) A shopping centre in the Central Zone, near the Patna Junction Railway Station which will form part of the Patna Improvement Trust's Scheme No. II.

The wholesale markets at Maroofgunj and Mansoorgunj in the Eastern Zone will continue to function as such but it is proposed to redevelop the area and provide proper parking as well as loading and unloading facilities. The present wholesale market in Bakargunj is proposed to be eventually shifted to the proposed shopping centre in the Western Zone. Most of the other shopping areas of the City are proposed to be retained and proper facilities for parking, etc. are proposed to be provided in all these areas.

Re-siting of Collectorate and District Courts

The Collectorate, the District Courts and the District and Divisional Administrative offices are proposed to be shifted to a 50 hectare new site, east of Rajendra Nagar. A separate reservation of about 10 to 20 hectares is proposed to be made in the same area for private housing for lawyers and their clerks and for rest-houses for litigants.

Artificial Lake

The Serpentine Channel which passes through the Western Zone, is proposed to be converted into an artificial lake. The Mangles Tank in the Eastern Zone is proposed to be improved upon and landscaped to form a part of the zonal park proposed in that area. Two other artificial lakes are proposed, one is the proposed Zoological Garden and another in the proposed zonal park in the Western Zone.

Milk Colony

Two milk colonies are proposed to be developed, one in the Eastern Zone and another in the Western Zone. Milk centres are proposed to be provided in all district centres and milk depots in every neighbourhood centre.

Utility Services

The entire city is proposed to be provided with an underground drainage and sewerage system.

Supplies of both water and electricity are proposed to be augmented to meet the present shortage and future demand in respect of both.

Archaeological Monument

All monuments of archaeological and historical interest are proposed to be provided with appropriate physical setting so that these may be properly protected and highlight the urban landscape.

Implementation

The Master Plan, which is prepared for a period of 20 years in the first instance, is proposed to be reviewed, revised and enlarged quinquenially, in the light of subsequent changes that may take place in the city.

It is further suggested that a 'forum' of all the agencies involved in the development of the City be set up at Government level to draw up five-year programmes with definite annual targets for all development works within the framework of the Master Plan. These targets shall have to be based on the combined resources available with the Government and the various agencies involved in the development of the City and also on the policies laid down and allocations made in the Government's Five Year Plans.[3]

When an existing area is redeveloped or a new area developed, certain zoning ordinance would be required to be promulgated to restrict the use of land and the type of structures, that may be permitted in different parts of the area. Such zoning ordinances, which will be framed by the Development Authority, have to be approved and enforced by the Government.

Patna's Beautification Plan

As a promising start to the much-avowed beautification plan of Patna, the foundation stone laying ceremony in Kadam Kuan at an impressive function organized here today by Bhagwan Buddha Smarak Vikash Samiti, the roads are being built from the MP's fund.

The deplorable condition of roads in an old colony like Kadam Kuan was really a matter of concern and added that it was not the case with Kadam Kuan only. Rather development has eluded in most parts of the State even after 52 years of Independence.

Bihar is underdeveloped because we have not solved any problem including waterlogging at Kankarbagh and at several other places.

The roads in Kadam Kuan and West Lohanipur would be rebuilt and the drainage system improved. "But what is more important is a change in the people's attitude. People often praise cities like Bangalore and Mumbai, but they do not realize the reasons behind it. People must change their mentality and understand the importance of civic sense."

Expressing his concern over the pathetic state of affairs in the field of education, how could a man like without a spine can move. "Education is the spine of society." Unfortunately, few people have broken the spine.

> "Development does not mean only road, drainage, water, etc. We should think about real development, emancipation of mind, and if

that is achieved, other things will follow the suit. In Bihar, most of the ills have been due to frequent changes of the Government.

Incidentally, only recently the urban authorities are showing interest in giving a facelift to the city.

The people's cooperation was imperative to give a facelift to Patna. "The problem of waterlogging is acute in Patna and a plan has been chalked out to deal with it. It is a Rs. 14 crore plan. In the first phase, Rs. 7 core will be sanctioned for the installation of 14 pumps and 14 more will be installed in the next phase."

Beautiful and Healthy

There are things like tree planting for instance. It is not necessary for the municipality or the corporation to plant the trees. All that we have to do is to encourage every school and very college which has even a very small plot of land, to plant tree, or to grow vegetables if they have some land of their own. Vegetables can grow, even in a small box. So all the time I want people what new things they can do so that the city is able to meet its need of cleanliness, more beautiful and more healthy. We should have some kind of competition between cities also. They did this in America for many years and we have it for our gardens here. If we can have some competition like this between cities of the same size, we can see who come up with the greatest number of ideas and who can improve the conditions of the destitutes.

NOTES

1. Study of City Master Plans, Uttar Pradesh, *Urban Research Bulletin*, National Institute of Urban Affairs, Vol. 1, No. 7-8, 1978.
2. R.L. Bawa, *Master Plan of Patna*, Government of Bihar, Local-shelf Government, 1954.
3. V.K. Bawa, "Some Organizational Problems of Implementing Metropolitan Master Plans," in *Settlement Systems of India* (ed. M. Alam and K.N. Gopi), Oxford & IBH Publishing Co., New Delhi, 1982, pp. 57-64.

38

STATUS OF TOWN PLANNING

IN TOWN and city planning—the urban problems are searched in terms of traffic congestion, parking problems, land utilization besides the problems related to human habitation, water and electricity supply, transport, communication and garden development and the mounting problems of unemployment. The survey of problems and the search for their solution has found a respectable place in any Town Plan or Master Plans of an area. In this way, the urban planning approached for beautifying the urban environment and making it more attractive for convenient living. The city planning authorities makes a differential design of its growth and development. The plan is always prepared for the known number of population, and the availability of site and resources.

Definitions of Town Planning

Town planning is an art and science of rearranging and beautifying the urban environment. It is concerned with the relationship of man and environment in terms of the influence of each other and guiding the growth of urban centres in a right direction. As a practical science town planning must be considered in close coordination with economics, social science, politics and statistics. As an art it is considered as the probable approach of the future society and directions of differential designs of urban development.

1. According to Thomas Adams—"Town Planning is a science, art and policy which fulfils the social and economic needs of the society for proper development and management of a particular size-class of towns. It considers the building arrangement and architectural style besides the provision of various services and set the direction of migration suited for an urban area. Town planning also considered the urban layout

arrangement, setting the conditions of land utilization, transport management and construction of buildings in a systematic, healthy and efficient manner.

2. Nelson P. Lewis—The Considered town planning as the future estimate of the urban society by a planner in relation to the multi-directional development of the environment and growth of local areas. Such an idea must be supported by the availability of resources, development of industries and commercial establishment.
3. According to Edward M.—The planning of streets, parks, community buildings, proper arrangement of services, besides ports, communication lines, legal consideration and local participation should be searched for proper development.[1]

Reframing Designs of Cities Development

During ancient period the urban areas were planned on certain principles whose size were either square, rectangular and cellular in type. In most of the cases the upper part were pyramidal shape. Such urban centres were found in Mohanjodaro and Harappa in the Indus Valley civilization of Pakistan. The cities planned in 400 B.C., e.g., Olynthus, Parise-milletus, were based on grid iron pattern. The plan of some of the urban centres of Aryan Period could be seen in Fig. 38.1. The detailed regulations for planned and fortified types of settlements are prescribed in the *Mansara Shilpashastra.* According to Havell and R.L. Singh, the Aryans lived in eight types of planned settlement—Dandaka, Sarvatobhadra, Nandyavarta, Padmaka, Swastika, Prastara, Karmuka and Chaturmukha. Location wise these towns are rectangular or square in shape and are similar in essential components. Each town is surrounded by a moat for Jefence purposes. There were generally four gates in the middle of the four sides, thus dividing the village into four quarters. The main gate of the town was occupied by a temple, a tank or a public hall. The four quarters were further sub-divided by straight streets, inhabited on the basis of the Varnas and professions. The best quarters were generally occupied by the Brahmans and the Kshatriyas.

Present Status

Urban planning refers to a science and art of ordering landuse, siting buildings and communication network to secure maximum practicable economic growth, convenience, beauty and safety. This is done by a

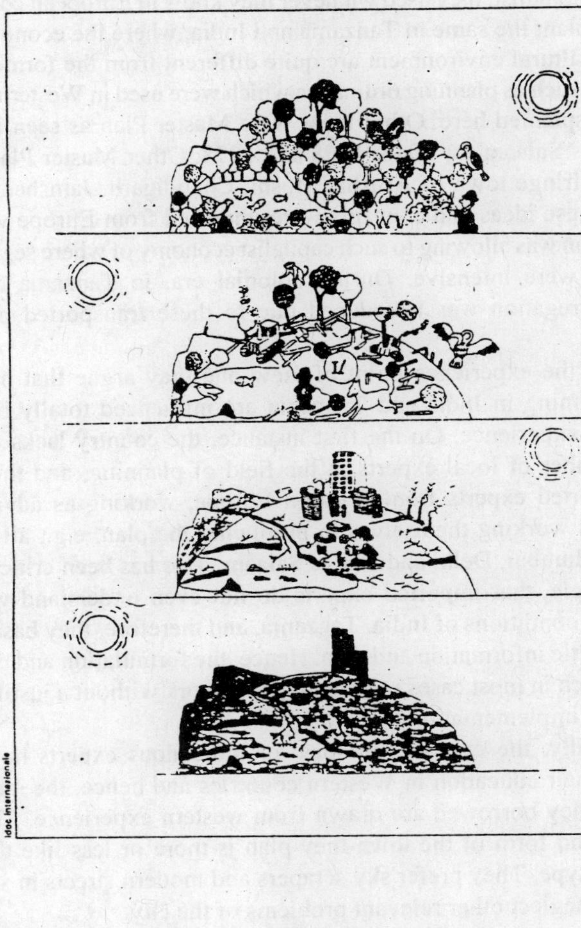

FIG. 38.1 : Planning of Urban Environment

section of people from the local government who draw up and implementing plan and programmes within the municipal boundaries. Urban planning, therefore, is geared to control and protect urban environment, and other related measures to ensure that there is smooth life, security and conveniences within the area.[2]

The influence of foreign planning ideas can be assessed as follows : that since the introduction of urban planning in Tanzania and India by foreign colonialist, they used whatever they know in European conditions and transplant the same in Tanzania and India, where the economic and physico-cultural environment are quite different from the former. Thus concept—such as planning ordinance which were used in Western Europe were transplanted here. Other ideas were Master Plan as seen in Delhi and Dar es Salaam for 1948, 1968 and 1978. Other Master Plans were made for fringe town of Arusha, Moshi, Chandigarh, Jamshedpur and others. These ideas and concepts were imported from Europe whereby the situation was allowing to such capitalist economy of where segregative measures were intensive. During colonial era, in Tanzania also the social segregation was introduced due to these transported planning ideas.[3]

From the experience point of view one may argue that most the urban planning in India and Tanzania are influenced totally by West European experience. On the first instance, the country lacks or have small number of local experts in the field of planning, and therefore, they imported experts from Western Europe, working as advisors or sometimes working themselves in producing the plan, e.g., all Master Plans of Mumbai, Delhi and Dar es Salaam. This has been criticized by many people, that imported experts do not even understand well the prevailing conditions of India, Tanzania, and therefore, they based their ideas on little information and data. Hence, the formulation and drawing a plan which in most cases remain on paper work without a justification for proper implementation.

Secondly, the case of few local or indigenous experts have also received their education in Western countries and hence, the ideas and concepts they borrowed are drawn from western experience. Thus the standard and form of the town they plan is more or less like those of European type. They prefer sky scrapers and modern streets in selected areas and neglect other relevant problems of the city.

Form of the city which follows foreign influence are as follows : neighbourhood unit strategy, green belt reserved, gardening, flaming C, B, D. and other high sounding projects such as sophisticated industries, luxurious air ports, etc. These forms of cities found in Western Europe of which the planners in India as our cities faced with the problems of squatter settlements, unemployment and under employment in unproductive sectors, etc.

Another influence of foreign planning in India includes : the way they form zone in the city. They are always maintaining social inequalities started by colonialists of where residential zoning was their main moto in urban planning. They maintain the zoning of high income, low density and low income/high density, of which the low density receives too much funds for social infrastructure and high density receives minimal funds. Therefore, social inequalities which are found in capitalist countries were borrowed and maintained even after independence and the declaration of building egalitarian society, zoning also is applied to the location of industries like those found in Europe where transportation is a big problem because industries are located far away from the residential areas.[4]

Urban planning must be realistic and practicable following the situation and data analysis of Tanzanian economy, and must be influenced by the needs of the areas. Also planners need to create an atmosphere which allows people's participation, and must make sure that all organs and government parasiticals who are supposed to implement, must be coordinated in all sections of the plan.

Planning is fraught because it is endless and infructuous in most of the cases. Any attempt to plan in the past and their failure has placed tremendous burden of physical and social regeneration upon urban society. Unchallenged application of the principle of laissez-fairs in the building of our cities has created a number of critical problems based upon observation of the physical plan, congestion, industrial dislocation and poverty. Urban authorities have been forced to assure an ever increasing responsibility for the welfare of their citizen, while municipal income has been threatened by decreasing valuations of houses and poor amount of taxes to meet the demands of urban society. Consequently, it is imperative that municipal officials should plan their activities with the objective of eventually resolving the conditions from which their problems emanate. This means that they should exercise foresight in dealing with the specific problems which arise day to day. In doing so they will be in need of the best available assistance from persons who could devote primary energy to study and think about the plan.

In principle, settlement planning has been widely accepted in American cities. The modern urban planning movement has sought to counteract the unsightliness and intolerable living conditions accompanying the rapid industrialization of the past. The ideals and energy of the public, planning consultants, commission members and far sighted public officials who supported and led the city planning movement have made an invaluable condition to the improvement of urban living. Their work provides incentive to several public sectors and has secured the need for planning.

In practice urban planning agencies have fallen far short of their potential usefulness. Two factors have combined to limit severally the work of local planning agencies :

(a) The planning commission should be set-up as an independent unpaid boards of the periphery of the urban administrative structure to prepare plan of the city;
(b) Planning is meant for ameliorating the physical structure of the settlement; and
(c) The second factor is a result of standardization of the plans to include only zoning of streets, parks, transit, transportation and civic amenities.

Out of these six point zoning has been the predominant concern of most city planning organization. These are due to lack of interest and financial support from officials who understands planning agency as a zoning body of public works. Hence, it has a little value in matters of social and financial issues which now occupy much of their attention.

The dimension of urban planning activity has expanded considerably during the last seven-decades (1930-2001) by including social and economic issues as well as in physical development of the community. Such plans are specially related with property inventories, housing surveys and the tax evading studies carried on with the aid of workers from relief rolls. It has been felt that planning should be a staff function attached to the potential extension areas to include all aspects of local government activities. The planning agency should take a comprehensive view of the future of community and should aid directly in integrating the necessary planning work of all departments, e.g., public health, welfare and finance, etc. This, however, does not imply that land planning and public works planning should be abandoned. The plans for physical development should be closely related to the plans for industrial, social and financial development of the urban community.

The judiciary have adopted an increasing liberal attitude towards the exercise of zoning as a part of comprehensive planning. This is very important for the municipal government as to warrant public control of private property under the supervision of police force.

The intermittent planning cannot adequately meet modern planning problems. The system of consultant planning has not been effective for long time. Both civic and public officials have tended to regard the plan prepared in this manner as the end product of the planning process. Planning poses responsibility on officials to formulate public policy. Urban planning gets changed continuously each and every year. It is a continuous process and lassaiz-fare work which will not serve the purpose. The coordination of planning is essential by different planning units of the municipality. In this way, the preparation of Master Plan is a useful device for the broad guidance of the planning agency.

As the independent citizen planning commission is not satisfactorily executing the work of planning for the present. This happens because the majority of city planning agencies are now inactive. Even in the larger cities many of the commissions forbidden their interest to zoning and public works. Planning body constitute the numbers of shop-keepers, architects, engineers, ward commissioners and pleaders. In general, these commission members have a limited social outlook and an inadequate grasp of planning.

Urban planning is one of the staff function and it should be attached to the executive officers of the municipality. The greater integration in administrative responsibility and authority on the hands of the chief has influenced the coordination of urban activities.

Planning agencies have been mostly handicapped by the lack of a full time executive officer. In most cases the consultant played a vital role in the city planning is no longer important because it has been replaced by resident planners and a fulfledged town planning organization. Recently, in most of the developed and developing countries of the world the federal government is spending more in the principal urban planning projects carried on with the assistance of State Government.

For the success of micro-level planning the perfect integration of local plans with the national plan is quite essential. Metropolitan planning is older and has been closer to the city planning movement. Thus any planning, for the city should be more realistic and it should be related with more general plans for the region, state and nation as a whole. Urban plan generates resources for the towns and city whereas the State plan generates resources for the surrounding countryside.

The Agenda of Urban Issues in the World

The urban inhabitants of the World numbered about 25 million in 1800 (or 3 per cent of today population of the world) has reached about 2500 million or 40 per cent in 1990 and it would be 55 per cent in the year 2001. Until the turn of 19th century, only Britain had an urban majority; but now all developed countries do, and a majority of developing countries should have made the same transition by early in the 21st century. The 200 years since 1900 will have encompassed this remarkable world demographic transition.

The speed of urbanization in developing countries today is mediocre level, but the numbers involved is unprecedented. Inhabitants of towns and cities increased from 300 million in 1950 to 1.8 billion of the eighties; if UN projections are right, there could be 4 billion by A.D. 2020. In Sub-Saharan Africa, where the level of urbanization is still low (25 to 27%), expected an spectacular growth in the next thirty years could produce urban population of great size—179 million in Nigeria, 58 million in Zaire, 47 million in Tanzania and 42 million in Kenya.[5]

The patterns of urbanization have completely changed from the past. Many more people are going to live in the larger cities. In the developing world of 1950, there were 31 cities of one million or more people, and five of 4 million or more; by 1985, the equivalent figures were 146 of one million and 28 of 4 million; projected for A.D. 2025 are 486 of one million and 114 of 4 million. In Sub-Saharan Africa, the share of the urban population living in cities of half a million or more people grew from 7 to 41 per cent in the year 1950.

The largest cities are, with a few exceptions, not growing, but their inhabitants are spreading out over much larger metropolitan regions so as population size increases and densities decline. The regions absorb the medium and large manufacturing plants along major highways, leaving small scale industries in the old cities.

The majority of city population now comes from the natural increase of the urbanities, exaggerated because of their youthful stage. However, immigration remains important in China and Sub-Saharan Africa. The experience of migrant is—contrary to popular opinion—not bad. They do improve the standard of living of themselves and their families, and there is no correlation between being a migrant and being unemployed, badly housed due to poverty.

The question is 'Why do cities grow and societies urbanize?" In the process of economic development, the physical concentration of workers makes a key contribution to raise the productivity of modern industry and services—and also of the labourers themselves. The growth of productivity in urban areas is possible only through the concentration of high quality infrastructure, both physical (power, roads, water supply drainage, etc.) and social (educational, health, facilities, etc.) besides an infrastructure which could not be evenly distributed throughout the country due to involvement of high cost.

The relationship between the urbanization and economic growth is very close—i.e., is 0.6557. There are 49 'numbered points' which represent the countries with 10 million or over population.

City growth is part of the creation of new patterns of territorial specialization which is a result of the growth and diversification of output. The poorer a country, the more important this is; the richer it is, the less the difference between town and country and the more workers can afford to live at considerable distances from the place of work.

Cities exist not only to provide a framework for production and population concentration but they play a key role in the transactional systems that span individual countries and the world—inflows of people, of finance, of information, and of commodities, through banking systems, markets and trading networks, through transport terminals (Air, Sea, Road and Rail) and their associated warehousing, through newspaper offices, publishing, radio and television, through telephone and Fax systems, and so on. This is because urban centres are such important junction points. They are inevitably the centres of power and government. For the same reasons they are usually the centres of high level education, research, health and cultural facilities as well.

Once congregated workers and their families form large markets for the produce of agriculture. Hence, urbanization forces the growth of agricultural productivity. In many developing countries of the World, the gap between rural and urban productivity is wide and growing. The World Bank estimates that 60 per cent of the output of developing countries is not produced in urban areas—and 80 per cent of the increase in output cities grow due to growing and changing demand of labour. The result of economic development itself and the changing composition of a growing output of goods and services force the cities to grow. Of course, people often have different motives for moving to cities. Some go for education, jobs, marriage migration while others for medical treatment. But in developing and underdeveloped countries search for work is the key force in urbanization. Cities in general do not grow because rural land is scarce, rural poverty is high or because the population is increasing fast.

The cities also concentrate great wealth which survives most notoriously along with the hiatus poverty. As urbanization takes place, the transfer of poverty from rural to urban areas will also occur. By the end of 20th century, it is expected that half the poor will remain only in urban areas (and half the urban children will be born to poor families) of developing countries. Low salary in urban areas is usually associated with excessive working hours (rather than unemployment), high costs of consumption and appalling standards of housing and services. The combination of poor health and poor diet makes workers exhausted, unable to work at normal speed, and they find it difficult to acquire the skills to make some improvement in the quality and quantity of goods. It is impossible for workers to attain the levels of productivity which command better incomes.

The urban health is better than rural, in poor urban localities is often much worse. It is even worse for the most vulnerable members of poor households—women, children, the elderly and the disabled. Girls born to poor families are likely to start off with a low birth weight due to poor nourishment as they grow and subject to early marriage and multiple births. They are heads of household and obliged to work for their families as well as keep house, the problems get compounded. For their children again of low body weight at birth, again poorly nourished with very poor health care, often obliged to start work at an early age or to run households which their parents can work. It is almost impossible for them to gain the education which might allow them as adults to escape from the vicious circle of poverty.

Poverty is a key issue facing the cities, but most visitors can see a whole agenda of issues. Traffic jams, air pollution, intermittent and contaminated water supply, inadequate waste disposal, poor health facilities and the development of slums and shanty towns. Workers spend an exhausting time in overcrowded buses and trains. Their families live in conditions which are a constant threat to sanitation in urban areas.

Even with these problems, the cities are relatively rich; and the level of employment and income grow. The task for public authorities is to translate this wealth into efficient cities, capable of fueling national development and meeting the needs of the society. The cities may provide the major part of national output—and revenue to the governments—but the public authorities are unwilling to employ these revenues to make cities more functional. Local authorities have quite often robbed of powers and morale by the State authorities and put more burden on municipalities but without financial help and institutional autonomy, any State governments intervene to prevent—electricity, telephone, water supply companies covering their costs, so they cannot invest in timely fashion as the city expands. The deficits in provision grow as the budgetary losses increase. In the case of water, households are then obliged to pay private water sellers for water which is often impure and costs between three and one hundred times the public supply, showing that it is not the poverty of the population which prevents the extension of public supplies to all the inhabitants. In areas of high population density the health dangers are especially great. The Peruvian cholera epidemic of January 1991 shows how potent was Lima's combination of contaminated water supplies, inadequate sewage disposal, and poor hygiene in handling the preparation of food. Nearly 313,000 people had been affected by the late 1991 and 3,288 had died.

The inability of governments to maintain existing public facilities, let alone expand them, has been a great problem, leading to prodigious waste of resources, and ultimately to capital consumption. At this juncture the older cities often lose half their water supply through inadequate maintained pipes; repairing the pipes produces much more water than any new investment. Poor maintenance of roads damages vehicles; the cost of maintaining the road is often much less than the losses from depreciated vehicles. Blocked drains undermine roads, leading ultimately to unpassable. In all places of the urban areas the picture is of waste, of extravagance in the use of resources, of an inability to conserve and economise. Economic crisis in Asian and Sub-Saharan Africa—only aggravates the problem by cuts in both the creation of new infrastructure and the fund for its maintenance. The Governments having such a poor record have ground for being more optimistic about the future requirements of people. The financial squeeze of the 1980s has had perverse effects which forced many governments to reconsider their role, especially centralization of powers and finance, and the stress on the public delivery of services in urban areas. At this point three types of reform are currently being employed. The first is administrative decentralization, with financial hacking, to a strengthened local government. The second is the concentration of the government on facilitating services by private firms, non-government or self-help organizations and households. The third is relaxation in the regulatory regime including the setting up of standards of provision so that services can be provided which could be approachable.

There is already a wealth of experiment which has emerged—the private collection of wastes in Ibadan and provision of public W.Cs in Calcutta, NGOs providing cheap sanitation and sewage services to poor households in Karachi, water supplies in Abidjan, health, training and educational facilities in Karachi. The greening of Nairobi allows the expansion of informal transport to poor areas in Manila. Childrens have been mobilized to give health education to their parents in the squatters of Mumbai to clean up Bangkok and Curitiba.

Due to the agenda and perspective of urbanization has grown more daunting, the change of public attitudes could be the means to engage many more people in the process of copying with the problems of urban growth. In case this happens, the cities can be made to work. They will become more productive, making a greater contribution to national development and the welfare of their inhabitants.

Development of Township by LIC

Since 1972 LIC has also taken up construction of 'Townships on a commercial basis, primarily for its policy-holders'. The houses or flats are sold to policy-holders on reasonable terms with a loan component either directly or through cooperative societies formed by the tenant owners. These projects will serve the needs of middle income and low income group of people.

The first such projects was taken up at Borivli, Mumbai. Here 2,500 tenements are living in a self-sufficient township with all modern amenities. This project estimated to cost about Rs. 6 crores. In 1st and 2nd phases altogether 1845 flats have been completed. Out of these, 1,286 flats have been sold to the policy holders, 135 flats to the employees, 296 flats are allotted as staff quarters and 128 flats are under the process of sale. The 3rd phase of 140 flats is now under construction. Another 500 to 600 flats are expected to be constructed by the time of project is completed. LIC has sanctioned so far a loan of Rs. 2.66 crore to as many as 1,183 policy-holders for the construction of houses at Borivli.[6]

The second township undertaken is at Jeevan Bima Nagar, Bangalore, which consists of 2,500 tenements. Here 468 flats have been completed and offered to policy-holders for sale. Another 323 row houses are nearing completion and 108 houses are under construction.

Similar township projects are being planned at Hyderabad, Ahmedabad, Kanpur, Ludhiana, Patna, Delhi and Mumbai (Bandra). At Delhi 5,000 to 6,00C tenements have been constructed within a period of three years over 40 hectares of plot at Patparganj.

Strategy for Sustainable Development of Urban Centres

Sustainable development means development of the economy of urban areas in such a manner so that urban environment could be protected without harming any sector. It could be managed through calculated growth policy of the economy and the urban society. Although development is an irreversible process and the same state cannot be achieved by means of reverting the process. Sustainable development also implies a future in which life is improved worldwide in urban areas through economic progress, protecting the scenario of local environment and the biosphere besides the mobilization of science to create new opportunities for the progress of urbanities.

Hence, sustainable development is a sure and calculated action not to be started hurriedly but it should be such that our economy could tolerate, technology will cooperate, environment could be protected, people achieve their target and the structure of society and the government may not distort the urban scenario whatever targeted.

The Concept

The concept of sustainability is variable in quality and not always very helpful. This concept was popularized by G.H. Brundtland, Prime Minister of Norway through Brundtland[7] Commission in the year 1987. An elaboration of the concept of sustainable development was made by D.W. Pearce,[8] E. Barbier and A. Markandya in London in the year 1989. The new substitution between resources and modes of capital input can achieve a sustainable outcome. This has been presented first of all by Robert Solow[9] in the year 1974.

There has been a temporal shift in the paradigm of development due to changing contemporary problems and planning methodologies. But, an equitable distribution of benefits of development both in social and geographical space has remained the focus of 'development strategies and models'. Keeping the contemporary problems and the idea of 'equitable distribution' in mind the study of urban geography of India with predominantly low levels of urban development, growth of slums and squatters, failure of electricity and dilapidated roads has been conducted to suggest measures to ensure availability of resources not only to the present generation of urban India but to future generation also. The parameters for this study may be considered as urban housing, slum improvement, improving the conditions of road, etc.

FOOTNOTES

1. A. Rimsha, *Town Planning in Hot Climates*, Mir Publishers, Moscow, 1976.
2. W.H. Clarie (Ed.), *Urban Planning Guide—ASCEM & R*, 49, New York, 1969.
3. J.B. Cullingwoth, *Town and Country Planning in England and Wales*, Allen & Unwin, 1967.
4. I.H. Seeley, *The Planned Expansion of Country Towns*, Godwin, 1969.
5. R.E. Dickinson, "Town Plans of East Anglia," *Geography*, Vol. 19, 1934.
6. R.L. Dwivedi, "Replanning an Existing City," *The National Geographer*, Vol. 5, 1962.
7. G.H. Brundtland, "Sustainable Development" in *World Commission on Environment and Development : Our Common Future*, Oxford University Press, London, 1987.
8. D.W. Pearce; Barbier, E. and Markandya, A., *Sustainable Development, Economics and Environment in the Third World*, Elward Elgar, London, 1989.
9. R. Solow, "The Economics of Resources and the Resources of Economics," *American Economic Review*, May 1984.

39

NATIONAL COMMISSION ON URBANIZATION AND URBAN DEVELOPMENT

THE Ministry of Urban Development has constituted a National Commission on urbanization to examine the status of urbanization in the country. This will identify priority action areas and formulate specific guidelines for an action plan for managing rapid urbanization. The Commission headed by the eminent architect Shri Charles Correa, submitted its report to the Government on 19th August, 1988 :*

(a) With the discussion of a wide spectrum of urban issues arising out of the inter play of factors like rapid population growth, inadequate infrastructure, financial constraints, scarce land resources, etc. The Commission has come up with specific suggestions on a board range of policies necessary to bring about more efficient urban settlements which could generate rapid economic growth with equity and social justice.

(b) The Report has been circulated to all the Union Ministries/ Departments, State Governments/UT Administration, concerned research institutions, professionals, etc. soliciting their comments.

(c) The Report has also been circulated amongst all the Members of Parliament. The Union Minister for Urban Development. Smt. Mohsina Kidwai made a statement on the report in both the Houses of Parliament on 7.12.1988.

(d) The Report was also discussed in the Conference of the State Ministers for Local Government and Urban Development on 14.12.1988. The resolutions that emerged therefrom have been sent to all the concerned agencies for necessary action.

1. Ministry of Urban Development, Report 1988-89, New Delhi, p. 12-14.

(e) A Seminar on Urban Poverty, a component of the Report, was organized in collaboration with UNICEF on 15-16th December, 1988.

(f) A Task Force was constituted to consider convergence of current urbanization programmes. The report is being processed expeditiously for decisions on various recommendations.

Terms of Reference of the National Commission on Urbanization

(i) To examine the status of urbanization in the country with reference to the present demographic, economic, infrastructural, environmental, physical, shelter, energy, communication, land, poverty, aesthetics and cultural aspects;

(ii) To identify priority action areas, make projections of future needs and estimate available resources;

(iii) To formulate and recommend basic guidelines for the specific action plan in each of the identified priority action areas;

(iv) To evolve and recommend policy frames and suggest basic approaches for the encouragement of manageable urbanization;

(v) To recommend an institutional framework for monitoring the effective implementation of the Commission's recommendations; and

(vi) To consider any other matter having a bearing on urbanization and make suitable recommendations thereon.

Salient Features of the Report of the National Commission on Urbanization

1. After discussing a wide spectrum of urban issues arising out of the inter-play of factors like rapid population growth, inadequate infrastructure, financial constraints, scarce land resources, etc. the Commission has come up with specific suggestions on a broad range of policy interventions necessary to bring about more efficient urban settlements which could generate rapid economic growth with equity and social justice. The strategy that has been proposed by the Commission involves the dispersal of urban population among a larger number of settlements by :

(i) Consolidating the economic base of cities and towns which have a potential for generating high rates of economic growth;

(ii) Accelerating the development of predominantly rural districts to arrest the relentless outmigration from the rural hinterlands to larger cities;

(iii) Giving an impetus by way of larger investment for the upgradation of infrastructure to sustain the growth of urban areas; and

(iv) Devising population control measures to stabilize the urban situation.

2. Keeping in line with the above strategy, the Commission has identified 329 urban centres whose economic and physical bases are to be consolidated, strengthened and expanded. These are described as GEMs (Generators of Economic Momentums), which fall into 49 Spatial Priority Urbanization Regions (SPURs). The Commission is of the opinion that for urbanization to play a major role in the development process, the Planning Commission and the State Governments should view urbanization in the total developmental context and allocate resources which will ensure optimum utilization of the natural and human resources within each of these SPURs.

3. To accelerate urban development, the Commission has recommended that the current share of about 4 per cent of the total plan allocation for the urban sector in the Five Year Plan be raised to 8 per cent. Half of this allocation should come from the Central Sector. To ensure predictable and adequate devolution of funds from the State Governments to local bodies, a Constitutional amendment is proposed for setting up quinquennial State Finance Commissions.

4. In the Indian context, the preponderance of the poor should be taken as the principal concern of urban planning. The physical planning of our cities must therefore, change dramatically to achieve the sensitivity to the requirements of low income groups and informal sector for shelter, employment, basic services and finance. To ameliorate urban poverty, self-employment of urban poor must be encouraged by an appropriate credit support programme supervised by an Urban Small Business Development Bank. The Commission has recommended further that four lakh urban youths should be selected from poor urban households every year and trained in various skills so that they can get gainful employment.

5. The Commission has recommended that the housing policy must aim at increasing the supply of serviced land and low-cost shelter, improving and upgrading slums and conserving the existing housing

stock. Apart from providing access to land, the housing programme must also provide for finance, infrastructure development and community facilities. Public agencies in the housing sector should be restructured for fulfilment of their new role as facilitators rather than builders of housing.

6. The Commission accords high priority to the supply of developed land in the urbanization process. State intervention in the urban land market is imperative for ensuring equitable access to land for the urban poor and to recycle existing land efficiently. It is felt that the Urban Land (Ceiling and Regulation) Act, 1976 has failed not only to transfer significant amount of vacant land to State agencies, but has led to an unwarranted increase in land prices. The Commission feels that the Act be amended radically and supplemented by taxation measures to encourage owners to develop the land for housing for lower and middle income groups and bring a larger supply of land into market.

7. The Commission refers to the deleterious effects of existing Rent Control legislations and recommends that Rent Acts should be modified to limit tenancy protection to the poor and the existing tenancies and to provide for annual revision of rent to reflect increases in the cost of living, the increases varying between residential and non-residential premises and houses above and below 80 square metres.

8. The Commission believes that India's architectural heritage is truly extraordinary and our towns and cities present unique characteristics worthy of preserving. The Commission, therefore, suggests that conservation should go beyond preservation of monuments and encompass the whole built heritage. Direct fiscal and other incentives should be offered as an encouragement to individuals to conserve places and sites.

9. For the efficient management of urban centres, the Commission has proposed a two-tier administrative set-up consisting of city corporation and local councils for cities with over 5 lakh population. The Commission suggests that the holding of elections for reconstituting a superseded local body within the specified period should be made mandatory and municipal electoral process brought under the umbrella of the State Chief Electoral Officer. It has proposed steps to improve the financial and managerial capacity of local bodies, besides larger access to budgetary and institutional resources.

10. The Commission has also recommended that for the formulation of future urbanization policies, there should be a National Urbanization Council at the national level with a counterpart of State Urbanization Council in each State.

Urban Development

Urban Development is a State subject and the Central Government performs an advisory and coordinating role besides providing technical and financial assistance for promoting orderly urbanization.

Centrally Sponsored Scheme for Integrated Development of Small and Medium Towns

In order to check the migration of the rural population to major metropolitan centres and also to diffuse the economic growth patterns so as to provide reasonable living standards in small and medium towns, the Government of India sponsored a scheme for the Integrated Development of Small and Medium Towns with a population of less than one lakh which has been in operation since December, 1979. Central assistance on matching basis is provided for works like land acquisition and development for residential schemes, traffic and transportation, markets and *mandis,* low cost sanitation, etc. in the selected towns. At present, each selected town under the scheme is eligible for an assistance on matching basis upto a maximum of Rs. 52 lakhs including a compulsory component of Rs. 6 lakhs on low cost sanitation. Besides, additional assistance for low cost sanitation to the extent of Rs. 8 lakhs will be available on matching basis, if opted for. The State Governments are also required to include components such as slum improvement/upgradation, low cost sanitation, preventive medical facilities, health care, parks and playgrounds.

Till the end of the Sixth Plan 235 towns benefited under the Scheme and a sum of Rs. 63.57 crores was provided as Central assistance. The Scheme has been continued during the Seventh Plan with an outlay of Rs. 88.00 crores part of which is to be utilized for funding the on-going schemes and the remaining amount is to be utilized for schemes to be taken up in 102 additional towns in various States/UTs. About 107 towns have been brought under the purview of the scheme and Rs. 63.49 crores have been made available as Central assistance to the States/Uts till 29.3.1989, including Rs. 15.84 crores released during 1988-89.

Environmental Improvement of Urban Slums

This State sector scheme, which is part of the Minimum Need Programme, envisages provision of basic amenities like water supply, sewers, storm water, drains, community baths and latrines, widening and paving of existing lanes and pathways, street lighting, etc. The Scheme has also been included in the latest 20-Point Economic Programme.

1. In the Sixth Plan, against a target of 10 million slum dwellers, 9.45 million slum dwellers were covered under the scheme. The Seventh Plan aims at covering 9 million slum dwellers at an estimated cost of Rs. 269.55 crores. A total of 5.73 million slum dwellers have benefited during the first three years of the Seventh Plan. During 1988-89 (upto February, 1989), benefit of the Scheme has been extended to 1.74 million slum dwellers as against the annual target of 1.48 million.
2. The scheme is monitored at the Central level through the Town and Country Planning Organization, a subordinate office of this Ministry.

Urban Transport

The concept of Urban Transport Consortium Fund has been approved in principle. An amount of Rs. 2 crores was provided in the Annual Plan 1988-89 for this purpose. Rail India Technical Economic Services (RITES), a public sector undertaking, has been entrusted with the task of preparing a detailed feasibility report for a Mass Rapid Transit System in Delhi and for taking up Traffic and Transport studies in Chennai and Bangalore.

Urban Ceiling Act

The Urban Land (Ceiling and Regulation) Act, 1976 came into force on 17th February 1976, in the States of Andhra Pradesh, Gujarat, Haryana, Himachal Pradesh, Karnataka, Maharashtra, Orissa, Punjab, Tripura, Uttar Pradesh and West Bengal and in all the Union Territories. The Act was subsequently adopted by the States of Assam, Bihar, Madhya Pradesh, Manipur, Meghalaya and Rajasthan. During the year 25,493.52 hectares of vacant land was acquired by various State Governments/ UTs. The State Government/UTs exempted 50,531.31 hectares of land from acquisition under Section 20 of the Act. During the year 2,396 schemes were approved by State Governments/UTs for the construction of 3,19,057 dwelling units. This involved land to the extent of 4,477.47 hectares.

Local Self-Government

The 24th Meeting of the Central Council for Local Government and Urban Development and the Executive Committee of the All-India Council of Mayors was organized by this Ministry in New Delhi on 14th February, 1989. The meeting was attended by the State Ministers for

Housing and Urban Development, Secretaries in-charge of the Housing and Urban Development and Municipal Administration Departments and the Mayors/Commissioners of the Municipal Corporations. The recommendations of the National Commission of Urbanization and the problems and constraints experienced by the local bodies were discussed in the meeting and seven resolutions were adopted.

1. The Ministry continued to assist the 3 Regional Centres for Urban and Environmental studies at Mumbai, Lucknow and Hyderabad in addition to the Centre for Urban studies, I.I.P.A., New Delhi, by giving grant-in-aid from the non-Plan and Plan funds. This Ministry is keen to ensure that the training needs of the Municipal employees all over the country are adequately met. For this purpose, a proposal to set-up an additional centre is under consideration of this Ministry. The regional centres organize training programmes in the field of Housing, Municipal Administration and Urban Development. The Centre also conduct specialized courses and workshops for the personnel engaged in the Centrally Sponsored Schemes of UBS and IDSMT. During the current financial year Rs. 45.60 lakhs have been provided for the purpose.
2. The Regional Centre at Hyderabad organized an International Training course on 'Planning and Management of Urban Development for overseas Administrators' from the 16th January to 26th February, 1989. Financial assistance is given for this course by the Ministry of Finance and External Affairs.
3. The Centrally sponsored scheme of Urban Basic Services, which was launched in the country w.e.f. 22.8.1986, will cover 200 towns in 36 selected districts for their development. The scheme is mainly designed to enhance the survival and development of children and women of urban low-income families in selected least-developed districts. The cost of the scheme is to be shared by the UNICEF, the State/UTs Government and the Central Government in the ratio 40 : 40 : 20 respectively. During the current financial year a provision of Rs. 100 lakh has been made for the scheme.

40

POLICIES FOR URBAN DEVELOPMENT

URBAN Oriented Policies (i.e., generally one-sided policies directed at the largest city or cities) have tended to dominate urban development in many LDCs. However, it may be suggested that many of these measures have not really been coherent polices at all but rather spontaneous or adhoc campaigns applied arbitrarily in the hope of achieving quick solutions to the mounting problems of our major cities.

Many of these Urban Oriented Policies have consisted of negative measures, either attempts to control immigration or to encourage migrants or other unwanted town dwellers to return to the countryside, i.e., various forms of influx controls and back to the land 'movements'. Some of the common problems are as follows :

At administrative level the policies required to effectively check and stem the inward flow of migrants which is very demanding and often beyond the capacities of implementing authorities. Apart from monitoring the arrival of migrants at the city's points of entry, resort has often been made to arbitrary 'roundup' which usually succeed in capturing a mere handful of those sought. Repatriation of unwanted migrants presents further problems in particular, finding an effective means of ensuring that they would not return home due to corruption and tax evasion.

Political

Negative policy operations are often arbitrary, inhuman and repressive whether the same may be clearance of slums and squatter areas or round ups of unemployed. Except in authoritarian states as South Africa, the extreme unpopularity of such measures often makes them politically unfeasible and the strong political will needed to deny poor rural dwellers access to the towns and poor urban dwellers the right to remain in town is often lacking over a sustained period.[1]

Consistency and Continuity

Partly for the reasons cited above, negative urban control measures are often put forward on the basis of campaigns rather than as part of a wider planned programme. A feature of this policy-making style is that while hectic activity might be a feature of early stages of the campaign, the momentum quickly diminishes and the application of measures relaxed so that there is no consistent commitment to such policies. After initial enthusiasm, policy application often becomes temporary and half-hearted. This may be revived at a later date. Stop-go co-ordination—Perhaps the main inherent failing is that urban—directed policies attack the problem at the receiving end rather than at the source. Most of these types of policies attempt to keep newcomers out of the capital as such urban dwellers may deemed a problem, without asking why these people keep flooding in and resort to irregular housing and employment. They also ignore the fact that, as Third World countries have reached a considerable size, a significant proportion of their growth is increasingly due to natural increase among existing city dwellers. Unless other positive policy measures are implemented which would improve the situation in source areas or alternative destinations, then coercive controls focussed on the major cities are unlikely to be effective in themselves. Urban policies in LDCs are likely to be more effective and socially beneficial if their focus is on managing rapid urban growth or increasing the attractiveness of secondary and lower order urban centres.

Three reasons are presented for the frequent failure of NUDP in general—not simply urban—oriented policies discussed above :

(a) Lack of high level political commitment to a better distribution of economic activity;
(b) Exclusive concentration on problems of urban decentralization, ignoring the fact that other national economic policies provide strong implicit incentives to locate in urban areas; and
(c) Stop-go application of policy instruments according to short term political and economic circumstances, e.g., business investment requires long-term planning. B.J.L. Berry stated that most urban policy is unconscious, partial, uncoordinated and negative.[2]

National Urban Development Policy (NUDP)

The formulation of Appropriate Policy : Appropriate NUDPs would eliminate the unintended and unwanted spatial effects of anti-national economic policies and would promote more efficient internal management of cities and improve national economic efficiency. Appropriate management of all cities is very important to the success of a decentralization policy. Planning for any area cannot be done effectively in isolation. In case of large cities, policies to stop population growth are not a substitute for policies that directly address the problems of congestion, pollution and the inadequate provision of services for large segments of the population. If cities other than the capital are not efficiently managed, their chances of attracting industries and deflecting rural-urban migrants as well as city migrants away from the capital region are very limited.[3]

In this context the recently enacted Urban Land (Ceiling and Regulation) Act, 1976 is important in that it helps to check racketeering and profiteering in land and sencering equitable distribution. The States where the Act is in force are already examining the surplus land that would be available and which could be allotted to the weaker sections of the society. With the enactment of this legislation it should be possible for the State Government to tackle the problem of slum and squatter settlements. To some extent if we would plan the development of these surplus lands on a long range basis and channelize the newly arriving settlers, it would help surely urban development.

The above solutions are not adequate. They help only in meeting the existing needs. But the problem of the urban poor will continue to be acute unless solutions can be found to reduce migration into urban areas, especially to metropolitan cities. It is here that a whole gamut of policy issues are involved. Urbanization as a process is unavoidable but it can be channelized.

While urbanization may be a necessary commitment of economic development, improving the rural situation and encouraging proper growth centre is also necessary for distributing equitably the fruits of development amongst the urban and rural sections of the population. The main emphasis, therefore, in economic planning will be to arrange distribution of economic activity specially, according to well chalked out regional plans and to encourage the growth of smaller towns and cities to deflect the migration of population to bigger urban areas and metropolitan cities. The aim is to reduce the complex problems of huge metropolitan cities to the extent possible and at the same to encourage growth of rural population. A proper urban development strategy is proposed to be evolved for this purpose.

The Central Government, during the Fifth Five Year Plan, has evolved a scheme for giving financial assistance to metropolitan cities and areas of national importance for development projects. The assistance from the Central Government supplements the financial resources raised by the state Governments. The draft Fifth Plan makes a provision of Rs. 250 crores for this purpose.

To limit population concentration in large centres decentralization measures must encourage the growth of farm sector as well as the growth of intermediate urban centres. Decentralization must be based on polices favourable to the farm sector because the stagnation of agriculture would slow the growth of provincial rural centres and small towns and would speed concentration in large cities. Decentralization policies also encourage the development of medium size urban centres with better growth potential because economic goods and services, financial flows and innovations circulate throughout the country via the system of cities. Secondary urban centres could be strengthened through appropriate transport policies.

At the time of preparing urban housing scheme, an urban environment should be created that will satisfy the utilitarian socio-cultural and aesthetic functions of urban society. These enter-related functions must find proper expression in the ultimate urban design of residential areas irrespective of caste and income groups that they will house.

The utilitarian function implies that the urban environment is a tool in the realization of human existence in the form of housing including the places for social and cultural facilities, recreation, circulation and the like. However, the effectiveness of environment will not be attained solely by the presence of such elements as dwelling units, roads, shops, schools, etc. but by the manner in which they are situated and by the layout of the surroundings.

The socio-cultural function of the urban environment lies in how the environment reflects the social and cultural needs of the residents. The urban environment must not only serve an utilitarian function, but it must also give expression to human existence at a reasonable socio-cultural level.[4]

The aesthetic function of urban environment relates to visual expression of the urban form as directed to form colour, mass and space, etc. The aesthetic experience of forms of housing must lead on one hand to the appreciation of the totality, namely the spatial composition, the relationship with the environment, the degree of openness or intimacy, etc. and on the other hand the components of which the overall residential complex is built up, namely the relation between high rise and low buildings the street designs, etc.

The ultimate urban design must relate to the totality of a settlement in relation to the city as a whole. The character of the urban environment as a whole will be determined by the choice of the form of housing that will be developed for adoption in the scheme. The form of housing plays an important part in urban design. For instance the use of low, medium and high rise residential buildings and the variety of ways in which they can be applied should express the dwelling requirements of the income groups to be housed in a form which not only fulfil the utilitarian requirements, but it also give expression to the contemporary social conditions and succeed in arousing aesthetic interest among the residents.

A careful assessment of the requirements—physical, functional and social, etc. from which the urban design should emerge is to be undertaken, rather than an attempt to apply what has been done elsewhere. However, it will be necessary to consider studies done elsewhere also so that the methodology and techniques employed on acquiring information and interpretation and translating information into required architecture could be evaluated for adoption in future as well.

Policies of Urban Planning in India

In India, in the First Five Year Plan, the problem of urbanization was identified in terms of increasing housing shortage due to heavy influx of population from rural to urban areas. The Second Five Year Plan marked the expansion of housing programme of the First Five Year Plan, besides slum clearance, slum improvement, land acquisition and development. During Third Five Year Plan it was recognized that 'housing policies need to be set in the larger context of economic development and industrialization. In this Plan, it was decided to prepare Master Plans and Regional Development Plans especially for the metropolitan and industrial cities. The Third Five Year Plan widened the horizon for urban development planning. In this Plan urbanization was recognized as an important aspect of socio-economic development besides stressing the development of urban services and amenities.

The most important policies for urban development are :

(a) New industries to be established away from large and congested cities;
(b) The rural and urban components in the community development projects should be knit together to strengthen the economic independence between the town and adjoining rural areas. The

Fourth Five Year Plan considered urban planning by the expansion of urban community development. During Fourth Plan the urban planners were considered about spatial implications and the problems of urbanization. This also concerns with the regional urban plans and regional extension of Master Plans. This Plan concerns about the operational strategy for urban development which may be a part of overall regional development. The Fifth Plan concerns about the growth of industrialization and urbanization and vice versa. During Sixth, Seventh and Eighth plans checking urban population growth, slum clearance, meeting the shortage of housing needs, creation of townships, colonies and housing apartment were planned to meet the urban crisis.

Urban Development Policies : Some Conclusions

With the increasing urbanization in the World the cities play a crucial role in national development, and spearhead economic growth. In the past, much has been talked about the problems caused by city development, i.e., the congestion, the plight of the urban poor, the degradation of the urban poor and the environment. The question is whether it is possible for the governments to bring about changes in the ways in which urbanization is taking place; changes of a kind that will make cities more efficient and a satisfactory place to live in.

The intervening in this process at the national level has recently been considered seriously in most countries of the World. Although economic planning and management has long been undertaken by the government at the central level but the planning and management of cities has usually taken place at the local level, with little inter-relationship between the two processes. It is now seen that national policies can have a profound effect on the ways in which cities grow. As for example, to encourage the growth of manufacturing, the export often determines the type, size and location of industries, and this in turn may emphasize the importance of some cities rather than others.

Recently, a number of countries have attempted to fill this policy gap by undertaking comprehensive national strategy of urban development. The purpose of such studies have broadly been threefold: to analyze the dynamics of the urban process in their respective countries, to identify the problem that has caused, and to devise policies and programmes which render more efficient urban areas with high growth potential, and ameliorate the plight of those places where the speed of urban growth has caused environmental degradation.

Now it is clear that an increasing awareness of the policies and programmes operate at two distinct levels. National governments can lay down a broad range of policy in areas of economic development, and make decisions on the location of major infrastructure investments, i.e., for roads, large scale industry, water supply, electricity and can legislate the basis upon which State Government can proceed. Most of the development that takes place within cities themselves, it has now recognized that the best arbiters for the choice of projects and for the priorities to be assigned to them, are the municipalities and corporations rather than the State Departments of the Government such as ministries of public works, education and health. Recently, the Health Department decides where and what to build, and when to built it, with the minimum of consultation of local dwellers. For the success of such a policy great effort is needed to upgrade local self-government so that they possess the skills and management capabilities required to plan and implement such programmes. The role of NGO's is also likely to be most helpful in assisting the people to articulate their needs and present them to the local administrative matters.

The purpose of national urban policy is to create a flexible framework within which cities can prosper and grow, for the economic and social well-being of the country as a whole. However, nowadays the 'megacity'—with more than 8 million people—in most of them the population is growing fast, and absorbing resources even to the urban based and this has deteriorated the situation. On the globe it appears that there is a definite correlation between the proportion of a country's population living in urban areas and its per capita Gross National Product. The urbanized areas are richer in comparison with others. The megacities do contribute greatly to the economic growth of the countries within which they are located. In Bangkok less than 10 per cent people of Thailand's resides and this contributes 40 per cent of the country's G.N.P. These days the recommendations tend to emphasize the need to facilitate the working of these cities, rather than attempt to migrate people out of urban places.

The image of these mega cities is most frequently conveyed in its vast squatter areas that are home for people who are destitute, and often ill-housed and ill-provided. Recent studies show that there is a greater capability to generate wealth, than had previously been possible. In urban places the basic lacunae is infrastructure, water supply, access paths, drainage channels, sewerage, garbage collection, etc. as well as clinics for the people, schools for the children and whereever such facilities are provided, the people have shown themselves capable of self-regeneration and tries to introduce modernity and novelty in the fields of economic development.

Urban policy concentrating the need to target resources into those urban areas where there is greatest potential for growth and need relief from chronic deprivation. Some of the policies lie behind the programmes of secondary city development which are actively supported by the World Bank and the regional development banks. Furthermore, by upgrading, secondary cities, there is also the strong possibility that they may become more attractive to the private investor who overwhelmingly prefers to locate in the cities of capital status. Some of these are basic issues of urban development in Africa and India where the pace of urban growth is at present highest in the world. In 1970 only about 16 per cent of Africa's and 18 per cent of India's people lived in urban areas. The United Nation estimates that this will grow to almost 30 per cent by the year 2000. Although at present the only African 'megacity' is Cairo but in India Mumbia, Calcutta and Delhi come under this category. The indications are that centripetal forces are causing African and Indian people to congregate more and more in major cities, and many capitals will have populations of two million or more by the end of this century. It is, therefore, a task for India and Africa at large to review the whole process of urbanization within their respective countries.

Some of these countries appears to have undertaken strategy at national level for preparing physical plans often with the assistance of the UNDP and UNCHS (United Nations Centre for Human Settlements). Such plans have objectives of the alleviation of poverty and the encouragement of local participation as in Kenya, Tanzania and India to encourage the expansion of secondary cities. An even spread of investment in urban and rural infrastructure as in Ghana and Malawi is a formidable spatial strategy for the development of urban and rural settlement. These objectives will also be followed in ongoing projects for Zimbabwe, which aimed at preparing a national spatial and physical plan for the country, and identifying specific investment projects and programmes within the basic framework of urban development.

The elements discussed above need to decentralize decision-making as far as possible to the local administration, and to relate plan preparation to the identification and eventual implementation of actual projects that are designed to upgrade the concerned urban areas.

The success of these initiatives will depend on the political will and administrative capability of the government to implement them. The processes which lead to the growth and development of cities is immensely complex, as it is the outcome of a huge range of both public and private endeavours, which are taking place within a very limited and congested space in urban centres.

The recent years experience shows that there is no 'quick fixes' of panaceas to guide the policy-maker in framing urban policy. In each country, the process of urbanization is an unique phenomenon with deep roots in the social, economic and political make up of the country as a whole. The outcome that a country can expect is the implementation of an urban policy is considerable as the cities play key role for its economic and social upgradation along with better human living.

FOOTNOTES

1. B.N. Sinha, "Urban Land Policy in India," *The National Geographical Journal of India*, Vol. XI, No. 3-4, 1965, pp. 131-43.
2. T.C.S. Clark's (Ed.), *Urban Policy Analysis*, Sage Publications, London, 1981.
3. B.D. Raheja, *Urban India and Public Policy*, Somaiya Publishers Pvt. Ltd., Mumbai, 1973, pp. 1-323.
4. H.K.L. Bhagat, "Housing : A National Task, *Yojana*, Vol. XX/22, December 1976, pp. 6-7.

BIBLIOGRAPHY

Abrams, C., "Urban Land Problems and Policies : Housing, Town and Country Planning," *Bulletin No. 7*, UNO, 1953.

Achuta Rao, T.N., "The Concept of Twin Towns," *Geographical Review of India*, 29, 104-107, 1967.

Adams, John S., "Residential Structure of Mid Western Cities," *Annals of the Association of American Geographers*, 60, 1970, pp. 37-62.

Adams, R.M., *The Origin of Cities*, Scientific American, 1960.

Ahmad, K.S., "Urbanisation Trends in West Pakistan," *Pakistan Geographical Review*, 21, (I), 1966, p. 1-20.

Ahmad, Nafis, "The Landscape of the Dacca Urban Area," *The Oriental Geographer*, VII(1), 1963, pp. 1-18.

Ahmad, Q., Indian Cities : Characteristics and Correlates, Research Paper No. 102, Chicago : University of Chicago Press, 1965.

Ahmad, Qazi S., "Distribution Pattern of Urban Centres in Pakistan," *Pakistan Geographical Review*, 22, (1), 1967, pp. 7-8.

Ahmed, E., "Distribution of Cities in India and Pakistan," *Geographical Outlook*, I, 1956.

———, "Origin and Evolution of Towns of the Uttar Pradesh," *Geographical Outlook*, I, 1956, 38-58.

———, *Geographical Essays on India*, Patna, 1954.

Ahmed, S., *Dacca : A Study in Urban History and Development*, Curzon Press, London, 1950.

Alam, S.M., "Masulipatam : A Metropolitan Port," *Indian Geographical Journal*, 34, 1959, 33-42.

———, "Urban Growth Trends and Patterns of Urban Development in Andhra Pradesh," *Studies in Applied and Regional Geography*, Ed. M. Shafi and M. Raza, Aligarh : Aligarh Muslim University, 1971, 114-41.

———, *Hyderabad-Secundrabad : A Study in Urban Geography*, Allied Publishers, Mumbai, 1965.

———, and W. Khan, *Metropolitan Hyderabad and Its Region*, Asia Publishing House, New Delhi, 1972.

Ali, S.M., "The Towns of Indian Desert," *Proceedings of International Geography Seminar*, 1956.

Anal, A.K. Singh, *Origin and Growth of Towns*, Concept Publishing Company, New Delhi, 1980.

———, "Begusarai : A Study in Urban Morphology," *The Deccan Geographer*, X, pp. 1-8.

Anantapadmanabhan, N. "Functional Classification of Urban Centres in Madras State," *Bambay Geographical Magazine*, 1965, 13.

Anderson, J., "On General Systems Theory and the Concept of Entropy in Urban Geography," Unpublished Paper, London : London School of Economics and Political Science, 1969.

Andrews, R.B., "Elements in Urban Fringe Pattern," *The Journal of Land and Public Utility Economics*, Vol. 18, 1942, pp. 167-83.

Anstead, R.D., "Sewage and City Geography," *Journal of the Madras Geographical Association*, 3, 1928, 119-36.

Applebaum, William, "A Technique of Constructing a Population and Urban Landuse Map," *Economic Geography*, 28, 1952, pp. 240-43.

Arunachalam, B., "Bombay City Stages of Development," *Bombay Geographical Magazine*, 3, 1955, pp. 34-9.

Aspinall, P.I. and I.W.R. Whitehand, "Building Plans : A Major Source for Urban Studies," *Area*, 12(3), pp. 199-203.

Aurousseau, M., "The Recent Contribution to Urban Geography—A Review," *Geographical Review*, XIV, 1924, pp. 444-55.

Badcock, B.A., "Central Place Evolution and Network Development in South Auckland, 1840-1968 : A System Analytic Approach," *New Zealand Geographer*, 26, 1970, 109-35.

Bagchi, K., "The Howrah Conurbation : A City in Urban Sprawl," *Geographical Review of India*, 28, 1966, 16-31.

Bagchi, K. and Usha Sen, "Giridih : Its Growth and Landuse," *Geographical Review of India*, 1963, 25.

Balchin, P.N. and J.L. Kieve, *Urban Land Economics*, Macmillan, London.

Banerjee, S., *Spatial Dimension of Urbanization in Relation to Development Planning in India,* New Delhi, Associated, 1968.

Barr, B.M., I.J. Lindsay and E.R. Reinett, "Patterns of Urban Spacing in the U.S.S.R. : Analysis of Order Neighbour Statistics in Two-Dimensional Space," *Journal of Regional Science*, 11 (1971), 211-24.

Bartholomew, H., *Landuse in American Cities : Harvard City Planning Studies*, Cambridge.

Barua, *Urbanization in North-East India*, New Delhi, 1977.

Basak, T.N., "Bochum—A Study of Urban Growth," *Geographical Review of India*, 38 (1), 1976, pp. 60-77.

Beavon, K.S.O., *Central Place Theory : A Reinterpretation*, Longman, 1977.

Berry, B.J.L., "Ribbon Development in the Urban Business Pattern," *Annals of the Association of American Geographers*, XLIX, 1959, pp. 145-55.

———, "Some Relations of Urbanization and Basic Patterns of Economic Development," in *Urban Systems and Economic Development*, Eugene, Oregan, 1962.

———, *Geography of Market Centres and Retail Distribution*, Englewood Cliffs, N.J. : Prentice Hall, 1967.

———, *The Human Consequences of Urbanization*, The Macmilan Press Ltd., 1973.

———, *Theories of Urban Location : An Introductory Essay*, Association of American Geographers, Commission on College Geography Resource Paper No. 1, Washington, 1968.

———, "Cities as Systems within Systems of Cities," *Papers of the Regional Science Association*, 13 (1964), 147-64.

Berry, B.J.L. and A.R. Pred, *Central Place Studies : A Bibliography of Theory and Applications*, including Supplement.

Berry, B.J.L. and W.L. Garrison, "A Note on Central Place Theory and the Range of Goods," *Economic Geography*, 34 (1958), 304-11.

———, "Recent Developments of Central Place Theory," *Papers of the Regional Science Association*, 4 (1958), 107-20.

———, "The Functional Bases of Central Place Hierarchy," *Economic Geography*, 34 (1958), 145-54.

Bhatia, S.S., "A Reconsideration of the Concept of the Primate City," *Indian Geographical Journal*, 37, 1962, 23-24.

———, "Historical Geography of Delhi," *Indian Geographer*, I, 1956, 17-43.

Bhatia, S.S. and A. Lal, "Geography of Delhi," *Science and Culture*, Vol. 29, (12), pp. 590-92.

Bhattacharya, A.N. and R. Lodha, *Evolution of Chittorgarh*, University of Udaipur Research Studies, 1962, 2.

Bhattacharya, N.D., "Geographical Zones of Murshidabad," *Indian Geographical Journal*, 36, 1961, 132-39.

———, "Murshidabad : A Study in Urban Geography," *National Geographical Journal of India*, 5, 1959, 38-51.

Bird, James, "Billingsgate—A Central Metropolitan Market," *Geographical Journal*, 1958, 124, pp. 464-75.

Bonine, M.E., "The Morphogenesis of Iranian Cities," *Annals of the Association of American Geographers*, 69(2), 1979, pp. 208-24.

Borchert, J.R., "America's Changing Metropolitan Regions," *Annals of the Association of American Geographers*, 62, 1972, 352-73.

———, "American Metropolitan Evolution," *Geographical Review*, 57, 1967, 301-32.

———, "The Urbanization of the Upper Midwest : 1930-1960," University of Minnesota, Upper Midwest Economic Study, Urban Report No. 2, 1963.

Bore Gowda, P., "Urban Sprawl in Bangalore," *National Geographical Journal of India*, XXV (3 & 4), 1979, pp. 194-99.

Bose, A., *Studies in India's Urbanization : 1901-1971*, New Delhi : Tata McGraw Hill, 1973.

———, *Urbanization in India : A Inventory of Source Materials*, Mumbai : Academic Books, 1970.

Bose, Ashish, *Bibliography on Urbanization in India*, Tata McGraw Hill, New Delhi, 1976.

Bose, Ashish, *India Urbanization, 1901-2001*, Tata McGraw Hill, New Delhi, 1978.

Bourne, L.S., "Forecasting Urban System : Research Design, Alternative Methodologies and Urbanization Trends with Canadian Examples." *Regional Studies*, 8, 1974, 197-210.

———, "Private Redevelopment of the Central City," University of Chicago, Department of Geography, Research Paper No. 112, Chicago : University of Chicago Press, 1967.

———, "Urban Structure and Landuse Decisions," *Annals of the Association of American Geographers*, 66, pp. 531-47.

———, *et al.*, (ed.), *Urban Future for Central Canada : Perspective on Forecasting Urban Growth and Form*, Toronto : University of Toronto Press, 1974.

Bourne, L.S. and Mackinnon (eds.), *Urban Systems Development in Central Canada,* Toronto : University of Toronto Press, 1972.

Breeze, Gerald, "Urban Development Problems in India," *Annals of the Association of American Geographers*, 53, 1963, 253-65.

Brodsky, Harold, "Land Development and the Expanding City," *Annals of the Association of American Geographers*, 63, pp. 159-66.

Bromley, R.J., "Markets in the Developing Countries : A Review," *Geography*, 56 (1971), 124-32.

Brush, J.E., "The Growth of Presidency Towns," Rutgers—The State University, Department of Geography, 1969 (unpublished paper).

Brush, J.E., "The Hierarchy of Central Places in South-Western Wisconsin," *Geographical Review*, 43, 1953, 380-402.

Brush, J.E. and H.E. Bracey, "Rural Service Centres in South-Western Wisconsin and Southern England," *Geographical Review*, 45, 1955, 559-69.

Brush, John, E., "Spatial Patterns of Population in Indian Cities," *Geographical Review*, 58, 1968, 362-91.

Brush, John E., "Morphology of Indian Cities," in Roy Turner, (ed.), *India's Urban Future*, Berkeley, University of California, 1962, 57-70.

Buckley, W., (Ed.), *Modern Systems Research for the Behavioural Scientist*, Chicago : Aldine, 1968.

———, *Sociology and Modern Systems Theory*, Englewood Cliffs, N.J. : Prentice Hall, 1967.

Bull, G.B.G., "Fieldwork in Towns : A Review of Technique," *Geography*, 49, 1964, pp. 206-21.

Burgess, E.W., *The Growth of the City*, The City, University of Chicago Press, 1925, pp. 47-62.

Burghardt, A.F., "The Location of River Towns in the Central Lowland of the United States," *Annals of the Association of American Geographers*, 49, (1959), 305-23.

Carruthers, Ian, "A Classification of Service Centres in England and Wales," *Geographical Journal*, 123, 1957, pp. 371-85.

Carter, H., "Urban Grades and Spheres of Influence in South West Wales : An Historical Consideration," *Scottish Geographical Magazine*, 71, 1955, 43-58.

———, *The Growth of the Welsh City System*, Cardiff, University of Wales Press, 1969.

———, *The Study of Urban Geography*, London : Edward Arnold, 1973.

Carter, H. and S. Wheatley, "Residential Segregation in Nineteenth Century Cities," *Area*, 12 (1), pp. 57-62.

Census of India, Paper No. 3, 1960, Study of Growth of Towns between 1901-1951, Delhi : Registrar General Office, 1960.

Chambers, W.T., "Geographic Areas of Cities," *Economic Geography*, VII, 1931 pp. 171-88.

Chang, S.D., "Some Aspects of the Urban Geography of the Chinese Hsien Capital," *Annals of the Association of American Geographers*, 51, 1961, 23-45.

———, "The Changing System of Chinese Cities," *Annals of the Association of American Geographers*, 66, 1976, 398-415.

Change, Sen-Dou, "Some Observations on the Morphology of Chinese Walled Cities," *Annals of the Association of American Geograhers*, 60, 1970, pp. 63-91.

Chapin, F.S., *Urban Landuse Planning*, Harper and Brothers, New York, XIV, 1957.

Chatterjee, A.B., "Cultural History of Howrah City," *Geographical Review of India*, XXVI, 1964, pp. 23-34.

———, "Demographic Features of Howrah City," *Geogaphical Review of India*, 1958, 20.

———, "Ecological Structure of Calcutta's Twin," *National Geographical Journal of India*, XI (2), 1965, pp. 59-62.

———, "Hinterland of a Symbolic City : A Case Study," *Geographical Review of India*, 27, 1965, 55-71.

———, "Industrial Landscape of Howrah," *Geographical Review of India*, 1963, 25.

———, *Howrah : A Study in Social Geography*, Kashipati-Bharati Studies, Calcutta, 1968.

———, "The Morphology of Howrah City," *Essays in Geography*, Calcutta, 1964.

Chatterjee, A.B. and P.N. Roy, "A Spatial Analysis of the Changing Livelihood Structure Around Calcutta," *National Geographical Journal of India*, XIX (2), 1973, pp. 92-97.

Chaudhuri, M.R. "Kulti-Burnpur—A Case Study of Growth and Problems of an Industrial Complex," *Geographical Review of India*, 29, 1967, 18-35.

Chettiar, C.M.R., "Growth of Modern Coimbatore," *Journal of the Madras Geographical Association*, 14, 1939, 101-16.

Childe, V. Gordon, "The Urban Revolution," in G.L. Possehl (ed.), *Ancient Cities of the Industries,* Vikas Publishing Houses, New Delhi, pp. 12-17.

Chisholm, M. and B. Rogers, ed., *Studies in Human Geography*, London : Heinemann, 1973.

Chittick, D. Growth, "Decline of South Dakota Trade Centres 1901-1951," South Dakota, State College, Agricultural Experiment Station, *Bulletin 448*, 1956.

Chorley, R.J. and P. Haggett, *Models in Geography*, London : Methuen, 1967.

Christaller, W., *Central Places in Southern Germany*, Prentice Hall, New Jersey, 1966.

Claire, Wm. H. (ed.), *Urban Planning Guide*, ASCE M & R, 49, ASCE, 1969, New York.

Clarke, C.G., "Urbanization in the Caribbean," *Geography*, 59, 1974, pp. 223-32.

Cohen, S.B. and G.K. Lewis, "Form and Function in the Geography of Retailing," *Economic Geography*, 43, 1967, pp. 1-42.

Colby, C.C., "Centrifugal and Centripetal Forces in Urban Geography," *AAAG*, XXIII, 1933, pp. 1-20.

Conzen, M.R.G., "Alnwick, Northumberland : A Study in Town-Plan Analysis," The Institute of British Geographer, *Transaction*, 27, 1960, pp. 27-81.

Corborg (ed.), *Proceeding of the I.G.U. Symposium in Urban Geography*, 1962, pp. 383-414.

Curry, L., "A Classical Approach to Central Place Dynamics," *Geographical Analysis*, 1, 1969, 272-82.

———, "Central Places in the Random Spatial Economy," *Journal of Regional Science*, 7, 1967, 217-38.

———, "The Random Spatial Economy : An Exploration in Settlement Theory," *Annals of the Association of American Geographers*, 54, 1964, 138-46.

Dacey, M.F., "A Note on the Derivation of Nearest Neighbour Distances," *Journal of Regional Science*, 2, 1960, 81-87.

———, "A Probability Model for Central Place Locations," *Annals of the Association of American Geographers*, 56, 1966, 549-68.

———, "Analysis of Central Place and Point Patterns by a Nearest Neighbour Method," *Lund Studies in Geography*, Series B, *Human Geography*, 24, 1962, 55-75.

———, "Some Properties of the Superposition of a Point Lattice and a Poisson Process," *Economic Geography*, 47, 1971, 86-90.

———, "The Spacing of River Towns," *Annals of the Association of American Geographers*, 50, 1960, 59-60.

Daula, Ashwini Kumar, "Plea for a Port at Paradip," *National Geographical Journal of India*, 6, 1960, 47-54.

Dann, R., "A Note on the Urban Geography of Malabar," *Journal of the Madras Geographical Association*, 6, 1931-32, 163-70.

Dann, R., "Urban, Geography of Trichinapaly and Srirangam," *Journal of the Madras Geographical Association*, 1933, 8.

———, "Urban Geography of Coimbatore 'District'," *Journal of the Madras Geographical Association*, 5, 1930, 130-33.

———, "Urban Geography with Special Reference to South Indian Cities," *Journal of the Madras Geographical Association*, 5, 1931, 149-59.

Datt, J.M., "Urbanization in Bengal," *Geographical Review of India*, 18, 1956, 19-23.

Datt, Kusum L., "Urban Zones of India," *National Geographical Journal of India*, 13, 1967, 97-109.

Davis, J.T., "Middle Class Housing in the Central City," *Economic Geography*, 41, 1965, pp. 238-51.

Davis, K. and H.H. Golden, "Growth and Functional Structure," *National Geographical Journal of India*, 22, 1976.

———, "Urbanization and the Development of Pre-Industrial Area," *Economic Development and Cultural Change*, 3, 1954, 3.

Davis, Kingsley and H. Hertz, *The Pattern of World Urbanization*, New York.

Davis, S.G., "The Rural-Urban Migration in Hong Kong and Its New Territories," *Geographical Journal*, 130, pp. 328-33.

Dayal, P. and B. Thakur, "Commercial Ribbons in Patna : Spatial Growth and Functional Structure," *National Geographical Journal of India*, XXII, (3 & 4), 1976, p. 164-77.

Deshpande, C.D., "Cities and Towns of Bombay Province : Aspects of Urban Geography," *Indian Geographical Journal*, 1941, 16.

———, "Visnagar," *Geographic Outlook*, 1956, I.

Dhabariya, S.S., "Evolution of Cultural Landscape in Udaipur City," *National Geographical Journal of India*, 10, 1964, 92-99.

———, "Udaipur : A Study in Urban Planning," *Annals of the Social Science* (Jodhpur), I, 1965.

———, "Blighted Areas and their Re-development," Academic Conference in Geography, Bhilwara, Rajasthan, 1965.

Dhar, H.K., "Role of Open Space in Residential Area," *Urban and Rural Planning Thought*, 1,2 1969, pp. 34-52.

Dickinson, R.E, "The Morphology of the Medieval German Town," *Geographical Review*, 35, 1945, pp. 74-97.

———, "Evolution of Cities in Maharashtra : An Analysis of the Phases of Urbanization," *The Indian Geographical Journal*, 45, 1970, 53-64.

———, "The Development and Distribution of the Medieval German Town," *Geography*, 27, 1942, 9-20, 47-53.

———, "The Distribution and Functions of the Smaller Urban Settlements of East Anglia," *Geography*, 17, 1932, 19-31.

———, "The Town Plans of East Anglia : A Study in Urban Morphology," *Geography*, 19, 1934, pp. 37-50.

———, *The West European City*, 1951, London.

Dickinson, R.E, "The Scope and Status of Urban Geography," *Land Economics*, 24, 1948, 221-38.

Dixit, K.R., "The Concept of Functional Zones," *National Geographical Journal of India*, 7, 1961, 191-200.

Doeppers, D.F., "Ethnic Urbanism and Philippine Cities," *Annals of the Association of American Geographers*, 64, 1974, pp. 549-59.

Doran, M.F. and R.A. Landis, "Origin and Persistence of an Inner-city Slum in Nassau," *Geographical Review*, 70, (2), 1980, pp. 182-93.

Dubey, K.K., "Use and Misuse of Land in Kaval Towns," Ph.D. Thesis, B.H.U., Varanasi (Unpublished).

Dutt, A.K., "An Analysis of Commutation to the Metropolis of Calcutta," *National Geographical Journal of India*, 10, 1964, 194-206.

———, "A Critique of Town Plans of Jamshedpur," *National Geographical Journal of India*, 5, 1959, 205-11.

———, "Daily Shopping in Howrah City," *National Geographical Journal of India*, 12, 1966, 163-77.

———, "Evolution of Jamshedpur City," *Indian Geographical Journal*, 41, 1966, 19-26.

———, "Shopping in Calcutta," *Town Planning Review*, January 1967.

———, "Umland of Jamshedpur," *Geographical Review of India*, 26, 1963.

———, "Inter-City Hierarchy of Central Places : Calcutta as a Case Study," *Professional Geographer*, 21, 1969, 18-23.

Dutta, A.K. and S.C. Chakraborty, "Reality of Calcutta Conurbation," *National Geographical Journal of India*, 9, 1963, 161-74.

Dwivedi, R.L., "Allahabad : A Study in Industrial Development," *National Geographer*, 4, 1961.

———, "Delimiting the Umland of Allahabad," *Indian Geographical Journal*, 39, 1964, 123-39.

———, "Origin and Growth of Allahabad," *Indian Geographical Journal*, 38, 1963, 16-32.

———, "Replanning An Existing City : Allahabad," *National Geographer*, 5, 1962.

———, "The Concept of Conurbation : A Review," *National Geographical Jounal of India*, 15, 1964, 45-55.

———, *Demographic Features of Allahabad City*, 27, 1965, 163-88.

Dwyer, D.G., "Urban Geography and the Urban Future," *Geography*, 64 (2), 1979, pp. 86-95.

———, *The City in the Third World*, Mcmillan, London.

Dziewonski, K., "Emerging Patterns of Urbanization in Poland," *Geographia Polonica*, 24, 1972, 21-29.

Dziewonski, K., "The Geographical Differentiation of Contemporary Urbanization," *Geographia Polonica*, 27, 1973, 31-41.

Edward, L.I., "The Geographer in Planning," *Professional Geographer*, XIV, 1962, pp. 84-98.

Ellefesen, Richard A., "City Hinterland Relationship in India," in Roy Turner, (ed.), *India's Urban Future*, Mumbai, Oxford University Press, 1962.

Epstien, Bart, J., "Geography and the Business of Retail Site Evaluation and Selection," *Economic Geography*, 47, 1971, pp. 192-99.

Falk, T., "Retail Change and Downtown Commercial Viability : The Case of Norrkoping, Sweden," *Geografiska Annaler Serries B, Human Geography*, 62 B (1), 1980, pp. 21-24.

Fano, P.L., "Organisaton City Size Distribution and Central Places," *Papers of the Regional Science Association*, 22, 1969, 29-38.

Ford, L.R., "The Urban House Type as an Illustration of the Concentric Zone Model : The Perception of Architectural Continuity," *Journal of Geography*, 140, 1974, pp. 29-39.

Forrester, J.O., *Urban Dynamics*, Cambridge : M.I.T. Press, 1969.

Friedman, J., "Two Concepts of Urbanization : A Comment," *Urban Affairs Quarterly*, I, 1966, 78-84.

Fuguitt, G.V. and N.A. Deeby, "Retail Service Patterns and Small Town Population Change : A Replication of Hassinger's Study," *Rural Sociology*, 31, 1966, 103-96.

Ganguly, Maya, "The Nature of Fringe Area of Calcutta," *Geographical Review of India*, 1967, 29, 53-59.

Garner, B.J., "Models of Urban Geography and Settlement Location," *Models in Geography*, Ed., R.J. Chorley and P. Haggett, London : Methuen, 1967.

Gauntia, R. and V.N.P. Sinha, "Trends in the Urbanization of the Chotanagpur Plateau," *The Deccan Geographer*, 7, 1969, 117-28.

Geddes, P., *Cities in Evolution*, London, 1955.

Getis, A., "Temporal Landuse Pattern Analysis with the Use of Nearest Neighbour and Quadrat Methods," *Annals of the Association of American Geographers*, 54, 1964, 391-99.

Ghosh, S., "The Urban Pattern of Calcutta," *Economic Geography*, 26, 1950, 51-58.

Gibbs, J.P., *Urban Research Method*, New Delhi, 1966.

Gilmore, H.W., *Transportation and Growth of Cities* : Chicago, Clencoe, 1953.

Gist, N.P., "The Ecological Structure of an Asian City : Madras," *Social Process*, Vol. 2, 1958, pp. 1-25.

Goodall, Brain, *The Economics of Urban Area*, Pergamon Press, London.

Gopal Krishan, "Spatial Analysis of Vegetable Supply of a Planned City : Chandigarh," *Indian Geographical Journal*, 38, 1963, 1-15.

Green, F.H.W., "Urban Hinterlands in England and Wales : An Analysis of Bus Services," *Geographical Journal*, 116, 1950, pp. 65-88.

Griffin, T.L.C., "The Evolution and Duplication of a Pattern of Urban Growth," *Economic Geography*, 41, 1965, pp. 133-56.

Guha, A., "Evolution of Serampur : A Study in Urban Geography," *The Earth*, 1964.

Guha, B., "Morphology of Burdwan," *The Earth*, 1964.

———, "The Rural Service Centres of Hooghly District," *Geographical Review of India*, 29, 1967, 47-52.

Guha, M., "Urban Regions of West Bengal," *Geographical Review of India*, 19, 1957.

———, "Modern Concept of Town Planning," *Observer*, 12, 1966.

———, "Morphology of Calcutta," *Geographical Review of India*, 15, 1953.

———, "Transport in and Around Calcutta," *Geographical Review of India*, 17, 1955, 4-8.

Gupta, A., "Nainital," *Geographical Review of India*, 16, 1954, 12-18.

Gupta, D.N., "The Community its Plan and the Planner," *Geographical Review of India*, 21, 1959, pp. 51-60.

Gupta, K.L. "Tribeni-Bansberia," *Geographical Review of India*, 22, 1960.

Gupta, S.L., "Faridabad : A Study in Industrial Growth," *Indian Geographical Journal*, 42, 1967, 23-28.

Gwynne, R.N., "City Size and Retail Prices in Less Developed Countires : An Insight into Primacy," *Area*, 10 (2), 1978, pp. 136-40.

Haggett, P., *Geography : A Modern Synthesis*, New York : Harper and Row, 1972.

———, *Locational Analysis in Human Geography*, London : Edward Arnold, 1965.

Harris, B., "Urbanization Policy in India," *Papers of the Regional Science Association*, 5, 1959, 181-203.

Harris, C.D., "The Cities of the Soviet Union," *Geographical Review*, 35, !945, pp. 107-21.

Harris, C.D. and E.L. Ullman, "The Nature of Cities," *Annals of the American Academy of Political and Social Science*, 242, 1945, 7-17.

Hartman, George W., "The Central Business District : A Study in Urban Geography, *Economic Geography*, 26, 1950, pp. 237-44.

Hartshorne, R., "The Twin City District : A Unique Form of Urban Landscape," *Geographical Review*, 22, 1932, pp. 431-40.

Harvey, D., *Explanation in Geography*, London : Edward Arnold, 1969.

Harvey, M.E., "The Changing Urban Network in Sierra Leone," in S.P. Chatterjee, (ed.), Selected Papers, *Population and Settlement Geography*, 1971, 104-20.

Hatt, Paul K. and J. Reiss Albert Jr., *Cities and Society*, The Free Press of Glencoe, Inc., New York, 1963.

Heght, Alfred, "Exploring the Urban Industrial Relocation Process," *The Journal of Geography*, 143, 1977, pp. 15-8.

Hester, J., *System Models of Urban Growth and Development*, Cambridge, Mass : M.I.T. Press, 1969.

Hill, P. and R.H.T. Smith, "The Spatial and Temporal Synchronization of Periodic Markets : Evidence from Four Emirate in Northern Nigeria," *Economic Geography*, 48, 1972, 345-55.

Hiltner, J. and B.W. Smith, "Intra-Urban Residential Location of the Elderly," *Journal of Geography*, 140, 1974, pp. 23-33.

Hirt, H.F., "Dravidian Temple Complex : A South Indian Cultural Dominant," *Bombay Geographical Magazine*, 8-9, 1961.

Hodder, B.W., "Some Comments on Markets and Market Periodicity," Markets and Marketing in West Africa, University of Edinburgh, Centre of African Studies, Mimeographed Seminar Report, 1966, University of Edinburgh, 1966, 97-106.

Holford, W.G., "Geography and Town Planning," *Geography*, 37, 1952, 191-98.

Honrao, M.S. and V.R. Prabhu, "Karwar : The Port Town," *National Geographical Journal of India*, 3, 1957.

Hort, J.F., "Urban Encroachment on Rural Areas," *Geographical Review*, 66, 1976, pp. 1-17.

Hoselitz, Bert F., "Cities of India and their Problems," A Review Article, *Annals of the Association of American Geographers*, 49, 1959, 223-31.

House, P.W., *The Urban Environmental System*, California : Sage Publications, 1973.

Houston, J.M., *A Social Geography of Europe*, London, 1953.

Hoyt, H., *The Structure and Growth of Residential Neighbourhoods in American Cities*, Washington, 1939.

Janaki, V.A. and M.C. Ghia, "The Tributary Area of Baroda," *Journal of Baroda University*, 11, 1962.

Janaki, V.A. and Z.A. Sayed, "The Geography of Padra Town," *Baroda University Geography Series*, I, 1962.

Jauhari, A.S., "Growth of Early Urban Settlements in the Sutlej-Yamuna Divide," *National Geographical Journal of India*, 8, 1962, 1-24.

———, "Post-Partition Expansion of Pre-Existing Towns in the Sutlej-Yamuna Divide," *National Geographial Journal of India*, 10, 1964, 100-26.

———, "Urban Settlements of the Sutlej-Yamuna Divide : Medieval and Later Historic Period Settlements," *National Geographical Journal of India*, 8, 1962, 114-35.

Jayaswal, S.N.P., "Sachendi : A Study of a Rural Service Centre," *Geographical Review of India*, 24, 1962, 46-51.

———, "Bindki : A Study in the Evolution of a Trade Centre," *National Geographical Journal of India*, 9, 1963, 187-200.

Jefferson, Mark, "The Distribution of the World's City Folks : A Study in Comparative Civilisation," *Geographical Review*, 21, 1931, 453.

———, "The Law of the Primate City," *Geographical Review*, 29, 1939, 226-32.

Johnson, E.A.J., *Market Towns and Spatial Development in India*, New Delhi : National Council of Applied Economic Research, 1965.

Johnson, James H., *Urban Geography : An Introductory Analysis*, Pargamon Press, London, 1977.

Johnson, L.J. "The Spatial Uniformity of a Central Place Distribution in New England," *Economic Geography*, 47, 1971, 156-70.

Johnston, R.J., "Regarding Urban Origin Urbanisation and Urban Pattern," *Geography*, 62, 1977, pp. 1-8.

Jones, R. (ed.), *Essays on World Urbanization*, London : George Philip, 1975.

Joshi, R.V., "Urban Structure in Western India, Poona : A Sample Study," *Geographical Review of India*, 14, 1952.

Kar, N.R., "Calcutta Conurbation, India : Regional Studies," *21st International Geographical Congress*, Ed. R.L. Singh, Calcutta : National Committee for Geography, 1968, 330-53.

Kar, N.R., "Pattern of Urban Growth in Lower West Bengal," *Geographical Review of India*, 24, 1962.

———, "Research Frontiers in Urban Geography : An Appraisal and Critique of Recent Trends in Quantification in Urban Geography," *Urban Affairs Quarterly*, 3, 1968, 37-68.

Kar, N.R., "Urban Characteristics of Calcutta," *Indian Population Bulletin*, 1960.

———, "Urban Hierarchy and Central Functions around the City of Calcutta and their Significance in Problems of Urban Geography," *Proceedings of International Geographical Union Symposium in Urban Geography*, Lund, Sewden, 1960.

———, *Calcutta Als Waltstadt*, Deutschen Geographentag, Berlin, 1959.

———, "Economic Characteristics of Metropolitan Sphere of Influence of Calcutta," *Proceedings of Summer School in Geography*, 1962.

Karan, P.P., "Patna and Jamshedpur," *Geographical Review of India*, 14, 1952.

———, "The Pattern of Indian Towns : A Study in Urban Morphology," *Journal of American Institute of Planners*, XXIII, 1957, pp. 70-75.

Karennvar, M.F., "A Quantitative Analysis of the Pattern of Urban Settlements in Karnataka State," *Geographical Review of India*, 38, 1976, pp. 78-84.

Karimi, S.M., "Late Medieval Towns of Bihar Plain (12th Century A.D. to Mid 18th Century)," *Journal of Bihar Research Society*, 56, 1970, 172-90.

———, "Origin and Evolution of Towns in Ancient Bihar," *Journal of Bihar Research Society*, 59, 1973, 75-92.

———, "Rajagriha : The Magadhan Capital : A Study in Ancient City Morphology," *Journal of Bihar Research Society*, 55, 1969, 69-89.

———, "Trends of Modern Urban Growth in Bihar Plain," *Geographical Outlook*, 7 (1970-71), 45-55.

———, *Urbanisation of Bihar Plain*, Aligarh Muslim University, Aligarh, 1967.

Kayastha, S.L., "Kandla : A Study in Port Development," *National Geographical Journal of India*, 9, 1963, 19-24.

———, "Kangra—A Himalayan Town," *National Geographical Journal of India*, 4, 1958, 89-94.

Kenyon, James B., "Landuse Admixture in the Built-up Urban Waterfront : Extent and Implications," *Economic Geography*, 44, 1968, 152-77.

Khan, F.K. and N. Islam, "High Class Residential Area in Dacca City," *The Oriental Geographer*, VIII, 1964, pp. 1-41.

King, K.J., "A Quantitative Expression of the Pattern of Urban Settlements in Selected Areas of the United States," *Tijdschrift Voor Economische on Sociale Geografie*, 53, 1962, 1-7.

King, L.J., "A Multivariate Analysis of the Spacing of Urban Settlements in the United States," *Annals of the Association of American Geographers*, 51, 1961, 222-33.

Kirk, W., "Town and Country Planning in Ancient India According to Kautilya's Arthasastra," *Scottish Geographical Magazine*, 94 (2), 1978, pp. 67-75.

Konge, G.H., "Functional Classification of Large towns of Karnataka," *The Deccan Geographer*, XVI, (2), 1978, pp. 450-60.

Krishan, Gopal and S.K. Agrawal, "Commercial Pattern of a Planned City : Chandigarh," *National Geographical Journal of India*, XIX (2), 1973, pp. 83-91.

Kulkarni, G.S., "Industrial Landscape of Greater Poona," *Bombay Geographical Magazine*, 8-9, 1961.

Kulkarni, K.M., *Urban Structure and Interaction : A Study of Nasik City Region*, Concept Publishing Company, New Delhi, 1997.

Kumbhat, P.C., "Jodhpur : A Study in Site and Evolution," *Indian Journal of Geography*, I, 1966.

Kuriyan, George, "Distribution of Population in the City of Madras," *Indian Geographical Jounral*, 16, 1941.

La Patra, J.W., "Applying the Systems Approach to Urban Development Stroudsburg," Dowden Butchinson and Ross, 1973.

Lacheve, R., "Networks and the Location of Economic Activity," *Papers of the Regional Science Association*, 14, 1965, 183-96.

Lahiri, T.B., "Calcutta Metropolitan District," Proceedings of Indian Science Congress, 1963.

———, "Service Costs and Size of Towns," *Geographical Review of India*, 29, 1940, 19-29.

———, "Urbanization Potential of Villages," *Geographical Review of India*, 28, 1966, 29-34.

Lal, A.K., *Urban Family : A Study of Hindu Social System*, Concept Publishing Company, New Delhi, 1992.

Lal, Nanhe, "Robertsganj : A Study in Urban Settlement," *National Geographical Journal of India*, 9, 1963, 201-10.

Lal, Parmanand, "Evolution of Port Blair : A Study in Urban Geography," *National Geographical Journal of India*, 8, 1962, 93-113.

Lall, A., "Some Aspects of Functional Classification of Cities and a Proposed Scheme for Classifying Indian Cities," *National Geographical Journal of India*, 5, 1959, 12-24.

Lall, A. and R. Thirtha, "Spatial Analysis of Urbanization in India," *Tijdschrift Voor Economische en Sociale Geografie*, 62, 1971, 234-48.

Latimer, E., *Handbook to Calcutta and Environs*, Calcutta : Oxford Book Company, 1963.

Lee, T. and I. Burton, "Future Water Supply for Indian Cities," *Deccan Geographer*, 1965-66, 3-4.

Lei.tnek, B.M., Charnews and J.V. Colter, "Commerical Factors in the Development of Regional Urban System : A Mexican Case Study," *Economic Geography*, 54, pp. 291-313.

Litovka, O.P., "Urbanisation in the USSR : Problems of Spatial Differentiation," *Soviet Geography : Review and Translation*, XXI (1), 1980, pp. 30-36.

Lösch, A., *Economics of Location*, New Haven : Yale University Press, 1954.

Lowenstein, L.K., "The Location of Urban Uses," *Land Economics*, Vol. 39, 1963, pp. 407-20.

Mabogunje, A.L., *Urbanization in Nigeria*, London : University of London Press, 1968.

———, "Evolution and Analysis of Retail Structure of Lagos, Nigeria," *Economic Geography*, 40, 1964, pp. 304-23.

Mahadev, P.D. and A. Ramesh, "Spatial Distribution of Hotel Industry in Mysore City," *National Geographical Journal of India*, 13, 1967, 208-13.

Mahadeva, P.D. and D.C. Jayasankar, "Concept of a City Region : An Approach with a Case Study," *Indian Geographical Journal*, 44, 1969, pp. 15-26.

Maher, C.A., "Spatial Patterns in Urban Housing Markets : Filtering in Toronto 1953-71," *The Canadian Geographer*, XVIII, 1974, pp. 108-24.

Mandal, R.B., "Urban Potentials in Bihar," *Indian Geographical Studies*, No. 6, Patna, 1976.

Mandal, R.B. and G.L. Peters, *Urbanization and Regional Development*, Concept Publishing Company, New Delhi, 1982.

Marshall, J.U., "The Location of Service Towns : An Approach to the Analysis of Central Place System," University of Toronto, Department of Geography, Research Paper No. 3, University of Toronto, 1969.

Mason, Greg, "The Deregulation of Urban Land Markets : A Note on Alternatives to Zoning," *The Annals of Regional Science*, XII (2), 1979, pp. 54-66.

Mathur, R.B., "Urban Spatial Planning in India : Some Misconceptions," *National Geographical Journal of India*, XX (4), 1974, 236-45.

Mathur, R.R., "Thana : A Study in Urban Landscape," *Bombay Geographical Magazine*, 1966, 14.

Mayer, Harold M., "Urban Geography," in Preston E. James and Clarence F. Jones (ed.), *American Geography : Inventory and Prospect*, Syracuse, Syracuse University Press, 1954.

Mayer, Harold M. and Clyde F. Kohn, *Reading in Urban Geography*, Central Book Depot, Allahabad, 1967.

McGoe, T.G., *The Urbanization Process in the Third World : Explorations in Search of a Theory,* London : G. Bell, 1971.

McGovern, P.D., "The New Towns of Scotland," *Scottish Geographical Magazine*, 84, 1968, pp. 29-44.

McLoughlin, J.B., *Urban and Regional Planning : A Systems Approach*, London : Faber, 1969.

McLoughlin, J.B. and J.N. Webster, "Cybernetic and General System Approaches to Urban and Regional Research : A Review of the Literature," *Environment and Planning*, 2, 1970, 369-408.

Medvedkova, O.L., "Components in the Evolution of Urban Systems : Theory and Empirical Testing," *Soviet Geography : Review and Translation*, XXI (1), 1980, pp. 15-29.

———, "Structural Characteristics of Cities as Multifunctional Centres," *Soviet Geography : Review and Translation*, XXI (5), 1980, pp. 263-83.

Mehrotra, S.N., "Urbanization of Madhya Pradesh," *Geographical Review of India*, 1961, 23.

Meier, R.L., *A Communication Theory of Urban Growth*, Cambridge : M.I.T. Press, 1962.

Mesarovic, M.D. and A. Reisman, (ed.), *Systems Approach and the City*, New York : Elsevier, 1972.

Misra, H.K., "The Size and Spacing of Towns in the Umland of Allahabad," *The Geographer*, XXII (1), 1975, pp. 45-55.

Misra, S.D., "A Note on the Socio-Historical Geography of Mathura," *National Geographical Journal of India*, 4, 1958, 189-99.

Mohanty, B., *Urbanization in Developing Countries : Basic Services and Community Participation*, Concept Publishing Company, New Delhi, 1997.

Mookerjee, S., "The Growth of Nagpur City," *Geographical Outlook*, 1958, 2.

Mookherjee, D.N., "Urban Pattern of Siliguri," *Geographical Review of India*, 1957, 19.

———, "West Bengal : Its Urban Character," *Geographical Review of India*, 1957, 19.

Mookherjee, Debnath, "The Concept of Urban Fringe and Its Delineation," *Geographical Review of India*, 25, 1963, 47-57.

———, "Urbanization Pattern in India : 1951-61," *Professioal Geographer*, 21, 1963, 308-14.

Morrill, R.L., "Simulation of Central Place Patterns Over Time," *Lund Studies in Geography*, Proceedings of the I.G.U. Symposium in Urban Geography, Lund 1960. Ed. K. Norborg Lund : Gleerups, 1962, 109-20.

———, "The Development of Spatial Distribution of Towns in Sweden : An Historical-Predictive Approach," *Annals of the Association of American Geographers*, 53, 1963, 1-14.

Mukerji, A.B., *Migration and the Spread and Growth of Urban Settlement*, Lund : Glerups, 1965.

"Modinagar : A Study in Urban Geography," *Geographical Review of India*, 1953, 15.

———, "Umland of Modinagar," *National Geographical Journal of India*, 8, 1962, 250-64.

Mukherjee, D., "Urbanization Pattern in India : 1951-61," *The Professional Geographer*, 21, 1969, 308-14.

———, "West Bengal : It's Urban Pattern," *Geographical Review of India*, 19, 1957, 67-72.

Mukherjee, M., "Functions and Functional Classification of Towns in Bihar," Unpublisheed Ph.D. Thesis, Patna University, Patna.

Muller, Edward K., "Selective Urban Centers of Ohio Valley (1800-1860)," *Geographical Review*, 66, 1976, pp. 178-99.

Mulvihill, D.F. and R.C. Mulvihill, *Geography, Marketing and Urban Growth*, New York : Van Nostrand, 1970.

Mumford, Lewis, *The City in History*, London, 1961.

Munsi, S.K., "Spatial and Temporal Disparities of Urban Growth in West Bengal," *Geographical Review of India*, 1972, pp. 339-53.

Murphy, R., "The City in the Swamp : Aspects of the Site and Early Growth of Calcutta," *The Geographical Journal*, 130, 1964, 241-56.

———, "Traditionalism and Colonialism : Changing Urban Roles in Asia," *Journal of Asian Studies*, 29, 1969, 67-84.

Murphy, Raymond E., *The American City : An Urban Geography*, McGraw-Hill Book Company, New York.

Muth, R.F., *Urban Economic Problems*, Harper & Row Publishers, New York, 1975.

Naqvi, H.K., *Urban Centres and Industries in Upper India, 1556-1803*, Mumbai, 1968.

Natesan, S., "The Early European Trade Centre of Malabar," *Journal of the Madras Geographical Association*, 6, 1931, 213-18.

Nath, V., "The Concept of Umland and Its Delimitation," *Deccan Geographer*, I, 1962.

Nelson, H.S., "The Form and Structure of Cities," Urban Growth Patterns," *Journal of Geography*, Vol. 58, 1969, pp. 198-207.

Newling, B.E., "The Spatial Variation of Urban Population Densities," *Geographical Review*, 59, 1969, pp. 242-52.

Nigam, M.N., "Evolution of Lucknow," *National Geographical Journal of India*, 6, 1960, 30-46.

———, "Functional Zones of Lucknow," *National Geographical Journal of India*, 10, 1964, 38-52.

———, "Gorakhpur : A Study in Historical Geography," *Geographical Thought*, 1967, 3.

Olsson, G., "Central Place Systems, Spatial Interaction and Stochastic Processes," *Papers of the Regional Science Association*, 18, 1967, 13-35.

Pacione, Michael, "Redevelopment of a Medium-sized Central Shopping Area : A Case Study of Cydehank," *Tidschriff Voor Economische on Sociale Geografie*, 71 (3), 1980, pp. 159-68.

Pahl, R.E., *et al., Structures and Processes of Urban Life*, Longeman, London, 1975.

Pal, A.K., Dutta, M.K. and P.K. Chakravarty, "Siliguri : A Case Study for Urban Redevelopment," *Geographical Review of India*, 27, 1965, 189-99.

Pal, S.K., "Changing Role of Urbanisation of Central Cities in India," *Geographical Review of India*, 30, 1970, pp. 97-112.

Paltu, "Spatial Segmentation of the Urban Housing Market," *Economic Geography*, 54, 1978, pp. 201-21.

Pandeya, P., *Impact of Industrialisation of Urban Growth : A Case Study of Chotanagpur*, Central Book Depot, Allahabad, 1970.

Panjabi, R.M., "Chandigarh : Indias Newest City," *Geographical Magazine*, 31, 1958, 401-14.

Pathak, C.R., "Urbanization Pattern in the Damodar Valley Region, India," *Metropolitan Growth and Planning*, (ed.), S.P. Chatterjee, Calcutta : National Committee for Geography, 1970, 70-79.

Pereira, A.B.M., "Human Geography of Mangalore and Environs," *Journal of the Madras Geographical Association*, 13, 1938, 373-94.

Possehl, Gregory, L. (ed.), *Ancient Cities of the Industries*, Vikas Publishing House Pvt. Ltd., New Delhi.

Pownall, L.L., "Low Value Housing in two New Zealand Cities," *Annals of the Association of American Geographers*, 50, 1960, pp. 439-60.

———, "The Retail Potential of Some Representative : New Zealand Towns," *Economic Geography*, 33, 1957, pp. 163-70.

Prabhu, V.R., "Dharwar : A City in Indian Urban Landscape," *Bombay Geographical Magazine*, I, 1953.

———, *Towns of Mysore State,* Calcutta, Asia Publishing House, 1964.

Prakasa Rao, V.L.S., "Town Planning," *Calcutta Geographical Review*, 1943, 5.

———, "Macro-Urban Analysis, Geographer's Contribution," *Bombay Geographical Magazine*, 10, 1962, pp. 5-12.

Prasad, A., "Double Towns of India : A Problem in Urban Morphology," *Geographical Outlook*, 1958, 2.

Pred, A., *The Spatial Dynamics of U.S. Urban Industrial Growth : 1800-1914*, Cambridge : The M.I.T. Press, 1966.

———, "Diffusion, Organisational Spatial Structure and City-System.," *Economic Geography*, 51, 1975, pp. 252-68.

Preston, R.E., "The Zone in Transition : A Study in Urban Landuse Pattern," *Economic Geography*, 42(3), 1966, pp. 236-60.

———, "Two Centrality Models," *Yearbook Association of Pacific Coast Geographers*, 32, 1970, 59-78.

Pyare Ram, "Functional Classification of Towns of Bundelkhand," *National Geographer*, XV (1), 1980, pp. 53-66.

Rafiullah, S.M., "A New Approach to Functional Classification of Towns," *Geographer*, 1965, 12.

Raheja, B.D., *Urban India and Public Policy*, Somiaya Publications Pvt. Ltd., Mumbai, New Delhi, 1973, pp.1-323.

Ram, L.N. and B. Thakur, "The Ecological Structure of Gaya," *Geographical Outlook*, 9 (1972-73), 89-99.

Ramanadham, V.V. and Y. Venkateswarlu, "Economic Aspects of Town Formations in Andhra Pradesh," *Indian Geographical Journal*, 1957, 32.

Rao, B.P., "Evolution of Vishakhapatnam," *National Geographical Journal of India*, 6, 1960, 241-59.

———, "Internal and External relations of the Port of Vishakhapatnam," *National Geographical Journal of India*, 12, 1966, 114-27.

Rao, V.E.N., "Conjeevaram—A Temple Town," *Journal of the Madras Geographical Association*, 1937, 12.

Rappoport, A., *Human Aspects of Urban Form : Towards a Man Environment Approach to Urban Form and Design*, Pergamon Press, London, 1977.

Ravindra, A., *Urban Land Policy : Study of Metropolitan City*, Concept Publishing Company, New Delhi, 1992.

Reddy, K.N., *Urban Redevelopment : A Study of High Rise Buildings*, Concept Publishing Company, New Delhi, 1992.

Reddy, N.B.K., "Urban Evolution, Growth Pattern and Urbanization Trends in the Krishna and Godavari Deltas." *The National Geographical Journal of India*, XVI (3&4) (1970), 270-87.

Reddy, S.N., "Vegetable Markets and Regional Relationship of Hyderabad City," *Geographical Review of India*, 23, 1961, 24-40.

Reeve, S.A., "Congestion in Cities," *Geographical Review*, 1917, pp. 278-93.

Renner, George, T., "Cities of Puget Basins," *Economic Geography*, 11, 1935, pp. 280-83.

Rimsha, A., *Town Planning in Hot Climates*, Mir Publishers, Moscow, 1976.

Rizvi, S.Q.A., "The Role of Rivers in the Evolution of Urban Centres," *Geographer*, 1956-57, 8-9.

Robson, B.T., "The Urban Environment," *Geography*, 60 (2), 1975, pp. 184-88.

———, *Urban Growth : An Approach*, London : Methuen, 1973.

Romesh, A., "Origin and Evolution of Ootacamund," *National Geographical Journal of India*, 10, 1964, 16-28.

Rothenberg, J., "Problems in the Modelling of Urban Development : A Review Article on Urban Dynamics," by J.W. Forrester, *Journal of Urban Economics*, 1, 1974, 1-20.

Rowland, D.T., "Theories of Urbanisation in Australia," *Geographical Review*, 67, 1977, pp. 167-76.

Rowland, Richard H., "Declining and Stagnant Towns of the USSR," *Soviet Geography : Review and Translation*, XXI (4), 1980, pp. 195-218.

Roy, B.K., "Urban Landuse Maps for Towns and Urban Agglomerations in Census of India," *National Geographical Journal of India*, XXIII, (3&4), pp. 160-65.

Russwurm, L.H., "Systems Stability and Urban Ecosystems," Mimeographed Paper, Waterloo : University of Waterloo, Department of Geography, 1972.

Sanders, Ralph A., "Bilevel Effects in Urban Residential Patterns," *Economic Geography*, 52, 1976, pp. 61-70.

Sargent, C.S. Jr., "Toward a Dynamic Model of Urban Morphology," *Economic Geography*, 68, 1972, p. 357-74.

Saxena, N.P. and R. Singh, "Diurnal Labour Mobility at Meerut and Zoning of Supply Area," *Geographical Observer*, I, 1965.

Sdasyuk, G.V., "Urbanization and the Spatial Structure of Indian Economy," Economic and Social-Cultural Dimensions of Regionalization, (ed.), A.C. Shekhar, Ministry of Home Affairs, *Census Centenary Monograph No. 7*, New Delhi : Office of the Registrar General, 1971, 297-327.

Sen, A.K., "Bankura : A Study of the Cultural Landscape of an Urban Area," *Geographical Review of India*, 18, 1956, 9-14.

———, "City Regions of India," *Geographical Review of India*, XIX, 1959, pp. 32-42.

———, "Techniques of Classifying the Functional Zone of City," *Geographical Review of India*, 21, 1959, 37-42.

Sengupta, S., *Residential Pattern of Suburbs*, Concept Publishing Company, New Delhi, 1995.

Sexena, Sudha, *Trends of Urbanisation in Uttar Pradesh*, Satish Book Enterprise, Agra, 1970.

Sharma, J.P., "A Note on the Evolution and Functional Morphology of a Temple Town of Himachal Pradesh : Chamba," *National Geographical Journal of India*, XXIV (1&2), pp. 77-85.

Shrivastava, V.K., "General Landuse Models : Method and Experience," *Uttar Bharat Bhoogol Patrika*, Vol. XIII, No. 1-2, 1977.

Siddiqui, N.A., "Composition of the Population of Moradabad City," *Geographer*, 1958, 10.

Simmons, J. and R. Simmons, *Urban Canada*, Toronto : Top Clark, 1969.

Simmons, J.W., "Changing Residence in the City : A Review of Intra-Urban Mobility," *Geographical Review*, LVIII (4), 1968, 622-51.

Sinclair, R., "Vön Thunen and Urban Sprawl," *Annals of the Association of American Geographers*, Vol. 57, 1967, pp. 72-87.

Singh, A.N., "Functional Zones of Itarsi Town," *National Geographical Journal of India*, 13, 1957, 89-96.

Singh, B., "Urban Geography of Orissa," *Indian Geographical Journal*, 32, 1957, 86-94.

Singh, H.H., *Kanpur : A Study in Urban Geography*, Varanasi.

Singh, Harihar, "Evolution of the Townscape of Jaunpur City," *National Geographical Journal of India*, 4, 1958, 35-46.

———, "Urban Landuse of Jaunpur," *National Geographical Journal of India*, 6, 1960, 115-23.

Singh, J.P., "Patterns of Landuse in Urban Areas : A Case Study of Shillong," *Indian Geographical Studies*, R.B. 6, 1976, pp. 25-32.

Singh, K.N., "Barkaj : A Study in the Changing Pattern of a Market Town," *National Geographical Journal of India*, VII (1), 1961, pp. 21-36.

———, "Changes in the Functional Structure of Some Small Towns in Eastern Utter Pradesh," *Indian Geographer*, 6, 1961, 21-40.

———, "Functions and Functional Classification of Towns in U.P.," *National Geographical Journal of India*, V (3), 1959, pp. 121-48.

———, "Spatial Pattern of Central Places in the Middle Ganga Valley, India," *National Geographical Journal of India*, 12, 1966, 213-26.

———, "The Territorial Basis of Medieval Town and Village Settlement in Eastern Uttar Pradesh, India," *Annals of the Association of American Geographers*, 58, 1968, 203-20.

———, "Central Places and Development Strategy in Varanasi Region : A Spatial-Functional Approach," in R.L. Singh (ed.), *Urban Geography in Developing Countries*, Varanasi : National Geographical Society of India, 1973, 257-75.

Singh, K.N. and R.L. Singh, "Morphology of the Twin-Township of Dehri-Dalmianagar," *National Geographical Journal of India*, III (3 & 4), 1957, pp. 169-79.

Singh, L., "The Umland of Agra Based on Bus Services," *National Geographical Journal of India*, 2, 1956, 149-53.

Singh, L.R., "The Role of Geographers in Town Planning," *National Geographer*, I, 1958.

Singh, Madhusudan, "Evolution of Meerut," *National Geographical Journal of India*, II, 1965, 144-58.

———, "Urban Field Meerut," *Deccan Geographer*, 1964, 2.

Singh, O.P., *A Study of Central Places in U.P.*, Banaras Hindu University, Varanasi, 1969.

Singh, Onkar, "The Trend of Urbanisation in Uttar Pradesh," *National Geographical Journal of India*, XII (3), 1967, pp. 141-57.

———, "Towns of Uttar Pradesh," Unpublished Ph.D. Thesis, Banaras Hindu University.

Singh, R.L., "Ballia : A Study in Urban Settlement," *National Geographical Journal of India*, II (1), 1956, pp. 1-6.

———, "Evolution of Bangalore City," *National Geographical Journal of India*, 7, 1961, 231-44.

———, "Faizabad : 'Faizabad-cum-Ayodhya," *National Geographical Journal of India*, 4, 1958, 1-6.

———, "Gorakhpur : A Study in Urban Morphology," *National Geographical Journal of India*, 1 (1), 1955, pp. 1-10.

Singh, R.L., "Mirzapur : A Study in Urban Morphology," *National Geographical Outlook*, I, 1956.

———, "Origin of Medieval Towns in the Saryupar Plain of U.P.," *National Geographical Journal of India*, IX (1), 1963, pp. 1-11.

———, "The Trend of Urbanization on the Umland of Banaras," *National Geographical Journal of India*, 1956, 2.

———, "Two Small Towns of Eastern U.P. : Sultanpur and Chumar," *National Geographical Journal of India*, 3, 1957, 1-10.

———, "Urban Hierarchy in the Umland of Banaras," *Journal of the Scientific Research*, Banaras Hindu University, 1955-56, 6.

———, (ed.), *Urban Geography in Developing Countries*, 21st International Geographical Congress, Varanasi : National Geographical Society of India, 1973.

———, *Bangalore : An Urban Survey,* The National Geographical Society of India, Varanasi, 1964.

———, *Banaras : A Study in Urban Geography*, Nand Kishore and Bros, Banaras, 1955.

———, (ed.), "Urban Geography in Developing Countries," Proceedings of I.G.U. Symposium No. 15, Varanasi, N.G.S.I., Varanasi.

Singh, R.L. and B. Mukherjee, "Functional Zones of Ranchi," *National Geographical Journal of India*, 3, 1957, 117-24.

Singh, R.L. and K.N. Singh, "Basic-Non-Basic Concept : Theory and Applications," Review article in *National Geographical Journal of India*, 12, 1966, 188-92.

———, "Evolution of the Medieval Towns in the Saryu-Par Plain of the Middle Ganga Valley," *National Geographical Journal of India*, 9, 1963, 1-11.

Singh, R.L. and S.M. Singh, "Mungra-Badshahpur : A Rural-Urban Settlement in the Ganga-Ghagra Doab West," *National Geographical Journal of India*, 6, 1960, 199-206.

Singh, R.L. and Ujagir Singh, "Road Traffic Survey of Varanasi," *National Geographical Journal of India*, 9, 1963, 139-60.

Singh, R.N. and Sahab Deen, "Occupatonal Structure of Urban Centres of Uttar Praesh Household Industry : A Case Study," *National Geographer*, XIV (2), 1979, pp. 159-74.

Singh, R.R.P., *Karanchaura : Atit Aur Vartman*, Bihar School of Yoga, 1978.

Singh, Ram Prabesh, "Bokaro : The Future Steel Town," *National Geographical Journal of India*, 7, 1961, 245-56.

Singh, S.N., *Urban Land Utilization in the City of Munger*, Ph.D. Thesis, Bhagalpur University, 1984

Singh, S.P., "Demographic Features of Lucknow," *National Geographical Journal of India*, 5, 1959, 157-75.

Singh, Satish Chandra, "Evolution of Azamgarh," *National Geographical Journal of India*, 9, 1963, 175-86.

Singh, Shiv Mangal, "The Stability Theory of Rural Central Place Development," *National Geographical Journal of India*, 11,1965, 13-21.

Singh, U., "Allahabad : A Study on Its Planning and Development," *National Geographical Journal of India*, 7, 1961.

———, "Bombay : Study in Historical Geography," *National Geographical Journal of India*, 6, 1960, 19-29.

———, "Cultural Zones of Allahabad," *National Geographical Journal of India*, 7, 1960, 95-104.

———, "Demographic Structure of Allahabad," *National Geographical Journal of India*, 4, 1958, 163-88.

———, "Evolution of Allahabad," *National Geographical Journal of India*, 4, 1960, 109-29.

———, "Functional Regions of Allahabad," *National Geographical Journal of India*, 5, 1959, 67-90.

———, "Geographical Analysis of Slum Area in Indian Cities with Special Reference to Kanpur," *National Geographical Journal of India*, 12, 1966, 178-87.

———, "Geographical Significance of Essential Services in Allahabad," *National Geographical Journal of India*, 6, 1960, 176-93.

———, "Growth of Transport and Communication in Allahabad," *National Geographical Journal of India*, 5, 1959, 186-204.

———, "Industrial and Commercial Landscape of Allahabad," *National Geographical Journal of India*, 6, 1960, 227-41.

———, "Kaval Towns : A Comparative Study in Functional Aspects of Urban Fringe," *Natonal Geographical Journal of India*, 1962, 8.

———, "Mughalsarai : A Study in Landuse," *National Geographical Journal of India*, 10, 1964, 136-45.

———, "The Calcutta Conurbation," *National Geographical Journal of India*, 4, 1958, 95-102.

———, "The Influence of Railways on the 'Kaval' Towns of U.P.," *Geographical Review of India*, 1960, 22.

———, "The Origin and Growth of Kanpur," *National Geographical Journal of India*, 5, 1959, 1-11.

———, "Umland of Allahabad," *National Geographical Journal of India*, 7, 1961, 37-51.

Singh, U., "Banaras," *Indian Geographical Journal*, 26, 1952, 26-33.

———, *Allahabad : A Study in Urban Geography*, Varanasi, B.H.U., 1966.

Sinha, B.N., "Urban Geography of Orissa," *The Indian Geographical Journal*, 32, 1957, 86-94.

———, "Urban Land Policy in India," *National Geographical Journal of India*, XI (3&4), 1965, pp. 131-43.

Sinha, R.L.P., "Optimum City Size," *Indian Geographer*, 1964, 9.

Sinha, R.P., "Dhanbad : A Study in Urban Geography," *National Geographical Journal of India*, 8, 1962, 284-300.

Sita, K., "Some Aspects of Urbanisation in South Konkan," *Geographical Review of India*, 42 (3), 1980, pp. 238-44.

Sivaramakrishnan, R.C.; B. Dasgupta and M.N. Buch, *Urbanisation in India : Basic Services and Peoples Participation*, Concept Publishing Company, New Delhi, 1997.

Skeldon, Ronald, "Recent Urban Growth in Papua New Guinea," *Australian Geographer*, 14 (5), 1980, pp. 267-77.

Skinner, G.W., "Marketing and Social Structure in Rural China," *Journal of Asian Studies*, 24, 1964, 3-43, 195-228, 363-99.

Smailes, A.E., *The Geography of Towns*, London : Hutchinson University Library, 1953.

———, "Urban Systems," Transactions and Papers, *Institute of British Geographers*, 53, 1971, 1-14.

———, *The Geography of Towns*, London, 1953.

Smith, P.J., "Calgary : A Study in Urban Pattern," *Economic Geography*, 38, 1962, pp. 315-29.

Smith, R.H.T., "Method and Purpose in Functional Town Classificaton," *Annals of the Association of American Geographers*, 55, pp. 539-48.

———, "The Functions of Australian Towns," *Tijdschrift voor Economische en Sociale Geografie*, 1965, 56.

———, "The Theory of Periodic Markets : Consumer and Trade Behavour," "Preconference Publication of Papers, University of Aterloo, Annual Meeting of the Canadian Association of Geographers, 1971, 183-89.

———, "West African Market Places : Temporal Periocity and Locational Spacing" in C. Meillssoux (ed.), *The Development of Indigenous Trade and Markets in West Africa*, London : Oxford University Press, 1971.

Solomon, R.J., "Procedures in Townscape Analysis," *Annals of the Association of American Geographers*, 56, 1966, pp. 254-68.

Spate, O.H.K., "Two Federal Capitals : New Delhi and Canberra," *Geographical Outlook*, I, 1956, 1-8.

Spate, O.H.K. and E. Ahmed, "Five Cities of the Gangetic Plain," *Geographical Review*, 40, 1950, 259-78.

Srinivasachari, C.S., "Growth of Madras City," *Journal of the Madras Geographical Association*, 1939, 14.

Stafford, Howard A., "The Functional Basis of Small Towns," *Economic Geography*, 39, 1963, pp. 165-75.

Stamp, L.D., "Land Reclamation in Relation to New Towns," *The City Planning*, No. 9, 1941, pp. 14-15.

Stanislawski, D., "The Origin and Spread of the Town Girid Pattern," *Geographical Review*, 36, 1946, pp. 105-20.

Subrahmanyam, N., "Aspects of Geography of Calicut," *Journal of the Madras Geographical Association*, 6, 1931, 171-78.

———, "Regional Distribution and Relative Growth of Cities in Tamilnadu," *Indian Geographical Journal*, 16, 1941, 71-87.

———, "Some Aspects of Growth of Greater Madras," *Journal of the Madras Geographical Association*, 13, 1938, 22-31.

———, "Problems of Suburban Municipality with Special Referene to Saidapet," *Journal of the Madras Geographical Association*, 15, 1940, 19-29.

Sundaram, K.V. and C.S. Chandrasekhara, "Urban Morphology and Internal Structure of Indian Towns—Some Case Studies," *National Geographical Journal of India*, 14, 1968, 55-65.

Swaminathan, E., "A Quantitative Approah to the Study of the Spatial Distribution and Functional Hierarchy of Shopping Centres in Coimbatore City, *National Geographical Journal of India*, XXV (3&4), 1979, pp. 231-53.

Tamaskar, B.G., "The Weekly Markets of Sagar Damoh Plateau," *National Geographical Journal of India*, 12, 1966, 38-50.

Taylor, G., "Towns and Townships in Southern Ontario," *Economic Geography*, XXI, 1945, pp. 86-96.

Terechow, W., "Soviet Urban Settlements : An Analysis of Chistering," Unpublished M.A. Thesis, Department of Geography, University of Chicago, 1965.

Thakore, M.P., "Sixteen Sites in Delhi," *Indian Geographer*, 1963, 8.

———, *Aspects of Urban Geography of New Delhi*, 1962, Ph. D. Thesis, London University.

Thakur, B., "Nearest Neighbour Analysis as a Measure of Urban Place Patterns," *Indian Geographical Studies*, Research Bulletin, I, 1973, 59-71.

Thakur, B., "Urban Dynamics and General Systems Theory," *Indian Geographical Journal*, 51, 1976.

———, *Urban Settlements in Eastern India*, Concept Publishing Co., New Delhi, 1980.

Thakur, B. and L.H. Russwurm, "Central Places in a South Western Ontario Urban Growth Corridor 1871 to 1971 : A Functional and Hierarchical Level Analysis," *The Canadian Geographers*, 21, 1977.

Thomas, E.N., "The Stability of Distance Population Size Relationships for Iowa Towns from 1900-1950," *Lund Studies in Geography, Series B. Human Geography*, 24, 1962, 13-30.

———, "Towards an Expanded Central Place Model," *Geographical Review*, 51, 1961, 400-11.

Thomson, J.M., *Great Cities and Their Traffic,* Penguin Books, New York, 1977.

Tiwari, A.K., "Mathura : A Study in Site and Evolution," *National Geographical Journal of India*, 9, 1963, 48-59.

Tiwari, A.R., "The Urban Regions of Agra," *The Agra University Journal of Research*, 6, 1958, 101-14.

Tiwari, P.S., *Urban Patterns in Madhya Pradesh*, Banaras Hindu University, Varanasi, 1969.

———, "Functional Pattern of Towns in Madhya Pradesh," *National Geographical Journal of India*, 14, 1968, 41-54.

Tuppen, John, "Redevelopment of Lyon City Centre," *Scottish Geographical Magazine*, 93, 1977, pp. 150-58.

Turner, Roy, (ed.), *India's Urban Future*, Mumbai.

Tyagarajan, V., "Tutikorin : A Town Study," *Indian Geographical Journal*, Vol. 14, 1941.

Upadhya, L.N., "Ajmer : A Study in City Morphology," Academic Conference in Geography, Bhilwara Rajasthan, 1965.

Van Suylichem, A.P.G.N., "Calcutta's New Harbour as a Magnet in the Urbanization Process in East India," *Tijdschrift Voor Economische en Sociale geografie*, 58, 1968, 326-34.

Varma, K.N., "Jabalpur : A Study in Urban Geography and Town Planning," *Deccan Geographer*, I, 1962.

Vasanta, A., "A Method of Delimiting Areas of Urban Concentration," *Indian Geographical Journal*, 1957, 32.

Verma, L.N., "Some Forms of Town Sites in the Aravallis," *Indian Geographical Journal*, 4, 1966, 52-63.

Verma, N.R. and Gopal Krishan, "Site Analysis of Hill Towns," *Transactions Institute of Indian Geographers*, 2 (1), 1980, pp. 101-10.

Wadhwa, K., *Urban Fringe Land Markets : An Economic Analysis*, Concept Publishing Company, New Delhi, 1992.

Wang, L.H., "Entropy Changes in Urban System Evolution : A Case Study of Malaysia," Unpublished M.A. Thesis, Department of Geography, McGill University, Montreal, 1970.

Warneryd, O., "A Systematic Approach to the Study of Inter-depenence in Urban Systems," *Urban and Regional Planning*, Ed., A.G. Wilson, London : Pion, 1971, 21-29.

Weaver, D.C., "Changes in the Morphology of Three American Central Business Districts 1952-1966," *The Professional Geographer*, 21, 1969, pp. 406-10.

Whebell, C.F.J., "Corridors : A Theory of Urban Systems," *Annals of the Association of American Geographers*, 59 (1969), 1-26.

Whitehand, J.W.R., "The Basis for Historico-geographical Theory of Urban Form," The Institute of British Geographers, *Transaction New Series*, 2, 1977, pp. 400-16.

Wilson, A.G., *Entropy in Urban and Regional Modelling*, London, Pion, 1971.

———, *Patterns and Processes in Urban and Regional System*, London, Pion, 1972.

———, *Urban and Regional Models in Geography and Planning*, John Wiley & Sons, London.

Winid, B., "Problems of Towns : Patterns in Developing Countries," *The Professional Geographer*, 19, pp. 15-16.

Woldenberg, M.J., "Energy Flow and Spatial Order : Mixed Hexagonal Hierarchies of Central Places," *Geographical Review*, 58, 1968, 552-74.

Yadav, C.S., (ed.), *Rural-Urban Fringe*, Concept Publishing Company, New Delhi, 1992.

———, (ed.), *Slums, Urban Decline and Revitalization*, Concept Publishing Company, New Delhi, 1992.

———, (ed.), *Urban Economics*, Concept Publishing Company, New Delhi, 1992.

———, (ed.), *Urban Planning and Policies*, Concept Publishing Company, New Delhi, 1992.

———, (ed.), *Urban Research Methods : Central Place Hierarchy and City Size Models*, Concept Publishing Company, New Delhi, 1992.

INDEX